ENERGY IN EUROPE: ECONOMICS, POLICY AND STRATEGY

ENERGY IN EUROPE: ECONOMICS, POLICY AND STRATEGY

FILIP L. MAGNUSSON
AND OSCAR W. BENGTSSON
EDITORS

Nova Science Publishers, Inc.
New York

For permission to use material from this book please contact us:
Telephone 631-231-7269; Fax 631-231-8175
Web Site: http://www.novapublishers.com

LIBRARY OF CONGRESS CATALOGING-IN-PUBLICATION DATA

Energy in Europe : economics, policy & strategy / Filip L. Magnusson and Oscar W. Bengtsson (editor).
p. cm.
ISBN 978-1-60456-829-5 (hardcover)
1. Energy policy--Europe. 2. Power resources--Europe. I. Magnusson, Filip L. II. Bengtsson, Oscar W.
HD9502.E8E518 2008
333.79094--dc22
2008023354

Published by Nova Science Publishers, Inc. ✢ New York

CONTENTS

PREFACE

Energy constitutes the motive force of the civilization and it determines, in a high degree, the level of economy development as a whole. Despite the increase use of different type of energy, particularly, renewable energy sources, fossil fuels will continue dominating the energy combinations in the world near future. However, oil reserves are declining and this situation would have a negative impact in the future economic development of many countries all over the world.

In Europe, the import energy dependency is rising. Unless Europe can make domestic energy more competitive in the next 20 to 30 years, around 70% of the European Union's energy requirements, compared to 50% today, will be met by imported products some of them from regions threatened by insecurity. Now, the energy requirements of the different countries are so high that, for the first time in the humanity's history, there is a need to consider different types of available energy sources and their reserves to plan the economic development of the countries. At the same time, there is also a need to use these sources in the most efficient possible manner in order to sustain that development.

The EU leads the world in demand management, in promoting new and renewable forms of energy. If the EU backs up a new common policy with a common voice on energy questions, Europe can lead the global search for energy solutions. However, EU must act urgently because it takes many years to bring innovation on stream in the energy sector, as well as to make productive the investments that are need to update the energy infrastructure in the region. This book provides leading-edge research on this field of study from around the globe.

Chapter 1 - Energy constitutes the motive force of the civilization and it determines, in a high degree, the level of economy development as a whole. Despite the increase use of different type of energy, particularly, renewable energy sources, fossil fuels will continue dominating the energy combinations in the world near future. However, oil reserves are declining and this situation would have a negative impact in the future economic development of many countries all over the world.

In Europe, the import energy dependency is rising. Unless Europe can make domestic energy more competitive in the next 20 to 30 years, around 60-70% of the European Union's energy requirements, compared to 50% today, will be met by imported products some of them from regions threatened by insecurity.

Now, the energy requirements of the different countries are so high that, for the first time in the humanity's history, there is a need to consider different types of available energy sources and their reserves to plan the economic development of the countries. At the same time, there is also a need to use these sources in the most efficient possible manner in order to sustain that development.

The EU leads the world in demand management, in promoting new and renewable forms of energy. If the EU backs up a new common policy with a common voice on energy questions, Europe can lead the global search for energy solutions. However, EU must act urgently because it takes many years to bring innovation on stream in the energy sector, as well as to make productive the investments that are need to update the energy infrastructure in the region.

Chapter 2 - The aim of this chapter is to assess whether long-term differences among the UK industrial subsectors are stochastic or deterministic. The analysis is implemented over the 1970-2004 time period. The authors discovered that the results are very influenced by whether one allows for deterministic structural breaks. Only when doing so, one can discover evidence of many the long-term differences being deterministic. This supports the adoption of policy instruments which are applied across productive sectors in a way which is not affected by the short-run evolution of the sectors.

Chapter 3 - The promotion of electricity from renewable energy sources (i.e., renewable electricity) is a general public policy priority for, both, the Member States and the European Commission. A Directive on the promotion of renewable electricity (Directive 77/2001/EC) has been in force since 2001, setting indicative targets for the share of electricity coming from renewables in 2010 in the Member States, and allowing these to apply the instrument they deem best in order to reach those targets, in line with the subsidiarity principle. An ambitious overall target for renewable energy of 20% of energy consumption has been set for the year 2020. This means that great efforts will have to be made by individual countries to reach this target, renewable electricity being the main element of this obligation (the others are heat and renewable fuels for transport).

The European Commission is about to propose a new Renewable Electricity Directive, which is likely to include flexibility provisions allowing the cost-effective attainment of those targets. One of the possibilities that are being considered is to allow countries to reach their individual targets by buying their renewable electricity deployment deficit to other countries which have a surplus (i.e., a renewable electricity deployment which is above their targets). This trade is likely to take the form of an exchange in guarantees of origin (GOs). Guarantees of origin are currently implemented in Member States to fulfil the Renewable Electricity Directive requirement that each country have a system that allows the tracing of the source of each kWh of renewable electricity and informs on this source.

This paper analyses the advantages and drawbacks of a system of trading renewable electricity attributes based on the GO concept and discusses several design alternatives and options to make such idea workable, effective and efficient.

Chapter 4 - The social costs of electricity generation may be classified into private costs and external costs or externalities. Private costs of electricity production range from the initial investment of capital in the power plant through purchase and transport of the primary energy source, infrastructures, etc. These are costs with known market prices. Related to the externalities component, environmental costs include the effect on atmosphere pollution, climate change and the impacts on landscape and the environment, while non-environmental external costs relate to socio-economic effects such as price and supply instability, energy dependence, etc. All these impacts have no market price, resulting in market failure and inefficient allocation of resources. In this chapter the author will show how the existing data on external costs could be used to define a public intervention to re-structure the energy pricing in Spain. The author will begin, therefore, by analysing private costs in electricity generation processes based on both renewable (wind power, hydro, and biomass) and non-renewable energy sources. They will also describe how private costs of electricity generated from biomass (biomass-e) are higher than with other sources, mainly due to the high supply costs of the biomass. The author will then analyze the price system of electricity generated from renewable energy sources (RES-e) in Spain and the author will explain how the market price does not always cover the private costs of production. The author will show how public intervention in the electricity market, through the design and implementation of price-premiums to RES-e production has encouraged the growth of wind power. This system has been highly ineffective in increasing the production of electricity from forest biomass, and in achieving regional and national objectives stated on energy plans. The chapter will finish with an estimation of the relative social benefits of biomass-e and concludes that the internalization of social benefits is a strong argument for a public policy to reinforce a higher penetration of this energy source in the Spanish electricity market.

Chapter 5 - This chapter deals with competition in the Spanish petrol retail market. The author will analyze whether regional price differences have been recently observed in the pricing of retail petrol. Retail petrol prices were regulated by a ceiling price system from June 1990 to October 1998. Such a price regulation imposed some degree of regional price uniformity nationwide. With the abolition of the ceiling price regulation, one would not expect uniform prices since there are no longer legal impediments to the existence of regional petrol price differences. The authors empirical findings show a more competitive scenario than the early years after the abolition of the ceiling price regulation, when political interference in price setting was documented. On the one hand, pre-tax petrol prices and retail margins in Spain have converged recently with European levels. On the other hand, retail petrol price differences exist among the Spanish regions. These differences are due both to different regional taxation regimes, and to the regional demand conditions which the service stations operate under.

Chapter 6 - By enacting the Electricity Market Law in 2001, Turkey reformed the electricity market and initiated to unbundle the state-owned monopolistic structure of the market by separating vertically. Thus, the regulatory endowment and the market structure before the reform fully changed. The regulatory reform has aimed to introduce competition to the market. As an independent regulatory agency with full regulatory discretions on the market, Energy Market Regulatory Agency (EMRA) was established. The aim of this chapter is to review the transition process to the reform, the effects of the regulatory reform on the new market structure and evaluate the current economic, political and legal institutional problems of the market.

Chapter 7 - Due to the constant growth of world population and the increasing needs in life quality, it is predictable that energy demand will have a great increase in the near future. Hence, it is foreseen that in the next decades, fossil fuels will continue to have a great contribution to energy production, particularly to power generation. World's economy depends very much on oil, not only because of its great contribution to the worldwide energy production, but also because it is the natural raw material for the production of many indispensable materials. Such a large dependency on oil has become threatening to economic stability, due to the limited lifetime of known oil reserves and to the constant rise of oil prices, which may compromise the sustainability of economic development. On the other hand, due to the threat of global warming and climate change, CO_2 and other greenhouse gases emissions have to be reduced drastically. Therefore, all the European Union countries have encouraged the use of renewable energies and the use of biomass and wastes for energy production, to reduce the dependency on imported fossil fuels and simultaneously to decreasing CO_2 emissions. To attain these goals, EU regulations, governmental political measures and research and development (R&D) strategies supported by research financing programs are needed.

As in years to come fossil fuels will continue to have a great contribution mainly to power generation, the new power plants should use advanced coal technologies like supercritical steam cycles, IGCC, natural gas based combined cycles, etc. However, to ensure sustainable security of energy supply in the future, several areas still need further scientific and technological developments, such as: 1) the use of oxygen for combustion and gasification, instead of air, 2) the production of H_2 from coal and biomass through gasification, 3), H_2 separation using membranes (dry processes), 4) production of liquids fuels, through methanol synthesis or Fischer-Tropsch synthesis, 5) CO_2 removal and storage at large scale, 6), pyrolysis of wastes to produce fuels or raw materials, 7) nuclear power, 8) biofuels production and 9) renewable energy. The main R&D needs of each process or technologies analyzed are underlined. The role of science is very important and crucial for the development of new robust technologies to ensure sustainable security of energy, but these goals can not be achieved without governmental political support, at national level and integrated in international wide policies.

Chapter 8 - Due to a number of structural market shifts, a new context for European gas regulation has arisen. In this new context, regulation must ensure adequate investments in a gas market that is dominated by producers, where political considerations are becoming more important, and where supply security and sustainability are top policy priorities. This requires a regulatory policy which emphasizes criteria like irreversible investments, risks, uncertainties, and governance. This contrasts starkly with the old situation where a main emphasis for regulation was to sweat existing assets.

This chapter's main question is whether existing gas regulation, which is based on neoclassical guiding principles, is adequate in this new context. This chapter argues it is not, because the neoclassical perspective does not explicitly acknowledge the above criteria. These criteria are explicitly treated by transaction costs economics. This chapter therefore argues from a theoretical perspective that inclusion of the transaction cost perspective will likely result in better outcomes in terms of market behavior. Hence, designing a proper regulatory regime in the new regulatory context requires that existing regulation be complemented with the transaction cost perspective. This implies a liberalized European gas market which is different from the one prescribed by the neoclassical perspective as well as the one envisaged in existing gas regulation.

Furthermore, due to its perfect competition guiding principles, existing regulation is tautological. That is, by assuming the eventual arrival of a perfectly competitive internal gas market, the option that the current shortcomings in the development of the internal gas market are attributable to a discrepancy between the existing regulation and the structural characteristics of the European gas market is ruled out by definition. The transaction cost perspective implies that the latter may well be the case, however.

Chapter 9 - In cold-climate countries, energy for heating of buildings is essential and heating options that interplay with the power system through electricity use or generation have potential for efficiency improvements. In Sweden, district heating is used extensively, especially in large buildings but to a growing extent also for small houses. Since rising electricity prices in recent years cause problems for the extensive use of electric heating, further switching to district heating should be a possible option. To be economically favourable, district-heating systems require a certain heat load density, which depends on heat production cost. New low-energy houses and energy-efficiency measures in existing buildings decrease the heat demand in buildings and, thus, in district heating systems. In addition, higher outdoor temperatures due to the enhanced greenhouse effect may reduce heat demand and district-heating sales substantially. Decreased heat demand should decrease the use of natural resources, such as fossil fuels, and enable redirections of bio-fuels into other fields of applications, such as production of automotive fuel. But the consequences of changed heat demand are complex when the heat is supplied from resources that can be seen as surplus, for instance heat from waste incineration or industrial waste heat. Increased electricity prices and policy instruments that promote renewable electricity generation have caused much interest in investments in new combined heat and power plants in the already well-developed Swedish district-heating systems. The opportunities for combined heat and power production rearrange the view of how the energy system should be evaluated. If district heating is produced in a combined heat and power plant, the heat demand enables electricity generation. Such effects can be studied with an energy system approach, which tries to go beyond the details in order to get a grip of the most important factors. Optimisation models have been used in several studies of large and complex energy systems. With the models, scenarios with changing policy instruments and changed consumer behaviour can be analysed. New applications for waste heat, such as using heat for absorption cooling, and switching from electricity and oil to district heating in industrial processes can also be studied. Energy efficiency measures as well as good conditions for efficient electricity generation, which can replace old and inefficient plants, are needed to reduce carbon dioxide emissions from the energy sector.

Chapter 10 - There is strong scientific evidence that the average temperature of the earth's surface is rising. This was a result of the increased concentration of carbon dioxide (CO_2), and other greenhouse gases (GHGs) in the atmosphere as released by burning fossil fuels. This global warming will eventually lead to substantial changes in the world's climate, which will, in turn, have a major impact on human life and the environment. Energy use reductions can be achieved by minimising the energy demand, by rational energy use, by recovering heat and the use of more green energies. This study was a step towards achieving this goal. The adoption of green or sustainable approaches to the way in which society is run is seen as an important strategy in finding a solution to the energy problem. The key factors to reducing and controlling CO_2, which is the major contributor to global warming, are the use of alternative approaches to energy generation and the exploration of how these alternatives are used today and may be used in the future as green energy sources. Even with modest assumptions about the availability of land, comprehensive fuel-wood farming programmes offer significant energy, economic and environmental benefits. These benefits would be dispersed in rural areas where they are greatly needed and can serve as linkages for further rural economic development. The nations as a whole would benefit from savings in foreign exchange, improved energy security, and socio-economic improvements. With a nine-fold increase in forest – plantation cover, the nation's resource base would be greatly improved. The international community would benefit from pollution reduction, climate mitigation, and the increased trading opportunities that arise from new income sources. The non-technical issues, which have recently gained attention, include: (1) Environmental and ecological factors e.g., carbon sequestration, reforestation and revegetation. (2) Renewables as a CO_2 neutral replacement for fossil fuels. (3) Greater recognition of the importance of renewable energy, particularly modern biomass energy carriers, at the policy and planning levels. (4) Greater recognition of the difficulties of gathering good and reliable biomass energy data, and efforts to improve it. (5) Studies on the detrimental health efforts of biomass energy particularly from traditional energy users. This study discusses a comprehensive review of biomass energy sources, environment and sustainable development. This includes all the biomass energy technologies, energy efficiency systems, energy conservation scenarios, energy savings and other mitigation measures necessary to reduce climate change.

Chapter 11 - Coal production has been in decline since 1980 in Western Europe; this decline has mainly resulted from the reduced consumption of low-quality lignite in Germany. Although some coal-fired power plants are earmarked for retirement in Europe, investment plans have been embarked on to refurbish the existing coal plants for the purpose of extending their operating lives. Coal is recognized as an indigenous energy base, and subsidies to the coal industry are authorized by the European Commission.

Since greenhouse gas, pollutant gases and dust are associated with coal utilization, the introduction of coal clean technologies is required to solve these environmental pressures. Furthermore, it is important to combine renewable energy with conventional energy (e.g. coal) in a proper manner. The EU has a binding target to increase the share of reneweable energy in total energy consumption in the EU to 20% by 2020, that is, the remaining 80% of required energy will have to come from conventional energy sources (e.g. coal).

A case study in Portugal shows that renewable energy (hydroelectricity production) is quite variable, and a model-based study in Ireland shows that the potential fuel and emissions savings are tempered by the inherent intermittence of wind and the increase in total generation costs (138 EUR per CO_2 ton).

The fluctuant supply of energy from renewable sources should be supplemented by conventional energy sources (e.g. coal-fired thermal plant) in order to meet the required demand. Indigenous coal has an important role in not only securing national energy but also measuring the variable supply of renewable energy, but the problem lies in the combination of renewable energy with coal-based energy production: although there is a reduction in the overall fuel bill when wind powered generation is introduced, this reduction is curtailed by the increased inefficiency of thermal power plants as a result of frequent start-ups and low loading levels. It is not advisable to separately consider renewable energy and conventional energy when determining how to realize a 20% share of renewable energy in total energy consumption in Europe. Renewable energy and coal (as a conventional energy source) are not mutually opposed but should coexist in the EU energy strategy. To this end, each new coal-fired power system is required to make up variable energy production from renewable sources, and this system is guided by the following conditions – frequent start-up with energy-saving measures, achievement of target efficiency directly after start-up, and stable efficiency at a low loading level.

Chapter 12 - European Union has set challenging energy targets for year 2020 including increasing the renewable energy utilization up to 20% of the final energy use. The European

Commission has also announced an ambitious strategic energy technology plan. The combined effect from these market stimulating and technology development measures could lead to cost reductions of new energy technologies and enhance significantly their market penetration. With this positive starting point in mind, the paper investigates further the commercialization process of new energy technologies and particular the effects from different strategies on the support level needed. The methodological frame employed is based on a learning curve approach and the corresponding analytical equations used to quantify the effects from policies are also derived. The results indicate that finding an optimal balance between technology push and market pull policies may lead to significant savings in the support needed. If the new technology is far off from a cost breakthrough then strong early RTD actions would well pay off compared to market stimulation measures only. Closer to the breakthrough, a shift to stronger market deployment action would be motivated. In case of indicative EU targets on photovoltaic technology, an intensive 5 years RTD action could lead to savings of tens of billions of € over a long run in the total costs of reaching the breakeven point. As another example, the EU renewables target could lead to a 15-20% cost reduction of wind power by year 2020.

In: Energy in Europe: Economics, Policy and Strategy ISBN: 978-1-60456-829-5
Editors: F. L. Magnusson and O. W. Bengtsson

Chapter 1

THE CURRENT SITUATION AND THE PERSPECTIVES OF THE ENERGY SECTOR IN THE EUROPEAN REGION

Jorge Morales Pedraza[1]
International Atomic Energy Agency (IAEA), Vienna, Austria

ABSTRACT

Energy constitutes the motive force of the civilization and it determines, in a high degree, the level of economy development as a whole. Despite the increase use of different type of energy, particularly, renewable energy sources, fossil fuels will continue dominating the energy combinations in the world near future. However, oil reserves are declining and this situation would have a negative impact in the future economic development of many countries all over the world.

In Europe, the import energy dependency is rising. Unless Europe can make domestic energy more competitive in the next 20 to 30 years, around 60-70% of the European Union's energy requirements, compared to 50% today, will be met by imported products some of them from regions threatened by insecurity.

Now, the energy requirements of the different countries are so high that, for the first time in the humanity's history, there is a need to consider different types of available energy sources and their reserves to plan the economic development of the countries. At the same time, there is also a need to use these sources in the most efficient possible manner in order to sustain that development.

The EU leads the world in demand management, in promoting new and renewable forms of energy. If the EU backs up a new common policy with a common voice on energy questions, Europe can lead the global search for energy solutions. However, EU must act urgently because it takes many years to bring innovation on stream in the energy sector, as well as to make productive the investments that are need to update the energy infrastructure in the region.

Keywords: e*nergy, Europe, oil, carbon, gas, hydropower, solar, wind, nuclear, geothermal, hydrogen, energy policy and strategy.*

---

[1] E-mail: jmorales_47@yahoo.com or jmorales47@hotmail.com.

I. INTRODUCTION

It is an undisputed reality that the energy production and their sustained growth constitute indispensable elements to guarantee the progress of any country. In other words, the energy constitutes the motive force of the civilization and it determines, in a high degree, the level of economy development of the different countries as a whole. For this reason, the use of all type of energy, particularly fossil fuels, is now an important element to ensure the economic development of all countries. According with the International Energy Outlook (IEO) of 2007, "fossil fuels (petroleum and other liquid fuels[2], natural gas, and coal) are expected to continue supplying much of the energy used worldwide". [14, page 2] Then the following question needs to be answered: There is enough fossil fuel reserves in the world to support the growing demand foresee in the future?

The American geologist Mr. K. Hubbert elaborated a statistic methodology to calculate oil world reserves. According with Mr. Hubbert projections "world oil reserves is around 2,1 billion of barrels". Other geologist predicted world oil reserves using a refined version of Mr. Hubbert's methodology. According with the projections of Mr. C. J. Campbell, of the Association for the Study of Peak Oil, "oil world reserves will be around 1,8 billion of barrels". The Old and Gas journal dated January 2007 "estimate oil world reserves in around 1,3 billions of barrels". Despite of the different levels of world's oil reserves predicted by different experts, one thing is true oil reserves are declining.

The situation regarding world coal reserves is very different. Coal world confirmed reserves are around 1 billion of tons, enough to satisfy the world energy demands at least for the next 155 years at present level of consumption. In the case of gas, there are enough reserves to satisfy the demand of gas in the coming decades. According with Oil and Gas journal dated January 1, 2007, the gas world reserves are estimate around 6,182,692 trillion of cubic feet. Developing countries, where live around three-quarter of the humankind, consume around 13% of the world energy, while the industrialized countries consume 87%. However, this last group of countries has to satisfy only the energy needs of around 18% of the population of the planet, while the other group 82% of it. It is estimate that near 1 600 million of people lives today without electricity in the world. It is expected that in 2030 around 1 400 million of people will continue to live without electricity.

Several countries in the developed world depend of the import of fossil fuel to satisfy its fundamental energy needs. In almost 67% of them, this dependence reached 70-75% of its energy needs. In Europe, the import energy dependency is rising. Unless Europe can make domestic energy more competitive in the next 20 to 30 years, around 60%-70% of the European Union's (EU) energy requirements, compared to 50% today, will be met by imported products from regions threatened by insecurity. [1, page 3]. In case of gas, Europe reliance on imports of this type of energy is expected to increase from 57% to 84% by 2030. In case of oil this increase would be from 82% to 93%. The EU holds extremely modest shares of world oil and gas reserves — around 1.2% and around 2.9%, respectively. Taking into account that oil and gas are the largest components in the EU's gross inland energy consumption, the EU is very vulnerable to markets and price trends.

---

[2] Petroleum and other liquids fuels include petroleum-derived fuels and non-petroleum-derived fuels, such as ethanol and biodiesel. Coal-to-liquids, and gas-to-liquids. Petroleum coke, which is solid, is included. Also included are gas liquids, crude oil consumed as a fuel, and liquid hydrogen

According to conservative calculations, more than 60% of the increase in the demand of world primary energy up to 2030 will come from Third World countries, mainly from the Asian region, particularly China and India. Because of this situation, "energy demand in the non-OECD Asia region is projected to grow at an average rate of 3.2% per year, more than doubling over the 2004 to 2030 period and accounting for more than 65% of the increase in energy use for the non-OECD region as a whole. In 2004, energy consumption in the countries of non-OECD Asia made up just over 48% of the non-OECD total; in 2030, its share is projected to be above 56%". [14, page 6]

Many of these countries have to dedicate around 50% of their total exports to import the energy resources that they needs to guarantee their economic development, spending around US$ 74 000 million for this concept. It is expected that this figure can reach US$ 170,000 million in the first decades of the 21th century.

In the other hand the participation of the OECD countries in the world energy demand is expected to diminish from 58% to 47% in 2030. In the cases of Russia and Central and Eastern Europe countries, the energy demand will decay until 10%.

It is important to single out that the industrialized nations of Europe are not free from of the negative effect of the limited oil reserves and the current level of oil price in their economies. This is despite of their high economic develop and the use of several sources of alternative energy with much more efficiency than almost all countries in the developing world. According to expert's projections oil and gas prices will continue to growth in 2008. In the specific case of oil, it is expected that the price could go beyond the psychological barrier of US$ 100 per barrel, creating a very difficult situation for the majority of the countries. However, if the USA carries out a military attack against Iran, then oil price could increase up to US$ 200 per barrel. With increasing global demand for fossil fuels, stretched supply chains and increasing dependence on imports, high prices for oil and gas are probably here to stay with important negative economic effects in most of the Third World countries. Just to confirm what has been said above "every 10% increase in the price of oil causes a loss of 0.5% of global GDP, this means approximately €255 billon", according with the International Energy Agency (IEA).

The limited reserves of fossil fuels, the negative impact in the environment due to the use of this type of fuels and the high price of oil, among others elements, are forcing almost all countries to look to other energy sources, like wind, solar, hydrogen, inclusive nuclear energy, as primary sources of energy.

The IEO fuel fossil projections and the level of its reserves raise serious concerns about the following four subjects: a) security of energy supplies; b) investment in energy infrastructure; c) environmental impact; and d) use and the unequal access of the world's population to modern energy.

Security of Supplies

Keeping in mind the weight of energy in any country's strategy for its sustained development, in particular for the European industrialized countries, the energy security has become one of the main concerns for their governments. For this reason, the world energy security was one of the main topics included in the G-8 Summit's agenda, celebrated from 15 to 17 of July 2006 in San Petersburg, Russia. The leaders of the G-8 countries agree that "the dynamic and sustainable development of our civilization depends on a secure access to energy sources". [10 , page 33] There should be no doubt that energy security, energy price, energy reserves, among other relevant issues will be in the agenda of future regional and international forums.

Investment in Energy Infrastructure

The investments in the current energy infrastructure in many countries in order to be able to take care of the expected demand of energy are enormous. In Europe alone, to meet expected energy demand and to replace ageing infrastructure, investments of around one trillion Euro will be needed over the next 20 years .[1, page 3]. It is expected that the EU invest "€100 billion in transmission networks and €340 billion in distribution networks for reinforcement, asset replacements and new connections over the three decades from 2001 to 2030". [16, page 6]

However, EU must act urgently because it takes many years to make productive the investments that are need to update the energy infrastructure in the region.

Environmental Impact

When politicians and experts discuss energy demand, energy supplies, level of reserves, energy price, investment in infrastructure, among others energy issues, the impact in the climate must be present. There is no doubt that the world climate is getting warmer. According to the Intergovernmental Panel on Climate Change (IPCC), "greenhouse gas emissions have already made the world 0.6 degrees warmer. If all countries take no action, especially the high consumer energy countries, then there will be an increase of between 1.4 and 5.8 degrees by the end of the century. All regions in the world – including Europe – will face serious consequences for their economies and ecosystems, if appropriate policies and strategies are not adopted and implemented as soon as possible, in order to reduce the negative impact in the climate that the use of some of the current energy sources has".

Access to Modern Energy

Table 1 shows the relationship between energy consumption and population.

Table 1. Relationship Between Energy Consumption and Population

Region	Energy Consumption %	Population %
USA and Canada	32	6
Europe and Japan	26	12
Former USSR countries	26	32
Other countries	13	50

From Table 1 can be easily seen that the most advanced countries from the economic point of view with only 18% of world population, are consuming 58% of world energy production, while the rest of the world with the 82% of the world population are consuming only 39% of world energy production. This situation must be changed in the future.

European Internal Energy Market

One of the current problems that Europe is facing is the following: Europe has not yet developed fully competitive internal energy markets. Only when such markets exist will EU citizens and businesses enjoy all the benefits of security of energy supply and lower prices. To achieve this aim, interconnections should be developed, effective legislative and regulatory frameworks must be in place and be fully applied in practice, and Community competition rules need to be rigorously enforced.

Furthermore, the consolidation of the energy sector should be market driven if Europe wishes to respond successfully to the many challenges it faces and to invest properly for the future. The EU has the tools to help. It is the world's second largest energy market in the world with over 450 million consumers.

EU countries are making good progress in several areas within the energy sector. The EU leads the world in demand management, in promoting new and renewable forms of energy, and in the development of low carbon technologies. If the EU backs up a new common policy with a common voice on energy questions, Europe can lead the global search for energy solutions [1, page 4].

In order not to lose the leadership of Europe in the energy sector, the Commission will organise a European Energy Technology Summit in the first half of 2009. The objective of this Summit is to bring together and engage all stakeholders in the entire innovation system, from industry to customers, as well as representatives of the European institutions, the financial community and other international partners. This event would be an opportunity to review progress, disseminate achievements and foster cross-fertilization between sectors.

New Initiatives

The EU is considering the implementation in 2008 of a group of new initiatives that are commented in details in section II. These initiatives are the following:

1) *European Wind initiative*: This initiative will focus on large turbines and large systems validation and demonstration (relevant to on and offshore applications).
2) *Solar Europe initiative*: This initiative will focus on large-scale demonstration for photovoltaic and concentrated solar power.
3) *Bio-energy Europe initiative*: This initiative will focus on "next generation" bio-fuels within the context of an overall bio-energy use strategy.
4) *European CO_2 capture, transport and storage initiative*: This initiative will focus on the whole system requirements, including efficiency, safety and public acceptance, to prove the viability of zero emission fossil fuel power plants at industrial scale.

5) *European electricity grid initiative*: This initiative will focus on the development of the smart electricity system, including storage, and on the creation of a European Centre to implement a research programme for the European transmission network.
6) *Sustainable nuclear fission initiative*: This initiative will focus on the development of Generation-IV technologies [27, page 10].

The energy requirements of the different countries are now so high that, for the first time in the humanity's history, there is a need to consider different types of available energy sources and their reserves to plan their economic development. At the same time, the use of these sources should be done in the most efficient possible manner in order to sustain that development. To achieve this goal in the particular case of the EU, it is indispensable to adopt and implement a correct European Energy Policy and Strategy. In sections III and IV, a summary of the European Energy Policy and Strategy recently adopted by the European Commission, as well as additional comments regarding these two subjects has been included.

II. The Use of the Different Energy Sources in the European Union

The different energy sources that the humanity has in their hands now to satisfy its energy needs are the following:

II.1-Fossil Fuels

II.1.1- Oil
II.1.2- Gas
II.1.3- Coa

II.2- Renewable Energies

II.2.1-Hydropower
II.2.2-Solar energy
II.2.3-Wind energy
II.2.4-Geothermal energy
II.2.5-Biofuels
II.2.6-Hydrogen energy

II.3-Nuclear energy.

General Overview

The general expert opinions in the field of energy all over the world are that the use of energy at world level will continue to increase gradually until 2030. Then, the main question that needs to be asked is the following: How much the energy demand will increase and how this demand is going to be met? According to studies made by the French Association of Oil Professionals, "it is expected that for 2030, the world energy demand will be double and it is probably that could be triple for 2050". Until 2030, the primary energy demands at world level are expecting to increase annually in 1.7%, which is somehow smaller than the world growth of 2.1% registered during the last three decades.

It is also expected that 90% of the increase in the world energy demand in the coming decades will be satisfied with fossil fuels. This means that around 15 300 million tons of oil equivalent will be consumed at world level in 2030.

The European Commission estimates that in 25 EU member states electricity demand has increased by an average of 1.7% per annum since 1990. In 15 of these EU's states, demand growth has reached up to nearly 1.9% annually. In the new EU member states demand initially exhibited limited growth (from 1990 levels), around only 0.2% annually, but in the second half of the last decade, it rose to around 1,6% per annum. The Commission expects that total electricity generation in 25 of the EU member states will grow by 1.4% annually from 2000 to 2030 (see Table 2).

It is important to consider the following reality when the future energy balance in Europe is analyzed. Europe is running out of indigenous energy resources in the form of fossil fuels at a time when a paradigm shift in energy prices are occurring. Today, the oil price is close to US$ 140 per barrel. The price of coal and gas is expected to growth also in the near future. Most observers agree that the era of cheap fossil fuels is over and signs are emerging that competition for ownership of oil and gas is becoming fiercer and will intensify heavily in the coming years. The era of energy uncertainty has come. [16, page 5]

Taking into account the above-mentioned situation, the role of hydropower in the world energy balance is expected to become increasingly important in the future. World production of hydroelectricity has grown steadily between 2.3% to 3.1% per year on average since 1980. The highest growth rates are expected to occur in developing or strongly industrializing countries with high, yet unexploited hydropower potential, particularly in Eastern Europe. In the case of Western Europe, most of the practical hydroelectric resources available have already been exploited and little further development is expected in the near future.

Since the oil crisis of the 70's, several European countries understood the urgent need to use nuclear energy as much as possible. Other adopts a different position.

Table 2. Electricity Generation

	TWh					% Annual Growth Rate				
	1990	2000	2010	2020	2030	90/00	00/10	10/20	20/30	00/30
NMS	317	324	392	498	551	0.2	1.9	2.4	1.0	1.8
EU-15	2,139	2,574	3,027	3,450	3,846	1.9	1.6	1.3	1.1	1.3
Total	2,456	2,898	3,419	3,949	4,397	1.7	1.7	1.5	1.1	1.4

Note: NMS means New EU member states.

Nuclear technology has been sufficiently used in the 60's and the 70's due to the construction of hundreds of nuclear power plants in the USA, Canada, Europe, Russia, among others. It is difficult to foresee the future of Europe without the use of this type of energy, despite the current opposition of the public opinion to use of nuclear energy for electricity production.

Regardless of the effort carried out up to now by many countries, in particular by the European industrialized countries in the development and use of renewable energy sources, the use of these technologies has not substantially modified up to now, the energy balance of these countries. Fossil fuel stay as the main energy sources for many of them, and it is expected that this situation will not change drastically before 2050. According to conservative projections, new technologies will begin to influence in the energy balance of the different countries, particularly in Europe, around 2050. To achieve this long–term goal, European government should adopt an appropriate energy policy and strategy that guarantee a successful development of these technologies.

In Figure 1 the foresee participation of the different types of energy in the generation of electricity until 2030 is clearly reflected.

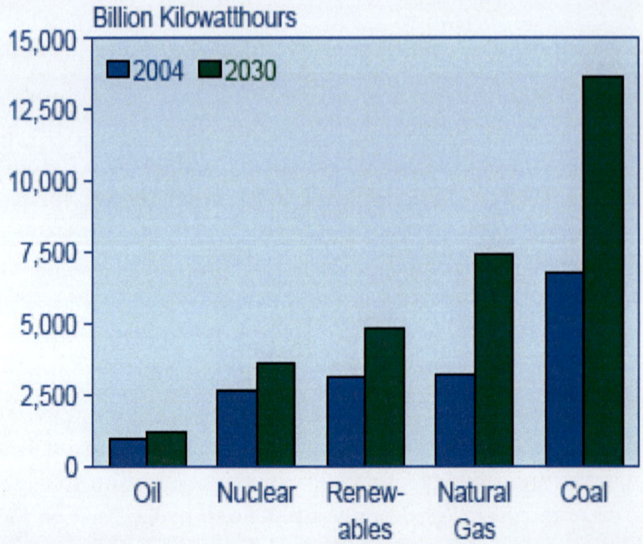

Sources: Derived from Energy Information Administration (EIA), Internatonal Energy Annual 2004 (May-July 2006), web site www.eia.doe.gov/iea.2030: EIA System for the Analysis of Global Energy Markets (2007), Figure 63, page 62

Figure 1. World Electricity Generation by Fuel. 2004 – 2030.

II.1- Fossil Fuels

II.1.1- Oil

Crude oil is a naturally occurring substance found trapped in certain rocks below the earth's crust formed more than 300 million years ago. It is a dark, sticky liquid which, scientifically speaking, is classed as a hydrocarbon. This means, it is a compound containing only hydrogen and carbon.

The first oil well was open in Pennsylvania, USA in 1859. Since that year, the oil consumption increased systematically in all countries. The world average level of oil consumption per habitant and per year is now around 1.6 ton. However, the level of consumption varies significantly between countries and regions. For example, in the USA and Canada average oil consumption per habitant per year is around 8 ton, this means 500 times world average. In Third World countries, that average level could be around 500 kg only, this means 0.3 times world average.

After so many years of discriminatory oil consumption in the world, now this type of fuel faces a serious problem that can limit, in a future not very distant, the possibilities of economic growth of many countries. This is due to the following reasons:

a) A decrease of the oil world reserves.
b) The current high level of consumption.
c) The ongoing high world price.
d) The negative impact on the environment.

As of January 1, 2007, proved world oil reserves were estimated on 1,317.4 billion of barrel. Of the world's total proved oil reserves, 78.3% is located in the Middle East, Canada, Venezuela and Russia. Countries with the largest crude oil reserves are show in the following table.

Table 3. Countries with Largest Crude Oil Reserves (Billion Barrels) January 2007

Country	Oil Reserves
Saudi Arabia	262.3
Canada	179.2
Iran	136.,3
Iraq	115.0
Kuwait	101.,5
UAE	97.8
Venezuela	80.0
Russia	60.0
Libya	41.5
Nigeria	36.2
United States	21.4
China	16.0
Qatar	15,2
Mexico	12.4
Algeria	12.3

Table 3. (Continued)

Brazil	11.8
Angola	8.0
Country	Oil Reserves
Norway	7.8
Azerbaijan	7.0
Rest of World	65.5
World Total	1,317.4

Source: International Energy Outlook 2007, Table 3, page 37.

It is expected that the increase in oil demand at world level will be around 1.6% annually in the period 2004-2030. This means an increase from 83,000 million of barrels a day in 2004 to 118,000 million barrels a day in 2030. It is expected, taking into account the current level of oil consumption in Europe that imports of oil will increase from 82% today to 93% in 2030.

Oil Demand in the Transport and Industrial Sectors

In the expected increase of oil energy demand, the transport sector will be responsible for 68% of it and the industrial sector of 27%. For this reason, it is projected that in 2020 the transport sector will displace the industrial sector as the one with the higher level in the final use of oil. In the case of Europe, it is predictable that the oil demand in the transport sector will growth to a slower rhythm, due to markets saturation. In the residential and service sectors it is expected that the growth in oil energy consumption will be around 1.7%, a little bit higher than in the industry, where the level of consumption it is expected to growth only 1.5%.

In the case of the generation of electricity, the situation at world level is expected to be the following: "Worldwide oil-fired generation is projected to increase by an average of 0.9% per year from 2004 to 2030; and in the OECD nations, it is projected to decline by 0.3% per year. Only the non-OECD Middle East region, with its ample oil reserves and a current one-third share of total electricity generation fueled by oil, is projected to continue relying heavily on oil to meet its electricity needs". [14, pages 62 and 63]

It is a fact that oil became, overnight, in the main source of energy at world level, with great impact in the development of the industrial and transport sectors. In the 30's oil covered 10% of the world energy needs. Forty years later, this percent was five times higher. Little by little, the oil displaced the coal in two of the activities in which the use of energy is fundamental: the production of heat and of electricity. The growth in the automobile and transport industry was a consequence, among other reasons, of a bigger use of the oil as energy source. There is not doubt that the impetuous development of Europe was achieved due, among other factors, to the massive use of oil in the industry and in the transport sectors and to the control that a reduced group of European and North American companies exercised in the extraction, sell and use of this type of fossil fuel.

In the 1960's the industrial sector consumed around 1,600 million of ton of oil. At the end of the 1970's this figure rose to 2,280 million of tons; this means an increase of 42.5% in only 10 years. In the same period, the role of coal in the world energy balance diminished from 57% to 25% and the gas increased from 12% to 23%.

With the significant rise in oil prices at the end of 2007 and during the first semester of 2008, the possible declination of oil reserves around 2025, and the growing concern about stable, secure and environmentally friendly energy supplies, the promotion of the use of bio-fuels in the transport sector should become a priority in the European political agenda. It is important to stress that today bio-fuels is the only type of energy that could significantly reduce oil dependence in the transport sector. For this reason, the European Commission is adopting additional measures in order to increase the participation of bio-fuels in the energy balance in the transport sector. However, the production of bio-fuels could has a negative impact in the production of cereals for food and in its price, affecting very seriously the already fragile economic situation of many Third World countries.

Oil Price Evolution

The vertiginous use of oil at world level was motivated by its low price during so many years. The fluctuations of the world oil price during the last century could be seen in the following table.

Table 4. Fluctuations of Oil Prices in the 20th Century

year	1900-1970	1979-1981	1985	1990-1999	2007 Projection	2007	2008
Price	< $2	> $35	~$10	Max $25	$40	+$98	+130

During the first seventy years the oil price stayed below US$ 2 per barrel; between 1979 and 1981 the oil price of barrel was above US$ 35, motivating a change in the energy strategy in Europe; in 1985 oil price collapsed reaching around US$ 10 per barrel; in the 90's oil price went up to a maximum of US$ 25. For 2007 the foresee oil price was US$ 40 per barrel, that is to say around US$ 58-59 less than the oil price of US$ 98-99 per barrel in November 2007. During the first semester of 2008 the oil price reached US$ 140 per barrel. Summing up can be stated that the oil price went up from US$ 2 per barrel in 1900 to up to US$ 140 per barrel in 2008; a maximum increase of US$ 138 in this period.

The Participation of Oil in the Energy Balance

According to some expert's predictions, it is expected that, with the current level of consumption, the oil will begin to present signs of exhaustion in some regions around 2025 (in some other regions even before). For example, oil production from Norway, OECD Europe's largest producer, appears to have peaked at about 3,4 million barrels per day in 2001, and it is projected to continue declining to about 1,4 million barrels per day in 2030 as the larger and older fields mature. Production from the United Kingdom, which peaked in

1999 at 3.0 million barrels per day, is projected to fall to 0.5 million barrels per day in 2030. [14, page 32]

Taking into account the analyses made by the Conservation Commission of the World Energy Conference, it is expected that for 2020 "the oil participation in the global energy balance will diminishes sensibly reaching half of the level reached at the beginning of the 1980's". It is very likely that oil participation in the global energy balance continues diminishing until reach 10% at some moment in the XXI century. The European countries will have a decisive influence in the reduction of oil participation in the global energy balance, if this group of countries continuing applying appropriate energy policies and strategies, increase the use of renewable energy sources and develop and use new technologies that make more efficient and less polluting the use of fossil fuels.

Annex 1 includes the world oil reserves and the reserves-to-production ratio by country for 2005.

II.1.2- Gas

Almost three-quarters of the world's natural gas reserves are located in the Middle East and Eurasia. Russia, Iran, and Qatar combined accounted for about 58% of the world's natural gas reserves as of January 1, 2007. Reserves in the rest of the world are evenly distributed on a regional basis. (See Table 5 and Figure 2). Despite high rates of increase in natural gas consumption, particularly over the past decade, most regional reserves-to-production ratios are substantial. Worldwide, the reserves-to-production ratio is estimated at 65 years. Central and South America has a reserves-to production ratio of about 52 years, Russia 80 years, and Africa 88 years. The Middle East's reserves-to-production ratio exceeds 100 years. [14, page 40]

Table 5. World Natural Gas Reserves by Country as of January 1, 2007

Country	Reserves (Trillion Cubic Feet)	Percent of World Total
World	6,183	100.0
Top 20 Countries	5,602	90.6
Russia	1,680	27.2
Iran	974	15.8
Qatar	911	14.7
Saudi Arabia	240	3.9
United Arab Emirates	214	3.5
United States	204	3.3
Nigeria	182	2.9
Algeria	162	2.6

Table 5. (Continued)

Country	Reserves (Trillion Cubic Feet)	Percent of World Total
Venezuela	152	2.5
Iraq	112	1.8
Turkmenistan	100	1.6
Kazakhstan	100	1.6
Indonesia	98	1.6
Norway	82	1.3
China	80	1.3
Malaysia	75	1.2
Uzbekistan	65	1.1
Egypt	59	0.9
Canada	58	0.9
Kuwait	55	0.9
Rest of the world	581	9.4

Source: International Energy Outlook 2007, Table 6, page 41

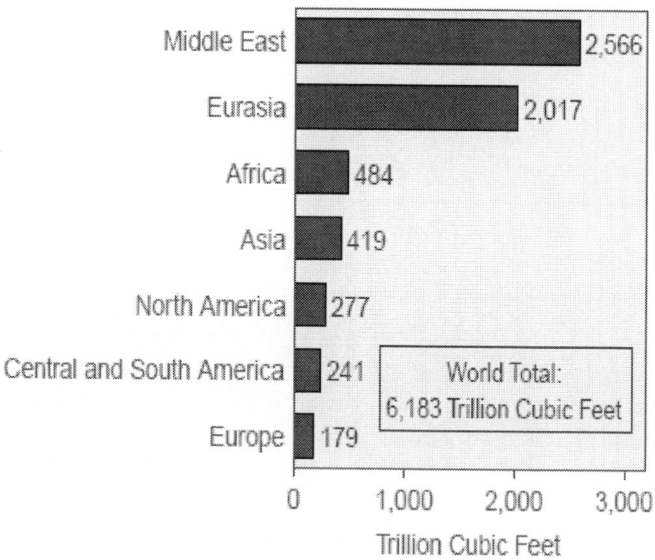

Source: International Energy Outlook 2007, Figure 43, page 4o; "Worldwide Look at Reserves and Production". Oil and Gas Journal, vol.104, No 47 (December 18, 2006), pp.22-23.

Figure 2. World Natural Gas Reserves by Region as of January 1, 2007.

Increased Use of Gas

Natural gas is the second-largest source of energy in Western Europe. Most of it is extracted from British, Dutch, Italian, Romanian, German and Danish fields, with additional gas imported from Russia, Norway, and Nigeria. Russian gas imports account for 26% of EU consumption, representing 40% of the imported gas consumed by households and businesses. In Central and Eastern Europe, Russian gas accounts for 87% of total imports and 60% of consumption. (See Table 6). Demand for natural gas is expected to rise significantly over the next 15–20 years, coincidentally with a steady depletion of reserves in EU countries. The consequence of this situation is an increase in the import of gas from countries outside the EU. For this reason, in an absolute terms Europe will be the region where the imports of natural gas will increase in the following decades. This increase will be carried out through different gas pipeline projects crossing different countries now under construction and due to the increment of liquefied natural gas trade. Reliance on imports of gas is expected to increase in Europe from 57% to 84% by 2030.

Table 6. Major European Recipients of Russian Natural Gas Exports in 2004

Rank	Country	Imports Bcf/year	% Domestic Consumption
1	Germany	1,110	44
2	Italy	777	29
3	Turkey	473	65
4	France	470	26
5	Hungary	378	72
6	Finland	269	100
7	Slovakia	261	100
8	Poland	258	60
9	Czech Rep.	240	82
10	Austria	201	63
11	Bulgaria	184	94
12	Romania	177	24
13	Former Yug.	74	-
14	Greece	74	92
15	Switzerland	18	17

Source: US Department of Energy, Energy Information Administration "Russia Natural Gas, Country Analysis Brief, January 2006

Increasing imports from some of the supplier countries would be a costly endeavor requiring massive investments to develop the necessary extraction, production, and transportation infrastructure. Piping gas from Algeria would require undersea pipelines, which are far more costly than overland pipelines. Norway's reserves are limited. For is reason, the EU is looking to tap other sources of natural gas, such as the Caspian Basin and Central Asia, but these regions, like the Middle East, are characterized by high levels of political instability and are thus less reliable as suppliers. As for liquefied natural gas (LNG) imports, they are economically viable only over large distances.

The world natural gas consumption is expected to increases by 1.9% per year on average; this means an increase from about 100 trillion cubic feet in 2004 to 163 trillion cubic feet in 2030. By energy source, the projected increase in natural gas consumption is second only to coal. In the case of Europe and other OCDE countries, natural gas is expected to be the fastest growing fuel source, with demand increasing at an annual average rate of 1.4%, from 18,8 trillion cubic feet in 2004 to 23.0 trillion cubic feet in 2015 and 26.9 trillion cubic feet in 2030. (See Figure 3).

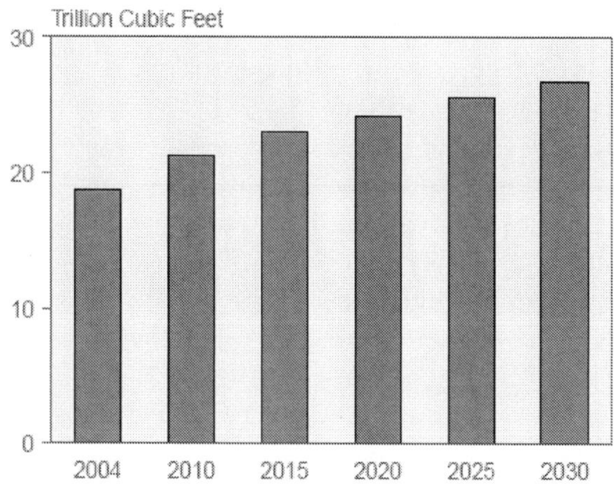

Source: 2004: International Energy Outlook 2007, Figure 48, page 42; Derived from Energy Information Administration (EIA). International Energy Annual 2004 (May-July 2006), web site www.eia.doe.gov/iea. Projections: EIA System for the Analysis of Global Markets (2007), Figure 48, page 45

Figure 3. Natural Gas Consumption in OECD, Europe, 2004-2030.

Growth in natural gas use for electricity generation is projected to account for the majority of total incremental gas use to 2030. Natural-gas-fired generation is less carbon-intensive than oil- or coal-fired generation and is expected to remain more cost-competitive than renewable energy, making natural gas the fuel of choice for new generating capacity in OCED- Europe.

Which are the main factors that are limiting an increase in the use of gas for electricity generation within the EU? One of these factors is the high levels of concentration and consequent concerns about wholesale market manipulation in many national gas markets.

Both gas and electricity markets are susceptible to concentration due to the existence of pre-liberalization monopolies and their natural characteristics. In particular, electricity is a product with low elasticity of demand, so that even with relatively high prices, customers do not reduce their consumption very significantly. This, the effects of congestion and the need for continual balance of supply and demand, increase the scope for market dominance. For gas, competition difficulties also have their roots in the structure of the gas industry outside the EU, as well as national production in member states. These problems are made worse due to the national scope of markets and the lack of integration. In addition, over time and provided there is fair acces to networks, new investment will contribute to eroding concentration. This would be particularly encouraged if incumbents or TSOs were to publish a list of suitable electricity generation or gas storage sites, and adopt mechanisms for releasing such sites to new investors. In the short term, market participants have identified a range of other possible measures that would also assist the process of moving to more competitive and contestable electricity and gas markets. These measures are: a) transparency; b) contract structures: and c) gas storage. [24, pages 15]

Transparency

One of the measures that were identified by market participants is transparency. The problem of concentration is made worse where dominant companies are not required to reveal information to other market participants. For example, wholesale price movements are often caused by variations in production or the use of import capacity by the largest electricity and gas companies. If smaller market participants are unable to track the underlying causes of changes in market price, they will be at a disadvantage. A higher degree of transparency would also allow for improved market surveillance. [24, page 15] The Commission intends to improve the transparency through either the adoption of new legislation or by modifiying existing Regulation (EC) No 1228/2003 or through using Regulation (EC) No 1775/2005.

Long-Term Gas Transmission and Downstream Contracts

The Commission has "repeatedly acknowledged the role of long-term contracts between external producers (i.e. upstream) and companies supplying customers in the EU. These long-term contracts reflect the need for upfront investments to be undertaken and have an important role to play as regards access to cost-effective energy inputs".

It is important to stress that the gas Regulation already "imposes strict use-it-or-lose-it conditions regarding transmission contracts. This includes contracts, which were concluded under Directive 91/296/EEC on the transit of natural gas through grids. These requirements, when combined with additional investment in gas networks, may help overcome the current blockages to meaningful competition. Further development of use-it-or-lose-it guidelines would also help competition develop more rapidly". [24, page 16]

Access to Gas Storage Facilities

According with reference 24, another factor which is affecting competition in the gas sector "is the limited availability of storage, which is often in the hands of the incumbent companies. Although storage is not a natural monopoly, facilities in certain locations may have a large impact on the functioning of the internal market. Voluntary guidelines for good practice for storage system operators (GPSSO) were agreed in 2004, but the findings of the ERGEG final report on monitoring the implementation of these guidelines showed a rather disappointing picture in terms of compliance".

The Commission therefore "intends to examine measures, which would best balance the need for effective access with maintaining incentives for new storage developments". [24, page 16]

Gas in the Industrial Sector

Natural gas remains a key fuel in the industrial sector. The industrial sector accounted for 44% of world natural gas consumption in 2004 and it is projected to account for 43% in 2030. With world oil prices expected to remain high relative to historical levels throughout the projection period, natural gas is projected to displace liquids in the industrial sector to some extent. Industrial use of natural gas is projected to increase at an average annual rate of 1.9% from 2004 to 2030, as compared with an average increase of 1.1% per year for liquids consumption in the industrial sector. (See Figure 4) [14, page 39]

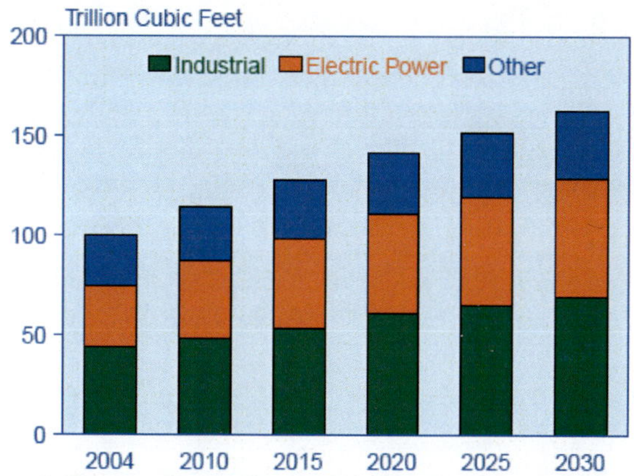

Source: 2004: International Energy Outlook 2007, Figure 40, page 39; Derived from Energy Information Administration (EIA). International Energy Annual 2004 (May-July 2006), web site www.eia.doe.gov/iea. Projections: EIA System for the Analysis of Global Markets (2007). Figure 40, page 39

Figure 4. World Natural Gas Consumption by End-Use Sector. 2004-2030.

Strict EU environmental regulations will force EU member states to replace high-emission fossil fuels, such as coal, with cleaner burning energy sources, such as natural gas. [7, page 1]. This change will reduce significantly the emission of CO_2 to the atmosphere in the coming years and will facilitate that the EU can accomplish their commitment with the Kyoto Protocol.

Summing up can be stated that in the EU, as well as in other countries and regions, gas remains a fuel of choice given its high efficiency and its effects in the environment. In other words, gas will be the fuel of choice between oil and carbon. However, even gas will need in the EU an investment of €150 billion on gas-fired power plant and an additional 220 billion of Euro on gas infrastructure, in order to coup with an increase in the participation of gas in the Europe's energy balance.

The world gas production by country and region is included in Annex 2.

II.1.3-Coal

The security and diversity of energy supply is causing growing concern in the EU. A renaissance in coal use, the major energy source 40-50 years ago, but currently with a modest and continuously declining contribution to the EU's energy supply, could potentially improve the energy balance of the EU.

The recent trends in coal markets make the long-term supply prospects for coal uncertain. The sharp increase in oil prices in 2005 and 2006, and the temporary cutback in natural gas supplies from Russia at the end of 2005 have boosted concerns about the security, diversity, reliability and affordability of energy supplies in the EU[3]. [15, page 6]

Coal was the main energy source not only in Europe but also worldwide until the 1960s. Owing to advances in oil extraction, conversion and application technologies, coal then began to lose market share to oil. The entry of natural gas and nuclear power into the energy market at the beginning of the 1970's put further pressure on coal. All these new energy sources were cleaner to use and in some cases even cheaper. Gradually, coal started to be perceived as a "dirty and old-fashioned fuel for use in poorer countries". As a result, despite the rising energy demand, gross coal consumption in the EU-15 has been declining since 1970, while the share of coal in gross inland energy consumption has more than halved, from more than 30% to approximately 15%. In contrast, coal retained a 25% share in gross inland energy consumption globally over the period 1970-2000. [15, page 7]

However, in recent years there is a renewed interest in the use of coal in the EU, as well as in other countries, for different purpose, based on a wide perception that coal is an abundant, widely available, cheap, affordable and reliable, energy source. The main reasons for this renewed interest are the following:

1) There is more coal than oil and gas worldwide, particularly in Europe. With current consumption trends, the reserves-to-production (R/P) ratio of world proven reserves of coal is higher than that of world proven reserves of oil and gas — minimum 155 years versus 40 and 65 years respectively (See Figure 5). World coal reserves are also more evenly distributed around the globe compared to oil and gas reserves. The geopolitical distribution of world coal reserves differs from that of oil and gas.

2) Historically, coal prices have been lower and more stable than oil and gas prices, owing to the more even spread of coal reserves and hence the smaller room for price manipulation. The EU has larger reserves of coal than of oil or gas, even though it does not hold a large share of world coal reserves (only 4.3% of the total world coal recoverable reserve). Consequently, import dependence on fossil fuels (i.e. mainly coal) is lower than the dependence on gas and oil. A more complete and efficient exploitation of indigenous coal reserves would reduce the EU's overall energy import dependence.

---

[3] In 2007 and 2008 the oil price increase significantly reaching more that US$ 140 per barrel

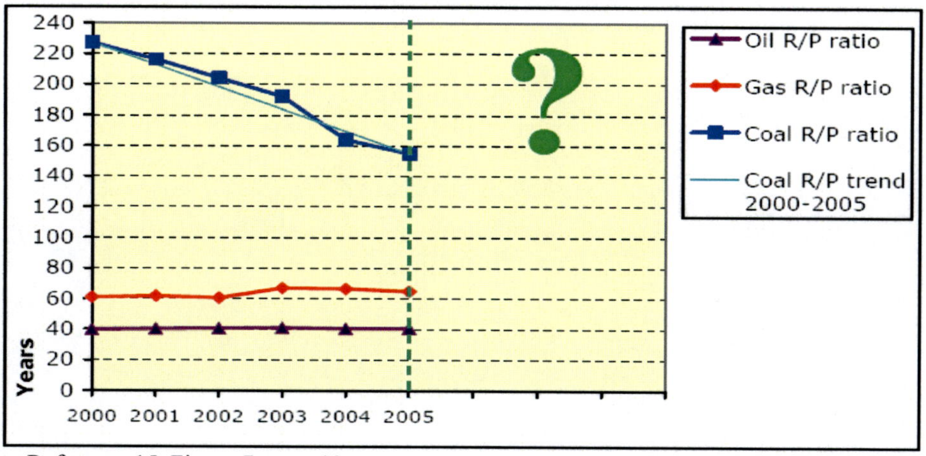

Source: Reference 15, Figure 7, page 10.

Figure 5. Oil, Gas and Coal Reserves-to-Production (R/P) Ratios in the World for 2000-2005

Increase in the Consumption of Coal

The development of novel and more environmentally friendly coal technologies, called Clean Coal Technologies, have change somehow the perception that coal is a "dirty and old-fashioned fuel for use in poorer countries". Designed to enhance both the efficiency and the environmental acceptability of coal extraction, preparation and use, these technologies are believed capable of bringing coal back into fashion. This is because the environmental concerns with coal are associated with the ways in which coal is used rather than with coal itself. Although some Clean Coal Technologies are still at the research and development stage, they are enjoying growing interest worldwide, particularly in the EU.

With the use of new technologies, the world consumption of coal is expected to increase from 25% to 28% in 2030. However, the increase in the production of coal will not take place in the EU. This main increase will take place mainly in areas outside Europe, where the extraction costs are smaller, like in Australia, China, India, Indonesia, North America and Latin America. In Europe, the use of coal will concentrate in the future mainly in the electricity generation. (See Annex 4) This increase will be, however, more slowly than the gas and oil.

Coal consumption in OECD-Europe is expected to declines by 1.7 quadrillion Btu (13%) from 2004 to 2030, according with the document IEO-2007. However, the region is and will continue to be a major market for coal. Coal consumption in OECD-Europe, at 13.1 quadrillion Btu in 2004, represented 28% of total OECD coal use. It is projected to drop to 11.5 quadrillion Btu in 2030, or 19% of total OECD coal use. The major coal-consuming countries of the region, all with consumption of 0,7 quadrillion Btu or more in 2004, include Germany, Poland, the United Kingdom, Spain, Turkey, and the Czech Republic. Although OECD-Europe relies heavily on imports of hard coal, low-Btu lignite represents an important domestically produced source of energy. In 2004, lignite accounted for 47% of the region's total coal consumption on a tonnage basis and 23% on a Btu basis. [14, page 52]

Participation of Coal in Electricity Generation

One of the main uses of coal as energy primary source is for electricity generation. Several hundreds of coal-fired plants has been built in the EU, but several of them need to be replace or refurnish, if coal is going to continue to be used with this purpose. Plans to replace or refurbish existing coal-fired capacity in a number of the countries of OECD -Europe are an indication that coal will continue to play an important role in the region's overall energy mix. In addition to some recent additions of coal-fired capacity (primarily in Turkey), electricity producers in Germany, Spain, France, Italy, Poland, the Czech Republic, and Slovakia have revealed plans to upgrade or replace existing coal-fired generating facilities over the next two decades. Power producers in Germany plan to build nearly 11 GWe of new coal-fired generating capacity by 2012, primarily to replace existing, less efficient, coal-fired capacity. A key incentive for the new coal builds in Germany is a provision guaranteeing carbon dioxide emission rights for the new capacity during the first 14 years of its operation. [14, page 53]

Coal is projected to provide larger shares of the total energy used for electricity generation worldwide in 2030 than in 2004. The coal energy share it is expected to increases, at world level, from 41% to 45%.

All these facts suggest that coal could have a future in the European energy mix, particularly in electricity generation, undergoing a real renaissance and becoming once again one of the fossil fuel options. Such a perspective could become reality. However, a few other very important facts that are often overlooked should be taken into account before this happens.

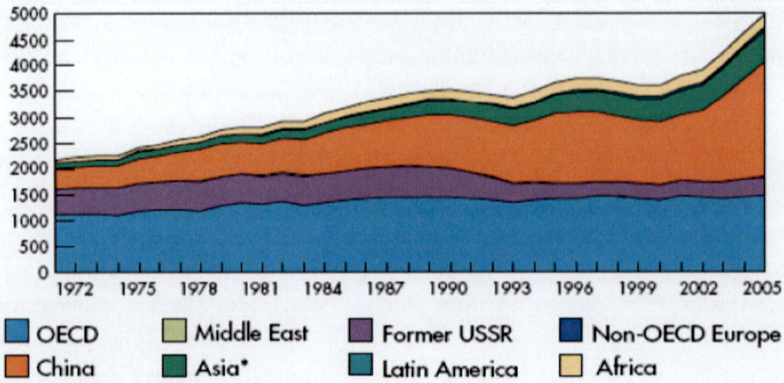

Source: Reference 15, Figure 23, page 24.

Figure 6. Absolute (Mt) and Relative (%) Evolution of World Hard Coal Production by Regions During 1971-2005

The recent growth in coal consumption has not been matched by a corresponding development of the supply base, unlike with oil and gas. Consequently, the coal R/P ratio, which had remained constantly above 200 years for several decades, plunged from 227 years in 2000 to only 155 years in 2005.

However, the use of coal in Europe in the near future should take into account that "coal is the fossil fuel with the highest carbon intensity, having carbon content of 50%-98%. Other coal components are hydrogen (3%-13%), oxygen, and small amounts of nitrogen, sulphur and other elements. Coal also contains different proportions of water and inorganic matter that remain as residue (ash) upon burning. These large variations in coal composition determine the type of coal available for different applications". (See Figure 8) [15, page 11]

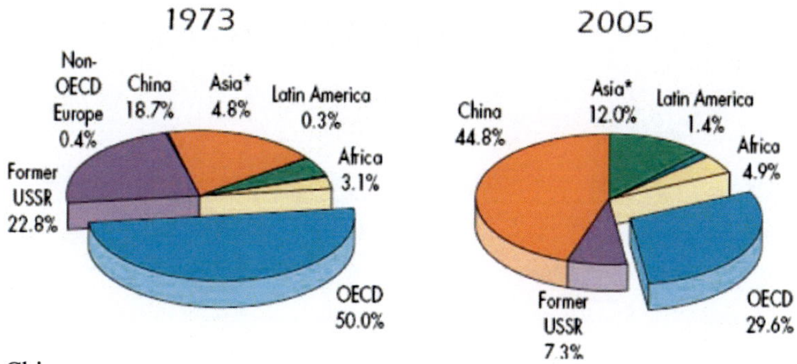

* Asia excludes China.

Source: Reference 15, Figure 23, page 25

Figure 7. Evolution of Coal Production by Regions During 1973-2005

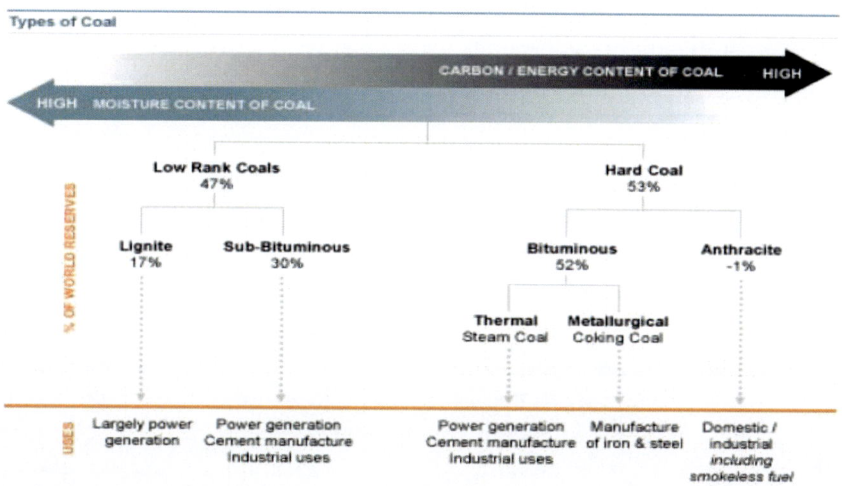

Source: Reference 15, Figure 8, page 11.

Figure 8. Types of Coal Depending on Carbon/Energy and Moisture Content, with their Shares in Total Coal Reserves and Typical Applications.

Figure 9 shows that the electricity generation sector is the main consumer of coal worldwide. This is not surprising as electricity and heat generation account for more than half of global coal demand. The metallurgical sector comes next, mainly for the production of pig iron and to a lesser extent steel.

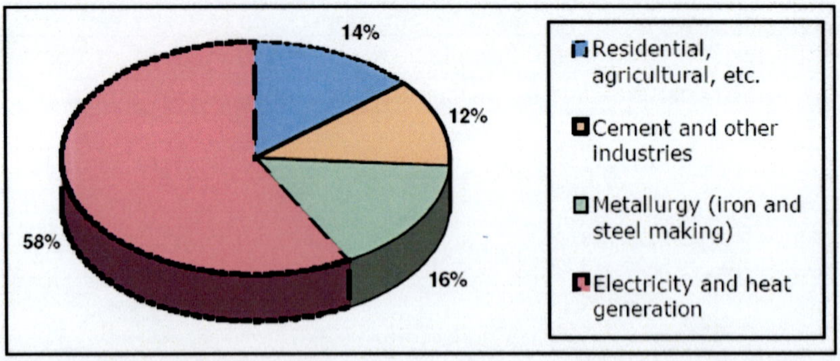

Source: Reference 15, Figure 9, page 12.

Figure 9. Main Applications of Coal

European Coal Production

At present, the prospects for European coal production are quite clear. Indigenous hard coal production in the EU will continue to decline (See Figure 10) for several reasons. Hard coal has been intensively mined in Europe for more than a century and the easier accessible deposits of good quality have already been exploited. As hard coal in Europe can be recovered mainly from underground deposits[4], European coal miners are forced to go for deeper and more difficult to recover reserves of poorer quality, which increases costs. European indigenous hard coal production is two to three times more expensive than imported coal. Some EU countries have therefore ceased hard coal production. In the countries where hard coal production still exists (mainly for socio-economic reasons), it is heavily subsidized[5], but the subsidies are gradually being phased out. The expiry of the European Coal and Steel Community in 2002 has accelerated this process. [15, page 25]

The case of lignite is different. The EU has greater reserves of lignite than of hard coal and reserves are available and exploited in a larger number of countries (Figure 11). Lignite in Europe is typically mined open-cut, which keeps extraction costs low. European lignite production is generally cost-competitive with imports of hard coal without subsidies. Consequently, lignite recovery in the EU will most likely survive, unlike hard coal production. Lignite represents an important energy source for the EU, as it helps to reduce its energy import dependence. Nevertheless, the R/P ratios (under current economic and regulatory conditions) in the major EU lignite producers are rather low, e.g., Germany - 33 years and Greece - 54 years. [15, page 26]

---

[4] With the exception of the UK, where hard coal is recovered also via surface mining.

[5] With the exception of the UK, whose indigenous hard coal production is generally competitive to imported hard coal.

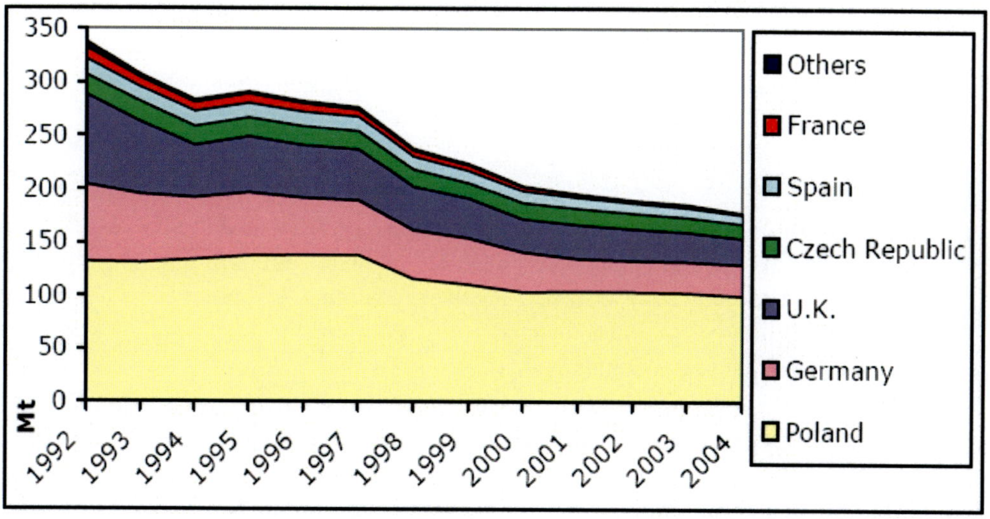

Source: Reference 15, Figure 24, page 25.

Figure 10. Hard Coal Production in EU-27 Over 1992-2004 (Mt).

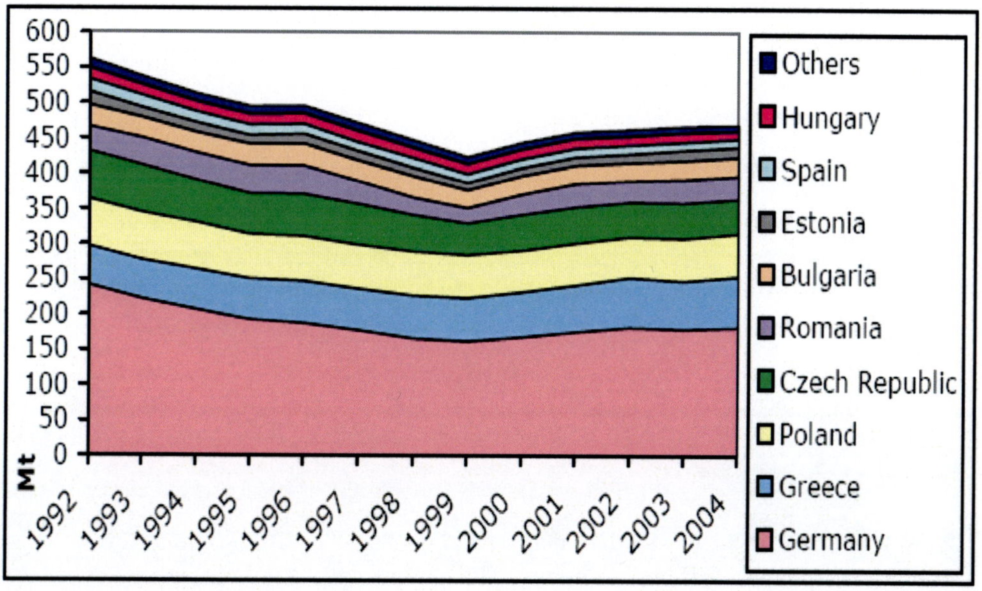

Source: Reference 15, Figure 25, page 26.

Figure 11. Lignite Production in EU-27 Over 1992-2004 (Mt) .

Three largest countries in the former Soviet Union (Russia, Kazakhstan and Ukraine) hold almost 20% of the world's hard coal reserves and 30% of brown coal reserves. Even so, they are not major players in the world coal market. After the substantial economic changes in the 1990s, Ukraine has become a net importer of coal, Russia has re-started exports just recently and only Kazakhstan has managed to keep a noticeable presence on the world market, but with a gradually declining volume and share. The future of coal exports from

Russia remains uncertain, but they will most likely play only a minor role in the coal market. The reasons for this assumption are the following:

1) Except for the coalfields currently in operation in the Kuzbass region (Western Siberia), the great majority of Russia's other coal reserves are located in the central and eastern parts of the country, away from both the main consuming centers in the western (European) part of the country and the seaports in the Far East. Bringing that coal to either destination involves expensive rail transport. Rail transport is also required for the currently operated Kuzbass coalmines. Consequently, Russia is a relatively expensive producer of coal competitive on the world market only when international coal prices are high.
2) Russia's railway system is generally in a poor state. There is a particular dearth of railway infrastructure in the low-populated Siberian regions, where the major coalfields are located. The seaport infrastructure is not any better either. Significant investment is needed to overcome these shortcomings, but the funds have not been secured.
3) Coal production and exports suffer from a generally low level of mechanization and productivity, as well as poor mine safety. [15, page 32 and 33]

Taking into account all the above factors, Russia no longer appears to be a coal– rich country. There are a few options that could be considered by the Russian government to increase coal exports but all of them need important investment.

The most feasible option, which however would not boost coal exports, would be to use the coal from remote fields for internal energy needs. This could be done either by building coal-fired power plants near the remote deposits and then transmitting or exporting the electricity ("coal-by-wire") or via gasification of the coal reserves with subsequent transhipment of the coal-derived gas. In both cases, significant investment is needed to improve and/or construct new transmission networks and/or pipelines. This option might be attractive for Russia from a strategic point of view, since it would release additional natural gas for export- maybe the country's most strategic export commodity. [15, page 33]

Low Coal Technologies

What the EU is doing to avoid increasing the emission of CO_2 while at the same time incrementing the use of coal for electricity generation in the coming decades? The EU is leading the world in responding to climate change by adopting targets and putting a price on coal through the Emissions Trading Scheme, as well as creating a truly internal energy market. The EU must act with equal determination and ambition on a policy for low coal technologies. These are the conditions to catalyse a new industrial revolution. In a coal constrained world, the mastery of technology will increasingly determine prosperity and competitiveness. If the EU falls behind in the intensifying global race to win low coal technology markets, the EU may need to rely on imported technologies to meet established targets, missing huge commercial opportunities for EU businesses. [27, page 4]

The transition to a low coal economy will take decades and touch every sector of the economy, but the EU cannot afford to delay action. Decisions taken over the next 10-15 years will have profound consequences for energy security, for climate change, for growth and jobs in Europe. The cost of action may be high, but the price of inactivity much higher. As an illustration of the scale of the problem, the Stern report (Stern Review on the Economics of Climate Change – UK HM Treasury) estimates that "the cost of action could be limited to around 1% of global GDP per year, while inaction could result in losing 5-20% of global GDP annually".

The vision is of a Europe with a thriving and sustainable economy, with world leadership in a diverse portfolio of clean, efficient and low-coal energy technologies as a motor for prosperity and a key contributor to growth and jobs. Low-coal technologies in general remain expensive and face market penetration obstacles. Energy efficient technologies tend to have high upfront costs, which deter market take-up. A twin-track approach is therefore needed: a) reinforced research to lower costs and improve performance; and b) pro-active support measures to create business opportunities, stimulate market development and address the non-technological barriers that discourage innovation and the market deployment of efficient and low coal technologies. [27, pages 4 and 5]

To achieve the 2050 vision and targets towards complete decarbonisation in the energy balance, the EU needs to develop a new generation of coal technologies through major breakthroughs. Even if some of these technologies will have little impact in the coming years, the EU should continue support these activities in order to ensure that they will be available as early as possible.

Member states have to deliver their contributions to the 20% targets agreed for 2020, and put their energy systems on a pathway toward decarbonization by 2050. A dedicated and substantial effort on energy technology can help achieve the targets in a way that maximizes the benefits for member states and limits the costs. The actions of the member states should aim at increasing investment and provide clear market signals to reduce the risks and stimulate industry to develop more sustainable technologies. In a world in which energy demand is still increasing and Europe's share of greenhouse emissions is set to fall from 15% to 10% by 2030, a global effort and cooperation are needed to address the global challenges. The EU need to take international cooperation on energy technology to a new level, in the same way as the Emissions Trade Scheme is being used to catalyse the development of a global cap and trdae system for carbon. If the EU is unable to create a global market appetite for low coal technologies and ensure their widespread take-up, then achieving the EU's ambitious goals could result in much wasted effort and resources – a high cost strategy for business and society. [27, pages 7 and 8]

What the EU could do to foster international cooperation in the use of low coal technologies in the future by other states? One of the actions that could be promoted is in the field of research; another is setting international standards to stimulate the global development, commercialization, deployment and access to low coal technologies. Other actions that could be implemented include, among other, the following:

a) Networking energy technology centres.
b) Setting up large-scale demonstration projects on technologies with the highest potential in those countries with the possibilities to implement them.
c) Increasing the use of innovative financing mechanisms, such as the Global Energy Efficiency and Renewable Energy Fund.

d) Reinforcing the use of the Kyoto Protocol mechanisms, notably the Clean Development Mechanism for investments in emissions reduction projects, if the post-2012 international agreement on further CO_2 reductions is reached.

The confirmed world coal reserves are around of one trillion of ton, which would be enough to cover the demand of all countries at the current consumption rhythms for no less than 155 years. Although coal deposits are widely distributed, 67% of the world coal recoverable reserves are located in four countries: United States (27%), Russia (17%), China (13%), and India (10%). The coal reserves by country are shown in the following table.

Table 7. World Recoverable Coal Reserves as of January 1, 2003
Billion of tons

Region/ Country	Hard coal	Sub-Bituminous*	Lignite*	Total
World Total	528.8	298.1	170.9	997.7
United States	123.7	110.3	33.5	267.6
Russia	54.1	107.4	11.5	173.1
China	68.6	37.1	20.5	126.2
India	99.3	0.0	2.6	101.9
Other Non-OECD Europe and Eurasia	50.1	18.7	31.3	100.1
Australia and New Zealand	42.6	2.7	41.9	87.2
Africa	55.3	0.2	*	55.5
OECD Europe	19.5	5.0	18.8	43.3
Other Non-OECD Asia	1.4	2.0	8.1	11.5
Brazil	0.0	11.1	0.0	11.1
Other Central and South America	8.5	2.2	0.1	10.8
Canada	3.8	1.0	2.5	7.3
Other[a]	1.8	0.4	0.1	2.3

Source: International Energy Outlook 2007, Table 8, page 50
*Brown coal

The world consumes much more hard coal than brown coal and the gap is growing continuously. In addition, the preference is naturally for coal that is easier (and cheaper) to recover. Without a corresponding increase in hard coal reserves, which will most likely be more difficult and more expensive to exploit than hard coal deposits in the past, the world is going to run out of higher-quality coal much earlier than it will of lower-quality coal. (See Figure 12)

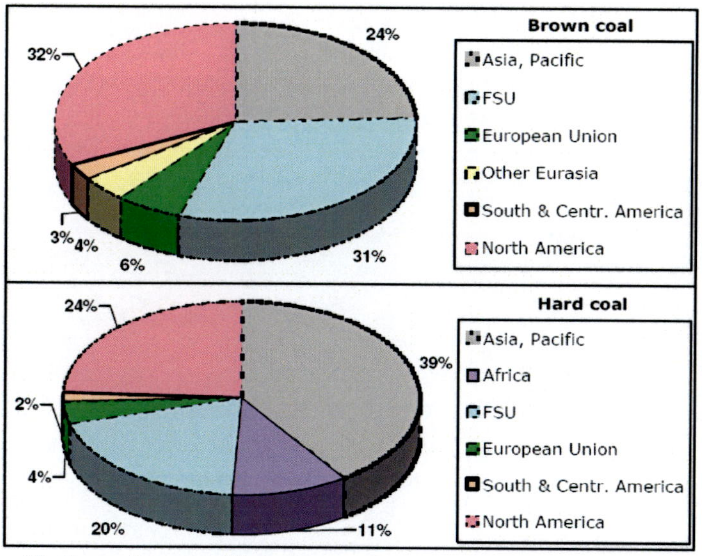

Source: Reference 15, Figure 12, page 15.

Figure 12. Breakdown of World Proven Brown and Hard Coal Reserves at the end of 2005 (%).

Table 8. World Share of the Top 5 and 10 Richest Countries in Hard and Brown Coal Reserves Worldwide (%).

Region	Country	Global Share	
		HardCoal	Brown Coal
North America	USA	23.3 (1)	31.4 (1)
South America	Brazil	-	2.4 (6)
	Colombia	1.3 (10)	-
	Kazakhstan	5.9 (7)	-
	Czech Rep.	-	0.8 (10)
Europe	Germany	-	1.5 (7)
and	Greece	-	0.9 (9)
Eurasia	Poland	2.9 (9)	-
	Russia	10.3 (4)	25.1 (2)
	Turkey	-	0.9 (8)
	Ukraine	3.4 (8)	4.2 (5)
Africa	South Africa	10.2 (5)	-
Asia	Australia	8.1 (6)	9.3 (4)
Pacific	China	13.0 (3)	12.2 (3)
	India	18.8 (2)	-
	Share of top 5 in world total	75.6	82.2
	Share of top 10 in world total	97.2	88.7

Source: Reference 15, Figure 13, page 16.

Note: The relative ranking is in brackets

From the above tables it is easily to confirm that the coal reserves in the EU are very small in comparison with the reserves in other regions in both types of coal. In the case of brown coal the EU reserves is only 6% of the world reserves (4th place) and only 2% in the case of hard coal (6th place).

The CO_2Emissions

An important element that should be kept in mind, when deciding which is the most appropriate place that coal should occupy within the world energy balance, is the negative effect that the use of this type of fuel has for man and the environment. This is particularly true, if we have to use more and more in the coming years of bad coal quality in the energy sector. While more and big coalfield are exploited the content of ashes and humidity will increase. The increment of ashes because of the burning of coal reduces the efficiency of the generating units, causes premature deterioration of some of the equipment used in the industry and has negative ecological and socio-economic effects.

The CO_2 emissions, as well as other harmful substances associated to the burning of coal, will not be able to be eliminated, even with the use of new technologies developed to reduce the mentioned negative effects. This situation could represent an important limitation in the participation of coal in the European and world energy balance in the coming decades.

In only one century we have discharged to the atmosphere as much CO_2 as other gases of green house effect that it concentration exceeds any previous level since the conquered of fire by the human being half million of years ago. All available models indicate that if the humankind does not radically reduce the emissions of gases of green house effect, then humankind is moving toward a catastrophe of disastrous consequences. This catastrophe can be irreversible from the human point of view, if urgent measures are not adopted and implemented by all countries. [6, page 54]

A study made by French specialists points out that "for Europe the growths in the consumption of coal will be above 40% for the coming years, and this increase would add, in an important way, the CO_2 emission to the atmosphere". However, to avoid that this would happens the European governments are taking measures to reduce the CO_2 emission in other activities. This foresees situation should be in the mind of politicians and executives when deciding the position to be given to coal in the world energy balance in the coming decades, particularly in the case of the European region.

It has been demonstrated that the main factor that contributes to the heating of the atmosphere is the CO_2 emission. Only the burnt of fossil fuels that takes place every year launched to the atmosphere around 20,000 million tons of this gas. An increment, as foresee in the use of coal in Europe and other developed countries as energy primary source as well, would increase significantly the quantity of CO_2 that would be emitted toward the atmosphere, even with the use of new technologies more efficient and less polluting. The emission of CO_2 can reach 9 thousand million of tons annually, mainly due to electricity generation. Also, the coal plants produced acid rain, smog, breathing illnesses, mercury contamination and contribute in a great extend to the emissions of gas with green house effect.

In general, the CO_2 emissions in the European countries and other countries of the OECD that signed the Kyoto Protocol could reach 12,500 million of tons for 2010, that is to say about 2,800 millions (29%) above fixed objective. It is important to single out that Russia and other countries of Central and Eastern Europe are in a different situation, since their CO_2 emissions are below of the assumed commitments. The level of CO_2 emissions by the main energy sources is shown in the following figure.

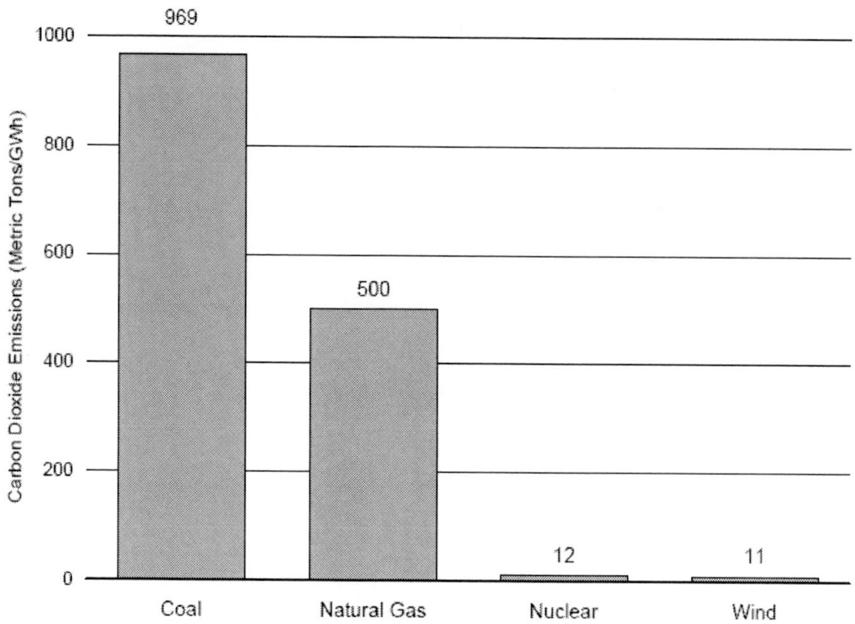

Source: "Life Cycle Energy Cost of Wind and Gas-Turbine Power" by White, Radcliffe and Kulcinski, University of Wisconsin, 1990 (http://fti.neep.wisc.edu/FTI/POSTERS/sww_energy_ctr.pdf) and CA-OWEE, Final Report, Dec 2001, p 7-1 (http://owe.starforze.net/ca-owee/indexpages/downloads/CA-OWEE_Complete.pdf).

Figure 13. Life Cycle Carbon Dioxide Emissions by Fuel.

In Annexes 3 and 4, the world coal production by region and the world flows by importing and exporting region for 2005, 2015 and 2030 are included.

II.2- Renewable Energies

The development of renewable energy, particularly energy from wind, hydro, solar, and geothermal is a central aim of the European Commission's Energy Policy. There are several reasons for this:

1) Renewable energy has an important role to play in reducing CO_2 emissions.
2) Increasing the share of renewable energy in the European energy balance enhances sustainability.
3) Helps to improve the security of energy supply by reducing the growing dependence of the region on imported fossil fuels.

Renewable energy sources are expected to be economically competitive with conventional energy sources in the medium to long term. The need for an EU support for renewable energy is clear. Not only some of the new technologies, especially wind energy, small-scale hydropower and solar thermal applications, are economically viable and competitive with respect other energy sources but European reserves of some of the fossil fuels are limited and located outside of the EU in some cases in instable political areas.

The EU sets out a strategy to double the share of renewable energies in gross domestic energy consumption in the region by 2010. This means to increase from the current participation of 6% in the region energy balance to 12%. The main elements of the EU strategy are the following:

a) Introduction of market measures in the internal regulatory and fiscal spheres.
b) Reinforcement of those EU energy policies, which allows an increase penetration of the renewable energies in the energy balance of the European countries.
c) Strengthening the co-operation between EU member states in the use of renewable energy sources.
d) Support measures to facilitate investment and enhance dissemination and information in the renewable energy fields.
e) A 4.9 % growth for renewable primary energy.

For 2020, it is expected that the share of renewable energies in gross domestic energy consumption in the EU be 20%, and a binding 10% target for the share of renewable energy in transport petrol and diesel, instead of a target of 5.5% previously approved. This is done for the following reasons:

1) The transport sector is the sector presenting the most rapid increase in greenhouse gas emissions of all sectors of the economy.
2) Bio-fuels tackle the oil dependence of the transport sector, which is one of the most serious problems of insecurity in energy supply that the EU faces.
3) Bio-fuels are currently more expensive to produce than other forms of renewable energy, which might mean that they would hardly be developed without a specific requirement. [22, page 8]

This target is truly ambitious and will require major efforts by all EU member states. The contribution of each member state to achieving the Union's target will need to take into account different national circumstances and starting points, including the nature of their energy mix. Member state should have the flexibility to promote the renewable energies most suited to their specific potential and priorities. The way in which member states will meet their targets should be set out in National Action Plans to be notified to the Commission. The Plans should contain sectoral targets and measures consistent with achieving the agreed overall national targets. In practice, in implementing their Plans member states will need to set their own specific objectives for electricity, bio-fuels, heating and cooling, which would be verified by the Commission to ensure that the overall target is being met. [3, pages 13 and 14]

To achieve a 20% share for renewable energies will result in an additional average annual cost of approximately €18 billion – around 6% extra on the EU's total expected energy import bill in 2020. However, this assumes oil prices of US$ 48 per barrel by 2020. If this price rose to US$ 78 per barrel, the average annual cost would fall to US$ 10,6 billion. With an oil price of around US$ 130 per barrel, the average annual cost will fall further. If a carbon price of more than €20 were factored in, the 20% would cost practically no more than relying on "traditional" energy sources, but create many jobs in Europe and develop new technology driven European companies.

When the share of renewable energy sources is analyzed within the EU energy balance in recently years, it is important to take into account the following. The increase in the consumption of renewable energy in 2005 has taken place in a context where total primary energy consumption for EU countries remained stable with respect to 2004. The renewable energy share could even have been greater if 2005 had not been a year that was so marked by a rainfall deficit that significantly slowed down hydraulic production in Latvia, Portugal, Slovenia, Spain, France and Italy. These are the European countries where hydropower production represents a significant share of the renewable energy contribution.

However, it is obvious that with the current rate of increase in the participation of the renewable energy in the EU energy balance, it will very difficult to reach the objective that was set for the EU countries. Under the above-mentioned situation it is expected only to reach a 9%-10% share of renewable resources in the European energy balance in 2010, this means 2%-3% lower that the target (12%). According to the European Wind Energy Association (EWEA) report of 2005, "the following countries could meet the established target for 2010: Germany, Denmark, Finland and Spain".

The main reason for the failure to reach the agreed targets for renewable energy - besides the higher costs of renewable energy sources today compared to "traditional" energy sources - is the lack of a coherent and effective policy framework throughout the EU and a stable long-term vision. As a result, only a limited number of member states have made serious progress in this area and the critical mass has not been reached to shift niche renewable production into the mainstream. [3, page 12]

The Commission estimates that "additional investments of €10 to €15 billion are required to achieve the 12% target the EU has set itself. These resources should come from both public and private sectors".

The main factors that should be taken into account in order to achieve the target goals are the following:

a) Using renewable energy today is generally more expensive than using hydrocarbons, but the gap is narrowing – particularly when the costs of climate change are factored in.
b) Economies of scale can reduce the costs for renewable energy, but this needs major investment today.
c) Renewable energy helps to improve the EU's security of energy supply by increasing the share of domestically produced energy, diversifying the fuel mix and the sources of energy imports and increasing the proportion of energy from politically stable regions as well as creating new jobs in Europe.
d) Renewable energies emit few or no greenhouse gases, and most of them bring significant air quality benefits.

The use of the new energy sources like solar, wind, geothermal, hydro, hydrogen, among others, is expected to suffers an increment of their use as primary sources of energy inside the world energy balance, as well as in Europe to an annually average rhythm of 3.3%. The wind energy will grow to a higher rhythm, especially in Europe and other countries of the OECD whose governments have adopted urgent measures to promote energy projects based on renewable energy sources. However, the scale use of some of these new energy sources present important limitations related with their geographical location, the intensity of the solar radiation and the meteorological conditions, among others.

According with the IEO 2007 document, "the renewable share of total world energy use it is expected to increases from 7% in 2004 to 8% in 2030". In the case of the EU, the Parliament "supported the proposal of a binding target to increase the level of renewable energy in the EU energy mix to 20% by 2020 as a good starting point, and considered that this target should be increased to 25% of the EU energy mix". [22, page 2]

Much of the growth in world renewable energy consumption is projected to come either from mid-to-large-scale hydroelectric facilities in non-OECD Asia and in Central and South America, where several countries have hydropower facilities planned or under construction.

Outside of Canada and Turkey, hydropower capacity is not expected to grow substantially in the OECD nations, because most hydroelectric resources in the region already have been developed or lie far from population centers. Instead, most of the increase in OECD renewable energy consumption is expected to be in the form of non-hydroelectric resources, such as wind, solar, geothermal, municipal solid waste, and biomass. [14, page 9]

According with the IAEA Nuclear Technology Review 2006[6], one of the main characteristics of the renewable energy sources is that this type of energy "is more evenly distributed over the world than oil reserves". For example, 60% of oil reserves are located in only 40 oil fields, often in African, Arabian and Central Asian states, prone to socio-political disruption.

One of the most important objectives of the EU in the next decades is an increase use of renewable energy for electricity production. The objectives of the European Directive on renewable origin electricity production, 21% of gross renewable origin consumption in 2010, does not take into consideration hydroelectric production coming from pump-storage installations functioning with power grid electricity. However, it is important to stress that if hydroelectric production could have been maintained at the level of 2004,, then the share of renewable electricity would have reached 14.6% in 2005, a little over the level reached in 2004. Four countries were particularly affected by drought in 2005: Portugal, Spain, France and Italy. For this reason, for these countries to maintain a share of renewable electricity around 14% efforts were made to develop other renewable sectors. This is especially truth in the case of Spain with a strong growth in its wind power sector.

Germany with a low hydroelectric potential, is a mainstay of renewable electricity production in Europe. In a very few years the country has succeeded to raise its share of renewable sectors up to a level very near to that of France. This is the result of ten years of policies in favor of electricity produced from the wind power and biomass sectors. The use of these two types of renewable energies are also at the origin of the increase in the share of renewable electricity in the Netherlands and Ireland. Summing up can be stated that in the case of the EU renewable energy sources are largely indigenous, they do not rely on the

[6] See reference 21, page 15

future availability of conventional sources of energy, and their predominantly decentralized nature makes EU countries economies less vulnerable to volatile energy supply. Consequently, they constitute a key element of a sustainable energy future. However, for renewables to become the "stepping stone" to reaching the dual objective of increased security of supply and reduced greenhouse gas emissions, a change in the way in which the EU promotes renewables is needed. Strengthening and expansion of the current EU regulatory framework is necessary. It is important to ensure that all member states take the necessary measures to increase the share of renewables in their energy mix. [22, page 3]

II.2.1. Hydropower

Hydroelectric power plants convert the kinetic energy contained in falling water into electricity. The energy in flowing water is ultimately derived from the sun, and is therefore constantly being renewed. Energy contained in sunlight evaporates water from the oceans and deposits it on land in the form of rain. Differences in land elevation result in rainfall runoff, and allow some of the original solar energy to be captured as hydroelectric power. Hydroelectric power plants can generally be divided into two categories:

a) "High head" power plants.
b) "Low head" power plants.

"High head" power plants are the most common and large hydroelectric power plants. Generally utilize a dam to store water at an increased elevation. The use of a dam to impound water also provides the capability of storing water during rainy periods and releasing it during dry periods. This results in the consistent and reliable production of electricity, able to meet demand. High head plants with storage are very valuable to electric utilities because they can be quickly adjusted to meet the electrical demand on a distribution system.

"Low head" power plants are those that utilize either a low dam or weir to channel water, or simply use the "run of the river". Run of the river generating stations cannot store water, thus their electric output vary with seasonal flows of water in a river.

"Pumped storage" is another form of hydroelectric power. Pumped storage facilities use excess electrical system capacity, generally available at night, to pump water from one reservoir to another reservoir at a higher elevation. During periods of peak electrical demand, water from the higher reservoir is released through turbines to the lower reservoir, and electricity is produced. Although pumped storage sites are not net producers of electricity - it actually takes more electricity to pump the water up than is recovered when it is released - they are a valuable addition to electricity supply systems. Their value is in their ability to store electricity for use later when peak demands are occurring. Storage is even more valuable if intermittent sources of electricity, such as solar or wind, are hooked into a system. [5, page 2]

Hydropower for Electricity Generation

Today, electricity generation from hydropower makes a substantial contribution to meeting the increasing world electricity demand. The installed hydropower capacity amounted to 22% (or around 700 GWe) of the total installed capacity for electricity generation.

Besides the fact that hydropower currently makes up a substantial share of the total amount of electricity generated in the world, the arguments for continued and increased utilization of hydropower are based on its advantages when compared to other sources of energy. The key positive characteristics of hydroelectricity can be summarized as follows:

a) Low-cost.
b) Effectiveness.
c) Sustainability.
d) Stored in large quantities.

The Utilization of Hydropower in Europe

The role of hydropower is expected to continue to play an important role in the world energy balance in the future. World production of hydroelectricity has grown steadily between 2.3% to 3.1% per year on average since 1980. Worldwide average growth rates of hydroelectricity generation in the future are estimated to be about 2.4% per year up to 2020. The highest growth rates are expected in developing or strongly industrializing countries with high, yet unexploited hydropower potential, particularly in Eastern Europe. About 22% of the total world demand for hydroelectricity is attributed to Western Europe. However, most of the practical hydroelectric resources available in the region have already been exploited and little further development is expected in the near future.

It is important to single out that since the 1970s annual energy production of some existing hydropower stations in Western Europe have decreased, in particular in Portugal, Spain and other Southern European countries. This reduction has been attributed to changes in average discharge; but whether this is due to temporary fluctuations or already the consequences of long-term changing climate conditions is not yet known. [17, page 8-1]

However, it should be noted that hydropower projects have also been known for their negative effects concerning environmental and social issues. [17, page 8-2] This negative effects must be considered when planning the construction of new hydropower plants, particularly in the EU.

The table below provides an overview of hydropower utilization in Europe. The share of hydropower within the total electricity generation varies considerably between countries, ranging from negligible to 99%. These differences in countries reflect both geographic and climatic constraints and suitability, as well as, to some degree, political guidelines in operation.[17, page 8-2]

Table 9. Hydropower Utilization in Europe 1998-2000

Energy Union		Annual generation of hydroelectricity[1]		Installed capacity for generation of hydroelectricity[2]				
				in total		by category		
		TWh	% of total electricity generation	GW	% of total capacity for electricity generation[3]	Run-of-river, incl. pondage GW	Reservoir and mixed pumped GW	Pure pumped storage GW
Austria	UCTE	42.2	78.0	10.9	67.7	5.5	5.4	0.0
Belgium	UCTE	1.7	2.1	1.4	8.9	0.1	0.0	1.3
Croatia	UCTE	5.8	59.0	2.0	57.0	n.a.	n.a.	n.a.
France	UCTE	66.9	13.3	24.3	22.1	10.8	11.6	1.9
Germany	UCTE	23.6	4.8	8.3	8.3	2.7	1.4	4.2
Greece	UCTE	4.1	9.0	3.0	32.9	0.7	2.3	0.0
Italy	UCTE	50.3	19.2	19.8	27.5	8.2	7.4	4.2
Luxembourg	UCTE	0.9	77.1	1.1	93.2	0.0	0.0	1.1
The Netherlands	UCTE	0.0	0.0	0.0	0.0	0.0	0.0	0.0
Portugal	UCTE	11.6	30.9	4.2	45.5	2.1	2.1	0.0
Slovenia	UCTE	3.5	28.0	0.8	32.1	n.a.	n.a.	n.a.
Spain	UCTE	31.4	16.1	16.3	35.4	6.1	7.7	2.5
Switzerland	UCTE	37.8	57.9	13.8	77.2	4.0	9.5	0.3
Yugoslavia, Macedonia, Bosnia-Herzegovina	UCTE	13.2	32.0	4.6	36.6	1.9	2.0	0.7
Czech Republic	CENTREL	2.3	3.4	2.0	14.3	0.2	0.7	1.1
Hungary	CENTREL	0.2	0.5	0.1	0.7	0.1	0.0	0.0
Poland	CENTREL	4.0	2.8	2.0	6.4	0.4	1.4	0.2
Slovakia	CENTREL	5.0	17.5	2.4	35.1	0.8	0.9	0.7
Denmark	NORDEL	0.0	0.1	0.0	0.1	n.a.	n.a.	n.a.
Finland	NORDEL	12.6	18.9	2.9	17.8	n.a.	n.a.	n.a.
Iceland	NORDEL	6.0	84.1	1.0	77.8	n.a.	n.a.	n.a.
Norway	NORDEL	122.1	99.4	27.6	98.9	n.a.	n.a.	n.a.
Sweden	NORDEL	70.4	46.8	16.2	48.5	n.a.	n.a.	n.a.
Albania	others	5	96	n.a.	n.a.	n.a.	n.a.	n.a.
Belarus	others	0	0	0	0	n.a.	n.a.	n.a.
Bulgaria	others	3	9	2	17	n.a.	n.a.	n.a.
Cyprus	others	0	0	n.a.	n.a.	n.a.	n.a.	n.a.
Estonia	others	0	0	0	0	n.a.	n.a.	n.a.
Ireland	others	1	5	0	0	n.a.	n.a.	n.a.
Latvia	others	4	77	n.a.	n.a.	n.a.	n.a.	n.a.
Lithuania	others	1	3	0	0	n.a.	n.a.	n.a.
Malta	others	0	0	n.a.	n.a.	n.a.	n.a.	n.a.
Moldova	others	0	7	n.a.	n.a.	n.a.	n.a.	n.a.
Romania	others	19	37	6	27	n.a.	n.a.	n.a.
Russian Federation (Europe and Asia)	others	158	20	44	22	n.a.	n.a.	n.a.
Turkey	others	42	39	10	44	n.a.	n.a.	n.a.
Ukraine	others	16	10	5	9	n.a.	n.a.	n.a.
United Kingdom	others	5	2	1	1	n.a.	n.a.	n.a.

Data sources: UCTE and CENTREL: UCTE 2001 (values of 2000); NORDEL: NORDEL 2000 (values of 1999); others EIA 1999 (values of 1998).

Data sources: UCTE and CENTREL: UCTE 2000 (values of 31.12.1998); NORDEL: NORDEL 2000 (values of 31.12.1999); others EIA 1999 (values of 01.01.1999).

Data sources: UCTE and CENTREL: UCTE 2000 (values of 31.12.1999); NORDEL and others as in 2.

N.a: No data available.

Note: For values of EIA 1999 it is not defined whether data included pumped storage.

Source: Reference 17, Table 8.1, page 8-3

When assessing the future of hydropower utilization in Europe, two main factors must be considered:

a) A change in exploitable river flow, mainly induced by climate change.
b) A change in the existing hydropower park of the individual countries.

In Eastern Europe and the former Soviet Union most of the development of hydroelectricity in the short to mid-term is expected to occur as expansion or refurbishment of existing hydroelectric plants, as the persistent economic problems interfere with the construction of new plants. In East-Central Europe, hydroelectricity already represents a substantial source of power in some countries such as Albania (96%), Croatia (59%) or Romania (37%) of total electricity generation.

Most of the potential for future hydropower expansion lies in Albania, Bulgaria and Romania, as well as in the former Yugoslav republics. However, despite a very large potential for future expansion, yet, these countries have found it difficult to secure financing for such projects.

The Nordic European countries generally show good suitability for hydroelectricity production, both from geographic and climatic aspects, and already have high shares of hydropower forming part of their energy supply. However, in the case of Sweden, for example, there is no scheme for building new large hydropower plants, as proposals for this meet strong opposition. Only old units are refurbished and some new small units are installed. Consequently, no major additions to the current production capacity are expected . Also for Norway, only small increases of the cumulative volume of power stations are expected. [17, page 8-4]

"Latvia, Portugal, Slovenia, Spain, France and Italy are the European countries where current hydropower production represents a significant share of the renewable energy contribution."

In conclusion, due to political and economical reasons the prediction of the future development of Europe's hydropower park is difficult task. Public opposition to large hydro schemes will probably result in very little new development of big dams and reservoirs. Small scale and low head hydro capacity will probably increase in the future as research on low head turbines, and standardized turbine production, lowers the costs of hydroelectric power at sites with low heads. New computerized control systems and improved turbines may allow more electricity to be generated from existing facilities in the future. As well, many small hydroelectric sites were abandoned in the 1950's and 60's when the price of oil and coal was very low, and their environmental impacts unrealized. Increased fuel prices in the future could result in these facilities being refurbished.

It is important to single out that there is a clear agreement or understanding within the EU that a dismantling of the existing hydroelectric plants is not an advisable option; they should continue to operate, and perhaps their number and capacities will be increased in the future.

II.2.2. Solar Energy

The massive use of solar energy is limited by the high costs of the solar collector and of the photoelectric cells. This situation makes, according with the UNESA Electricity

Generation Perspective for 2030, "solar energy the most expensive type of energy in Spain (380 Euro per MWe/h) and for this reason less competitive than other renewable and non-renewable energies available in the country (wind 72 Euro per MWe/h and nuclear 36 Euro per MWe/h)". The situation describe could be the same for other EU countries.

After having been blocked for a long time in its development, the concentrating solar power (CSP) sector is undergoing a full revival today due to falling costs, technologies that are more effective and policies sensitive to environmental questions. Technically, CPS consists in focusing sunlight to heat a fluid to a sufficiently high temperature to produce electricity. CPS use concentrated solar radiation as a high temperature energy source to produce electrical power.

These clean energy technologies are appropriate for applications where direct solar radiation is high. The many types of systems under development (including parabolic troughs, power towers, and dish/engine systems, see Figures 14 and 15) for different markets vary according to the concentration devices, energy conversion methods, storage options and other design variables. Exploitation of solar energy requires sunlight conditions that are particular to certain regions of the world only. In Europe, the best zones for solar power development plans are in the Mediterranean areas covering Spain, the south of France and Italy.

Source: Concentrating Solar Power from Research to Implementation, European Commission, page 14

Figure 14 . Solar Collector Elements of a Parabolic Through System.

The Use of Solar Energy in the EU

The USA, the birthplace of this sector, today concentrates practically all of the installed CSP capacity in the world with 355 MWe in California. Four American states are particularly involved in developing new projects: Nevada with 64 MWe, California where two contracts have been signed to develop 800 MWe between 2008 and 2011, Arizona and New Mexico.

Source : Concentrating Solar Power from Research to Implementation, European Commission, page 13
Figure 15. Solar Power Plant.

In Europe, the sector's rebirth is now being carried onward by Spain, which set an objective of 500 MWe for 2010. Wiser from the experience obtained from its first three CSP sites located in Almeria, Spain inaugurated it's first commercial power plant in Seville in 2006 with 11 MWe, with the next one expected for 2008 in Grenada (Andasol 1 with 50 MWe). Around additional 700 MWe are on the drawing boards. Spain has had a significant increase in its national market with, according to the Institute for Energy Savings and Diversification (Idae), "an additional capacity of 14,5 MWe bringing Spain's total installed capacity up to 51,8 MWe". The Idae has plans to reach the 500 MWe mark by the year 2010.

In France, the sector continues to mobilize actors even though all of the sites that were developed in the early 1980s have now been converted into research centers. In this way, the CNRS (National Center of Scientific Research) has launched the Pégase Project- a new installation coupled with a gas turbine is going to be erected on the historical site of the Thémis solar tower power plant. Experimentation should begin in the year 2009, if financing follows. A second project, named "ThemDish", provides for the installation of five Eurodish parabolas, development by Germany, at the beginning of 2008. In general, France has a solar power capacity installed of 33 MWe. It is expected that the solar power capacity in France reach 120 MWe in 2010 (85 MWe for the French overseas departments and 35 MWe for metropolitan France).

Germany continued to be the leading photovoltaic market in the world in 2005, far ahead of Japan and the USA, with, according to the new German Solar Industry Association (BSW), 603 MWe installed during that year. Germany now represents 85,8% of the total capacity installed in the EU. This unashamed success has inspired other countries to set up the conditions necessary for rapid development of their photovoltaic sectors. In Germany, the purchase price guaranteed until 2007 made possible the operation of a very high number of installations and it is more and more probable, taking the investments made by the German industry into consideration, that an incentive system would be renewable by the government.

With 6,8 MWe solar power capacity installed in 2006-2007, Italy occupies the third place in Europe in the use of solar energy. The Italian decision to increase the participation of solar energy in the country's energy balance is good news because it will create a big new photovoltaic market. The Netherlands occupies the second place in the use of solar energy in Europe.

The Situation of the EU Photovoltaic Market

The EU photovoltaic market reached the limits of the sector's procurement capacity for the first time in 2007. Photovoltaic industrialists would have been able to produce many more modules if it had not been for the current shortage of silicon, the principal raw material of solar cells. In 2005, the European photovoltaic market reached 644 MWe, bringing the cumulated total of installed European capacity to 1 791,7 MWe. This capacity consists essentially in grid-connected applications (solar roofs and facades and photovoltaic power plants) with 94.4% of installed capacity.

The prospects of growth in the photovoltaic market are now good. Silicon producers have finally responded to the expectations of the photovoltaic industry by announcing new production capacities. These extensions have reassured the photovoltaic industry, which in turn has responded by massively investing in new production capacities, in phase with ever-greater demand. This increase in demand remains, however, dependent upon the political will to develop this market at the national level.

Considering these elements a new estimate of European installed capacity of 6,000 MWe in 2010 was set up. The ambition of the Sustainable Energy Europe campaign is "to install an additional 1,500 MWe up to 2008". In the following table the installed photovoltaic capacities of the EU is shown.

Table 10. Installed Photovoltaic Capacities in theEU (in MWp)

	2004 Cumulated Power			2005 Cumulative Power		
	On-grid	Off-grid	Total	On-grid	Off-grid	Total
Germany	908,000	26,000	934,000	1,508,000	29,000	1,537,000
Spain	23,900	13,400	37,300	37,600	14,200	51,800
Netherlands	44,310	4,769	49,079	45,857	4,919	50,776
Italy	18,700	12,000	30,700	25,200	12,300	37,500
France	8,000	18,300	26,300	12,967	20,076	33,043
Luxemboug	23,200	0,000	23,200	23,266	0,000	23,266
Austria	16,493	2,687	19,180	21,126	2,895	24,021
United Kingdom	7,386	0,778	8,164	9,786	0,878	10,664
Greece	1,257	3,288	4,544	1,412	4,032	5,444
Sweden	0.194	3,672	3,866	0,254	3,983	4,237
Finland	0,193	3,509	3,703	0,223	3,779	4,002
Portugal	0,500	2,200	2,700	0,600	2,700	3,300

Table 10. (Continued)

	2004 Cumulated Power			2005 Cumulative Power		
	On-grid	Off-grid	Total	On-grid	Off-grid	Total
Denmark	2.035	0,255	2,290	2,335	0,295	2,630
Belgium	1,210	0,053	1,263	1,712	0,053	1,765
Cyprus	0,255	0,090	0,345	0,490	0,135	0,625
Czech Republic	0,269	0,147	0,416	0,380	0,150	0,530
Poland	0,069	0,165	0,234	0,085	0,232	0,317
Ireland	0,000	0,100	0,100	0,000	0,300	0,300
Slovenia	0,006	0,094	0,100	0,118	0,098	0,216
Hungary	0,055	0,083	0,138	0,085	0,091	0,176
Slovakia	0,000	0,060	0,060	0,000	0,060	0,060
Lithuania	0,000	0,017	0,017	0,000	0,017	0,017
Malta	0,006	0,000	0,006	0,015	0,000	0,015
Latvia	0,000	0,004	0,004	0,000	0.005	0,005
Estonia	0,000	0,002	0,002	0,000	0,003	0,003
Total EU	**1 056,038**	**91,673**	**1,147,710**	**1,691,511**	**100,201**	**1,791,712**

Source: EuroObserv'ER 2006.

II.2.3. Wind Energy

The energy sector of today faces a triple challenge, how to tackle climate change and meet rapidly increasing demand for energy while ensuring the security of its supply? Wind energy can be a significant part of the answer, if sufficient support and increased political will are applied to its development.

The European Commission's May 2004 Communication neatly encapsulates the benefits of wind and other renewable energy sources: "Developing Europe's potential for using renewable energy will contribute to security of energy supply, reduce fuel imports and dependency, reduce greenhouse gas emissions, improve environmental protection, decouple economic growth from resource use, create jobs, and consolidate efforts towards a knowledge based society". [23, page 2]

Wind energy involves no fuel price risk, as there is no fuel. Wind is an endless resource that only requires increased research for its effective extraction, accurate prediction, optimized and intelligent integration, and a market penetration large enough to "absorb" weather systems. Wind energy is one of Europe's largest indigenous energy resources, bigger than oil, coal and gas together. Wind can reduce energy import dependence as part of a more diverse and modernized European electricity supply. If Europe is to have secure electricity supply at stable prices, diversity, mitigating reliance on non-indigenous energy sources is essential. [23, page 15]

Europe is a world leader in renewable energy and in the most promising and matures renewable technology: winds power. Wind energy is a relatively young but rapidly expanding industry. Over the past decade, global installed capacity has increased from 2,500 MWe in 1992 to just over 48,000 MWe at the end of 2006, an annual growth rate of around 30%. (See Figure 16)

Wind energy will not only be able to contribute to securing European energy independence and climate goals. In the future, it could also turn a serious energy supply problem into an opportunity for Europe, in the forms of commercial benefits, technology research, exports and employment. Without reliable, sustainable, and reasonably priced energy, there can be no sustainable long-term growth and Europe will become economically disadvantaged. The fact that the wind power source is free and clean is, of course, economically and environmentally significant but the more fundamental point at issue is that the cost of the electricity is fixed, once the plant has been built. The long-term implication is that the economic future of Europe can be planned on the basis of known and predictable cost of electricity, derived from an indigenous energy source free of all the security, political, economic and environmental disadvantages associated with oil and gas. [16, page 5]

Wind energy technology has made major progression since the generation of wind turbines from the early 1980s. Twenty-five years of technological progress have resulted in today's wind turbines being a state-of-the-art modern technology - modular and quick to install. At a given site, a single modern wind turbine annually produces 180 times more electricity and at less than half the cost per kWh than its equivalent twenty years ago.

The Situation in the Use of Wind Energy in Europe

Installed wind power in Europe in 2006 stands over 48,000 MWe. If positive policy support continues to develop, the European Wind Energy Association (EWEA) has projected that "wind power will achieve an installed capacity of 80,000 MWe in the EU by 2010. By 2020, this figure is expected to increase to 15%, with wind power providing energy equal to the demand of more than 195 million European household consumers".

If the increase use of wind power projections were correct, then this would mean an annual business worth of some €80 billion. Wind provides now around 3% of European power needs, but is capable of delivering 15% by 2020 and in excess of 20% by 2030.

It is important to stress that 80% of today's world capacity is located in just five countries, Germany, Spain, Denmark, the USA and India, three of them in the European region. In terms of overall capacity, Germany will continue as European leader, with a total capacity installed of 20,622 MWe followed by Spain with 11,615 MWe and Denmark with 3,136 MWe. In Germany, since 2000, electricity produced from renewable energy sources is given priority for connection to the grid, the priority of purchase, of transmission, and for payments. German grid operators are obliged to feed in electricity produced from renewable energy and buy it at a minimum price within their supply area. In order to allow for technological progress and continuous cost reduction, the compensation rates are subject to nominal annual digression.

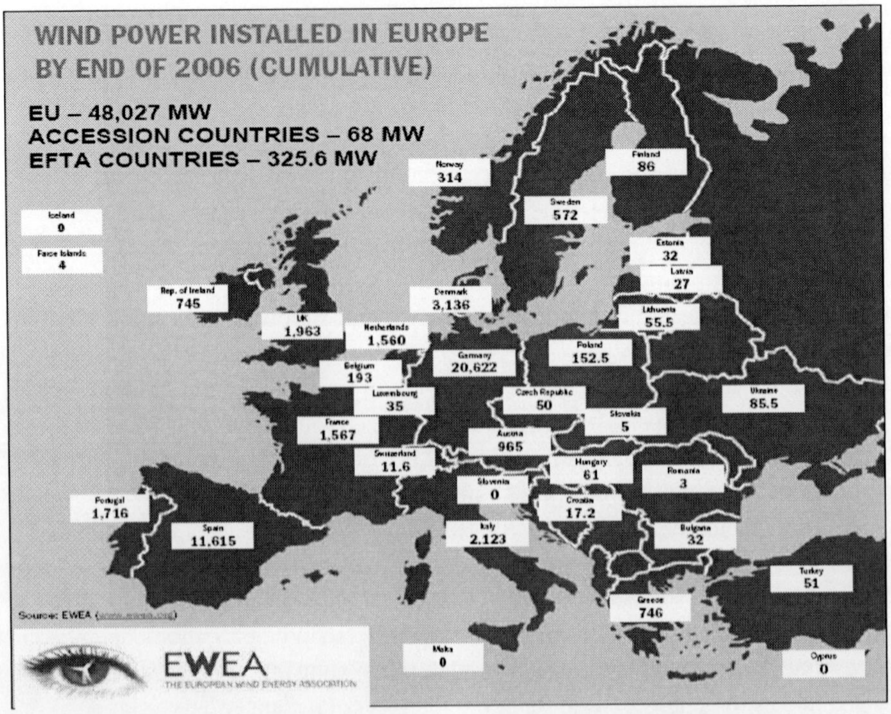

EU CAPACITY (MW)

Countries	Total end 2005	Installed 2006	Total End 2006
Austria	819	145.6	965
Belgium	167.4	26.3	193
Bulgaria*	10	22	32
Cyprus	0	0	0
Czech Republic	28	22	50
Denmark	3,128	11.5	3,136
Estonia	32	0	32
Finland	82	4	86
France	757	810	1,567
Germany	18,414.9	2,233.1	20,622
Greece	573.3	172.5	748
Hungary	17.5	43.40	61
Ireland	495.5	249.9	745
Italy	1,718	417	2,123
Latvia	27	0	27
Lithuania	6.4	49.05	55.5
Luxembourg	35.3	0	35
Malta	0	0	0
Netherlands	1,219	356	1,560
Poland	83	69.3	152.5
Portugal	1,022	694.4	1,716
Romania*	1.69	1.3	3
Slovakia	5	0	5
Slovenia	0	0	0
Spain	10,028	1587.16	11,615
Sweden	509.5	62.15	572
UK	1,332	634.4	1,963
EU-15	**40,301**	**7,404**	**47,644**
EU-10	**199.2**	**183.7**	**383**
EU-25	**40,500**	**7,587.9**	**48,027**
EU-27	**40,511**	**7,611**	**48,062**

Accession countries (MW)

	End 2005	Installed 2006	End 2006
Croatia	6	11.2	17.2
Turkey	20	30.85	51
Total	**26**	**42.05**	**68.2**

EFTA countries(MW)

	End 2005	Installed 2006	End 2006
Iceland	0	0	0
Liechtenstein	0	0	0
Norway	267	47	314
Switzerland	11.6	0	11.6
Total	**278.6**	**47**	**325.6**

Other countries (MW)

	End 2005	Installed 2006	End 2006
Faroe Islands	4	0	4
Ukraine	77.3	8.3	85.6
Total	**81.3**	**8.3**	**90.6**

	End 2005	Installed 2006	End 2006
TOTAL EUROPE	40,898	7,708.4	48,545

Note: Due to previous-year adjustments, project decomissioning of 61 MW and repowering, and rounding, the 2006 end-of-year cumulative capacity total does not exactly match the year-end 2005 total plus the 2006 additions.

* Bulgaria and Romania are members of the European Union from 01/01/2007

Source: EWEAFigure 16. Wind Power Installed in Europe up to 2006

For onshore wind farm development in Germany, the amount of sites commissioned is decreasing while repowering is proceeding at a slow pace. Repowering could play a stronger role in Germany today, but numerous Federal States have issued recommendations concerning vicinity and height restrictions and therefore hindering the use of a large number of suitable sits can no longer be used for the installation of modern turbines.

A recent study published by the Ministry of the Environment includes a projection for 1,100 MWe offshore wind capacity by 2010, and as much as 12,000 -15,500 MWe by 2020. More than 60,000 MWe of projects have been proposed by various companies and consortia. Eleven projects were licensed by the end of 2005 in the North and the Baltic Seas by the national maritime authority. The north German state of Schleswig-Holstein has 1 800 MWe of installed wind capacity, enough to meet 30% of the region's total electricity demand.

Denmark's 3 136 MWe of wind capacity meet, in 2006, 20% of its electricity needs, the largest share in any country. It ranks fifth in the world in installed capacity. Denmark is also the global leader in offshore wind power installations, with more than 400 MWe of existing capacity.

The year 2005 saw the consolidation of a reliable support system in Spain's electricity market. One major step for renewable energy was the approval of the Renewable Energy Plan for 2005-2010 in Spain (REP) in the Council of Ministers in August 2005. The REP had sets new targets for the installation of new wind energy capacity and for electricity production using this renewable source. The purpose is to achieve the 12% target of primary energy consumption from renewable sources. The REP also aims to steadily combat increasing CO_2 emissions, which are now 30% higher than previously forecasted in order for Spain to fulfill its Kyoto commitments. In the determination of the specific new targets, the wind energy potential in Spain is clearly recognized with 20,155 MWe. Spain holds the second position in the EU for installed wind power capacity with 11,615 MWe. In 2006 the output of Spanish wind farms exceeded the figure of large hydroelectric power plants.

According to the European Commission's Renewable Energy Directive, "at least 22 % of the Italian electricity supply should come from renewable sources by 2010". Both geothermal and hydroelectric energy is widespread in Italy but they have reached their saturation levels and have limited possibilities for further development. Wind and biomass energy are the only renewable energy sources, which can be exploited to reach the EU target in reasonable time scale and at competitive cost. The growing Italian market has brought about the arrival of several new domestic players, as well as growing interest from foreign developers, even if the wind resource is rather limited.

Potential sites outside the traditional areas in the Southern Italian mainland are more and more at stake. However, the main barriers to the development of the wind sector remain the regional authorization hurdles, local acceptance and grid connection difficulties.

United Kingdom has the best wind regime of any country in Europe, but the growth of its market has been hampered in the past by a mixture of opposition to development at a local level and lack of clear government policy. The UK government has a target for 10% of the country's electricity supply to be provided by renewable sources in 2010, and wind energy is expected to facilitate the achievement of this target. According to the British Wind Energy Association's latest annual survey "roughly 3,965 MWe of capacity on and off-shore is consented, but has not yet gone forward to construction and a very large capacity onshore - equivalent to approximately 6% of the UK total electricity supply is awaiting determination".

However, the UK is moving fast to develop its offshore wind capacity. In April 2001, it accepted bids for sites designed to produce 1,500 MWe of wind-generating capacity. In December 2003, the government took bids for 15 additional offshore sites with a generating capacity that could exceed 7 000 MWe. Requiring an investment of over US$ 12 billion, these wind farms off the east and northwest coasts of England, the north coast of Wales, and in the shallow waters of the Thames estuary, could satisfy the residential electricity needs of

10 million of the country's 60 million people. This goal could be affected because of the following. According with recently press reports, the UK government is reconsidering the establishment of new wind farms in the costal zones of Wales due to the opposition of the air force. Study carried out by the air force in the Northumberland and Scottish regions, had demonstrated that the establishment of new wind farms in these regions would interfere with the work of the military radar in operation in these zones, putting in danger the security of the British island.

The current amount of wind farms in the UK are 165. These wind farms are producing sufficient energy to satisfy the needs of more than a million of homes. The production of energy by these wind farms avoids the emission to the atmosphere of around five million tons of carbon dioxide each year.

The Use of Wind Energy for Electricity Production

The penetration levels to be reached in different countries regarding the use of wind energy for electricity production will require that decision makers and stakeholders in the electricity sector work together to make the necessary changes to the grid infrastructure in Europe. This infrastructure has been constructed and operated in the last century with large centralized coal, hydro, nuclear and, more recently, gas fired power plants in mind.

Until the 1980s, electricity generation, distribution, grid reinforcement, grid extensions, and electricity selling were undertaken by national, vertically integrated monopolies, which were granted exclusive rights and mandates to finance investments and research in new capacity and technologies through state subsidies and levies on electricity bills. As Europe is moving in the direction of more liberalized power markets, those options are no longer available and new technologies are facing a more challenging environment on the path to market penetration and maturity. [16, page 6] In recent years, Europe has spent around 19 billion of Euro to develop renewable energy technologies.

According with the EWEA report of 2005 "by 2010, in terms of capacity and power generation, wind power can provide:

a) Annual electricity generation of 167 TWh, equivalent to 5.5% of European electricity demand, or the power needs of 34 million European households (86 million people).
b) 28% of all new installed generation capacity over the period 2001 – 2010.
c) 10.6% of overall European installed electricity generation capacity.

In terms of sustainability and savings, successful development of wind power could:

a) Deliver 50% of the EU Renewable Energy Directive target.
b) Meet more than 30% of the EU Kyoto Protocol commitment.
c) Save annually 109 million tons of CO_2.
d) Cumulative CO_2 savings of 523 million tons (period 2001 – 2010).

In terms of fuel and external cost savings, and investment in Europe, wind power would mean:

a) Cumulative avoided fuel costs of €13,2 billion .

b) Annual avoided external costs of €1,8 - €4.6 billion .
c) Cumulative avoided external costs of €9,4 - €24 billion over the period 2001 – 2010.
d) An investment value of €49 billion over the period 2001-2010".

The Role of the Wind Energy in the Europe Energy Balances

When analyzing the role of wind energy in Europe energy balances for the future an important element need to take into account: Wind cannot be analyzed in isolation from the other parts of the electricity system, and all systems differ. The size and the inherent flexibility of the power system are crucial aspects determining the system's capability of accommodating a high amount of wind power.

Experience has shown that combining a diverse mix of creative demand and supply solutions allows large wind power penetration in an electricity grid without adverse effect. [9, page 5].

Three major trends have dominated the economics of grid connected wind turbines in recent years:

a) The turbines have grown larger and taller. The average capacity of all the turbines installed in Germany and Denmark increased from roughly 200 kW in 1990 to almost 1.5 MW during 2002.
b) Turbine efficiency has increased. A mixture of taller turbines, improved components and better sitting has resulted in an overall efficiency increase of 2% - 3% annually over the last 15 years.
c) Investment costs have decreased. The average cost per kW of installed wind power capacity varies from €900 /kW to €1 100/kW. The turbine comprises about 80% of this total cost.

The other principal cost element is operation and maintenance costs, including repairs and insurance, and this cost can account for 20% - 25% of total production costs per kWh, at some €1,2 /kWh over the lifetime of the turbine. Manufacturers aim to shrink these costs significantly through development of new turbine designs requiring fewer regular service visits and, consequently, reduced downtime. [9, page 7]

Wind Energy and the Impact in the Environment

Besides the positive effects of wind energy on the environment, especially in the replacing of fossil fuels and consequent reduction in CO_2 and other emissions, wind energy potentially has negative environmental impacts. These can be divided into impact on ecology, and visual and other-sense impacts. Into the first category fall:

a) Potential impacts on the populations, for example, collision, diversion, habitat disturbance, impact on migration routes.
b) Effects on benthic organisms and fish, as well as on sea mammals for offshore applications.

The second category includes:

a) Acoustic sound emissions.
b) Visual impact.
c) Shadow effects and “flicker”.
d) Safety.

Wind Energy and the Public Opinion

Let consider now the public opinion regarding the use of wind energy in Europe. A summary of public opinion surveys carried out in a number of countries, including Spain, the UK, Denmark, Germany and Sweden, shows substantial majorities in favor of wind energy, especially when compared with existing fuel sources. In Germany, for instance, a 2002 poll showed 86% in favor of increasing wind’s contribution to the energy mix. The surveys also show that local approval rises once a wind farm starts operating, and is increased with local involvement and good planning guidelines and integration into the landscape.

In some countries, direct involvement of the public in wind energy projects has helped to encourage public acceptance. For example, in Denmark about 150,000 families own wind turbines or shares in wind turbine co-operatives.

Wind turbines are a relatively new feature in the European countryside, and will become more visible as more wind farms are built. Consultation with, and acceptance by, local communities is essential, especially in rural areas where particularly high value is placed on the landscape amenity. Acceptance is more likely where there has been a clear assessment of the impacts of a proposed wind farm, and the mitigation measures have been properly explained. Once sited, however, existing activities such as agriculture can continue operating alongside wind farms. [9, page 9]

Summing up can be said that wind power has many benefits that make it an attractive source of power for both utility-scale and small distributed power generation applications. The beneficial characteristics of wind power include:

1) *Clean and inexhaustible fuel.* Wind power produces no emissions and is not depleted over time. A single one MWe wind turbine running for one year can displace over 1,500 tons of carbon dioxide, 6.5 tons of sulfur dioxide, 3,2 tons of nitrogen oxides, and 60 pounds of mercury (based on the USA average utility generation fuel mix).
2) *Local economic development.* Wind plants can provide a steady flow of income to landowners who lease their land for wind development, while increasing property tax revenues for local communities.
3) *Modular and scalable technology.* Wind applications can take many forms, including large wind farms, distributed generation, and single end-use systems. Utilities can use wind resources strategically to help reduce load forecasting risks and stranded costs.
4) *Energy price stability.* By further diversifying the energy mix, wind energy reduces dependence on conventional fuels that are subject to price and supply volatility.

5) *Reduced reliance on imported fuels.* Wind energy expenditures are not used to obtain fuels from abroad, keeping funds closer to home, and lessening dependence on foreign governments that supply these fuels. [26, page 4]

Today, wind power installed in Europe is saving over 50 million tons of CO_2 every year. If current trends in growth continue, by 2010, wind energy will save 109 million tons per year, the equivalent of more than 30% of the EU's total Kyoto Protocol obligation. [23, page 14]

Lastly, it is important to stress that wind energy capacity installed in Europe in the past 15 years is the equivalent of 50 coal-fired power stations, but with half of the price.

II.2.4- Geothermal Energy

Geothermal energy is energy stored in the form of heat beneath the surface of the solid earth. The use of geothermal energy depends of the temperature of the geothermal fluid. If this temperature is below 180°, then the most efficient use of the geothermal sources is in domestic heating or for heating the water. If the temperature of the geothermal fluid is 180° or above, then the energy associated with the geothermal fluid can be used for electricity generation.

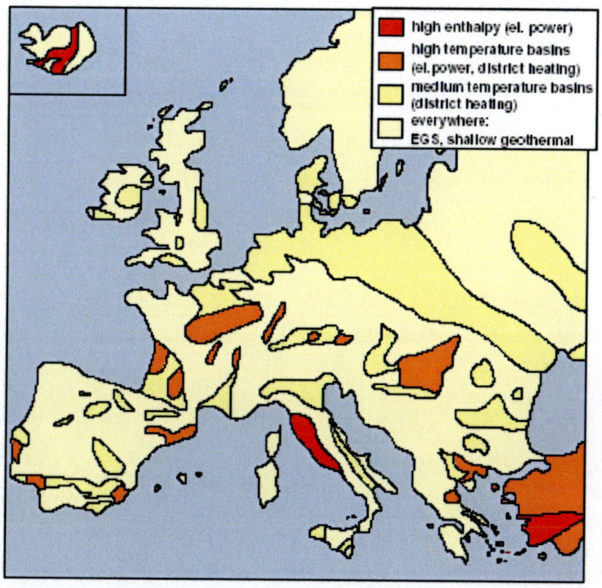

Sources: Reference 2, Figure 1, page 1.

Figure 17. Main Basins and Geothermal Resources of Europe.

Geothermal energy has been used for district heating in France during the Middle Ages, and for extracting various borate compounds at Lardarello, Italy, starting in the 1700's. More recently, extensive direct heat utilization projects have been undertaken in many European countries, and electric power developed extensively in Italy and Iceland. Geothermal heat pumps became extensively used in Austria, Switzerland, Germany and Sweden. [2, page 1]

Europe has significant geothermal resources both in volcanic and sedimentary basin environment (See Figure 17).

According to a study carried out by SHELL that investigates the temperatures at 5000 m depth (Figure 18), it can be observed that on the whole territory of the continental Europe there are favorable temperature anomalies, especially throughout the large basins.

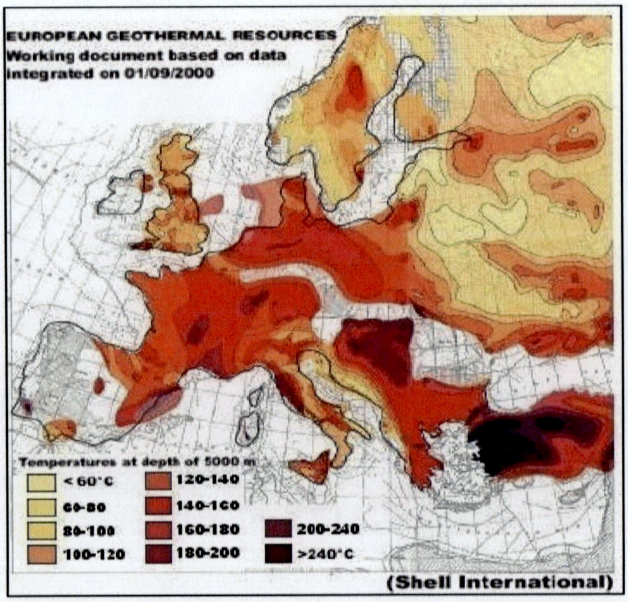

Source: Reference 2, Figure 3, page 2.

Figure 18. Temperature Distribution at 5000 m Depth

The Utilization of Geothermal Energy in Europe

In some regions of Europe, geothermal power plants already substantially contributes to an environmentally friendly and sustainable energy supply, using existing technologies of exploiting steam and hot water reservoirs. This is, for example, the case of Italy, the Azores and other islands of volcanic origin in Europe including, last but not least, Iceland[7], among others countries. In South-East Europe, Turkey and the Caucasian region, reservoirs may contribute to a sustainable energy supply. The status of geothermal power production is presented in Table 11. Distribution of installed capacities is given in Figure 19.

7 In Iceland , geothermal energy will be one of the two pillars upon which a renewable energy supply will be built.

Table 11. Geothermal Power Production in Europe (from Bertani 2005, Rybach 2006) Updated with Available Data May 2007

Country	Installed Capacity (MWe)	Running Capacity MWe)	Annual Energy Produced (GWh/y)	Number of Units	% of National Capacity	% of National Energy	2007 Data
Austria	1.2	1.1	3.2	2	Negligible	Negligible	
Germany	0.2	0.2	1.5	1	Negligible	Negligible	
Iceland	202	202	1,483	19	13.7	17.2	
Italy	810.5	711	5,200	32	1.0	1.9	X
Portugal San Miguel Island	16	13	90	5	25*	n/a*	
Turkey	30	30	108	2	Negligible	Negligible	X
Total in Europe	1059.9	957.3	6,885.7	61	-	-	
France Guadeloupe island	15	15	102	2	9*	n/a*	
Russia Kamtchatka	79	79	85	11	Negligible	Negligible	X
Grand Total	**1,153.9**	**1,051.3**	**7,072.7**	**74**	-	-	

*) Local capacity (Azores islands, Guadelupe)

Sources: Reference 2, Table 1, page 3

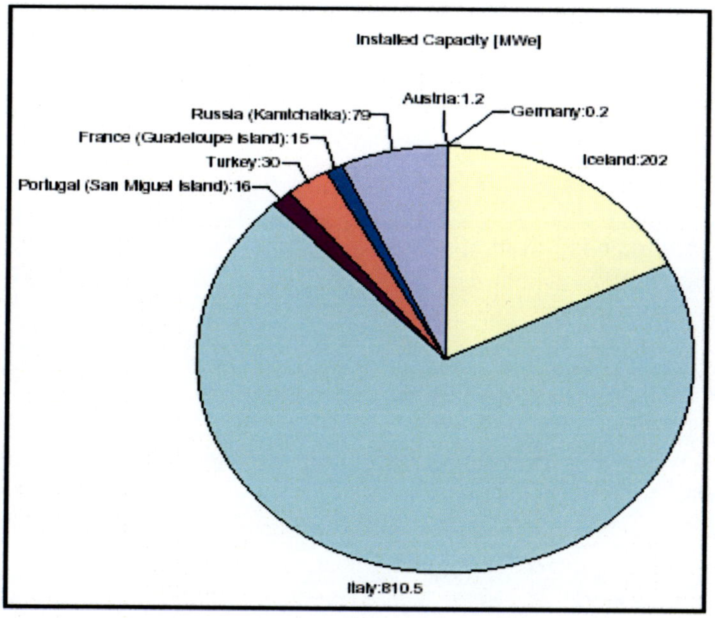

Sources: Reference 2, Figure 5, page 3.

Figure 19. Distribution of Installed Geothermal Power Capacities

In Germany, geothermal power generation is done through the use of binary cycle technology. Since November 2003, a 0,2 MWe pilot power plants using this process is exploited at Neustadt-Glewe and another 20 MWe (4 or 5 power plants) is currently in the planning and construction stage, chiefly in Southern Bavaria. The most advanced project is that of Unterhaching. This small city, located to the southwest of Munich, has a 122°C hot water deposit at a depth of 3,446 meters depth with a mean flow rate of 150 l/s The considered system consists of a 41 MW combined heat and power (CHP) plant that could be able to develop an electrical capacity of 3.5 to 4 MWe. [2, page 3]

Geothermal electricity production in Iceland has increased significantly since 1999, with the installation of new plants in Svartsengi, Krafla and Nesjavellir, up to the present value of 202 MW. An additional 30 MW single flash unit at Nesjavellir is being constructed. Italy hosts the two main high temperature geothermal deposits in Europe (810.5 MWe out of 1,059.9 MWe). Italy's geothermal installations are concentrated around three sites: Larderello (563 MWe), Travele-Radicondoli (160 MWe) and Monte Amiata (88 MWe). Italy installed 10 new power plants over the last five years with a total capacity of 264 MW. In Portugal, exploitation of geothermal energy to produce electricity has been developed on the volcanic archipelago Antics and Sanner of the Azores, or more precisely on the Sao Miguel Island. This island has five geothermal power plants achieving a total capacity of 16 MWe, one on the Pico Vermelho site (3 MWe) and 4 binary cycle power plants on the Ribeira Grande site (13 MWe). These power plants supply 25% of Sao Miguel's needs in electricity. A new 10 MW power plant is currently under construction. It will replace the old Pico Vermelho power plant . A high temperature geothermal field has also been identified on Terceira Island. A drilling program has been set up with the view of installing 12 MWe to supply 50% of the island's needs. [2, page 3 and 4]

The electricity generation in Turkey has been increased to 30 MWe with the addition of the Aydin-Salavatli binary cycle geothermal power plant, adding a 10 MWe installed capacity to the existing Kizildere geothermal power plant (20 MWe installed capacity). France started up its second geothermal power plant in 2004 on the Bouillante site, i.e., an additional 10 MW (14.7 MWe in total) that could produce an additional 72 GWh per year. Furthermore, the Bouillante 3 feasibility study, launched in 2003, could result in a third power plant with more than 10 MWe capacity. In metropolitan France, the experimental Soulz-sous-Forêt deep geothermal energy program continues. The first phase of the project (1999-2004) that consisted in constructing a high-depth heat exchanger (well drillings, stimulation) was accomplished as scheduled. The second phase of the project (2004-2007) should permit construction of a 5 to 6 MW pilot unit. Would tests prove conclusive, a 25 MW industrial prototype could be built in 2010. [2, page 4]

In Russia, despite of its vast territory has only two sites that can economically be exploited. In the European part of Russia there is no site that can be used for the production of electricity through this type of energy source.

The geothermal power production in Russia is located in the Kamchatka Peninsula and Kuril Islands. At present, three stations are in operation in Kamchatka: Pauzhetka geothermal power plant (11 MWe installed capacity) and two Severo-Mutnovka geothermal power plants (12 and 50 MWe respectively). Another geothermal power plant of 100 MWe is being commissioned on the same site. Two small geothermal power plants are in operation in Kuril's Kunashir Island, and Iturup Island, with an installed capacity of 2.6 MWe and 6 MWe respectively. Hungary is also considering electric power production from geothermal energy.

The national oil company MOL plans installations for 12-14 MWe within the next years. First well tests have been finished early in 2007 at the drilling site in Iklódbördoce. [2, page 4]

Some developments have been completed in the geothermal energy sector in Greece. New geothermal fields have been located, studied and reported. The installed capacity in the country for the direct utilization of geothermal energy could be approximately 75 MWe.

New Developments and Future Achievements

In the geothermal power sector, the main new developments that can be expected include:

1) Improved energy conversion efficiency for geothermal power plants, adapted to the reservoir temperatures on site for conventional turbines.
2) Successful demonstration of Enhanced Geothermal Systems (EGS) on key sites, like Soultzsous-Forêts in France, and the dissemination of the technology used in this site to other sites and regions.
3) Increased overall efficiency in geothermal CHP plans.
4) Improvement of exploration methods, installation technologies, and system components (pumps, pipes, turbines, etc.)

The future development of the geothermal heating sector includes:

1) Improved site assessment, exploration and installation for shallow systems, and the dissemination of successful approaches applied by some countries to the whole EU.
2) Further increase in the efficiency of ground source heat pumps, optimized system concepts, application of advanced control systems, and improved components and materials such as compressors, refrigerants, pipes, etc.
3) Construction of new district heating networks and optimization of existing networks and plants, in particular in East/South Eastern Europe and Turkey.
4) Increased application of innovative concepts related with the use of geothermal energy in agriculture, aquaculture, industrial drying processes, etc.
5) The use of geothermal energy for de-icing and snow melting of roads and airport runways, as well as for seawater desalination process.

II.2.5- Bio-fuels

As part of its Energy Policy for Europe, the Commission is committed to encouraging the production and use of bio-fuels by proposing to set a binding minimum target for bio-fuels of 10% of vehicle fuel by 2020.Transport in Europe is constantly growing. Cars and trucks provide such socioeconomic advantages that life without them today is difficult to imagine. However, the situation is becoming untenable. Transport produces nearly a third of CO_2 emissions – the greenhouse gas most responsible for climate change – and emissions are expected to grow significantly. Moreover, the transport sector relies 98% on oil, a largely imported fossil fuel, which is set to become more expensive as reserves become depleted. The EU is proposing an immediate response to this situation: it is encouraging the replacement of diesel and petrol with bio-fuels. These are clean, renewable fuels, which are produced from organic material. The development of the sector will also create jobs and open up new markets for agricultural production. Bio-fuels also contribute to solving broad common problems, such as the diversification of energy sources and the fulfillment of Kyoto Protocol commitments. [19] However, the massive use of bio-fuels could create other problems that could affects many countries, especially in the Third World without energy resources.

In the bio-fuels Commission's directive adopted in 2003, Europe set itself the objectives of replacing 2% of petrol and diesel for transport by bio-fuels by 2005, and 5.75% by 2010. The 2005 target was not met. However, substantial progress can be expected by 2010, but not enough to achieve the target established. The Commission therefore proposes reinforcing the legislative framework, with a 10% minimum for the market share of bio-fuels in 2020.

The most common bio-fuels today are bio-diesel (made from oleaginous plants such as rapeseed and sunflower) and bio-ethanol (produced from sugar and starch crops such as beet or cereals). These two liquid transport fuels have the potential to replace diesel and petrol on a large scale. They can be used in the engines of modern cars (unmodified for low blends, or with cheap modifications to accept high blends) and distributed via existing infrastructures. Research is under way to develop "second-generation" production techniques that can make bio-fuels from woody material, grasses and some additional types of waste.

However, the use of bio-diesel and bio-ethanol as bio-fuels in the transport sector is not exempt of some negative economic impact in many countries. The increasing use of bio-fuels in the EU and USA in the future will raise price and will reduce world supply of some important agriculture products, such as cereals, which is one of the basic components of the food basket in many developing countries even in a scale higher than the price increase which is now affecting cereals and other agriculture products. This situation cannot be ignored by world leaders, if the international community is committed to reduce hunger in the developing world and to preserve international security and stability. In Table 12 progresses in the use of bio-fuels in the EU member states covering the period 2003-2005 is shown.

Table 12. Progress in the Use of Bio-fuels in EU Member States 2003-2005

Member state	Bio-fuel share 2003	Bio-fuel share 2004	Bio-fuel share 2005	National tentative target 2005 (%)
Austria	0.06	0.06	0.93	2.50
Belgium	0.00	0.00	0.00	2.00
Cyprus	0.00	0.00	0.00	1.00
Czech Republic	1.09	1.00	0.05	3.70[1]
Denmark	0.00	0.00	No data	0.10
Estonia	0.00	0.00	0.00	2.00
Finland	0.11	0.11	No data	0.10
France	0.67	0.67	0.97	2.00
Germany	1.21	1.72	3.75	2.00
Greece	0.00	0.00	No data	0.70
Hungary	0.00	0.00	0.07	0.60
Iceland	0.00	0.00	0.05	0.06
Italy	0.50	0.50	0.51	1.00

Table 12. (Continued)

Member state	Bio-fuel share 2003	Bio-fuel share 2004	Bio-fuel share 2005	National tentative target 2005 (%)
Latvia	0.22	0.07	0.33	3.00
Lithuania	0.00	0.02	0.72	2.00
Luxembourg	0.00	0.02	0.02	0.00
Malta	0.02	0.10	0.52	0.30
The Netherlands	0.03	0.01	0.02	2.00[2]
Poland	0.49	0.30	0.42	0.50
Portugal	0.00	0.00	0.00	2.00
Slovakia	0.14	0.15	No data	2.00
Slovenia	0.00	0.16	0.35	0.65
Spain	0.35	0.38	0.44	2.00
Sweden	1.32	2.28	2.23	3.00
UK	0.026[3]	0.04	0.18	0.19[4]
EU-25	0.5%	0.7%	1.0% (estimate)	1.4%

Note: National Reports Under the Bio-fuels Directive

[1]: 2006.

[2]: 2006.

[3]: 0.03% in volume terms, equating to 0.26% in energy content, assuming 100% bio-diesel.

[4]: 0.3% in volume terms, equating to 0.19% in energy content, assuming 50-50% split between bio-diesel and bio-ethanol

II.2.6- Hydrogen Energy

Hydrogen has been known for along time as an energy carrier (not an energy source[8]), as well as the chemical atom most widely available in the world. Its high energetic content (1 kg corresponding to 3.5 liters of petrol) quickly led to its recognition as an ideal fuel in applications where weight was more important than volume, such as for zeppelins and more recently as a fuel for spacecraft. The burning of hydrogen under appropriate conditions in combustion engines or gas turbines results in very low or negligible emissions. [18, page 8]

One of the most significant global trends arriving in the near future is a shift away from fossil fuels and towards hydrogen. According with some expert's opinion, the future world economy will be powered by hydrogen, not by oil (the so-called "hydrogen economy").

The Hydrogen Economy: Reasons for it Adoption

The hydrogen economy is important for the advancement of humanity for several reasons:

8 On the Earth, hydrogen is not an energy source, only a means of storing and transporting energy.

a) *Pollution:* Burning fossil fuels generates alarming levels of pollution that affect every living organism on the planet. Fossil fuels also contribute to global pollution through oils spills, oil extraction, oil refining, and other processes.
b) *Global warming:* Although this topic is aggressively debated in these days, there is growing consensus that the burning of fossil fuels contributes strongly to global warming. The natural consequences of global warming are quite severe:

    - Rising oceans.
    - Disappearing coastlines.
    - Mass extinction of ocean life.
    - Severe and unpredictable climate change.
    - A sharp increase in natural disasters.

c) *Control of resources:* This is perhaps one of the most damning aspects of our global dependence on oil. Because oil is so vital to the economies of nations, its control escalates to "national security" priority. Accordingly, the 20th century (and now, the beginning of the 21st century) witnessed unprecedented death and destruction in the form of military conquest primarily motivated by control of resources. There is little doubt that today's military actions in the Middle East are largely motivated by oil interests. A shift away from the oil economy to a clean, renewable and widely available energy source would ease tensions that have historically surrounded the control of limited resources.
d) *Limited supply*: Fossil fuels are, indeed, running out. There is a finite supply of oil to be found on the planet, and once that oil is consumed, it simply cannot be recreated without waiting hundreds of thousands of years for nature to create more.

Beyond the problems with the oil economy, there are additional reasons why a hydrogen economy offers unprecedented benefits to the quality of life of people everywhere:

1) *Hydrogen is everywhere:* Hydrogen is in water and can be found in abundance at the bottom of the ocean in frozen gas hydrates. Hydrogen also can be found in natural gas, petroleum, and the byproducts of microbial activity. One of the important features of hydrogen is that it is not limited to a few geographic regions of the planet, making it a resource that automatically reduces geopolitical tension over the control of limited oil resources.
2) *Hydrogen is clean:* Through fuel cell technology, hydrogen can be converted to electricity with no harmful waste products. Hydrogen does not pollute cities, rivers, streams or oceans and does not cause global warming. Shifting to a hydrogen economy could save millions of lives each year in terms of human health effects alone, not to mention its effects on the health of the planet and its various forms of life.
3) *Gas hydrates are abundant:* At the bottom of the colder regions of the world's oceans, gas hydrates are plentiful. These are frozen ice-like crystals of frozen hydrogen. They are found off the coasts of Canada, Japan, Alaska, Russian, China, Iceland and the countries of Northern Europe. Technology now exists to harvest

these gas hydrates, store them at liquid nitrogen temperature, and easily convert them into usable hydrogen gas by allowing them to melt at normal atmospheric pressure. The entire process is clean, energy efficient, and technically feasible. The available supply of gas hydrates is enormous, far exceeding the known supplies of all fossil fuels on the planet.

4) *Hydrogen is renewable:* Unlike fossil fuels, hydrogen is renewable. Converting hydrogen gas to electricity in fuel cells does not "destroy" the hydrogen; it just alters the state of the hydrogen. As a result, hydrogen molecules can be used repeatedly to store and release electrical potential. For example, solar panel electrodes immersed in water cause the water to give off hydrogen gas. When that hydrogen gas is fed into a fuel cell, the byproduct is water. No hydrogen is destroyed in the process; it is simply transformed. In this way, hydrogen operates like a battery that transforms energy from the sun into usable electricity.
5) *Hydrogen solves serious global problems:* By shifting to a hydrogen economy, we will simultaneously solve a long list of problems tied to the oil economy (pollution, limited resources, global warming, etc.) while creating new opportunities with hydrogen (clean, renewable, plentiful energy, etc).

The Use of Hydrogen in the EU

Fuel cells using hydrogen can eliminate all polluting emissions, as the only by-product resulting from the generation of electricity is water. With fuel cells now being under growing development and deployment, hydrogen has found the perfect complementary converting process to generate electricity, which can then be immediately used in stationary, mobile or transport applications. Furthermore, hydrogen offers the possibility, depending on the production method used, to drastically reduce or avoid emissions, especially carbon dioxide (CO_2), in the whole production and distribution cycle.

Using hydrogen as secondary energy carrier can allow the flexible introduction of the most diverse renewable energies (primary sources) including wind, solar, hydro, geothermal, and biomass, by separating the primary energy production issue from the storage and distribution issue. [18, page 8].

In less than 20 years, and taking into account the current oil crisis and the climatic changes, the hydrogen could become an important new energy source, with multiples uses, according with a study carried out by French scientists of the Atomic Energy Commission (AEC). In about two or three years it is expected that portable telephones will be equipped with hydrogen's batteries more effective than the ones in use now. Electricity produced by hydrogen will also be able to be used in the near future for heating of houses and buildings. In the framework of the AEC research programs, a school in Saint-Pierre in the centre-west of France will be equipped from September of 2008 with the first generator for the production of electricity and heating using hydrogen. Another project with the French automobile manufacturer PSA Peugeot-Citron has allowed the production of hydrogen's batteries, compatible with the vehicles technology.

Research on hydrogen as future energy source is very advanced in Japan, Germany and United States. General Motors predicts the production of competitive and non-pollutants vehicles that will work with hydrogen by 2010, or by 2012 at the latest. The European Union has now embarked on a series of further demonstration projects grouped under the initiative Hydrogen for Transport. Around 200 hydrogen-powered vehicles will be demonstrated over the next three years. The aim is: to improve vehicle efficiency and infrastructure reliability; to facilitate the understanding of our citizens and our decision makers regarding hydrogen; and to prepare even larger demonstration projects necessary to bridge the gap between the future state of technology and the market. [8, page 22]

The Difficulties of Using Hydrogen as Energy Source

However, progresses achieved in the use of hydrogen as energy sources should not mean that the difficulties and obstacles for the development of this new type of energy at industrial level are over. Still there is a need to reduce production costs and the creation of adequate infrastructures of transport, storage and distribution before hydrogen can be used worldwide.

Limiting Factors in the Use of Hydrogen

One of the current limiting factors in the use of hydrogen as energy source is, according with AEC's experts, the weight and cost of the hydrogen's batteries, particularly due to the use of platinum, which is a very expense raw material. Materials used in the fuel cell system are not very expensive but required significant research investments in order to improve fuel cell durability, hydrogen storage densities and simplify the integration of these systems in existing cars. Industrial processes and volumes are not yet there to optimize production costs, and it will certainly take years if not decades to get there. However, in the last seven years, the weight and cost of the special badges of hydrogen's batteries diminished in 90%, making possible the use of hydrogen as an important energy source in the near future.

Another limiting factors in the use of hydrogen worldwide as energy source is the need to use other energy source to produce it. A key attraction of hydrogen as an energy vector is that can be produced from a variety of sources, including renewables, nuclear and fossil sources. The production of hydrogen today is mainly performed by steam reforming, partial oxidation of gaseous or liquid fuels or the gasification of coal. Electrolysis is used when a small amount of pure hydrogen is required at a specific site. The purification of hydrogen rich gases is an important step in improving the quality of hydrogen produced, depending on eventual use. Certain fuel types require very high purity hydrogen. Distribution of hydrogen is done through pipelines, or using trucks carrying hydrogen in high-pressure gas cylinders or cryogenic tanks. The latter involves an energy-intensive liquefaction step, though the energy required just to compress gaseous hydrogen is itself significant. In the short- to-medium term, the lack of readily available non-fossil sources means that the bulk of hydrogen produced will come from fossil fuels, firstly without carbon capture and sequestration (CCS), and then with CCS in the medium term. The long-term goal is to produce hydrogen from indigenous carbon-free and carbon-lean energy sources. [8, page 10]

It is important to stress that if we use fossil fuels with this purpose, then the production of hydrogen will not be clean, because CO_2 will be release to the atmosphere. The only possibility to produce hydrogen without the emission of CO_2 is to use renewable energies, which could not be always available and competitive or nuclear energy. In the short term, hydrogen can be produced economically by electrolysis of water using off-peak nuclear power. In the future hydrogen can be produced also by thermo-chemical conversion of water using high-temperature nuclear reactors.

There is already today a major worldwide industry that provides hydrogen as a chemical used in making nitrogen fertilizers and in converting low-grade crude oils into transport fuels. But this hydrogen is made from natural gas, giving rise to emissions of CO_2. To make hydrogen cleanly and on a large scale, two nuclear- powered processes are conceivable. In the short term, hydrogen can be produced economically by electrolysis of water using off-peak nuclear power. In future, a major possibility is direct thermo-chemical conversion of water using high-temperature reactors [25, page 35].

The shift to a hydrogen economy will, of course, require a major support from government in the form of mandated usages of this new type of energy, or by limiting the use of carbon or any other energy sources more contaminant from the point of view of the environment.

Iceland is implementing a project replacing the fuel of its cars and their fishing fleet by hydrogen, bearing in mind that this country is rich in geothermal and hydroelectric sources. In Luxembourg, three fuel cell buses run on the city's streets as part of the Clean Urban Transport for Europe (CUTE) Project, an initiative of the EU. The project that runs from 2003 until the end of 2006 has been developed in order to gain experience on how fuel cell technology really works on a day-to-day operation in a challenging environment. The same type of project run in the Netherlands as part as the EU initiative in the transport sector.

In other parts of the world, there is an increase in the use of renewable sources to produce hydrogen at moments of excess of electricity generation, with the purpose to extract later on its energy in deficit generation situations.

The Production of Hydrogen in the EU

Approximately 600 billion Nm3 of hydrogen is produced every year through well-established commercial processes for the hydrocracking of oil, the production of ammonia and hydrogenation of edible fats, 95% of which is captive (consumed on-site). Significant progress is needed for hydrogen to become a widely available "consumer fuel" which meets the previously mentioned energetic, environmental and production cost constraints.

II.3. Nuclear Energy

There should be no doubts that nuclear energy could contribute to alleviate European energy supply dependency, local air pollution and global climate change. However, during the consideration of the possible role to be played by nuclear energy in the European energy balance in the future, the following problems need to be thoroughly considered and favorable solved:

a) Management of the radioactive waste, particularly high –level nuclear waste.
b) Proliferation security.
c) Operational safety of the nuclear power plants.
d) Economic competitiveness.

e) Public acceptance.

Nuclear power is cost competitive with other forms of electricity generation, except where there is direct access to low-cost fossil fuels. Fuel costs for nuclear plants are a minor proportion of total generating costs, though capital costs are greater than those for coal and oil fired plants. In assessing the cost competitiveness of nuclear energy, decommissioning and waste disposal costs should be taken into account.

If no actions are taken by the EU to solve the above-mentioned problems, then the participation of nuclear energy for electricity generation in the region will decline notably due to the following two reasons:

1) There are only a few new nuclear reactors under construction in Europe in the coming years[9]. However, this situation is changing due to the position adopted recently by UK, Italy and other EU countries in favor of constructing new nuclear power plants.
2) The closure of several nuclear power plants, due to the extinction of their exploitation license, if extension of these licenses are not requested and approved in the coming years.

Cost Associated with the Use of Nuclear Energy for Electricity Generation

The relative costs of generating electricity from coal, gas and nuclear plants vary considerably depending on location. Coal is, and will probably remain, economically attractive in countries such as China, the USA and Australia with abundant and accessible domestic coal resources as long as carbon emissions are cost-free. Gas is also competitive for base-load power in many places, particularly using combined-cycle plants, though rising gas prices have removed much of the advantage. Nuclear energy is, in many places, competitive with fossil fuel for electricity generation, despite relatively high capital costs and the need to internalize all waste disposal and decommissioning costs. If the social, health and environmental costs of fossil fuels are also taken into account, nuclear is outstanding. [4, page 1]. However, the increase cost due to delay in the construction of a nuclear power plants is an important element that need to be taken into account when the overall cost associated with the use of nuclear energy for electricity generation is analyzed.

The future competitiveness of nuclear power will depend substantially on the additional costs, which may accrue to coal generating plants and the cost of gas for gas-fired plants. It is uncertain how the real costs of meeting targets for reducing sulphur dioxide and greenhouse gas emissions will be attributed to fossil fuel plants. [4, page 4]

One of the important components of the cost associated with the construction of a nuclear power plant is the decommissioning cost. These costs are about 9-15% of the initial capital cost of a nuclear power plant. But when discounted, they contribute only a few percent to the investment cost and even less to the generation cost. The back-end of the fuel cycle, including

[9] Finland and France are the two only countries in Western Europe that are constructing nuclear power plants in this moments. Bulgaria, Romania, Czech Republic, and Lithuania are constructing nuclear power plants or are in the final stage of planning for the construction of new nuclear power plants. UK Ais planning to construct up to 10 new nuclear power plants in the future.

spent fuel storage or disposal in a waste repository, contributes up to another 10% to the overall costs per kWh, - less if there is direct disposal of spent fuel rather than reprocessing. Nuclear is favorable because of the large standardized plants used. [4, page 3] French figures published in 2002 (Euro-cents/kWh) are shown in the following table.

Table 13. Cost of Kilowatt Per Hour of Three Types of Energy

Type of Energy	Cost Euro cents/kWh
Nuclear	3.20
Gas	3.05 – 4.26
Coal	3.81 – 4.57

Source: Reference 4, page 3

The French Energy Secretariat in 2003 published updated figures for new generating plant. The advanced European PWR (EPR) would cost €1 650- €1 700per kilowatt to build, compared with €500- €550 for a gas combined cycle plant and €1 200- €1 400 for a coal plant. The EPR would generate power at 2.74 cents/kWh, competitively with gas, which would be very dependent on fuel price. Capital costs contributed 60% to nuclear's power price but only 20% to gas's. While the figures are based on 40-year plant life, the EPR is designed for 60 years. [4, page 8] There should no doubt that the extended life of the EPR will have an important impact in the level of the nuclear power price.

The IAEA and the World Energy Outlook for 2006 has foresee that "the participation of nuclear energy in the country's energy balance will drop from 16% to 8-10% in 2030, if no decision is adopted by Europe, Canada and the USA to build more nuclear power plants for electricity generation in the coming years". The major increase in nuclear energy capacities in the near future are foresee in China[10], Japan, India, Russia, and the Republic of Korea. It is expected that these five countries and the USA, in case that this country approve the construction of new nuclear power plants, will have around two-third of the world nuclear energy capacity in 2030, an increase of 16%-17% from the current level (50%).

The major reduction in the use of nuclear energy for electricity production will occur in Europe, despite recently announcement by the UK government to built up to 10 new nuclear power plants in the coming years and the constructions of nuclear power plants in other European countries. One of the scenarios that have been considered regarding the participation of nuclear energy for electricity generation is the following: The electricity production by nuclear energy will drop from 29% in 2005 to 12% in 2030. In another scenario, this reduction could be from 29% to 20% in the same period. The foresee reduction in the use of nuclear energy for electricity production in Europe in the coming years would be from 131 GW to 74 GW. However, this scenario could change if more European countries follows the UK position regarding the use of nuclear energy for electricity production in the future.

---

[10] At the end of 2007, China and France signed an agreement for the construction of nuclear power plants for US$ 11,8 billion up to 2026. This is the major contract signed by the French company Areva up to now for the construction of nuclear power plants. It is important to stress that for China, nuclear energy is a fundamental option to satisfy the growing demand of energy to sustain the high economic growth of this country.

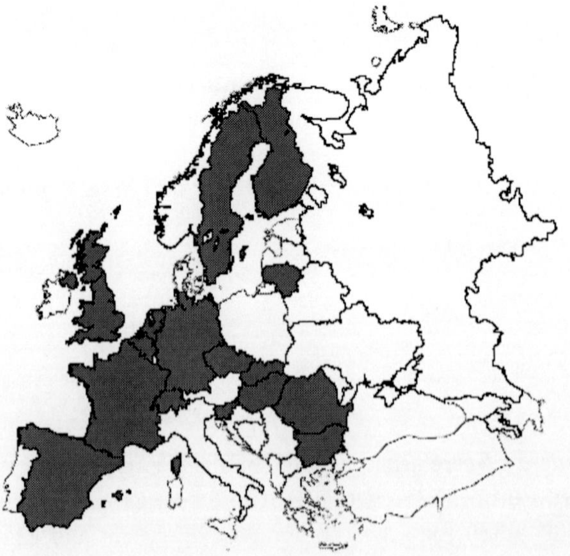

Note: 16 countries today produce nuclear energy (in red) and 20 countries do not (in white).

Figure 20. Nuclear Power in Europe. Reference 29, Figure 1, page

The Situation of the Use of Nuclear Energy in the World and in Europe

Worldwide, according with IAEA Nuclear Technology report for 2007, "there are 435 nuclear power plants operating at the end of 2006 totaling 369,682 MWe of generating capacity and supply about 15% of the world's capacity". (See Tables 14 and 15)

Table 14. Reactors in Operations*.
Country Reactors in Nuclear Electricity Supplied Total Operating Experience Operations Through 2006 Through 2006

	Number of units	Total MWe	TW-h	% of Total	Years	Months
Argentina	2	935	7.15	6.93	56	7
Armenia	1	376	2.42	41.95	32	8
Belgium	7	5,824	44.31	54.43	212	7
Brazil	2	1,901	12.98	3.31	31	3
Bulgaria	2	1,906	18.15	43.64	141	3
Canada	18	12,610	92.43	15.81	528	7
China	10	7,172	51.81	1.93	66	7

Table 14. (Continued)

	Number of units	Total MWe	TW-h	% of Total	Years	Months
Czech rep	6	3,523	24.50	31.48	92	10
Finland	4	2,696	22.0	27.99	111	1
France	59	63,260	429.82	78.07	1,523	2
Germany	17	20,339	158.71	31.82	700	5
Hungary	4	1,755	12.51	37.70	86	2
India	16	3,577	15.59	262	267	7
Japan	55	47,587	291.54	29.97	1.276	8
Korea	20	17,454	141.18	38.64	279	8
Lithuania	1	1,185	7.94	72.30	40	6
Mexico	2	1,360	10.40	4.86	29	11
Netherlands	1	482	3.27	3.47	62	-
Pakistan	2	425	2.55	2.74	41	10
Romania	1	655	5.18	9.0	10	6
Russia Fed	31	21,743	144.64	15.91	901	4
S. Africa	2	1,800	10.07	4.41	44	3
Slovakia	6	2,442	16.34	56.06	112	6
Slovenia	1	666	5.29	40.26	25	3
Spain	8	7,450	57.43	19.82	245	6
Sweden	10	9,097	65.05	47.98	342	6
Switzerland	5	3,220	26.37	37.41	158	10
UK	19	10,965	69.39	18.40	1,400	8
Ukraine	15	13,107	84.91	47.53	323	6
USA	103	99,257	788.31	19.42	3,188	2
TOTAL	435	369,682	2,660.86	18.76	12,599	1

Source: IAEA

*According with the latest IAEA information, there are 449 nuclear power plants in operation at the end of 2007.

According to the above IAEA information, "twenty-seven nuclear power plants were under construction in the world in 2006". In Asia the number of nuclear power plants under construction was 15, representing 55.6% of the total. Western Europe has 135 operating nuclear power plants and only one under construction in Finland, representing 3.7%[11]. Russia

---

[11] France are constructing one nuclear power plant in Flamanville, Normandy. The construction started in December 2007

has 31 nuclear power plants in operation and 5 under construction, representing 14.8% of the total, and Eastern Europe has 39 operating nuclear power plants and 5 under construction, representing 18.5%. France produces 78% of its electricity in 58 nuclear power plants, the highest level in the EU[12]. In the USA, 105 nuclear power plants cover 20% of the electricity generation.

Table 15. Reactors Under Construction in 2006

Country	No of Units	Total (MWe)
Argentina	1	692
Bulgaria	2	1,906
China	4	3,610
Finland	1	1,600
India	7	3,112
Iran	1	915
Japan	1	866
Republic of Korea	1	960
Pakistan	1	300
Romania	1	655
Russian Federation	5	4,525
Ukraine	2	1,900
Total	27	23,641

Source: IAEA.

China has nine nuclear power plants in operation and four under construction and hopes to incorporate 21 new nuclear power plants up to 2020 and another 20 later on. This is until now the most ambitious nuclear program for generation of electricity under discussion and in implementation in the world.

In 2006 were in the European region 192 nuclear power plants in operation and 11 under construction in 18 countries. Overall, there was a net increase in capacity of 1,418 MWe in 2006, taking into account new nuclear power plants connected to the grid and plant retirements (See Tables 14 and 15).

In Western Europe, capacity has remained relatively constant despite nuclear phase-outs in Belgium (which passed a phase-out law in January 2003), Germany and Sweden. The most advanced planning for new nuclear capacity was in Finland[13]. The Russian Federation continued its program to extend licenses at 11 nuclear power plants. [12, page 2]. Recently, the UK government has announced the construction of 10 nuclear power plants in the coming years, in order to balance the reduction in the extraction of oil in the North See.

---

[12] See foot-note 6

[13] See foot-note 6

During the next two decades, the European region might need between 200 and 300 GWe of new generating capacity. Replacement of the existing infrastructure could double this amount. The increase in the demand of energy in the European region in the coming years could be a very difficult goal to be achieved, if the European countries decided to use mainly clean energy to reduce the emission of CO_2, as well as other gas of green house effect, and if nuclear energy is excluded. Has been demonstrated that the uses of nuclear energy is environmental clean, that the safety of the nuclear power plants operation are very high and there is a safe and proven technology available for the disposal of the nuclear waste.

The Future Participation of Nuclear Energy in Electricity Generation

There is no doubt that wind and solar have their place in electricity generation, "but because they are intermittent and unpredictable they simply cannot replace big base load plants like coal, nuclear and hydroelectric. Natural gas, a fossil fuel, is too expensive already and its price is too volatile to risk building big base load plants. Given that hydroelectric resources are largely built to capacity, nuclear is by elimination the best option. It's that simple. [20, page 57]

When analyzing the future participation of nuclear energy in the electricity production in Europe the following elements should be kept in mind:

1) Nuclear technology has reached its majority age. Is a safe and non- pollutant environment technology. In the 2006, more than 30 countries were using nuclear energy for electricity production, 18 of them in the European region (including Russia, Ukraine and Lithuania).
2) The supply of fossil fuels could not be enough to support the growing energy needs of many countries for the second half of this century.
3) A safety nuclear culture has been developed at world level, elevating the safety of the current nuclear power plants in operation.
4) A new generation of nuclear reactors will be available with a high-level of safety than the ones now in operation.
5) The appropriate technologies exist for the safe treatment and for the final disposal of the nuclear waste, specially the high-level nuclear waste.

According with UNESA Electricity Generation Perspective for 2030, "the nuclear energy is the cheapest type of energy for electricity generation and the most effective and clean in front of the climatic change".

The acceptance by the public opinion of Finland and Sweden of the final disposal of nuclear waste in the country, and the technological advances reached by the USA on this important subject, allows a little bit more optimistic view about the future of nuclear energy for electricity production in the European region. However, according with the IAEA Nuclear Technology Review of 2006, "Finland, Sweden and the USA are furthest along in developing final high nuclear waste repositories, although none is expected to be in operation much before 2020". [13, page 11]

The Projection of the Contribution of Nuclear Energy in the Global Energy Balance

The IAEA revised projection of the contribution of nuclear energy in the global energy balance for 2030 is shows in Table 16.

If the world is to meet even a fraction of the economics aspirations of the developing world, energy supplies must expand significantly. If the increase needs of the European countries, as well as other industrialized countries are considered as well, then the world energy supplies must expand even further.

Table 16. Low and High Estimates of Total Electricity Generation and Contribution by Nuclear Power (Low estimate: first row ; High estimate: second row)

Region		2004 Total Elect. GW(e)	Nuclear GW(e)	%	2010* Total Elect. GW(e)	Nuclear GW(e)	%	2020* Total Elect. GW(e)	Nuclear GW(e)	%	2030* Total Elect. GW(e)	Nuclear GW(e)	%
North America		1055	111.3	10.6	1099 1155	116 117	11 10	1194 1279	118 128	10 10	1318 1422	115 145	8.7 10
Latin America		264	4.1	1.6	303 350	4.1 4.1	1.4 1.2	383 543	6.1 6.1	1.6 1.1	483 828	5.8 15.0	1.2 1.8
Western Europe		724	125.1	17.3	762 816	119 125	16 15	842 951	97 130	11 14	940 1118	79 145	8.5 13
Eastern Europe		466	49.4	10.6	469	48	10	505	64	13	543	66	12
					496	51	10	605	78	13	736	97	13
Africa		105	1.8	1.7	115 135	1.8 1.8	1.6 1.3	143 207	2.1 4.1	1.5 2	181 316	2.1 9.3	1.2 3.0
Middle East and South Asia		284	3	1	331 370	9 10	2.8 2.8	430 555	15 27	3.6 4.9	556 811	18 43	3.2 5.3
South East Asia and the Pacific		143			169 184			213 270	0.9 0.9	0.4 0.3	264 391	0.9 3	0.3 0.8
Far East		651	72.8	11.2	685 840	82 85	12 10	804 1167	113 142	14 12	937 1589	131 183	14 11
World Total	Low estimate High estimate	3693	367.5	10	3934 4347	380 395	10 9.1	4515 5576	416 516	9.2 9.3	5223 7210	418 640	8 8.9

In the following figure (See Figure 21) the historical growth in worldwide nuclear generating capacity since 1960 plus high and low projections by the IAEA as updated in 2005 is shown.

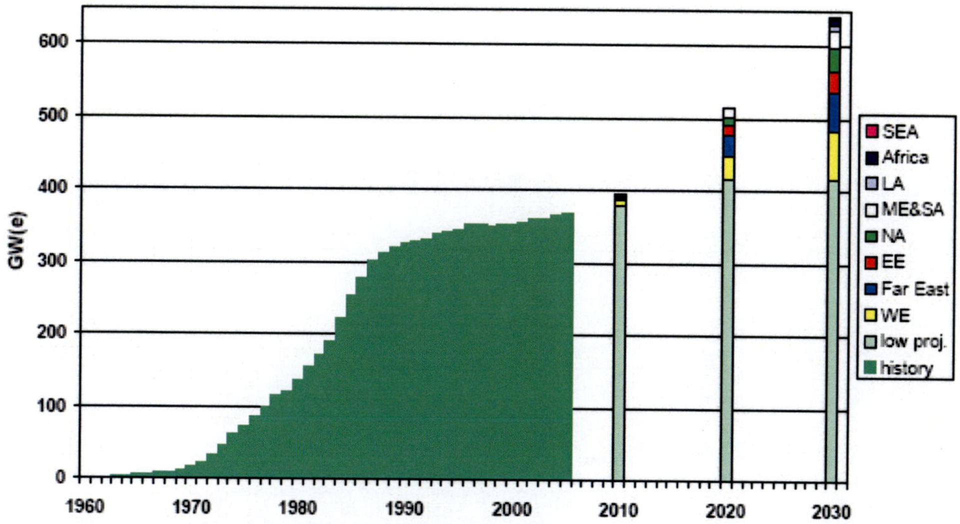

Note: Installed nuclear power generating capacity worldwide. The dark green bars show historical growth form 1960 through 2005. The light green bars show the IAEA's latest low projection for 2010, 2020 and 2030. The other colors show how much of the difference between the IAEA's low projections is attribute to different regions of the world. WE refer to the Western Europe region and EE to Eastern Europe.

Source: Reference 13, Figure A-1, page 7

Figure 21. Historical Growth in Worldwide Nuclear Generating Capacity Since 1960 Plus High and Low Projections

Limiting Factors for the Use of Nuclear Energy in Europe

One of the limiting factors in raising the use of nuclear energy for electricity production in Europe is the increase number of power plants that are reaching the point where choices must be made between extensions of their useful live and decommissioning. In the following figure, the aggregated capacity of nuclear reactors in Europe per age of operation is shown.

Other limiting factors are:

a) The inability of designers and operators to improve significantly the competitiveness of nuclear power plants while meeting increasingly stringent safety requirements.
b) The opposition of public opinion to the use of nuclear energy for electricity production.
c) The problem causing by the final disposal of nuclear waste, particularly high-level nuclear waste.
d) The high construction cost.
e) The time needed for the construction of the current type of nuclear power plants.

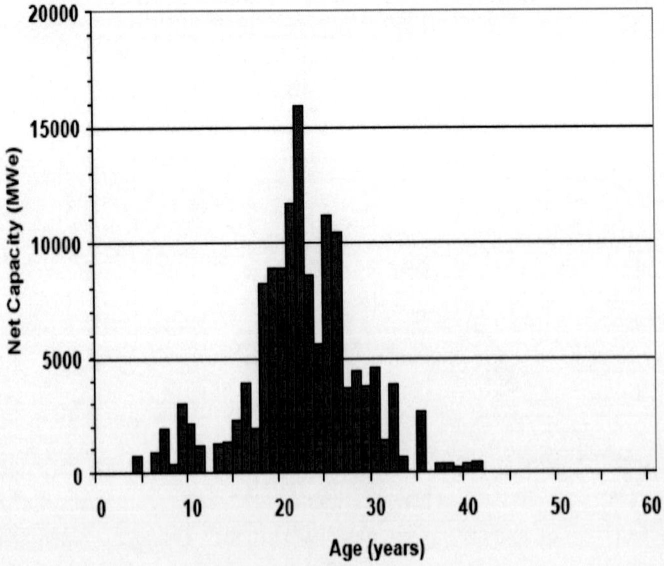

Source: IAEA-PRIS (2006), data from February 2006.

Figure 22. Aggregated Capacity of Nuclear Power Plants in Europe Per Age of Operation

Existing well-run operating nuclear power plants continue to be a generally competitive profitable source of electricity, but for new construction, the following factors should be considered:

1) First, it depends on the alternatives available. Some countries are rich in alternative energy resources, others less so.
2) Second, it depends on the overall country electricity demand and how fast it is growing.
3) Third, it depends on the market structure and investment environment.

Regarding market structure and investment environment, it is important to stress that nuclear power's front-loaded cost structure is less attractive to a private investor in a liberalized market that values rapid returns, than to a government that can look longer-term, particularly in a regulated market that assures attractive returns. Private investments in liberalized markets will also depend on the extent to which energy related external costs and benefits (e.g. pollution, GHG emissions, waste and energy supply security) have been internalized. In contrast, government investors can incorporate such externalities directly into their decisions. Also important are regulatory risks.

Political support for nuclear power varies across countries, and, within a given country, it can change over time. An investor must weigh the risk of political shifts that might require cancellation of the project midstream or introduce delays and costs that would reduce an originally attractive investment. Different countries also have different approval processes. Some are less predictable than others and create greater risks, from the investor's perspective, of expensive interventions or delays. [21, page 8]

The Present and Future Use of Nuclear Energy in the European Region

In Europe, several countries, including Sweden, Germany, Italy, Austria and Belgium, have passed different laws that oblige their governments to phase-out nuclear power plants in the coming years, or are prohibiting the operation of these plants in the country. In line with nuclear phase-out policies in Germany and Sweden there were two retirements, one in Germany (Obrigheim) and one in Sweden (Barsebäck-2).

Despite of the phase-out of nuclear plants laws adopted by Belgium and Sweden, it is expected that for 2030, some of the current reactors in both countries will still be in operation. The reduction in the use of nuclear energy for electricity production in Germany, Sweden and Belgium will be around 35 GWe.

In Germany, despite of a widespread public opposition, polls indicate that 80% of Germans oppose the renewal of the country's nuclear power reactors. The current government appears to be shifting its policy on nuclear power adopted by the previous one, which ordered the phasing out of the country's nuclear power industry. The present German government favor the maintenance of the nuclear option for electricity production, as a means of lessening Germany's reliance on Russian oil and gas. If nothing is done in the coming years to keep the nuclear energy option open in Germany, then it is foresee that in 2030 only one of the current 17 nuclear reactors units will be in operation, representing a reduction of 94.2%.

In UK, the current government openly expressed its support for a new generation of British nuclear reactors, taking into account the declining of oil and gas reserves in Britain's North Sea fields, and the growing instability of imports from some other regions. As many as 10 new reactors are under consideration.

At the same time, other European governments approved a lifetime extension of two nuclear power plants in operation. These plants are one in the Netherlands (Borssele) and one in the UK (Dungeness-B1 and Dungeness-B2). The Swedish Government approved up rates of 15 MWe for Ringhals-1 and 3, and a similar request for a 250 MWe up rate at Oskarshamn-3 has been supported by regulators and is awaiting final government approval. Additional up rate requests were presented for 120 MWe at Forsmark-1, 120 MWe at Forsmark-2 and 170 MWe at Forsmark-3.

Finland began constructing a third-generation pressurized water reactor last year to come on-line in 2009[14], while France has approved a similar plant[15] and another pilot plant by 2020. Bulgaria is expected to award a contract for construction of two plants. Romania has restarted work on a plant abandoned 15 years ago, while the Czech Republic predicts the construction of two more by the end of the decade. [28, page 1]

Currently, EU generates around 30% of its electricity from nuclear power stations. This proportion should be maintained or increased in order to meet the 2020 EU target, and the increasing demand.

As the demand for electricity continues to increase and the drive for sustainable development gain momentous, the need to exploit energy sources with limited impact (in particular to meet commitments made in connection with the Kyoto Protocol) could revitalize the nuclear power option. A number of countries announced plans for significant expansion, specifically China, India, Japan, Pakistan, the Russian Federation, and the Republic of Korea.

---

[14] It is expected that the reactor could come on-line by 2011 due to some delay in the construction process

[15] See foot-note 11

In addition, Argentina, France, South Africa, Ukraine and the USA have expressed an intention to expand their existing programmes[16] [11, page 1]

In conclusion, during the consideration of the possible role of nuclear energy in the European energy balance in the future, it is important to take into account the following elements:

1) Renewable energy sources have the potential to play a much larger role than they presently do, but until now and in the near future the participation of the renewable energy sources will not have a significant impact in the world energy balance.
2) There has been significant progress in renewable energy technology in recent years. For example, wind energy is now widely recognized as a viable option for electricity generation in several European countries, as well as in other countries. However, problems associated with the climate, public opinion, structure of the generation electricity system, the efficiency of the turbines technology available, among others factors are limiting the use of the wind energy balance worldwide.
3) There are in general no supply problems in the use of renewable energy sources, but not in all countries available renewable energy sources can be used without limitation and difficulties.
4) Over the last 15 years, there has been a steady increase in energy production from renewable energy sources (around 30%). In the wind energy sector the recent increase is more than spectacular (2,000% in 10 years). However, the increase in the use of renewable energy sources worldwide is still very low.
5) Most renewable energy sources produce little, or no greenhouse gases, so do not contribute to climate change.
6) Despite of the efforts made by several EU countries, targets for steadily increasing the share of renewable energy sources have not been achieved.
7) The development of new sites, in particular for hydroelectricity, often meets with strong local resistance.
8) There are obstacles of a structural nature – especially for the generation of electricity that need to be overcome, in order to increase the use of renewable energy sources.
9) Several renewable energy sources are more expensive than conventional fuels.

Due to the opposition to the use of nuclear energy for electricity production by the public opinion in several UE countries, the future of nuclear energy in the region is somehow unpredictable and will depends on the solution of the following problems:

a) Change in the public opinion position in some countries that is again the use of nuclear energy for electricity production.
b) Safe management of nuclear waste, particularly of high-level nuclear waste.
c) Economic viability of new nuclear power plants generation.
d) Safety of the reactors currently in operation in Eastern Europe.
e) Production of new inherent safe nuclear reactors.
f) Proliferation requirements.

---

16 Others countries have also expressed their intentions to expand their current nuclear programmes or initiate new ones in the coming years.

g) Reduction of global warming.

None of the above-mentioned topics are new and some of them are under discussion at the European Commission level for more that 20 years and in some cases even longer. Millions of Euro and in some instances, billions of Euro, have been spent in order to find an acceptable solution to some of these problems.

Fusion Technology

By far the largest financial contribution of Europe into research in the nuclear area is into fusion technology. Collaboration started in 1978. The following countries are now involved in the development of fusion technology for electricity production: China, , Japan, the Republic of Korea, the Russian Federation and the USA. The EU is also participating in thi activity. The above-mentioned countries and the EU are cooperating in building the International Thermonuclear Experimental Reactor (ITER) in Cadarache, France, with the support of the IAEA. This reactor will enabled Europe to assume a leading position worldwide in the development of the fusion technology..

However, the use of fusion rectors to produce electricity is not expected to be ready before 2020 and commercially available not before 2050, or even later. This will happens if, and only if, all difficulties and technological problems now facing the use of this technology could be overcome, in an acceptable manner, in the coming years. This is something that is very difficult to predict now. For this reason, the use of the fusion technology for electricity production is a long-term solution not only for the EU but for other countries as well.

Fission Technology

In the field of fission technology, which is the technology now use for the construction of the current generations of nuclear reactors, the position of the EU is that the industry should take care of the design of more efficient and inherent safe reactors. There are some progresses in this direction, which could represent an acceptable solution for the foresse increase in the use of nuclear energy for electricity generation in several EU countries. It is expected that when new nuclear power plants will be required, and this will not take several years, there will be several attractive designs already on the market using this type of technology.

The future reactors are expected to have the following advantages over the present generation of nuclear power plants:

1. Lower investment costs and construction times.
2. Simpler designs.
3. Modular units.
4. More passive safety features.
5. Proliferation issues.

Safety of Nuclear Reactors

One of the important issues that are limiting the wider use of nuclear energy for electricity production in Europe is the current level of safety of some of the nuclear reactors now in operations.

The European Commission started to discuss safety issues in 1975, when nuclear power programs promoted by different European states had progressed and diverged along very different routes, including different legal frameworks, types and number of reactors, as well as national systems regulating them in force in several countries. The safety requirements for the safe operation of nuclear reactors in Western Europe were also different to the ones applied in the past in Central and Eastern Europe, which make even more difficult to find a common acceptable solution to the level of safety to be applied to nuclear reactor produced in Russia and other Eastern and Central Europe countries.

The above-mentioned situation force, for the first time in the history, to the EU to start in the 1990's the process of carrying out an evaluation of nuclear safety in Eastern and Central Europe, particularly in those countries requesting to entry in the EU. The debates on nuclear safety in the context of the enlargement of the EU raised vital questions about what are the Western standards for nuclear safety. The objective to be reached within this debate is to ensure that all European states would have equivalent levels of nuclear safety. This by no means imply that the operator of nuclear reactors should not retain the primary responsibility for the nuclear safety of the nuclear reactors under the control of the national regulatory bodies, but to have common nuclear safety standards that should be applied by all European countries.

Over the past 20 years, reactor safety has improved significantly, both in and outside Europe. No early generation Soviet (Chernobyl-type) Light Water Graphite Reactors (LWGRs) are in operation in Europe, and the present cohort of European reactors has had a good overall safety record. Since 1986, accident probabilities have decreased substantially, because of improvements in reactor technology, peripheral equipment and operation practices. Most of the European reactors are equipped with confinement domes, ascertaining that, in the occurrence of an accident, radioactive material are not released to the external environment and consequences are controlled to a minimum.

Man-machine interactions in plant operation have been considerably perfected, and a better safety culture has been established through the creation of an international "early notification system", obliging operators to report any incident using the International Nuclear Event Scale (INES) of the IAEA. Continued efforts in maintaining and elaborating high safety standards are among the actions carried out by the operators of nuclear power plants for an expansion of nuclear power in Europe.

Furthermore, the EU is in the process of creating new directives to further improve reactor operation safety, develop regulatory safety oversight, and orchestrate this presently largely national affair on a European level. Among the issues also addressed are the ascertaining of sufficient funds for the complete decommissioning of power plants, the exchanging of best operation practice for existing installations, the maintaining of high safety standards for plants whose operation licenses are extended, and the providing of transparency for citizens. These measures and new reactor types need to continue complying with cost requirements in order to ascertain nuclear energy's economic competitiveness.

Despite of the positive general situation regarding the level of safety of the nuclear power reactors now operating in the EU describe above, the IAEA continue registering different types of incidents in some of these plants but without any serious consequences for the health of the people or the environment.

Public Opinion and the Problem of Nuclear Waste

Another important issue that need to be considered, in the framework of a future wider use of nuclear energy for electricity production in Europe, is the public opinion. In October of 2001, a survey was carried out within the EU that shows that the majority of people would support the nuclear energy option, if and only if problems related with the final disposal of high-level nuclear waste, and the safety of the nuclear reactors are properly solve. Only in Austria, people against the use of nuclear energy for electricity production outnumber those who agree.

Findings of a recent survey, conducted among 18,000 citizens of 18 countries representing the major regions in the world, show that 62% believe that existing nuclear reactors should continue to be used, but 59% are not favorable of building new nuclear plants. Also, when citizens from the current 25 member states of the EU were asked what national governments should focus on in order to reduce its energy dependency, only 12% answered that first the use of nuclear energy should be further developed. As the impacts of climate change and the vulnerability of the European economy to foreign fuel imports become more evident, it is likely that the gradual shift in public opinion of the last decade will further develop towards less skepticism or in favor of nuclear energy. [29, page 9]

Progress on these issues will have an important positive impact in the public opinion towards nuclear power for electricity generation. For example, once in Europe an underground permanent storage site is in operation and proves to safely contain high-level radioactive waste, the public may start accepting it as a satisfactory solution for the back-end of the fuel cycle. While many European governments delay on this subject, progress on deep geological disposal has been made in Finland, France, and Sweden. The Finnish government has made a decision to start building a final repository for spent nuclear fuel in 2011 near Olkiluoto, which would be in operation around 2020. Based on studies performed between 1991 and 2005, the French government has initiated in 2006 a debate with Parliament on which solution to choose for the long-term disposal of spent nuclear fuel. Sweden presented in 2007 a plans to make a site proposal based on ongoing geological investigations at two candidate locations.

The main issue concerning underground storage remains uncertainty about the integrity of spent fuel canisters, and whether the isolation offered by geological formations will be sufficient over a period of thousands of years. The fear is that canisters, as a result of corrosion, after a long time will start to leak and consequently contaminate groundwater. This mostly explains the hesitation towards final waste disposal implementation by governments and the skeptical attitude of the general public. [29, page 10]

Undoubtedly, the biggest issue concerning radioactive waste management is the one related with the high-level of nuclear waste. All other aspects related with low and middle level of nuclear waste management have already reached the industrial stage and to large majority of the waste has already been disposed of, without any major problems.

Two main channels exist through which the problem of high-level of nuclear waste could be mitigated: "reducing the radioactive lifetime of the long-living isotopes the waste contains by transmutation processes, and organizing its disposal regionally through Internationally Monitored Waste Repositories (IMWRs)". The European Commission is preparing legislation that creates a regulatory framework for EU states to undertake concrete and timely action for the development of permanent, underground and aboveground, disposal facilities. [29 , page 11]

This Directive would require EU member states to prepare a detailed strategy or work for radioactive waste and report on this to the Commission at regular intervals. It would also seek to impose a relatively strict timetable on progress towards sitting and developing geological and other repositories. In this way, the Community will advance rapidly lives towards an acceptable–and fully transparent–solution to the issue of high-level nuclear wastes.

The technologies for the safe disposal of high-level of nuclear waste have been tried and tested in the USA, Sweden and other countries. Research and development will continue to refines models and concepts associated with this important and delicate issue for the public opinion.

Decommissioning of Nuclear Power Plants

According with the IAEA Annual report for 2006, "roughly three quarters of the world's operating reactors are 20 or more years old. This means that decommissioning decisions and needs are likely to grow in importance over the next two decades. The Agency provides information and guidance to its member states on strategic, methodological and technological approaches to decommissioning, as well as on when to choose decommissioning over license renewal. As of 2006, nine power plants around the world had been completely decommissioned, with their sites released for unconditional use. Seventeen plants have been partially dismantled and safely enclosed, 30 are being dismantled prior to eventual site release and 30 are undergoing minimum dismantling prior to long-term enclosure. The Agency also assisted several of its member states in their efforts to dismantle nuclear power plants. For example, at the Ignalina 1 plant in Lithuania, the Agency's efforts were focused on upgrading local capabilities and, in doing so, facilitating Lithuania's coordination with major international donors for this project. At the severely contaminated A-1 nuclear power plant in Slovakia, the Agency's decommissioning assistance focused on the development of remote viewing equipment and remotely operated tools, both of which are essential given the difficulty of access to a number of components and areas". [11, pages 3 and 4]

There are three main strategies for decommissioning of nuclear facilities:

1) Immediate dismantling, this means the dismantled of the facility after the removal of the nuclear materials and waste.
2) Deferred dismantling, this means keeping the facility in a situation of safe enclosure for many years, after which the facility is dismantled.
3) Entombment, this means keeping the facility encapsulated until the nuclear materials are no more harmful from the radioactive point of view.

One of the main elements that need to be considered in any decommissioning strategy is the decommissioning costs, which includes dismantling, decontamination operations, and waste management costs.

How to finance the decommissioning costs associated with a nuclear facility? One of the possibilities is that the funds to cover these costs could come from resources collected during the operational life of the nuclear facility. These resources should be available when needed. However, this is not always possible. In case of a premature closure of a nuclear facility, the level of the resources collected and available could not be enough. The lack of sufficient funds could be a situation that may force to change the decommissioning strategic to be used in a specific nuclear facility.

The availability, or non-availability, of a waste disposal facility is a significant factor in the selection of an appropriate decommissioning strategy and the cost associated with it. This is often a matter for national policy or government, and beyond the control of operators. If no repository is available, radioactive waste must be processed and stored until the appropriate repository is available.

In this case, the costs associated are likely to be highly dependent on the type of nuclear facility, the site where the storage facility is located and the characteristics and amount of nuclear waste to be storage.

Lastly, relevant knowledge and technical information about nuclear facilities is of prime importance for their safe decontamination and dismantling. Conservation of such knowledge is a key consideration in any decommissioning strategy.

Summing up can be confirmed that in Germany 17 nuclear power plants and prototype reactors, 31 research reactors and critical assemblies, as well as 9 fuel cycle facilities have been permanently shut down. Two of the power reactors, 21 of the research reactors and critical assemblies and four of the fuel cycle facilities have now been decommissioned, and the sites of the two power reactors restored and released from regulatory control. The other power reactors are currently in safe enclosure or are being dismantled, and their sites will be returned to a condition suitable for unrestricted reuse.

In Belgium, a small prototype PWR is in the process of being dismantled and decontaminated, and laboratories used previously for nuclear R&D have been decontaminated, released from radiological control and are now being used for conventional research.

Programs in France and in the UK are quite extensive in scale. Ten gas-cooled, graphite moderated power reactors have been shut down (six in France and four in the UK), including a fast breeder reactors

However, other countries with relative young nuclear programs, like Finland, the Czech Republic and Hungary have no decommissioning programs underway and none foreseen for some years (except maybe for research facilities).

III . European Energy Policy

The continuous increment of the electricity demand obeys to the fact that this type of energy has become the main form of energy in many industrial processes, mainly in the European region. At the same time, this situation provokes an increase in the consumption of fossil fuels by the EU countries, as well as of an increase of the negative impact in the environment. It is expected that the electricity grow more quickly than any other source of energy, reaching 1.5%-2.4 % annually during the coming years.

Taking into account the above-mentioned growth the question to be answered is the following: What to do in order to continue producing electricity and to satisfy other energy requirement, with the purpose to sustain a fast economic development of the EU reducing, at the same time, the consumption of fossil fuels and the negative impact that this increment will has in the environment?

Following the adoption of the 1997 Treaty of Amsterdam, the EU adopted an energy policy based on three core principles, namely:

1) Environmental protection, in both energy production and energy use, in order to maintain ecological and geophysical balances in nature.

Energy accounts for 80% of all greenhouse gas (GHG) emission in the EU; it is at the root of climate change and most air pollution. The EU is committed to addressing this by reducing EU and worldwide greenhouse gas emissions at a global level to a level that would limit the global temperature increase to 2°C compared to pre-industrial levels. However, current energy and transport policies would mean EU-CO_2 emissions would increase by around 5% by 2030 and global emissions would rise by 55%. [3, page 3]

To reduce the level of CO_2 emission the European Commission is promoting the use of renewable energy sources in all European countries. Europe is increasing the use of clean sources of energy, particularly renewable energy for electricity generation, but the level reached until now is still not enough to have a positive impact in the energy balance of several European countries.

2) Security of supply, which aims to minimize risks and impacts of possible supply disruption.

Europe is becoming increasingly dependent on imported hydrocarbons. With "business as usual", the EU's energy import dependence will jump from 50% of total EU energy consumption today to at no less 60% in 2030[17]. (See Figure 23).

The new EU Energy Policy has, as is main purpose the reduction, to the minimum, of the dependency of the region to external energy sources, mainly oil and gas. This could be achieved by promoting the use of renewable energy sources available in the region. Another purpose of the EU Energy Policy is the diversification of its energy sources.

3) Competitive energy systems in order to ensure low cost energy for producers and consumers.

The EU is becoming increasingly exposed to the effects of price volatility and price rises on international energy markets, as well as the consequences of the progressive concentration of oil reserves in few hands. Providing that the right policy and legislative frameworks be in place, the Internal Energy Market could stimulate fair and competitive energy prices and

---

[17] The EU dependency of imported energy could be 60-70% in 2030.

energy savings, as well as higher investment. However, all the conditions to achieve this do not yet exist. This situation prevents the EU economy from receiving the full benefits of energy liberalization. A longer time horizon in the area of carbon constraints is required in order to promote the necessary investments in the electricity sector.

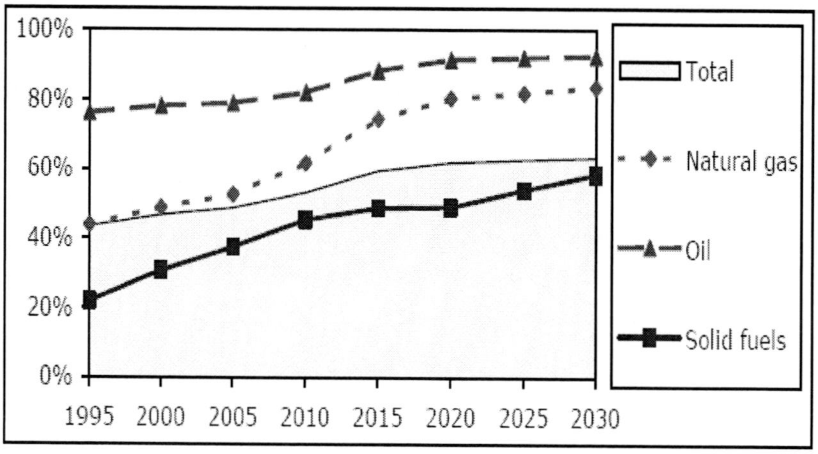

Source: Reference 15, Figure 5, page 9.

Figure 23. Retrospective (1995-2000) and Projected (2005-2030) Import Dependence of EU-27, Total and by Fuels (%).

Boosting investment, in particular in energy efficiency and renewable energy should create jobs, promoting innovation and the knowledge-based economy in the EU. The EU is already the global leader in renewable technologies, accounting for a turnover of 20 billion of Euro and employing 300 000 people. It has the potential to lead the rapidly growing global market for low carbon energy technologies. In wind energy, for example, EU companies have 60% of the world market share. Europe's determination to lead the global fight against climate change creates an opportunity to drive the global research agenda. All options should be kept to ensure the development of emerging technologies. [3, page 4]

Main Elements of the European Energy Policy

The European Energy Policy should be underpinned by:

- An EU objective in international negotiations of 30% reduction in greenhouse gas emissions by developed countries by 2020 compared to 1990. In addition, 2050 global GHG emissions must be reduced by up to 50% compared to 1990, implying reductions in industrialized countries of 60-80% by 2050.
- An EU commitment now to achieve, in any event, at least a 20% reduction of greenhouse gases by 2020 compared to 1990. [3, page 5]

The application of the new European Energy Policy should reduce the CO_2 emissions in some 2,150 million of tons by 2030, that is to say 16% below previously forecasts. This level of emission is equivalent to the CO_2 current emissions of Germany, the UK , France and Italy all together.

Meeting the EU's commitment to act now on greenhouse gases should be at the center of the new European Energy Policy for three reasons:

a) CO_2 emissions from energy make up 80 % of EU-GHG emissions, reducing emissions means using less energy and using more clean locally produced energy.
b) Limiting the EU's growing exposure to increased volatility and prices for oil and gas.
c) Potentially bringing about a more competitive EU energy market, stimulating innovation technology and jobs.

The EU Energy Policy should be prepared based on the realities of the different countries and should contemplate, among others, the following elements:

1) The increase in the saving and in the efficiency of the use of the different energy sources now available in the region. Since the 1970's the European countries have been guaranteeing an economic growth sustained without significant increments in the energy consumption. This achievement is the result not only of a more rational use of the energy but also to a more efficient use of it. In the 1970's and 1980's the growth of the European countries was, as average of 2.4% annually, being the increment of the energy requirements of only 0.4%.
2) Relevant technological advance that can be able to revolutionize the conversion, transport and storage of energy, as well as to elevate it efficiency.
3) The impact in the environment of the use of the different energy sources.
4) An increase in the resources allocated for research and development of the different renewable energy sources and for the discovery of new and even more efficient energy sources.
5) An increase of the international cooperation between the European countries and other countries, with the purpose to join effort in the use of the renewable energy sources now available in the world in the most efficient manner. The search for new more efficient energy sources should also be a subject of international cooperation.
6) An increase of the investments in each one of the links of the energy supplies chain.

Energy security

Energy security has passed to occupy the first place inside the Europe's Energy Policy. In the framework of the application of this policy, the European countries will.

a) Pay higher attention to the maintenance of the security of the international marine routes.
b) Consider the most efficient form of diversifying the use of the different types of fuels and of the geographical sources of the same ones.
c) Reduce the energy demand growth and the gradual substitution of the use of fossil fuels by new, more efficient and less polluting energy sources.

To achieve an energy security it is indispensable, according to the G-8 leaders:

a) To increase the transparency, the forecast possibility and the stability of the world market.
b) To improve the investment climate in the energy sector.
c) To promote the efficiency and the energy saving.
d) To diversify the mixture of energy.
e) To guarantee the physical security of the critical energy infrastructure.
f) To reduce the energy shortage.
g) To approach the climatic change and the sustainable development in the most effective manner. [10, page 33]

New Energy Advanced Technologies

The use of advanced technologies by the European countries has increased the efficiency in the electricity production using fossil fuels. The use of these technologies has produced an important saving of primary energy for electricity production. It is expected to increase this efficiency in 30% in the coming years, if all the investments that are required for energy conservation in the European region are available.

The application of the new European Energy Policy, as well as a bigger use of the new energy advanced technologies (liquid gas, fuel cells, carbon capture and fuel storage), could produce a saving of 9% of the energy demand projected for 2030. The reduction of CO_2 could even be higher if all European countries are supported the change of the most polluting fuels to less polluting ones. The biggest reduction in CO_2 would take place in the electricity generation sector, due to a growth in the use of renewable energy sources, and to a reduction in the energy demand that are expected to take place in coming years.

The Community's external energy policy should ensure the common voice of the EU in support of intensifying its relationship with its energy partners, with a view to further diversifying sources and routes, strengthening partnership and cooperation and focusing on the reduction of greenhouse gas emissions, renewables and increasing energy efficiency. Third countries should be able to benefit from the promotion of renewables in the EU through the supply of bio-fuels and other bio-liquids, which meet sustainability requirements, or the supply of renewable electricity from neighboring countries. While in principle, no trade restrictions should apply to renewable energy imports or exports, the Community must ensure that a level playing field is afforded to all renewable energy producers, in and outside of the Community. As this proposal sets ambitious objectives for member states and its industry to meet, the issue of legal framework of third countries must be addressed. [22, page 4]

Finally, the priorities to be pursued by an effective external EU Energy Policy during the next three years are:

1) The EC and its member states should be a key driver in the design of international agreements, including the future of the Energy Charter Treaty and the post-2012 climate regime.

2) EU energy relations with its neighbors are fundamental to European security and stability. The EU should aim to build up a wide network of countries around the EU, acting on the basis of shared rules or principles derived from the EU Energy Policy.
3) To enhance relations with our external energy suppliers, further developing comprehensive partnerships based on mutual interest, transparency, predictability and reciprocity.
4) To continue to develop closer energy relations with other major consumers, in particular through IEA and G8, or through intensified bilateral cooperation.
5) Develop the use of financial instruments, via enhanced co-operation with the EIB[18] and the EBRD[19] and the establishment of a Neighborhood Investment Fund, to enhance the EU's energy security.
6) To improve the conditions for investments in international projects, working for example to secure a clearly defined and transparent legal framework and appointing European coordinators to represent EU interests in key international projects.
7) Promote non-proliferation, nuclear safety and security, in particular through a reinforced cooperation with the International Atomic Energy Agency. [3, page 18]

The detailed manner in which these objectives will now be pursued, as discussed in detail by the European Council at the Lahti Summit and the December 2006 European Council is contained in reference 3. In addition, however, the Commission considers that two additional priority actions should now be pursued:

- A comprehensive Africa-Europe Energy partnership. The importance of Africa as an energy supplier has increased greatly in recent years, but its potential is still greater. The dialogue should include security of supply, technology transfer in renewable energy, sustainable exploitation of resources, transparency of energy markets and respect for good governance. The dialogue should be launched through a joint event at the highest level.
- An international agreement on energy efficiency.[3, page 19]

IV. European Energy Strategy

Developing a European Energy Strategy is being a long-term challenge for the EU Commission. This needs a clear but flexible framework: clear in that it represents a common approach endorsed at the highest level, flexible in that it needs periodic updating. Six priority areas have been identified and included in the current European Energy Strategy contained in reference (1). These priorities areas are the following:

1. Energy for growth and jobs in Europe: completing the internal European electricity and gas markets.

---

[18] European Investment Bank

[19] European Bank for Reconstruction and Development

2. An Internal Energy Market that guarantees security of supply: solidarity between member states.
3. Tackling security and competitiveness of energy supply: towards a sustainable, efficient and diverse energy mix lives.
4. An integrated approach: to tackling climate change.
5. Encouraging innovation: to strategic European energy technology plan.
6. Towards to coherent external energy policy.

According with the paper titled A European Strategy for Sustainable, Competitive and Secure Energy of 2006, a “sustainable, competitive and secure energy will not be achieved without open and competitive energy markets, based on competition between companies looking to become European-wide competitors rather than dominant national players. A truly competitive single European electricity and gas market would bring down prices, improve security of supply and boost competitiveness. It would also help the environment, as companies react to competition by closing energy inefficient plant” [1, page 5]

The following core areas were included in the above-mentioned paper.

IV.1. Energy for Growth and Jobs in Europe: Completing the Internal European Electricity and Gas Markets

IV.1.1. An European Grid

A European grid code could encourage harmonized, or at least equivalent, grid access conditions. This would take the form of common rules on regulatory issues that affect cross border trade [1, page 6].

The establishment of a real and effective internal European electricity and gas markets, with a common European grid, is one of the priorities that the European Commission has established in the field of energy.

Without a common European energy markets and a common electricity grid, the following problems, among others, will be present:

a) Energy price would not go down.
b) Difficult to increase the saving and the efficiency of the use of the different energy sources now available in the region.
c) Some European countries would not benefit with the existence of major European grids of the region.
d) It would be extremely difficult to help some countries, in case of energy supply disruption, due to their isolation from major European grids.

IV.1.2. A Priority Interconnection Plan

At the Barcelona European Council in 2002, the Heads of State and Government of the EU agreed, “to increase minimum interconnection levels between member states to 10%”. There can be no truly competitive and single European market without additional physical capacity: this is particularly vital for countries such as Ireland and Malta, or for the Baltic States, which remain an “energy island”, largely cut off from the rest of the Community. Equally, additional electricity interconnection capacity is necessary between many areas, particularly between France and Spain with the purpose to permit real competition between these two countries to develop [1, page 6].

At the same time, it is also important to develop offshore wind power in Northern Europe, the construction of the Nabucco pipeline, bringing gas from the Caspian to central Europe, among others additional electricity interconnection capacities.

IV.1.3. Investment in Generation Capacity

The EU will need substantial investment in building new generation capacity over the next 20 years in order to replace ageing electricity generation capacity and to meet electricity demand. This includes capacity to deal with peaks, and additional capacity to replace the ageing generation of nuclear reactors actually in operation in Germany, UK, Sweden, among other countries.

It is extremely important, to prevent disruptions during high demand period and to serve as back up for intermittent renewable energy sources, to have the necessary reserve in electricity generation capacity to deal properly with any situation that could occur related with electricity generation.

IV.1.4. Boosting the Competitiveness of European Industry

One of the most important objectives of the internal energy market is to promote the competitiveness of EU industry and thus contribute to growth and jobs. Industrial competitiveness requires a well-designed, stable and predictable regulatory framework respectful of market mechanisms. Energy policy therefore needs to favor cost-effective options and be based on a thorough economic analysis of different policy options and their impact on energy prices. Secure availability of energy at affordable prices is crucial. Integrated and competitive electricity and gas markets with the minimum of disruption are essential. [1, page 7]

Boosting competitiveness of the European industry will not only contribute to growth and create jobs but also reduce energy price for the consumers. This is an important objective to be achieved of the European Energy Strategy.

IV.2. An Internal Energy Market that Guarantees Security of Supply: Solidarity Between Member States

IV.2.1. Enhancing Security of Supply in the Internal Market

To enhance security of supply the following actions need to be taken:

1) The establishment of a European Energy Supply Observatory as soon as possible to monitor the demand and supply patterns on EU energy markets, identifying likely shortfalls in infrastructure and supply at an early stage and complementing, on an EU level, the work of the International Energy Agency.
2) Improved network security through increased collaboration and exchange of information between transmission system operators in defining and agreeing common European security and reliability standards.

3) With respect to the physical security of infrastructure, two main actions merit further consideration. Firstly, a mechanism could be developed to prepare for and ensure rapid solidarity and possible assistance to a country facing difficulties following damage to its essential infrastructure. Secondly, common standards or measures might be taken to protect infrastructure. [1, page 8]

Under the current international situation, the physical protection of energy infrastructure against terrorist activities is one of the most important tasks that should be carried by the European governments, in order to avoid disruptions of energy supply and generation.

IV.2.2. Rethinking the EU's Approach to Emergency Oil and Gas Stocks and Preventing Disruptions

Oil is a global market and major supply disruptions, even if local or regional, require a global response. Recent experience has raised important questions, including whether Europe's gas stocks can meet the challenge of shorter-term supply disruptions. This review should also considered whether the appropriate signals are being given to encourage the necessary investment in Europe's gas and electricity markets in the years ahead, including investments in security of supply and infrastructure to enable mutual assistance. This could include a new legislative proposal concerning gas stocks to ensure that the EU can react to shorter-term emergency gas supply disruptions in a manner that ensures solidarity between member states, whilst taking account of the different potential for storage in different parts of the EU. [1, pages 8 and 9]

IV.3. Tackling Security and Competitiveness of Energy Supply: Towards a More Sustainable, Efficient and Diverse Energy Mix

Each member state and energy company should choose its own energy mix. However, choices made by one member state inevitably have an impact on the energy security of its neighbours and of the Community as a whole, as well as on competitiveness and the environment. For example decisions to rely largely or wholly on natural gas for power generation in any given member state have significant effects on the security of supply of its neighbours in the event of a gas shortage. Decisions by member states relating to nuclear energy can also have very significant consequences on other member states in terms of the EU's dependence on imported fossil fuels and CO_2 emissions. For this reason, the decision on which is the best energy mix for a given European country should be thoroughly studied, and should considering not only it possible energy options but the impact of its decision in the neighbouring countries and in the whole EU as well.

The Strategic EU Energy Review would offer a clear European framework for national decisions on the energy mix. It should analyze all the advantages and drawbacks of different sources of energy, from indigenous renewable energy sources such as wind, bio-fuels, small hydro and energy efficiency to coal and nuclear, and the knock-on effects of these changes for the EU as a whole. It might be appropriate to agree an overall strategic objective, balancing the goals of sustainable energy use, competitiveness and security of supply. This would need to be developed on the basis of a thorough impact assessment and provide a benchmark based on which the EU's developing energy mix could be judged and would help the EU to stem the increasing dependence on imports. [1, page 9]

During these analyzes governments should have in mind the following two important elements:

- Fossil fuels sources are declining and as consequences fossil fuels prices will continue to increase in the following decades.
- Some renewable energy sources will take more time to development and to have a significant impact in the energy balance of several European countries.

IV.4. An Integrated Approach to Tackling Climate Change

Effective action to address climate change is urgent and the EU must continue to lead by example and, above all, work towards the widest possible international cooperation. Europe needs to be ambitious and must act in an integrated manner that promotes the EU's energy objectives.

The EU is already at the forefront of approaches to decouple economic growth from increasing energy consumption. Its action has combined robust legislative initiatives and energy efficiency programs with encouragement to competitive and effective renewable energy. However, the EU's commitment to fighting climate change is a long-term one. Action on renewable and energy efficiency, besides tackling climate change, will contribute to security of energy supply and help limit the EU's growing dependence on imported energy. [1, page 10]

In any future energy strategy to be adopted by any given country, the impact of the implementation of the adopted strategy in the world climate should be thoroughly considered. In today's world, it is impossible to adopt an energy strategy without having a complete assessment study of the impact that thisstrategy will have in the environment.

IV.4.1. Making More From Less: Leading on Energy Efficiency

An effective energy efficiency policy does not mean sacrificing comfort or convenience, nor does it mean reducing competitiveness. In fact an effective policy in this area means the opposite; making cost-effective investments in order to reduce the waste of energy, thereby increasing standards of living and saving money, and using price signals, that would lead to more responsible, economical and rational use of energy. [1, page 10]

Taking into account the current level of world energy reserve, especially oil and good quality coal reserves, it is extremely important to increase, as much as possible, the efficiency in the use of all available fossil fuel sources, mainly in the transport sector and in the industry. The purpose should be to produce more with less energy and to reduce to the minimum the negative impact in the environment.

IV.4.2. Increasing the Use of Renewable Energy Sources

Since 1990, the EU has been engaged in an ambitious and successful plan to become world leader in renewable energy. In 1997, the EU started working towards a target of a 12% share of renewable energy in its overall mix by 2010, a doubling of 1997 levels. Since then, renewable energy production has increased by 55%. Nevertheless, the EU is set to fall short of its target. The share of renewable energy is unlikely to exceed 10% by 2010. Only a limited number of member states have made serious progress in this area and the critical mass has not been reached to shift niche renewables production into the mainstream. [3, page 12]

The EU's renewable energy market has an annual turnover of €15 (half the world market), employs some 300,000 people, and is a major exporter. Renewable energy is now starting to compete on price with fossil fuels. [1, page 11]

However, the penetration of the renewable energy sources in the world energy balance, particularly in the European energy balance, will be a slow process. The main reason for this is, besides the higher costs of renewable energy sources today compared to "traditional" energy sources, the lack of a coherent and effective policy framework throughout the EU and a stable long-term vision. The renewable energy sources will have a significant impact in the world and European energy balance not before 2030 or even beyond. However, to ensure that the renewable energy sources become an important element of the EU energy balance in the future, the EU should adopt and implement a clear, effective and long-term vision policy.

IV.5. Encouraging Innovation: A Strategic European Energy Technology Plan

There should be no doubt that innovation is a pre-condition to ensure the economic development of a country or region. At the same time can be said that there is no economic development without energy. Based on what has been said before, it is easy to conclude that energy innovation, particularly energy technology innovation, is crucial to ensure the economic development of any country or region, as well as to deliver security of supply, sustainability and industrial competitiveness.

Energy related research has contributed strongly to energy efficiency (e.g. in car engines) and to energy diversity through renewable energy sources. However, the magnitude of the challenges ahead requires increased efforts. This necessitates a long-term commitment. As an example, research has allowed efficiency of coal power stations to be improved by 30% in the last thirty years. Further technological developments would see significant reductions in CO_2 emissions. Research can also bring commercial opportunities. Energy efficient and low carbon technologies constitute a rapidly growing international market that will be worth billions of Euro in the coming years. Europe must ensure that its industries are world leaders in these new generations of technologies and processes. How to achieve this goal? This goal could be achieved by the adoption of an appropriately resourced Strategic Energy Technology Plan within the EU. This Plan "should accelerate the development of promising energy technologies, but should also help to create the conditions to bring such technologies efficiently and effectively to the EU and the world markets. Research in areas of high-energy use, housing, transport, agriculture, agro-industries, and materials, should be addressed as soon as possible". [1, page 13]

Europe should also invest in other possible future forms of energy and in certain technologies, such as hydrogen and fuel cells, carbon capture and storage, large-scale renewable technologies as well as even longer-term prospects such as methane hydrates. Consideration should be given on how to mobilize the resources of the European Investment Bank to promote close to market R&D in this area and how to enhance cooperation in areas of global concern.

Actions to accelerate technology development and drive down the costs of new energy technologies must be complemented by policy measures to open the market and to ensure the market penetration of existing technologies that are effective in addressing climate change. Competing against entrenched technologies and huge locked-in investments in the current energy system largely based on fossil fuels and centralized generation, new technologies face high entry barriers. The EU Emissions Trading Scheme, green certificates, feed-in tariffs and other measures can ensure that the implementation of environmentally friendly energy production, conversion and use is financially viable. Such measures can provide powerful policy signals to the market and create a stable climate in which industries can take the long-term investment decisions required. The Intelligent Energy-Europe Program will also provide the necessary tools and mechanisms to overcome the non-technical barriers to the take up of new and effective energy technologies. [1, page 14]

IV.6. Towards a Coherent External Energy Policy

Why is so important for the EU to have a coherent external energy policy? According with the document titled "An European Strategy for Sustainable, Competitive and Secure Energy" (reference 1), the reason is the following: "The energy challenges facing Europe need a coherent external policy to enable the region to play a more effective international role in tackling common problems with energy partners worldwide. A coherent external policy is essential to deliver sustainable, competitive and secure energy. It would be a break from the past, and show member states' commitment to common solutions to shared problems. The first step is to agree at Community level on the aims of an External Energy Policy and on the actions needed at both Community and national level to achieve it. The effectiveness and coherence of the EU's external energy policy is dependent upon the progress with internal policies and, in particular, the creation of the internal market for energy". [1, page 14]

The following topics are essential elements of a coherent energy external policy for the EU.

IV.6.1. A Clear Policy on Securing and Diversifying Energy Supplies

Europe is running out of indigenous energy resources in the form of fossil fuels, and has not yet a significant participation of renewable energy sources in its energy balance. Some countries within the EU depend strongly of a very limited number of external energy suppliers, situation that could become extremely danger in case of any disruption of energy supplies due to political or economic instability.

A clear policy on securing and diversifying energy supplies "is necessary both for the EU as a whole and for specific member states or regions, and is especially appropriate for gas". To this end, it is important to propose "clearly identified priorities for the upgrading and construction of new infrastructure necessary for the security of EU energy supplies, notably new gas and oil pipelines and liquefied natural gas (LNG) terminals, as well as the application of transit and third party access to existing pipelines". Examples include independent gas pipeline supplies from the Caspian region, North Africa and the Middle East into the heart of the EU; new LNG terminals serving markets that are presently characterized by a lack of competition between gas suppliers, and Central European oil pipelines aiming at facilitating Caspian oil supplies to the EU through Ukraine, Romania and Bulgaria. The new EU-Africa

Strategy, envisaging interconnections of energy systems as a priority area, could also help Europe to diversify its oil and gas supply sources. [1, page 15]

IV.6.2. Energy Partnerships with Producers, Transit Countries and other International Actors

The EU and its energy partners are interdependent. This is reflected at bilateral and regional level in a number of specific EU energy dialogues with a number of producer and transit countries. Equally, energy issues are a growing feature of the EU's political dialogues with other major energy consumers, such as the USA, China and India, including through multilateral forum like the G-8. [1, page 15]

The EU should take the lead and reinforce this dialogue with other countries in order to reach agreements that ensure setting up appropriate priorities for the upgrading and construction of new infrastructure necessary for the security and diversification of EU energy supplies.

IV.6.3. Dialogue with Major Energy Producers/Suppliers

The EU lack of significant fossil fuels reserves and renewable energy sources available in the region does not represent yet an important component of the energy balance of the EU. For this reason, the EU needs to import an important amount of energy from other countries to satisfy its energy needs. To ensure a stable supply of energy, it is indispensable to develop and maintain a good relationship with major energy suppliers.

The EU has an established pattern of relations with major international energy suppliers, including OPEC and the Gulf Cooperation Council. A new initiative is particularly opportune with regard to Russia, the EU's most important energy supplier. The EU, as Russia's largest energy buyer, is an essential and equal partner in this relationship. The development of a common external energy policy should mark a step change in this energy partnership at both Community and national level. A true partnership would offer security and predictability for both sides, paving the way for the necessary long-term investments in new capacity. It would also mean fair and reciprocal access to markets and infrastructure including, in particular, third party access to pipelines. Work should start on an energy initiative based on these principles. [1, page 15]

IV.6.4. Developing a Pan-European Energy Community

In line with the European Neighborhood Policy and its Action Plans (and in addition to the current work undertaken through Partnership and Cooperation Agreements and Association Agreements), the EU has for some time been engaged in widening its energy market to include its neighbors and to bring them progressively closer to the EU's internal market. Creating a "common regulatory space" around Europe would imply progressively developing common trade, transit and environmental rules, market harmonization and integration. This would create a predictable and transparent market to stimulate investment and growth, as well as security of supply, for the EU and its neighbors. Existing political dialogues, trade relations and Community financing instruments can be further developed and, for other partners, there is potential for new agreements or other types of initiative. [1, page 16]

For example, by building on the Energy Community Treaty with partners in South-East Europe, as well as the development of the EU-Maghreb electricity market and the EU-Mashrek gas market, a pan-European Energy Community could be created both through a new Treaty, and through bilateral agreements.

Certain essential strategic partners, including Turkey and Ukraine, could be encouraged to join the South East European Energy Community Treaty. The Caspian and Mediterranean countries are important gas suppliers and transit routes. Algeria increasing importance as a gas supplier to the EU could point to a specific energy partnership.

In addition, as one of the EU's most important strategic energy partner, attention should be given to facilitating Norway's efforts to develop resources in the high north of Europe in a sustainable manner, as well as facilitating its entry into the South East Europe Energy Community. This framework would also offer a clearer framework to promote best long-term use of Community investment through Trans-European Energy Networks and their extensions to third country partners and to maximize the impact on energy security of EU resources devoted to the energy sector in third countries. [1, page 16]

The EU should encourage other initiatives in order to enlarge its energy market. This enlarges should include other neighbors countries with the purpose to bring them progressively closer to the EU's internal market.

IV.6.5. Reacting Effectively to External Crisis Situations

Consideration should be given on how best to react to external energy crises. Recent experiences with respect to both oil and gas have shown the need for the Community to be able to react quickly and in a fully coordinated manner to such events. [1, page 16]

The establishment of a monitoring mechanism to provide early warning and to enhance response capabilities in the event of an external energy crisis is an appropriate mechanism that could be used y the EU to react in an effective manner to external crisis.

IV.6.6. Integrating Energy iInto other Policies with an External Dimension

At the political level, a common European external energy policy will permit a better integration of energy objectives into broader relations with third countries and the policies, which support them. That means increasing the focus in relations with global partners facing similar energy and environmental challenges – such as the USA, Canada, China, Japan and India – on issues such as climate change, energy efficiency and renewable energy sources, research and development of new technologies, global market access and investment trends, with better results in multilateral forum such as the UN, the IEA and the G-8. If these countries reduce the use of fossil fuels, it will also be beneficial for Europe's energy security. The EU could significantly step up bilateral and multi-lateral cooperation with these countries with the objective of encouraging the rational use of energy worldwide, of reducing pollution and encouraging industrial and technological cooperation on the development, demonstration and deployment of energy efficient technologies, renewable energy sources and clean fossil fuel technologies with carbon capture and geological storage. [1, pages 16 and 17]

In particular, greater efforts need to be made towards widening the geographic scope of the EU Emissions Trading Scheme and, as mentioned above, as a first step, the EU should propose and promote an international agreement on energy efficiency. In addition, more focus could be given to technological cooperation, in particular with other energy consuming countries.

Similarly, there is scope to make better use of trade policy tools to promote goals such as non-discriminatory energy transit and the development of a more secure investment climate. The EU should press for a better respect of existing WTO rules and principles in this field, and bilateral or regional initiatives should build on these. Such agreements can include provisions on market opening, investment, regulatory convergence on issues such as transit and access to pipelines, and competition. Reinforced market-based provisions on energy and trade–related energy issues would thus be incorporated in the EU's existing and future agreements with third countries. [1, page 17]

Summing up can be stated that the new European Energy Strategy is driven primarily by the need for the European powers to reduce their dependence on unstable oil and gas imports. Europe, with its limited and dwindling hydrocarbon energy sources, is highly sensitive to the resource wars that have erupted across the globe since the collapse of the Soviet Union in 1991.

As the USA has launched military adventures in several unstable regions in order to control the main hydro and carbon supplies and transit routes of the world, Europe has found itself increasingly at risk of having its energy needs curtailed. The insecurity of the European powers has been intensified by the resurgence of Russian power due to current high-energy prices. The strategy adopted is largely aimed at enabling Europe to reduce dependence on Middle Eastern and Russian oil and gas and includes a series of binding targets that should be met by the 27 EU members for 2020. These targets are the following:

1. To reach 10% of automobiles to be powered by bio-fuels
2. To reach 20% of all energy (including electricity, heating and transport fuel) produced from "coal-neutral" sources such as solar, wind and nuclear energy, and the development of "coal burying," involving reducing atmospheric coal emissions from fossil fuel burning power stations by disposing of it by other means.

The EU Energy Strategy has been widely greeted as a noble effort to respond to the threats of global warming, but some experts consider that this strategy has a limited impact on the environment. For instead, the target to have 10% of automobiles powered by bio-fuels by 2020 will do very little to reduces current coal emissions due to an increase number of car that it is expected to be running the European highways by that year. The main beneficiary of such target would be, in the opinion of some experts, the European agribusiness, which can turn over large extension of land to growing fuel crops that are likely to be highly subsidized by the UE in the future. This situation will increase the price of certain crops used to produce bio-fuels and this will increase the price of certain foods that are essential to feed the population of most of the developing countries. The first signal of the price increase in certain crops used to produce bio-fuels is already increasing the price of foods at international level. This tendency will continue in the future is no urgent actions are taken by the international community to stop it.

Another important element of the UE Energy Strategy that needs to be considered is the deregulation of the continent's energy markets in order to liberalize the energy sector. Some experts have the opinion that with this strategy the EU governments are serving the interests of the capital finance sector, which are looking where to invest with high rate of return.

The energy debates within the EU should take into account that the current energy demand is covered by 41% of oil, 22% of gas, 16% of coal (hard coal, lignite and peat), 15%

of nuclear and 6% of renewable. If nothing is done, it is expected that the total energy picture in 2030 will be continued dominated by fossil fuels: 38% of oil, 29% of gas, 19% of solid fuels, 8% of renewable and barely 6% of nuclear (See Figure 24)

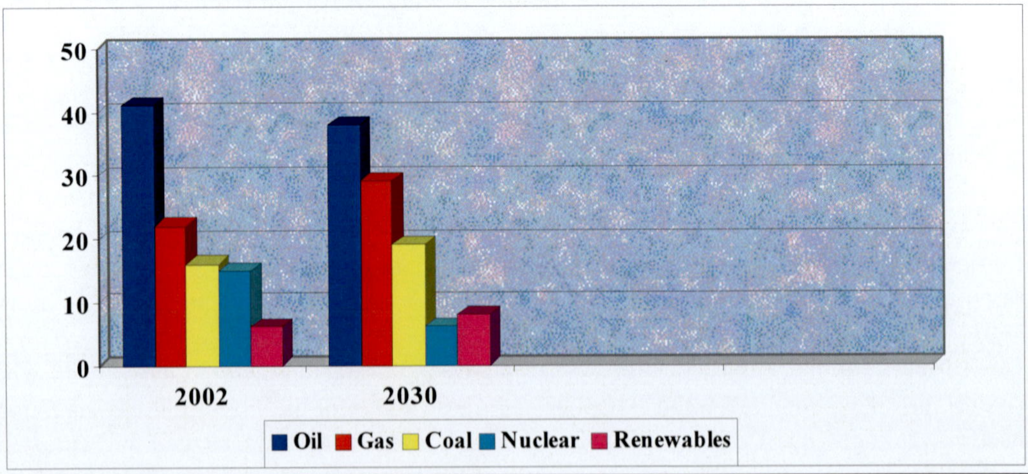

Figure 24. The Present and Future Situation of the Different Types of Energy in the Energy Balance in the EU

V. CONCLUSIONS

This paper presents a summary of the energy realities facing Europe in the 21st century. Three main objectives have been identified within the Europe's Energy Policy:

1. Sustainability.
2. Competitiveness.
3. Security of supply.

To achieve these objectives, it is important to put them in an overall framework within a European Energy Strategy.

The main elements of the European's Energy Strategy have been single out in this paper. A group of concrete actions has been identified in order to meet these three objectives. These are the following:

1) The EU needs to complete the internal gas and electricity markets.

The EU needs to ensure that its internal energy market guarantees security of supply and solidarity between member states. Concrete measures should include:

a) A review of the existing Community legislation on oil and gas stocks, to focus them on today's challenges.
b) An European Energy Supply Observatory, enhancing transparency on security of energy supply issues within the EU.

c) Improved network security through increased cooperation between network operators and possibly a formal European grouping of network operators.
d) Greater physical security of infrastructure, possibly through common standards.
e) Improved transparency on energy stocks at the European level.

2) The Community needs a real Community-wide debate on the different energy sources, including costs and contributions to climate change to be sure that, overall, the EU's energy mix pursues the objectives of security of supply, competitiveness and sustainable development.
3) Europe needs to deal with the challenges of climate change.
4) An European Strategic Energy Technology Plan, making best use of Europe's resources, building on European technology platforms and with the option of joint technology initiatives or joint undertakings to develop leading markets for energy innovation need to be adopted and implemented.
5) A common external energy policy in order to react to the challenges of high and volatile energy prices, increasing import dependency, strongly growing global energy demand and global warming. The EU needs to have a clearly defined external energy policy and to pursue it at the same time at both national and Community level, with a single voice.

To achieve this goal the following actions need to be implemented:

a) Identifying European priorities for the construction of new infrastructure necessary for the security of EU energy supplies.
b) Developing a pan-European Energy Community Treaty.
c) A new energy partnership with Russia.
d) A new Community mechanism to enable rapid and coordinated reaction to emergency external energy supply situations affecting EU supplies.
e) Deepening energy relations with major producers and consumers.
f) An international agreement on energy efficiency.

ACKNOWLEDGEMENT

I would like to thank Mag. Lisette Morales Meoqui, from Dialog and Konzept, Financial Services, Vienna, Austria, for her support in the preparation of this paper and for her valuable comments and suggestions during its revision and to Eng. Hector Espejo from the Argentinean Association on Non-destructive Testing for his valuable comments and suggestions.

Annex 1 . Oil production and reserve-to production ratio 2005

Country	2005 Production (Million Barrels per Day)	2005 Share of World Production (Percent)	Reserve-to-Production Ratio (Years)
Saudi Arabia	9.55	13.3	75
Russia	9.04	12.6	18
United States	5.18	7.2	11
Iran	4.14	5.7	83
China	3.61	5.0	14
Mexico	3.33	4.6	12
Norway	2.70	3.7	9
Nigeria	2.63	3.6	37
United Arab Emirates	2.54	3.5	106
Kuwait	2.53	3.5	110
Venezuela	1.98	2.7	107
Iraq	1.88	2.6	168
Algeria	1.80	2.5	18
United Kingdom	1.65	2.3	7
Brazil	1.63	2.3	18
Libya.	1.63	2.3	65
Canada.	1.28	1.8	10
Angola	1.26	1.7	12
Indonesia	1.07	1.5	12
Kazakhstan	1.05	1.5	23
Qatar	0.84	1.2	50
Oman	0.77	1.1	19
Malaysia	0.75	1.0	11

Annex 1. (Continued)

Country	2005 Production (Million Barrels per Day)	2005 Share of World Production (Percent)	Reserve-to-Production Ratio (Years)
Argentina	0.70	1.0	10
India	0.66	0.9	22

Sources: 2005 Production: Energy Information Administration, Short-Term Energy Outlook (October 2006).

Reserves: "World- wide Look at Reserves and Production," Oil and Gas Journal, Vol. 104, No. 47 (December 18, 2006), pp. 24-25.

Annex 2.World Natural Gas Production by Region and Country, 2004-2030

(Trillion Cubic Feet) Region/Country	2004	2010	2015	2020	2025	2030	Average Annual Percent Change 2004-2030
OECD North America	269	28.1	28.2	29.3	29.2	29.6	0.4
United Statesa	19.0	19.5	19.7	20.9	20.7	20.7	0.3
Canada	6.5	6.8	6.4	6.0	5.9	6.0	-0.3
Mexico	1.5	1.8	2.0	2.4	2.6	3.0	2.7
OECD Europe	11.4	11.7	11.2	10.7	10.5	10.1	-0.4
OECD Asia	1.6	2.2	3.1	3.8	4.3	4.7	4.2
Japan	0.1	0.1	0.1	0.1	0.1	0.1	0.4
South Korea	0.0	0.0	0.0	0.0	0.0	0.0	—
Australia/New Zealand	1.5	2.1	3.0	3.7	4.2	4.6	4.3
Total OECD	39.9	42.0	42.5	43.8	44.1	44.5	0.4
Non-OECD Europe and Eurasia	28.6	33.2	36.4	39.5	42.4	45.2	1.7
Russia	22.4	24.9	27.4	30.0	32.6	35.2	1.7
Other	6.3	8.3	9.1	9.5	9.8	10.0	1.8
Non-OECD Asia	10.5	13.6	16.4	19.1	22.2	25.2	3.3
China	1.4	2.5	3.1	3.5	4.0	4.3	4.1

Annex 2. (Continued)

(Trillion Cubic Feet)							
							Average Annual Percent Change
India	1.0	1.5	1.7	2.1	2.4	2.5	3.5
Other non-OECD Asia	8.1	9.6	11.5	13.5	15.8	18.4	3.1
Middle East	9.9	13.8	17.4	201	21.8	24.1	3.3
Africa.	5.3	7.8	9.5	11.1	13.0	15.1	4.0
Central and South America	4.5	5.8	70	7.7	8.4	9.2	2.7
Brazil	0.3	0.6	0.7	0.8	0.9	1.0	4.1
Other Central /South America. .	4.2	5.3	6.3	6.9	7.6	8.2	2.5
Total Non-OECD	58.9	74.3	86.7	97.4	107.8	1188	2.6
Total World	98.9	116.3	129.2	141.2	151.9	163.3	1.9

Includes supplemental production or forecast discrepancy. For details, see Energy InformationAdministration (EIA), Annual Energy Outlook 2007, p. 159, Table A13, "Natural Gas Supply, Disposition, and Prices." Note: Totals may not equal sum of components due to independent rounding. Sources: 2004: EIA, International Energy Annual 2004 (May-July 2006), web site www.eia. doe.gov/iea. 2010-2030: United States: EIA, Annual Energy Outlook 2007, DOE/EIA-0383(2007) (Washington, DC, February 2007), web site www.eia.doe.gov/oiaf/ aeo. Others: EIA, System for the Analysis of Global Energy Markets (2007).

Annex 3 . World Coal Production by Region, 2004-2030

(Quadrillion Btu)							
							Average Annual Percent Change
Region	2004	2010	2015	2020	2025	2030	2004-2030
OECD North America	24.6	27.0	28.3	29.4	33.3	37.0	1.6%
United States	22.8	24.6	25.8	26.7	30.4	33.9	1.5%
Canada	1.5	1.9	2.1	2.2	2.4	2.5	2.0%
Mexico	0.2	0.4	0.4	0.4	0.5	0.5	3.3%
OECD Europe	7.9	8.0	7.7	7.1	6.6	6.6	-0.7%

Annex 3 .(Continued)

OECD Asia	8.1	9.7	10.4	11.2	11.9	12.7	1.7%
Japan.	0.0	0.0	0.0	0.0	0.0	0.0	—
South Korea	0.1	0.1	0.1	0.1	0.1	0.1	0.9%
Australia/New Zealand	8.1	9.6	10.2	11.1	11.9	12.6	1.7%
Total OECD	40.6	44.7	46.4	47.7	51.9	56.2	1.3%
Non-OECD Europe and Eurasia . . .	10.0	11.5	12.5	13.3	13.6	13.7	1.2%
Russia	5.9	7.1	7.5	7.9	8.2	8.5	1.4%
Other	4.1	4.4	5.0	5.3	5.4	5.3	0.9%
Non-OECD Asia	55.2	70.6	82.4	94.7	105.4	116.7	2.9%
China	43.0	55.4	64.6	74.3	83.4	93.4	3.0%
India.	7.3	8.1	9.5	10.8	12.0	13.0	2.3%
Other	4.9	7.1	8.3	9.6	10.0	10.3	2.9%
Middle East	*	*	*	*	*	*	-1.2%
Africa	5.9	7.1	7.7	8.0	8.6	8.9	1.6%
Central and South America	1.8	2.7	3.1	4.0	4.3	4.3	3.4%
Brazil	0.1	0.1	0.2	0.2	0.2	0.2	3.3%
Other	1.7	2.5	2.9	3.8	4.1	4.1	3.4%
Total Non-OECD	72.8	91.9	105.7	120.1	131.9	143.7	2.6%
Total World	113.4	136.6	152.1	167.7	183.8	199.9	2.2%

*Less than 0.05 quadrillion Btu

Note: With the exception of North America, non-seaborne coal trade is not represented in the IEO2007 cases. As a result, the projected levels of production assume that net non-seaborne coal trade will balance out across the world regions. Currently, a significant amount of non-seaborne coal trade takes place in Eurasia, represented by exports of steam coal from Kazakhstan to Russia and exports of coking coal from Russia to Ukraine.

Sources: 2004: Energy Information Administration (EIA), International EnergyAnnual 2004 (May-July 2006), web site www.eia.doe.gov/iea. Projections: EIA, System for the Analysis of Global Energy Markets (2007) and National Energy Modeling System run IEO2007.D032707B

Annex 4 . World Coal Flows by Importing and Exporting Regions, Reference Case 2005, 2015, and 2030

(Quadrillion Btu)												
Exporters	Importers											
	Steam				Coking				Total			
	Europe (a)	Asia (c)	America	Total (b)	Europe (a)	Asia (c)	America	Total (b)	Europe (a)	Asia (c)	America	Total (b)
			2005				2005				2030	
Australia	0.08	2.57	0.15	2.81	0.79	2.26	0.22	3.27	0.86	4.83	0.37	6.07
United States	0.08	0.03	0.36	0.47	0.44	0.10	0.22	0.77	0.52	0.13	0.58	1.23
South Africa	1.54	0.10	0.03	1.70	0.02	0.00	0.00	0.03	1.56	0.10	0.03	1.73
Eurasia	0.92	0.29	0.00	1.22	0.08	0.10	0.00	0.18	1.00	0.40	0.00	1.40
Poland	033	0.00	0.00	0.34	0.04	0.00	0.00	0.04	0.37	0.00	0.00	0.37
Canada	0.00	0.02	0.01	0.04	0.24	0.38	0.11	0.72	0.24	0.41	0.12	0.76
China	0.07	1.62	0.01	1.70	0.00	0.14	0.00	0.14	0.07	1.76	0.01	1.84
South America	0.86	0.00	0.77	1.63	0.00	0.00	0.00	0.00	0.86	0.00	0.77	1.63
Vietnam	0.00	0.28	0.00	0.28	0.00	0.00	0.00	0.00	0.00	0.28	0.00	0.28
Indonesia/Others	0.42	2.06	0.08	2.57	0.00	0.46	0.00	0.46	0.42	2.52	0.08	3.03
Total	4.30	6.98	1.41	12.75	1.61	3.44	0.56	5.60	5.91	10.41	1.97	18.35
							2015					
Australia	0.00	3.33	0.00	3.33	0.64	3.20	0.27	4.10	0.64	653	0.27	7.43
United States	0.07	0.03	0.19	0.29	0.33	0.03	0.35	0.71	0.41	0.06	0.54	1.00
South Africa	1.69	0.31	0.13	2.13	0.02	0.00	0.02	0.04	1.71	0.31	0.15	2.17
Eurasia	1.35	0.54	0.00	1.89	0.08	0.23	0.00	0.31	1.43	0.77	0.00	2.20
Poland	0.13	0.00	0.02	0.15	0.03	0.00	0.00	0.03	0.16	0.00	0.02	0.18
Canada	0.00	0.00	0.00	0.00	0.29	0.54	0.13	0.96	0.29	0.54	0.13	0.96
China	0.00	1.07	0.00	1.07	0.00	0.03	0.00	0.03	0.00	1.10	0.00	1.10
South America	1.42	0.00	1.25	2.66	0.00	0.00	0.00	0.00	1.42	0.00	1,25	2.66
Vietnam	0.00	0.16	0.00	0.16	0.00	0.01	0.00	0.01	0.00	0.16	0,00	0,16
Indonesia/Others	0.00	3.66	0.13	3.79	0.00	0.43	0.00	0.43	0.00	4.09	0,13	4,22
Total	4.67	9.10	1.70	15.48	1.39	4.46	0.76	6.61	6.06	13,56	2,47	22,09

Annex 4 . (Continued)

(Quadrillion Btu)												
Exporters	Importers											
	Steam				Coking				Total			
	Europe (a)	Asia (c)	America	Total (b)	Europe (a)	Asia (c)	America	Total (b)	Europe (a)	Asia (c)	America	Total (b)
							2030					
Australia		4.17	0.00	4.17	0.84	3.88	0.41	5.13	0.84	8,05	0,42	9,31
United States	0.00	0.01	0.16	0.17	0.26	0.04	0.46	0.76	0.26	0,05	0,62	0,93
South Africa	1.44	0.73	0.18	2.36	0.00	0.00	0.02	0.02	1.44	0,73	0,21	2,38
Eurasia	1.25	0.66	0.00	1.90	0.16	0.27	0.00	0.43	1.41	0,93	0,00	2,33
Poland	0.07	0.00	0.03	0.10	0.01	0.00	0.00	0.01	0.08	0,00	0,03	0,11
Canada	0.00	0.00	0.00	0.00	0.28	0.66	0.11	1.04	0.28	0,66	0,11	1,04
China	0.00	1.07	0.00	1.07	0.00	0.03	0.00	0.03	0.00	1,10	0,00	1,10
South America	1.58	0.00	2.19	3.77	0.00	0.00	0.00	0.00	1.58	0,00	2,19	3,77
Vietnam	0.00	0.80	0.00	0.80	0.00	0.01	0.00	0.01	0.00	0,80	0,00	0,80
Indonesia/Others	0.00	3.90	0.35	4.26	0.00	0.43	0.00	0.43	0.00	4,33	0,35	4,68
Total	4.33	11.35	2.93	18.61	1.55	5.31	1.00	7.86	5.88	16,65	3,93	26.47

A) Europe/Mediterranean, including coal shipments to the Middle East and Africa. B) In 2005, total world coal flows include a balancing item used to reconcile discrepancies between reported exports and imports. The 2005 balancing items by coal type were 1.062 quadrillion Btu (steam coal), 0.003 quadrillion Btu (coking coal), and 0.065 quadrillion Btu (total). C) Includes 0.37 quadrillion Btu of coal for pulverized coal injection at blast furnaces shipped to Japanese steelmakers in 2005. D) Coal exports from South America are projected to originate from mines in Colombia and Venezuela. E) Includes shipments from other countries not modeled for the projection period. The 2005 exports from other countries by coal type were 0.08 quadrillion Btu (steam coal), 0.03 quadrillion Btu (coking coal), and 0.11 quadrillion Btu (total). Notes: Data exclude non-seaborne shipments of coal to Europe and Asia. Totals may not equal sum of components due to independent rounding. Sources: 2005: SSY Consultancy and Research, Ltd., SSY's Coal Trade Forecast, Vol. 15, No. 2 (London, UK, August 2006); and Energy Information Administration, Quarterly Coal Report, October-December 2005, DOE/EIA-0121(2005/4Q) (Washington, DC, March 2006). 2015 and 2030: Energy Information Administration, National Energy Modeling System, run IEO2007.D032707B.

Annex 5 . World Carbon Dioxide Emissions by Region, 1990-2030

(Billion Metric Tons)									
Region	History		Projections					Average Annual Percent Change	
	1990	2004	2010	2015	2020	2025	2030	1990-2004	2004-2030
OECD	11.4	13.5	14.1	14.7	15.2	15.9	16.7	1.2%	0.8%
North America	5.8	6.9	7.3	7.8	8.2	8.8	9.4	1.3%	1.2%
Europe	4.1	4.4	4.5	4.6	4.6	4.6	4.7	0.5%	0.3%
Asia	1.5	2.2	2.3	2.4	2.4	2.5	2.6	2.5%	0.6%
Non-OECD	9.8	13.5	16.8	19.2	21.6	23.9	26.2	2.3%	2.6%
Europe and Eurasia	4.2	2.8	3.1	3.3	3.5	3.7	3.9	-2.8%	1.2%
Asia	3.6	7.4	9.7	11.4	13.1	14.8	16.5	5.2%	3.1%
Middle East	0.7	1.3	1.6	1.8	2.0	2.1	2.3	4.4%	2.3%
Africa	0.6	0.9	1.1	1.3	1.4	1.5	1.7	2.5%	2.3%
Central and South America. .	0.7	1.0	1.2	1.4	1.6	1.7	1.9	3.1%	2.3%
Total World	21.2	26.9	30.9	33.9	36.9	39.8	42.9	1.7%	1.8%

Sources: 1990 and 2004: Energy Information Administration (EIA), International EnergyAnnual2004 (May-July 2006), web site.www.eia.doe.gov/iea. 2010-2030: EIA, System for the Analysis of Global Energy Markets (2007).

About the Author

Jorge Morales Pedraza has a University Diploma in Mathematics and in Economy. He was Ambassador for more than 25 years and University Professor in Mathematics and Invited Professor for International Relations. He worked as Senior Manager in the International Atomic Energy Agency (IAEA) in the Director's Office. He is the author and co-author of more than 30 papers already published by the United Nations and by other international publisher houses, and the author or coauthor of several chapters of different books related with the peaceful uses of nuclear energy, non-proliferation, disarmament, arms control and security.

REFERENCES

[1] *A European Strategy for Sustainable, Competitive and Secure Energy*, Brussels, 8.3.2006, COM (2006) 105 final, pages 3- 11 ,13-17.

[2] Antics, M and Scanner, B., Status of Geothermal Energy Use and Resources in Europe, *Proceedings European Geothermal Congress 2007 Unterhaching*, Germany, 30 May-1 June (2007), pages 1, 3, and 4.

[3] *An energy policy for Europe*, Brussels, 10.1.2007, COM (2007) 1 final, pages 3, 4, 5, 12, 13, 14, 18 and 19.

[4] Australian Uranium Association, The Economics of Nuclear Power, *Briefing Paper* 8, 2007, pages 1, 3, 4 and 8.

[5] Baird, S., MBA, *Hydroelectric power*, page 2.

[6] Barré, B., *IAEA Bulletin* Vol. 45/2, 2003, page 54.

[7] Cohen, A., *The North European Gas Pipeline Threatens Europe's Energy Security*, Heritage Foundation, page 1.

[8] *European Fuel Cell and Hydrogen Projects 2002-2006*, Directorate-General for Research Sustainable Energy Systems, 2006, pages 10 and 22.

[9] Group of authors; *Wind Energy. The facts. An analysis of the wind energy in the EU-25*; European Wind Energy Association, 2002, pages 5, 7 and 9.

[10] *IAEA Bulletin* Vol. 48/1, 2006, G-8 Summit, San Petersburg, Russia, 15-17 July 2006, page 33.

[11] *IAEA Annual Report for 2006*, GC (50)/4, pages 1, 3 and 4.

[12] *IAEA Annual Report for 2003*, GC (48)/3, page 2.

[13] *Nuclear Technology Review 2006*, IAEA document GC (50) INF/3, 2006, page 11.

[14] I*nternational Energy Outlook 2007*, DOE/EIA-0484 (2007), Energy Information Administration, Office of Integrated Analysis and Forecasting, U.S. Department of Energy Washington, DC 20585, May 2007, pages 2, 6, 9, 32, 39, 40, 52, 53, 62 and 63

[15] Kavalov, B. and Peteves, S.D., *The future of coal,* DG JRC Institute for Energy, 2007, pages 6, 7, 11, 25, 26, 32 and 33.

[16] *Large-scale integration of wind energy in the European power supply: analysis, issues and recommendations*", (December 2005), EWEA report, pages 5 and 6.

[17] Lehner, B., Czisch, G., Vassolo, S., *Hydropower potential today and in the future*, Center for Environmental Systems Research, University of Kassel, Germany Institut für Solare Energieversorgungstechnik (ISET), Kassel, Germany, pages 8-1, 8-2 and 8-4.

[18] Maio, P., *The Hydrogen Economy, Energy*, 2005, pages 8.

[19] Memo/07/5, 10 January 2007, Brussels.

[20] Moore, P., Director Scientific, Greenspitit Enterprise, Vancouver, Canada, *IAEA Bulletin* Vol. 48/1, 2006, page 57.

[21] *Nuclear Technology Review 2006*, IAEA, GC (50)/INF/3, 3 July 2006, page 8.

[22] *On the promotion of use of energy from renewable sources, proposal for a Directive for the EU Parliament and the Council*, Brussels 23.01.2008 COM (2008) pages 2, 3, 4 and 8.
[23] *Prioritizing Wind Energy Research: Strategic Research Agenda of the Wind Energy Sector*, EWEA, July 2005, pages 2, 14 and 15.
[24] *Prospects for the internal gas and electricity market COM* (2006) 841 final, Brussels, 10.1.2007, pages 15 and 16.
[25] Ritch III, J., Nuclear energy at the moment of truth, *IAEA bulletin 44/2*, 2002, page 35.
[26] Reeves, A., Beck, F., Executive Editor; *Wind Energy for Electric Power*, page 4 .
[27] Towards a low carbon future, *A European Strategy Energy Technology Plan COM* (2007) 723 final, Brussels, 22.11.2007, pages 4, 5, 7, 8,, 10. .
[28] Thurner, B., *Europe Debates Nuclear Energy*, Washington, USA, 2006, page 1.
[29] Van der Zwann, B., *Prospects for Nuclear Energy in Europe*, Energy research Centre of the Netherlands (ECN), Amsterdam, The Netherlands, pages 9, 10 and 11.

In: Energy in Europe: Economics, Policy and Strategy ISBN: 978-1-60456-829-5
Editors: F. L. Magnusson and O. W. Bengtsson

Chapter 2

INDUSTRIAL ENERGY INTENSITIES IN THE UK: A PRELIMINARY ANALYSIS

Paolo Agnolucci
Environment Group Policy Studies Institute, London, UK

ABSTRACT

The aim of this chapter is to assess whether long-term differences among the UK industrial subsectors are stochastic or deterministic. The analysis is implemented over the 1970-2004 time period. We discovered that the results are very influenced by whether one allows for deterministic structural breaks. Only when doing so, one can discover evidence of many the long-term differences being deterministic. This supports the adoption of policy instruments which are applied across productive sectors in a way which is not affected by the short-run evolution of the sectors.

1. INTRODUCTION

Energy consumption, like the consumption of any productive factors, is determined by the final product and the technology used in the production process. Clearly, the energy intensity, i.e., the ratio between energy consumption e_t and economic activity y_t, of, say, the chemical sector will be different from that of, say, the construction sector. The existence of long-term differences among the energy intensities of different sectors is simply a fact – see Figure 1. This study assesses whether long-term differences among the energy intensities of the British industrial subsectors are deterministic or stochastic. The properties of the long-term differences, i.e., differences which persist across time, are related to the properties of short-term random differences. If these are transitory, long-term differences will be deterministic. If short-term random differences are persistent, they accumulate across time and long-term differences will be stochastic. A number of articles use the term convergence to indicate that long-term differences across sectors with respect to a certain variable are deterministic. As this terminology is not common in the energy literature, this paper avoids this use of the word convergence as some readers might be wrongly mislead to think that the levels of energy intensities in different sectors become gradually similar. This is a statement which is not supported by any evidence (see Figure 1) and, as far as the authors of this paper are aware, by any theoretical proposition.

This paper is structured as follows. Section 2 describes the approach used in this study and its motivation. Section 3 describes the dataset used in this study. Section 4 briefly reviews a number of univariate unit root tests and applies them to the dataset. After stressing the limitations of panel tests, Section 5 applies a limited number of tests. Like in the case of the univariate tests, one can conclude that long-term differences among the energy intensities of industrial subsectors are stochastic. However, the opposite conclusion can be drawn, if one takes into account breaks in the series, as done in Section 6. Despite the limited length of the sample, tests allowing for one or two breaks often point at long-term differences being deterministic. This is confirmed by a panel unit root test allowing for structural breaks.

2. Methodological Approach and Motivation of this Study

Energy intensity is a ratio which has received considerable attention in the academic and policy making community. From a policy point of view, energy intensity is related to energy efficiency, i.e., the focus of a number of policies adopted in several countries (World Energy Council 2004) and a priority for the European Commission and the member states (DTI 2006). Given the downward trend in energy intensity in the industrial sector (see the Odyssee database and Figure 1 below) policy markers might have a special interest in the time pattern of energy intensities of a number of sectors or equivalently in their long-term differences with a benchmark. This special interest could be motivated by the fact that these sectors are considered energy inefficient, compared to international benchmarks, or politically-sensitive as they attract a great deal of attention from the public.

This paper follows the approach of Carlino and Mills (1993) where series $Y_{i,t}$ are assessed relative to a benchmark $Y_{b,t}$. The dependent variable is therefore $y_{i,t}^{d} = Y_{i,t} - Y_{b,t}$ rather than the levels $Y_{i,t}$. One can further assume that the difference from the benchmark consists of a time invariant differential x_i, a linear trend t, and the random short-run difference v_t. Breaks in the deterministic components can also be accommodated by the variable m_i. Long-term differences between a series and the benchmark can be written as

$$y_{i,t}^{d} = x_i + v_0 + \beta_i t + \delta_i m_i + v_t \tag{1}$$

The nature of the differences among the series can be assessed by testing for the presence of unit roots in $y_{i,t}^{d}$. If there is a unit root, long-term differences are stochastic, as short-term random differences accumulate across time. If there is no unit root, short-term random differences dissipate and long-term differences are deterministic.

The nature of the long-term differences among industrial subsectors affects policy-making with regard to the timing and frequency of policy interventions and with regard to the choice among policy instruments. Suppose that long-term differences among industrial sectors are stochastic. As mentioned above, this implies that short-term random difference accumulated across time. When policy makers have an interest in the series following a certain time pattern, they will need to intervene any time a random shock push the series into a direction opposite to that desired by policy makers. On the other hand, when long-term differences are deterministic, policy makers do not need to intervene at any single minor departure of the series from the desired time pattern, as random shocks will naturally dissipate across time. In fact, if they want to affect the evolution of the series, policy makers will need to implement prolonged policy measures able to affect the deterministic component. Also the choice among policy instruments is affected by the nature of the difference among energy intensity among industrial subsectors. When long-term differences are stochastic, policy mechanisms affecting economic actors in a uniform fashion would seem a rather blunt instrument. Ideally, policy instruments would need to take into account the evolution of the series and the peculiarity of the sectors. On the other hand, when long-term differences are deterministic, a uniform policy instrument not subject to periodical reviews would seem suited to meet the policy-makers' objectives. It goes without saying that periodical reviews of these instruments can still be needed due to changes in the information set or in the policy aims.

3. DATA DESCRIPTION

Annual data on energy consumption is taken from IEA (2005), which provides information for the sectors listed in the first column of Table 1 for the 1970-2004 timespan. Data on the GVA have been sourced from ONS (2006a) and ONS (2006b)[1].

Table 1. Industrial sectors modelled in this study

<table>
<tr><td>Industrial subsectors</td><td>ISIC code</td><td>Acronym</td></tr>
<tr><td>Mining and Quarrying</td><td>13 and 14</td><td>MIN</td></tr>
<tr><td>Food and Tobacco</td><td>15 and 16</td><td>FT</td></tr>
<tr><td>Textile and Leather</td><td>17, 18 and 19</td><td>TXT</td></tr>
<tr><td>Wood and Wood Products</td><td>20</td><td rowspan="2">WPP</td></tr>
<tr><td>Paper, Pulp, and Printing</td><td>21 and 22</td></tr>
<tr><td>Chemical and Petrochemical Industry</td><td>24</td><td>CHE</td></tr>
<tr><td>Non-Metallic Minerals</td><td>26</td><td>NMM</td></tr>
<tr><td>Construction</td><td>45</td><td>CON</td></tr>
<tr><td>Iron and Steel</td><td>27.1 + 27.31</td><td rowspan="5">OTH</td></tr>
<tr><td>Non-Ferrous Metals</td><td>27.2 + 27.32</td></tr>
<tr><td>Transport Equipment</td><td>34 + 35</td></tr>
<tr><td>Machinery</td><td>28, 29, 30, 31 and 32</td></tr>
<tr><td>Non Specified Industry</td><td>25, 33, 36 and 37</td></tr>
</table>

---

[1] Energy and GVA data are available separately for the sectors Wood and Wood Products, and Paper, Pulp and Print. The decision to consider these two sectors together is related to the quality of the data on energy consumption in the Wood and Wood and Products. Energy Consumption decreases from 127 TTOE in 1991 to 21 TTOE in 1992 while GVA decreased from 2783 to 2751 million pounds. The breaks in the series disappear when the energy consumption from Wood and Wood Products is added to Paper, Pulp and Print due to the considerable higher consumption of the latter.

The energy intensities can be observed in Figure 1. As can be seen, the Construction, Wood, Pulp, Paper and Print, Food and Tobacco, Textiles, Machinery and Transport Equipment have lower energy intensities than the average of the industrial sector. The Chemical and the Non-Mineral Metal sectors have the highest energy intensities. Figure 2 displays the difference between the sectoral energy intensities and the average of the industrial sector. The time series of some sub-sectors presents a number of spikes mainly caused by sudden changes in the sub-sectors[2]. In the Food and Tobacco sector, the increase between 1980 and 1990 is due to energy intensity in the sector decreasing at a lower rate than in the other industrial sectors. The increase in the textile sector at the end of the sample is caused by the fact that a decrease in GVA is not matched by a corresponding change in energy consumption. Similarly to the Food and Tobacco sector, the time series for the Chemical sector in Figure 2 switches between two levels, namely the pre-1982 and the post-1989 levels. Energy Intensity in this sector almost halves between 1982 and 1989 – see Figure 1. As this reduction exceeds the decrease in the industrial sector, the time plot in Figure 2 presents a downward trend. Finally, the decrease in the time series for the Non-Metallic Mineral sector observed in the first few years and in the last 10 years of the sample is mainly due to changes in the energy consumption in the sector.

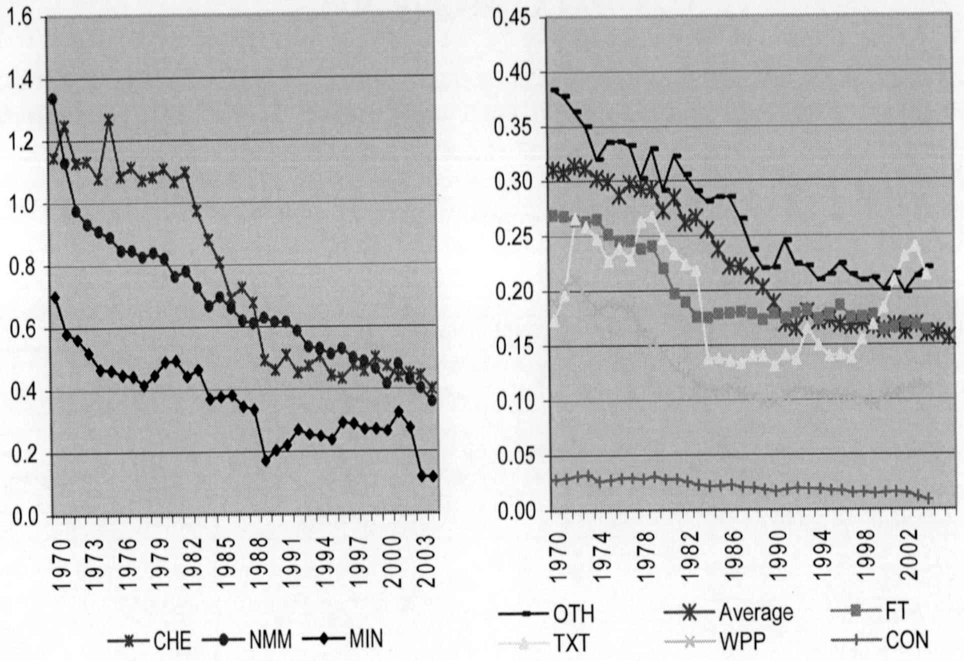

Figure 1. Energy intensity of the industrial subsectors assessed in this study. The list of sectors described by the acronyms in the figure column can be seen in Table 1.

[2] Those occurring in 1970-72 and 1984 in the textile sector and in 1989 and 2003 in the mining sector are all caused by sudden changes in energy consumption unmatched by changes of the GVA. A sudden decrease in the energy consumption is also the cause of the marked decrease in the construction sector in the last two years of the sample.

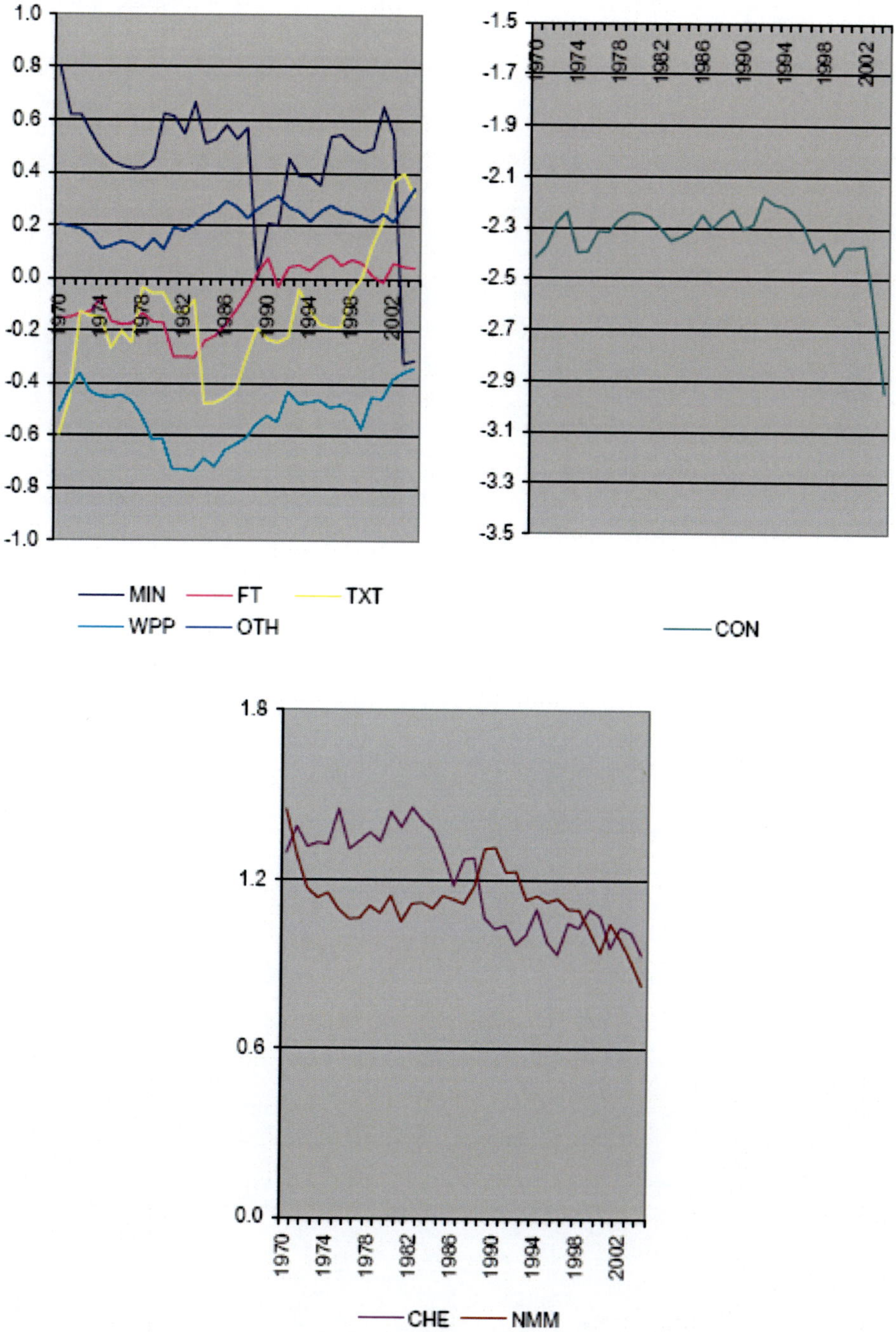

Figure 2. Difference between energy intensity in the subsectors assessed in this study and the energy intensity in the industrial sector as a whole. The list of sectors described by the acronyms in the figure column can be seen in Table 1.

While gradual changes in the level or trend of long-term difference could be explained by structural change in the sectors or by changes in the level or trends of factors affecting the energy intensity (e.g. energy price), it is worth stressing that a number of the sudden changes described above are likely to be caused by measurement errors and other limitations in the dataset. While we bear this point in mind, using a dataset published by international institutions seems more advisable than embarking on a subjective data adjustment process guided by uncertain and somewhat arbitrary criteria.

4. Univariate Testing

The presence of unit roots is tested by running the ADF, the PP test and a number of other tests. As both the PP and ADF tests suffer from low power, especially when the value of ρ is close to one or when the series is trend-stationary (see among others, DeJong et al. (1992) and Campbell and Perron (1991)), two strategies have been followed in the literature to increase the power of the tests. The first approach exploits the identical covariance structures of forward- and backward-looking finite order stationary AR models; the other recurs to a non-OLS estimator in order to estimate the deterministic components of the univariate process. Among the tests using the first approach, the DF^{MAX} (Leybourne 1995) can be implemented by running the ADF test on the ordinary realization of the time series and on its inverse representation $\mathbf{y}_{\mathbf{rt}} = (y_t, y_{t-1}, ..., y_2, y_1)$, and taking the value of the statistic $DF_{\max} = \max\{DF_f, DF_r\}$, where the former is the ordinary ADF test and DF_r is the test on $\mathbf{y}_{\mathbf{rt}}$. The DF^{RMA}, the DF^{GLS} and the M tests are three types of tests using a non-OLS estimator to compute the deterministic components in the series. In the DF^{RMA} test (Shin and So 2001) the deterministic terms are estimated recursively rather than over the whole sample[3]. In the DF^{GLS} (Elliott, Rothenberg and Stock 1996), GLS detrending procedure is used to increase the power of unit root tests[4]. The logic of ERS (1996) is extended by Ng and Perron

---

[3] For example, in the case of a model with intercept only and no lagged first differences to take into account for serial correlation, the test statistics is $\hat{\rho}_r = \sum_{t=2}^{n}(y_t - \bar{y}_{t-1})(y_{t-1} - \bar{y}_{t-1}) \Big/ \sum_{t=2}^{n}(y_{t-1} - \bar{y}_{t-1})^2$ where $\bar{y}_t = t^{-1}\sum_{i=1}^{t} y_i$ is the recursive mean. Serial correlation is accommodated by adding lagged first differences $\Delta y_{t-1} \ldots \Delta y_{t-p+1}$, $t = p+1, \ldots, n$.

[4] The DFGLS is obtained by running the ordinary DF test with y_t^d in the place of y_t, i.e., $\Delta y_t^d = \rho_0\, y_{t-1}^d + \sum_{j=1}^{p}\gamma_j\, \Delta y_{t-j}^d + \varepsilon_t$ where $y_t^d \equiv y_t - \hat{\delta}' z_t$ and $\hat{\delta}$ are obtained by regressing $\mathbf{y}_{\boldsymbol{\rho}}$ on $\mathbf{z}_{\boldsymbol{\rho}}$ $\mathbf{y}_{\boldsymbol{\rho}} = (y_1, (y_2 - \rho\, y_1), \ldots, (y_T - \rho\, y_{T-1}))'$ $\mathbf{z}_{\boldsymbol{\rho}} = (z_1, (z_2 - \rho\, z_1), \ldots, (z_T - \rho\, z_{T-1}))'$. The term $\mathbf{z}$ contains either an intercept or a trend and an intercept. The implementation of the test requires the specification of the value $\bar{\rho}$. ERS (1996)

(NP) (2001) to the modified version of the tests developed in Phillips (1987), Phillips and Perron (1988) and Bhargava (1986) and the Point Optimal test proposed in ERS (1996).

Considering the limited power of unit root tests and the fact that in the testing framework the null hypothesis is accepted unless there is strong evidence against it, it seems sensible to exchange the role of the null and alternative hypotheses. In theory, by carrying out stationarity tests, such as the KPSS test Kwiatkowski, Phillips, Schmidt and Shin (KPSS) (1992) and comparing the results with those from unit root tests, it is possible to collect firmer evidence on the degree of integration of the series. In practice, the results from stationarity tests are often highly affected by the bandwidth lag used to compute the long-run variance.

The results from the tests discussed above are shown in Table 2. In the case of the $\overline{M}^{GLS}$ tests, i.e., MZ_{α} , MZ_t, MSB and MP_T in the table, GLS detrended data are used to construct the series on which the test is run and the autoregressive spectral density estimator s^2_{AR} , as suggested in ERS (1996) and NP (2001). The Modified Akaike Information Criterion (MAIC) has been used to determine the lag in the estimation of s^2_{AR} and in the unit root regressions. When using the MAIC, the maximum lag was set at *l*12. The PP and KPPS tests were run for a bandwidth going from *l*0 to *l*12. The deterministic components in the ADF and DF^{GLS} tests were chosen according to the value of the Schwarz and Akaike Information Criterion (SC and AIC, respectively). The specification selected for the ADF test is also used for the KPSS, PP, DF_{MAX} and the DF^{RMA} tests while that selected for the DF^{GLS} test is also used for the $\overline{M}^{GLS}$ tests. In the case of the KPSS tests, Table 2 reports the value of the statistics computed with 4 lags, i.e., equal to $4(T/100)^{2/25}$, as suggested in Schwert (1989) or Newey West (1994), and also the range of the lags leading to a certain conclusion on the dynamic properties of the series. For example, in the case of energy intensity in the chemical sector, the value of the statistic computed with 4 lags is 0.10. The use of any number of lags between 4 and 9 leads to the acceptance of the null hypothesis. However, this can be rejected at the 90%, 95% and 99% confidence level when the statistic is computed with two or three lags, one lag and without lags.

Overall, there is agreement on the results estimated over the two samples. In the construction sector, however, energy intensity is I(1). The results from the KPSS test are influenced by the bandwidth lag, a finding also reported in Pesaran et al. (2006) . However, in a handful of cases the null hypothesis of stationarity cannot be rejected for a rather wide range of lags, more specifically the Food and Tobacco (when a trend is included), the Chemical and the Non-Metal Minerals and the Other sectors.

---

recommend $\bar{c}$ = - 7 in the model with an intercept only and $\bar{c}$ = - 13.5 in the case of a linear trend; $\bar{c}$ is equal to $T(\bar{\rho}-1)$.

Table 2. Results from the unit root tests listed in the first column of the table. The list of sectors described by the acronyms in the first row of the table can be seen in Table 1. Key: ADF = Augmented Dickey Fuller; PP = Phillips Perron; MZ_α = Modified Z_α test; MZ_t = Modified Z_t test; MP_T = Modified Point Optimal test; MSB = Modified R_1 statistic. The symbols $^{(+)}$, $^{(*)}$ and $^{(**)}$ indicate statistical significance at the 10, 5 and 1% significance level. In the case of the KPSS tests, the symbol X indicate I and T indicate the model selected by the information criteria, i.e., intercept only (I) or an intercept and trend (T). Both models are reported when the information criteria were inconclusive on the choice of the model.

	MIN		FT		TXT		WPP		CHE		NMM		CON		OTH	
ADF	-1.99/-2.43	I/T	-1.05/-1.83	I/T	-1.58	I	-1.11	T	-1.91	T	-2.15	I	0.06	T	-2.39	T
PP	-1.96/-2.43	I/T	- 1.14/-1.97	I/T	- 1.80	I	- 1.26	T	-3.01	T	-2.42	I	1.44	T	- 2.42	T
MZ_α	- 10.69	T	-5.24	T	-5.84	T	-3.31	T	-4.67	T	0.07	I	-31.19	T	-8.03	T
MZ_t	- 2.08	T	-1.62	T	-1.66	T	-1.17	T	-1.52	T	0.02	I	-3.45	T	-1.92	T
MSB	0.19	T	0.31	T	0.28	T	0.35	T	0.33	T	0.35	I	0.11	T	0.24	T
MP_T	9.58	T	17.38	T	15.51	T	25.14	T	19.45	T	12.76	I	5.58	T	11.57	T
DF^{GLS}	-2.58	T	-1.80	T	-1.90	T	-1.25	T	-1.56	T	-0.54	I	-1.29	T	-2.34	T
DF^{MAX}	-1.99/-2.43	I/T	-1.05/-1.63	I / T	- 1.58	I	- 1.11	T	-1.20	T	-1.56	I	0.06	T	-2.35	T
DF^{RMA}	-1.36/-1.45	I/T	-0.29/0.51	I / T	-0.05	I	0.86	T	1.00	T	0.52	I	1.67	T	-0.62	T
KPSS	0.41$^{(*)}$/ 0.06 (**) 0-1 and 9 / (x) 0-9 (*) 2-8 / na	I/T	0.55$^{(*)}$/ 0.11 (**) 0-2 / (**) 0 (*) 3-5 / (*) 1-2 (+) 6-8 / (+) 3 (x) 9 / (x) 4 - 9	I/T	0.44$^{(*)}$ (**) 0-1 (*) 2-3 (+) 4-9	I	0.16$^{(*)}$ (**) 0-2 (*) 3-5 (+) 6-9	T	0.10 (**) 0 (*) 1 (+) 2-3 (x) 4-9	T	0.33 (**) 0 (*) 1 (+) 2-3 (x) 4-9	I	0.19$^{(*)}$ (**) 0-2 (*) 3-9	T	0.10 (**) 0 (*) 1 (+) 2 (x) 3-9	T

5. Panel Testing

In order to discern whether the results in Table 2 are caused by the relatively low power of unit roots tests, one could run panel unit root tests. Among the published papers adopting this approach, one can mention Chen and Lee (2007), Lee (2005), Mehrara (2006), Narayan and Smyth (2007) and Tauchmann (2006). Using panel unit root tests seems to be a rather accepted methodological choice in the profession. The fact that the tests in Sections 5.1 and 5.2 are now incorporated in Eviews – a commercial econometric software – is likely to contribute to an increase in the number of papers adopting this methodology.

5.1. Homogenous Tests

Hadri (2000) extends the KPSS statistic to panels with iid disturbances across cross-sections i and time periods t. Both the intercept and the trend are allowed to vary across cross-sections. The null hypothesis of stationarity can be tested by the statistic $LM = \frac{1}{N}\sum_{i=1}^{N}\left(\frac{T^{-2}\sum_{t=1}^{T} S_{it}^2}{\hat{\sigma}_{\varepsilon}^2}\right)$ where $\hat{\sigma}_{\varepsilon}^2 = \frac{1}{NT}\sum_{i=1}^{N}\sum_{t=1}^{T}\hat{\varepsilon}_{it}^2$ and $S_{it} = \sum_{j=1}^{t}\hat{\varepsilon}_{it}$, $\hat{\varepsilon}_{it}$ being the residuals from the regression of the time series on the deterministic components of the series. It is also shown that $Z = \frac{\sqrt{N}(LM - \xi)}{\zeta} \Rightarrow N(0,1)$. Z can be computed from the KPSS statistic for the single series η_{iT}, more precisely $Z = N^{-1/2}\sum_{i=1}^{N}\left(\frac{\eta_{iT} - \xi_T}{\zeta_T}\right)$. Heteroschedasticity and serial correlation can be accommodated by computing the variance of each series, $\hat{\sigma}_{\varepsilon,i}^2$. The test of Hadri (2000) has been extended by Hadri and Larsson (2005) to models allowing heterogeneity under the alternative hypothesis[1]. In the energy literature, the test in Hadri (2000) has been applied by Lee (2005) and Chen and Lee (2007).

Levin, Lin and Chu (LLC) (2002) proposes a standardised t-statistic obtained from the pooled regression $\tilde{e}_{it} = \rho\tilde{v}_{it-1} + \tilde{\varepsilon}_{it}$ where $\tilde{e}_{it} = \hat{e}_{it}/\hat{\sigma}_{\varepsilon i}$ and $\tilde{v}_{it} = \hat{v}_{it}/\hat{\sigma}_{\varepsilon i}$, $\hat{e}_{it}$ and $\hat{v}_{it-1}$ being the residuals from $\Delta y_{it} = \sum_{l=1}^{p_i}\hat{\pi}_{il}\Delta y_{it-l} + \hat{\alpha}_i d_t + e_{it}$ and $y_{it-1} = \sum_{l=1}^{p_i}\tilde{\pi}_{il}\Delta y_{it-l} + \tilde{\alpha}_i d_t + v_{it-1}$, d_t a vector of deterministic variables and $\hat{\sigma}_{\varepsilon i}$ the regression standard error in an ordinary ADF regression. The standardised t-statistic can be written $t_{\rho}^{*} = \frac{t_{\rho} - N\tilde{T}\hat{S}_N\,\hat{\sigma}_{\tilde{\varepsilon}}^{-2}(s.d.(\hat{\rho}))\mu_{m\tilde{T}}^{*}}{\sigma_{m\tilde{T}}^{*}}$ where $\mu_{m\tilde{T}}^{*}$ and $\sigma_{m\tilde{T}}^{*}$ are the mean and standard

---

[1] Hlouskova and Wagner (2006) conclude that the non-parametric estimation used to correct serial correlation in Hadri and Larsson (2005) is too imprecise to deliver improved results. The advantages of using the test in Hadri and Larsson (2005) are apparent only when no serial correlation is present.

adjustment, computed through a Monte Carlo simulation. However, as pointed out in Breitung (2000), adjusting the statistics for the bias caused by the presence of a constant or time trend implies a serious loss of power. The LLC test has been applied to energy data by Lee (2005) and Mehrara (2007).

Breitung (2000) provides a class of statistics that do not require the bias adjustment which lowers the power of the tests. In the case of models with an intercept only or with a common linear trend, the statistics are easily obtained by subtracting the first observation instead of the individual mean. In the case of individual-specific linear trends one needs to transform the variables y_{it} and Δy_{it} so that the usual t-statistic without bias correction can be used. In the case of AR(1) models, given $\mathbf{y_i} = [\Delta y_{i1}, \Delta y_{i2}, \ldots, \Delta y_{iT}]$ and $\mathbf{x_i} = [y_{i0}, y_{i1}, \ldots, y_{iT-1}]$, the statistics are constructed from the transformation $\mathbf{y}_i^* = \mathbf{Ay}$ and $\mathbf{x}_i^* = \mathbf{Bx}$ where the $\mathbf{A}$ and $\mathbf{B}$ are such that $E(y_{it}^* x_{it}^*) = 0$. Higher order processes can be easily accommodated – see Breitung (2000). The t-statistic for $H_0 : \rho = 0$ in the pooled regression $y_{it}^* = \rho x_{it}^* + \varepsilon_{it}^*$ t = 2,3 , …, T-1, called the UB statistic, has normal limiting distribution.

5.2. Tests Allowing for Heterogeneity

The test in Im, Pesaran and Shin (IPS) (2003) is based on the average of the t-statistics from individual ADF regressions on the N cross-sections, i.e., $t_{NT} = \frac{\sum_{i=1}^{n} t_{iT_i}(p_i)}{N}$ where p_i denotes the lag order in the ADF regression. IPS (2003) provides the expectation and variance of the statistics needed to obtain a standardised statistic, i.e.,

$$W_{\bar{t}_{NT}} = \frac{\sqrt{N}\left(\bar{t}_{NT} - \frac{1}{N}\sum_{i=1}^{n} E(\bar{t}_{iT})\right)}{\sqrt{N^{-1}\sum_{i=1}^{n} Var(\bar{t}_{iT})}} \Rightarrow N(0,1).$$

Applications of the tests in IPS (2003) in the energy literature can be found in Chen and Lee (2007), Lee (2005), Mehrara (2007) and Narayan and Smyth (2007).

Another approach to building panel unit roots allowing for heterogeneity requires the combination of the p-values from the results of univariate unit roots. One advantage of this approach is that the value of any type of tests on univariate series can be combined by using the statistics below. One of these tests has been presented independently by Maddala and Wu (1999) and Choi (2001). As originally proposed in Fisher (1932), given a continuous univariate test statistic, $-2\ln p_i$ is distributed as a χ_2^2. If u_{it} is independent from u_{js} $\forall$ $t \neq s$ and $i \neq j$, $\lambda = -2\sum_{i=1}^{N} \ln p_i \Rightarrow \chi_{2N}^2$. Choi (2001) proposes the inverse normal test, $Z = \frac{1}{\sqrt{N}}\sum_{i=1}^{N} \Phi^{-1}(p_i)$, originally discussed in Stouffer et al. (1949). Z $\Rightarrow$ N(0,1). Both statistics converge to zero under the null hypothesis; under the alternative hypothesis λ diverges to + ∞ while the other two diverge to - ∞. A number of simulations have been undertaken to compare the tests mentioned above, with the one in Hlouskova and Wagner (2006) maybe being the most exhaustive. Overall, it is difficult to draw general conclusions with the exception maybe of the Hadri test which was found to be very oversized[2].

5.3. Tests Allowing for Cross-Sectional Correlation

A limitation of the panel tests discussed so far is that they assume independency of the members of the panels. Cross-section dependency can involve cointegration across the cross-sections. Banerjee et al. (2005) argue that panel unit root tests may be severely biased if the panel units are cross-cointegrated, namely if under the null hypothesis (of unit roots) one or more linear combinations of the series are stationary. When cross sections are not co-integrated, Lyhagen (2000) shows how a common stochastic trend affects the distribution of a number of tests and inflates their empirical size. It is worth mentioning that in the presence of cross-section cointegration the common trends may be identified as common factors and removed from the data when common factor tests are employed (Breitung and Pesaran 2005). If the remaining idiosyncratic component is stationary, tests will tend to indicate stationarity, even though all panel units are non-stationary. The time series assessed in this study are likely to be correlated due to the fact that each of the variables in the panel share one common component by construction, i.e., the energy intensity of the industrial sector. Another reason for expecting cross-sectional correlation is related to the fact that integrated sectors are subject to shocks which propagate across sectors.

The effect on cross-sectional correlation has been an issue discussed at length in the literature since O'Connell (1998) showed that the size of the test in Levin and Li (1992) is much bigger in the presence of constant cross-sectional correlation (i.e., it can be expressed by a single parameter multiplying a matrix with 1 in the cells off the main diagonal). However, this type of correlation can be tackled by demeaning the data, i.e., subtracting the cross-sectional average from each observation - see among others IPS (2003), LLC (2002) and Smith, Leybourne, Kim and Newbold (2004). Bornhorst (2003) argues that the results in O'Connell (1998) can be obtained only when tests are run without demeaning the data.

Another option which can be applied to more general types of correlation consists of using the feasible generalised least squares (SUR) method, i.e., estimating the covariance matrix so as to make the time series free from cross-sectional correlation (O'Connell 1998).

---

2 When the model comprises an intercept only, the IPS and the UB tests show acceptable empirical size regardless of the span of the time series in the case of serially uncorrelated innovations. When serial correlation due to negative MA errors is present, the size of all tests diverges to 1 for a sufficiently high number of cross-sections. When the model comprises intercept and trend, in the case of serially correlated innovations, UB and the LLC outperform the other tests in terms of the size of the tests. In the case of the model with intercept only, LLC is the test with highest power for small values of T, e.g. T =10; for higher values of T, i.e., T =25 UB has the highest power; for T approaching 100, there is no difference between the power of the test. In the case of the models with intercept and trend, there are panel dimensions and values of the parameters on the AR and MA components for which any of the five tests has highest power. Stationarity tests have acceptable size only when the DGP has innovations close to being white noise. Stationarity tests have also been found to have high power. Unfortunately, in empirical applications, researchers are unable to distinguish if rejections are due to high power or high size.

The SUR technique can be used if N is small (less than 10) and T is relatively large (Breitung and Pesaran 2005). SUR-based tests can handle only cross-sectional correlation in the stationary components (Wagner 2005). Lopez (2006) combines the SUR procedure with the GLS detrending of ERS (1996). The test is named DF-GLS-SUR. The DF-GLS-SUR performed rather well when compared with some of the tests based on the common factor approach – see below. Another test, the SURADF of Breuer, McNown and Wallace (2002), is able to identify the series in a panel which are I(1) and those which are stationary. Although N equations are tested separately, the SURADF model exploits the information in the error covariance to produce potentially more powerful tests. However, the test is able to take into account cross-section correlation at the cost of the critical values being dataset-specific, i.e., they need to be computed through simulation. The power of the test is very dependent on the timespan of the series and mildly on the cross-correlation in the data. Considering the timespan of the economic series normally available to researchers, it is fair to conclude that at least quarterly data are needed to make the application valuable. Ford et al. (2006) show that the outcome of the SURADF tests are highly sensitive to the selection of panel members, arguably a criticism which can apply to any panel study (Breuer et al. 2006). Also the approach suggested by Maddala and Wu (1999) requires a fair bit of computation as cross-correlated errors are accommodated by the bootstrap method to get empirical distributions of the test statistics to make inference. Chen and Lee (2007) have applied the Maddala and Wu (1999) procedure to the energy consumption per-capita of a number of countries.

Another approach assumes that cross-section correlation enters the model through common factors. Given $y_{it} = \alpha_i + y_{it}^0$ where $y_{it}^0 = \rho_i y_{it-1}^0 + \varepsilon_{it}$ and $\varepsilon_{it} = \lambda_i' F_t + u_{it}$, F_t are the common factors and λ_i' are the loading parameters. After estimating the common factors, the time series is multiplied by a projection matrix yielding de-factored data (projecting method). Otherwise, de-factored series can be obtained by subtracting the factors estimated from the data (subtracting methods). Given de-factored time series, test statistics can be obtained by averaging Dickey-Fuller tests (heterogeneous tests) or by pooling time series (homogenous tests). The test in Phillip and Sul (PS) (2003) uses projecting and averaging. The tests in Moon and Perron (2004) use projecting and pooling. The test in Bai and Ng (2003) uses subtracting and averaging. All tests use OLS as the estimation method[3].

The most general results have been obtained in Bai and Ng (2004) where any number of stationary or integrated factors can be allowed. The authors developed tests for the number of common trends contained in the common factors and panel unit root tests based on the Fisher principle. Estimation of the common factors in Moon and Perron (2004) is similar to Bai and Ng (2004a) although all common factors are assumed to be stationary. Two pooled *t*-type tests are proposed, the difference consisting of the variance and serial correlation corrections applied. The test in Phillips and Sul (2003) differs from those in Moon and Perron (2003) as only one factor is permitted which is also assumed to be iid ~ N(0, 1) across time. Pesaran (2006) extends the IPS test to allow for the presence of one stationary factor. The test can be easily implemented by adding the first difference and the level of the cross-sectional averages

3 The effect of more powerful estimators (e.g. WS-LS, GLS and RMA) on the power of the tests has been discussed in Jang and Shin (2005). Estimation methods affect power performances of the tests based on projection but not that of the tests based on subtraction. Jang and Shin (2005) recommend the use of tests based on subtracting and OLS. Either pooling or averaging can be used.

time to the ADF regression, hence the name cross-sectionally augmented Dickey-Fuller (CADF) regression, i.e.,

$$\Delta y_{it} = a_i + \rho_i y_{it-1} + b_i \bar{y}_{t-1} + d_i \Delta \bar{y}_t + \varepsilon_{it} \quad (2)$$

where $\bar{y}_t = N^{-1}\sum_{i=1}^{n} y_{it}$ and $\Delta\bar{y}_t = N^{-1}\sum_{i=1}^{n} \Delta y_{it}$; $\bar{y}_{t-1}$ is used as a proxy for the unobserved factor f_t. In case of serially uncorrelated errors, lagged values of Δy_{it} and $\Delta\bar{y}_t$ need to be included. The cross-sectional augmented version of the IPS tests (CIPS) is obtained by averaging, i.e., $CIPS = \frac{1}{N}\sum_{i=1}^{N} CADF_i$, where CADFi is the cross-sectionally augmented Dickey-Fuller statistic. The limiting distribution is non-standard; critical values can be found in Pesaran (2006).

When there are no a priori reasons to believe that the cross-sectional dependence is restricted to stationary components, it would seem sensible to use tests in Bai and Ng (2004), (Wagner 2005). However, these tests are severely distorted in finite samples (Gutierrez 2005). From Monte Carlo simulations in Gutierrez (2005) and Moon and Perron (2005) one can conclude that the performance of tests allowing for cross-section correlation depends on the underlying, unknown, DGP. However, for panels of limited cross-sectional dimension, the test in Moon and Perron (2004a) shows a reasonable combination of size and size-adjusted power.

The CIPS test might be preferred in the presence of one common factor considering that it can be computed in a straightforward manner (Gegenbach et al. 2005). In multi-factor settings tests from Moon and Perron (2004a) tests are to be preferred, as the cross-section component is not eliminated by the Pesaran (2006) procedure, although the loss in power when two common factors are observed is rather limited. In applied work, the various tests could be carried out in a sequence, starting with the test in Pesaran (2006) when one believes that there is a single common factor present in the data or starting with those in Moon and Perron (2004a) when two common factors are likely to be present. When the unit root hypothesis is not rejected, one can check whether the unit roots are either in the common factors, or in the idiosyncratic components, or in both by using the tests in Bai and Perron (2004) (Gegenbach et al. 2005).

5.4. Empirical Results

The table below summarises the results of the panel unit root tests computed in this study. The bandwidth lag in the Hadri and LLC tests and in the tests based on the PP tests (Fisher Chi and Choi Z in the table) has been selected by using the Newey-West automatic selection criterion. The tests above have also been computed by varying a common lag length across cross-sections in order to assess the sensitiveness of the outcome to different bandwidth lags. The values of the Hadri test do show considerable variation, although the decision on rejecting the null hypothesis of stationarity is not affected. The value of the other three tests is rather stable. Among the tests not allowing for cross-section correlation only the Breitung tests (in the case of the intercept model) point at the series being stationary. In the case of the statistics which are distributed normally, very high (positive) values are taken by tests with non-stationarity under the null (see LLC, Choi Z and IPS tests and the Breitung test in the model with intercept and trend). Also in the case of the CIPS test, the statistics are not statistically significant.

Given the results so far, a testing strategy focussed on cross-sectional correlations would require the implementation of the other tests allowing for cross-correlation described above. However, this strategy is not followed in this study. In the data discussion in Section 3, it was noticed that a number of series present breaks in the deterministic components. For this reason, rather than running more tests allowing for cross-section correlation, it seems more appropriate to assess the results from tests, univariate and panel, allowing for breaks.

6. The Role of Structural Breaks

The results from unit root tests can be misleading when a structural break goes unaccounted, the so-called Perron-effect (Perron 1989). Considering the time series in Figure 2, it is therefore unclear if the failure to reject the null hypothesis is due to the process being I(1) or to the presence of structural breaks. The results of unit root tests are also affected when series are generated by an I(1) process with a break in the deterministic terms, as the size markedly increases and the series is likely to be considered I(0), i.e., the so–called converse Perron effect (Leybourne et al. 1998).

Table 3. Results from the panel tests implemented in this study. The tests listed in the first column of the table are all unit root tests with the exception of the Hadri Z test which is a stationary test, i.e., the series is assumed stationary under the null hypothesis. The symbols (+), (*) and (**) indicate statistical significance at the 10, 5 and 1% significance level

	Intercept	Intercept and Trend
Homogenous tests		
LLC	2.39	2.53
Breitung	- 2.056 (*)	1.42
Hadri Z	5.48 (**)	4.75 (**)
Heterogeneous tests		
PP Fisher Chi	9.87	12.40
PP Choi Z	1.46	1.53
ADF Fisher Chi	8.68	8.67
ADF Choi Z	1.50	1.46
IPS	1.35	1.46
Tests allowing for cross-section dependence		
CIPS	- 1.20	- 1.82

If the level of energy consumption is determined by economic activity and energy price, any change in the correlation between the omitted energy price and the variables in the unit root regressions is likely to cause a break in the regression, an issue discussed in more depth in Agnolucci (2007). Breaks are likely to happen both under the null and under the alternative hypothesis. Hence, it becomes important to assess the dynamic properties of the series when allowing for structural breaks.

6.1. Univariate Tests

A straightforward approach to tackling the size distortion and the reduced power in unit root tests on series presenting structural breaks is to explicitly model the breaks as in Perron (1989) where breakdates are supposed to be known and breaks are considered as the outcome of exogenous events. On the other hand, in Zivot and Andrews (ZA) (1992) the breaks are considered a realisation from the distribution tails and the breakdate is chosen by a data-dependent criterion. For each observation in $\Lambda = [0+\varepsilon; \mathrm{T}-\varepsilon]$, where ε is a trimming factor, ZA (1992) propose to perform ADF tests on an ADF regression augmented with the break in the deterministic components[4]. In the case of breaks occurring in both the intercept and slope (Model C), the testing regression is equal to

$$y_t = \mu^C + \theta^C DU_t + \beta^C t + \gamma^C DT_t + \rho^C y_{t-1} + \sum_{i=1}^{k} c_i \Delta y_{t-i} + \varepsilon_t \quad (3)$$

where $DU_t = 1_{(t>T_B)}$, i.e., $DU_t = 1$ if $t \geq T_B + 1$ and 0 otherwise, $DT_t = t - \mathrm{T_B}\, 1(t > T_B)$; $\mathrm{T_B}$ is the time of the break. In the case of breaks in the intercept only, Model A, $DT_t = t - \mathrm{T_B}$ is omitted. The breakpoint is then fixed at the date that gives the least favourable results to the null hypothesis. More formally,

$$t^*_{\hat{\rho}^{\mathrm{i}}}[\lambda^{\mathrm{i}}_{\mathrm{inf}}] = \inf_{\lambda \in \Lambda} t_{\hat{\rho}^{\mathrm{i}}}(\lambda) \quad \mathrm{i} = \mathrm{A, C} \quad (4)$$

where $t_{\hat{\rho}^{\mathrm{i}}}(\lambda)$ is the t-test and $\lambda = T_B / T$. Another algorithm has been proposed in Perron (1997). Another statistic based on ADF regressions is proposed in Murray and Zivot (1998) and discussed in Sen (2003). The endogenous break methodology has been extended by Lumsdaine and Papell (LP) (1997) to the case of two breaks. The test for the model with two breaks in the intercept and trend is carried out by running

---

[4] A number of the more powerful tests discussed above have been applied to the case with structural breaks (Leybourne et al. 2005 and Perron and Rodriguez 2003). The KPSS test has been extended to allow breaks by a number of authors, e.g. Busetti and Harvey (1998), Lee and Strazicich (2001a), Kurozumi (2002) and by Carrion-i-Silvestre and Sanso (CS) (2005) to the case of 2 breaks. However, as far as the authors of this paper are aware, they have never been applied in the energy sector and they are not further discussed. Stationary tests are also very sensitive to the bandwidth used to estimate long-run covariance.

$$\Delta y_t = \mu + \beta t + d_1 DU_{1t} + d_2 DU_{2t} + \delta_1 DT_{1t} + \delta_2 DT_{2t} + \rho y_{t-1} + \sum_{i=1}^{k} c_i \Delta y_{t-i} + e_t$$

where $DT_{it} = t - TB_i \ \ 1(t > TB_i)$. The test statistic (4) is estimated for all feasible couples of breakpoints (λ_1, λ_2) with the restriction $\varepsilon < \lambda_i < 1 - \varepsilon$ and $\lambda_2 > \lambda_1 + 1$.

It is worth mentioning that the ability of ADF-based tests to accept the null hypothesis and estimate the right breakpoint has been seriously put in doubt by Lee and Strazicich (2001) when a break occurs under the null. In Monte Carlo simulations $t^*_{\hat{\rho}}\left[\lambda_{\inf}\right]$ tended to select T_B-1 as the breakpoint under both stationary and non-stationary DGPs, where T_B is the real breakdate. When a break occurs under the null hypothesis, the size of the test markedly increases. In addition, searching for a minimum will also tend to increase the size of the tests proposed in ZA (1992) or LP (1997). In order to account for this type of problem Lee and Strazicich (LC) (2004) extend the Lagrange Multiplier procedure of Schmidt and Phillips (1997) to DGPs with structural breaks. Given a DGP written in the unobserved components representation, $y_t = \delta' Z_t + X_t$ and $X_t = \beta X_{t-1} + \varepsilon_t$, where Z_t contains deterministic variables, the test statistic can be obtained from the regression

$$\Delta y_t = \delta' \Delta Z_t + \phi \widetilde{S}_{t-1} + u_t \tag{5}$$

where $\widetilde{S}_t = y_t - \widetilde{\psi}_x - Z_t \widetilde{\delta}$ t = 2,...., T. $\widetilde{\delta}$ is the vector of coefficients from the regression of Δy_t on ΔZ_t and $\widetilde{\psi}_x$ is the restricted MLE of $\widetilde{\psi}_x = (\equiv \psi + X_0)$ given by $y_1 - Z_1 \widetilde{\delta}$. The LM statistic is given by the t-statistic testing $\phi = 0$ in (6). To correct for autocorrelated errors LS (2004) include augmented terms $\Delta \widetilde{S}_{t-j}$ j = 1, 2, … k in the testing regression , where the number of lags is chosen by the t-sig procedure. The LM unit root test has been extended to the case of two breaks by Lee and Strazicich (2003) (LC). The main advantage of the LM-based tests is that breaks are allowed under the null hypothesis. In the case of the ADF-based test, no break is assumed under the null, as the critical values of test statistic would depend on the location and magnitude of the break (LC: 2003: 1083). On the other hand, the LM statistics are invariant to the location and magnitude of the break. Unlike the LP test, LM unit root tests will not diverge, i.e., to spuriously reject the null hypothesis in the presence of breaks under the null even when the breaks are large. However, the power is negatively affected by the size of the break under the alternative hypothesis, i.e., the bigger the break, the bigger the decrease in power of the LM test when compared to the ZA and LP tests.

6.2. Panel Tests

Structural breaks can be allowed in the test suggested by Harris et al. (2005), which is an extension of the univariate stationarity test introduced by Harris, McCabe and Leybourne

(2003) and in the test in Carrion-i-Silviestre et al. (2005) which extended the test in Hadri (2000) to consider structural change[5]. The latter allows for individual intercepts, temporal effects and structural breaks, i.e., shifts in the mean and/or in the time trend. The breakpoints and the number of breaks are allowed to vary across time series. Structural breaks may have different effects on each individual time series. The test in Carrion-i-Silviestre et al. (2005) has been recently applied by Chen and Lee (2007) to study the energy per capita consumption of a number of countries. In this paper this test is not employed as the test can be extremely oversized, ij presence of serial correlation, especially when T is low (50 observations), N is big or when breaks occurred in the intercept and linear trend (model C).

Im, Lee and Tieslau (ILT) (2005) propose a panel test based on the univariate tests discussed in Schmidt and Phillips (1992) as extended to DGPs with structural breaks by Lee and Strazicich (2003). As noted by the authors, allowing for breaks in the ADF-type tests, like in the IPS test, requires the expected value and variances of the univariate statistics for all different shift locations in the sample, as the statistics depend on the nuisance parameters indicating the location of structural changes. On the other hand, the LM test is not affected asymptotically by the presence of breaks. The panel LM test is equal to the average of the univariate statistics for each time series, i.e., $\bar{t}_{LM,NT}(p) = \frac{1}{N}\sum_{i=1}^{N} t_{LM,iT}(p_i)$, where p_i denotes the number of lags needed to take into account serial correlation in the series. Given the expected value $E[\tau_{LM,T}(p_i)]$ and the variance $V[\tau_{LM,T}(p_i)]$ of each $\bar{t}_{LM,iT}$, the null hypothesis can be tested by computing the standardised statistic

$$\Gamma_{LM}(p) = \frac{\sqrt{N}\left[t_{LM,NT}(p) - 1/N\sum_{i=1}^{N} E(\tau_{LM,T}(p_i))\right]}{\sqrt{1/N\sum_{i=1}^{N} V[\tau_{LM,T}(p_i)]}}$$

which is distributed N(0,1) as N increases for finite T. While invariance to breaks is valid only asymptotically, simulation results in ILT (2005) show that that no serious size distortions or power loss occurs when one mistakenly controls for non-existent shifts. On the other hand, loss of power from ignoring existing structural shifts are very serious. Modelling structural shifts whenever their existence is questionable seems to be a reasonable strategy ILT (2005: 413-4). Narayan and Smyth (2005) is an example of the application of the Γ_{LM} test, apparently computed with the assumption of no serial correlation in the series.

---

[5] The statistics to test the null hypothesis of stationarity can be built from the univariate scaled autocovariance of order k, $C_{ik} = (T-k)^{-1/2}\sum_{t=k+1}^{T} z_t z_{t-k}$. The panel test is simply the sum of the univariate statistics, i.e., $C_k = \sum_{i=1}^{N} C_{ik}$. C_k is asymptotically normal when suitably studentised. Determinist regressors, break dummies included, can be allowed for in the testing procedure. Given $y_{i,t} = \boldsymbol{\beta}_i^i \mathbf{x}_{i,t} + \mathbf{z}_{i,t}$, one needs to regress the time series on the deterministic components so as to obtain the OLS-residuals $\hat{z}_{i,t}$ and build the statistic above by using the standardised residuals $\tilde{z}_{i,t} = \hat{z}_{i,t} / s_i$ where s_i is the standard deviation of $\hat{z}_{i,t}$.

6.3. Empirical Results

Results from the tests discussed in this section differ somewhat from those presented in Sections 4 and 5. As can be seen in Table 4, in 38% of the industrial sectors long-run differences can be considered deterministic when using the ZA test. In the case of the LM test allowing for a break, only in two instances long-term differences among sectors can be considered deterministic in the longer sample. Similarly to what occurred in the case of the ZA tests, model selection on the basis of the information criteria is sometimes inconclusive, although this does not influence the properties of the long-term differences. In these cases both models have been reported in the tables.

Table 4. Results for the ZA test and the LM test with one break. The symbols $^{(+)}$, $^{(*)}$ and $^{(**)}$ indicate statistical significance at the 10, 5 and 1% significance level. The list of sectors described by the acronyms in the first column of each table can be seen in Table 1. For each test the table shows the value of the test statistic, the estimated breakdate and the model selected by the two information criteria. The letters A and C indicate the model with a break in the intercept and with a break in the intercept and trend, respectively. Both models are reported when the information criteria were inconclusive on the choice of the model

	ZA	Breakdate	Model	LM – 1 break	Breakdate	Model
MIN	- 3.76	1998	C	-3.87	1998	C
FT	- 3.52	1986	A	-3.42	1990	C
TXT	-5.43 $^{(**)}$ / 5.69 $^{(**)}$	1983	A / C	-3.46	1983	C
WPP	-4.08 / - 4.17	1977 / 1980	A / C	-2.88 / - 3.14	1991 / 1979	A / C
CHE	-5.41 $^{(**)}$	1988	A	-3.92 $^{(+)}$	1988	A
NMM	-5.67 $^{(**)}$	1988	C	-2.77	1989	C
CON	-3.53	1993	C	-6.27 $^{(*)}$	1998	C
OTH	-3.55	1980	A	-2.86	1980	A

Table 5. Results for the LP test and the LM test with two breaks. The symbols $^{(+)}$, $^{(*)}$ and $^{(**)}$ indicate statistical significance at the 10, 5 and 1% significance level. The list of sectors described by the acronyms in the first column of the table can be seen in Table 1. For each test the table shows the value of the test statistic, the estimated breakdates and the model selected by the two information criteria. The letters AA and CC indicate the model with two breaks in the intercept and with two breaks in the intercept and trend, respectively. Both models are reported when the information criteria were inconclusive on the choice of the model

	LP	Breakdates		Model	LM – 2 breaks	Breakdates		Model
MIN	- 5.30	1988	1998	CC	-4.02	1996	1998	CC
FT	- 6.45	1980	1995	CC	-5.86 (**)	1979	1990	CC
TXT	- 8.16 $^{(**)}$	1983	1996	CC	-2.90	1977	1983	AA
WPP	- 5.80	1980	1995	CC	- 3.06 / -4.56	1977 / 1977	1991 / 1990	AA / CC
CHE	- 6.85 $^{(*)}$	1984	1988	AA	-5.54 $^{(**)}$	1985	1988	AA
NMM	- 6.57 $^{(+)}$	1976	1988	CC	-5.68 $^{(*)}$	1976	1987	CC

Table 5. (Continued)

	LP	Breakdates		Model	LM – 2 breaks	Breakdates		Model
CON	- 3.65	1982	1993	CC	-6.57 (**)	1991	1998	CC
OTH	- 3.68	1980	1998	CC	-3.03	1980	1983	AA

In the case of the LP test, 37% of the series can be considered deterministic. All series which appear stationary according to the LP were also stationary according to the ZA tests. By comparing Table 4 and Table 5, one can notice that in many instances the breakpoint identified by the ZA test is one of the two breakpoints identified by the LP test. When using the LM test allowing for two breaks - see Table 5 - in 50% of the cases the long-term differences can be considered deterministic. The fact that the rejection rate is higher in the case of the LM than in the case of LP contradicts the expected better size of the LM in the presence of breaks under the null, i.e., more “wrong” rejection for the LP test, and the better power of the LP in presence of breaks under the alternative, i.e., more “right” rejection for the LP test.

With regard to breakpoints, in six instances the LM with one and two breaks select a common breakdate. Some agreement on the breakdates can also be noticed across LM- and ADF-based tests. Half of the series have a breakpoint in common when tests allowing for one break are employed. When allowing for two breaks, the two tests agree on one of the two breakpoints in 50% of the series.

With regard to the properties of the long-term differences, the LM with two breaks and the LP agree in five instances. Both tests agree on the long-term differences being deterministic in the Chemical and Non-Metallic Minerals sectors. In the Mining, the Wood, Paper, Pulp and Printing sectors, and the Other sector both tests point at the long-term differences being stochastic. Also the ZA test and the LM test allowing for a break agree in five instances with regard to the properties of the long-term differences.

In the absence of a specific criterion, the choice between the LP and LM specification can be carried out on the basis of the information criteria. Table 6 reports the result of this exercise. Long-run differences among the series are stochastic only in three instances. It is fair to conclude that the application of univariate tests allowing for structural breaks has reversed the conclusions of the univariate and panel unit root tests discussed in Sections 4 and 5. Application of the panel LM test of ILT (2005) supports the results of univariate tests. Long-term differences among the energy intensity are deterministic, as the value of the statistic $\Gamma_{LM}(p)$ is -9.86.

7. Conclusions

This paper has assessed whether long-term differences among energy intensities of industrial subsectors are stochastic or deterministic. By applying what can be considered a limited selection of tests when compared to those presented in the literature, this paper has shown how the conclusions from the tests can be contradictory. The spirit of our conclusions is therefore similar to Gaffeo et al. (2005) where inference appears to be strongly dependent on the type of tests used, so that conclusions reached at early stages of the testing procedure can be discarded at subsequent stages.

Table 6. Model specification selected by the information criteria. The symbols (+), (*) and (**) indicate statistical significance at the 10, 5 and 1% significance level. The list of sectors described by the acronyms in the first column can be seen in Table 1. For each test the table shows the value the test statistic, the estimated breakdates and the model selected by the two information criteria. The letters LP and LM indicate whether the specification from the LP test or the LM test with two breaks has been selected. The letters AA and CC indicate whether the model with two breaks in the intercept and with two breaks in the intercept and trend ahs been selected. In the case of the Other sector (OTH) information criteria are inconclusive on the choice of the model. As this does not affect the conclusions from the tests, only one model is presented.

	Statistic	Breakdates		Model
MIN	- 5.30	1988	1998	LP / CC
FT	-5.86 (**)	1979	1990	LM / CC
TXT	- 8.16 (**)	1983	1996	LP / CC
WPP	- 5.80	1980	1995	LP/ CC
CHE	-5.54 (**)	1985	1988	LM / AA
NMM	- 6.57 (+)	1976	1988	LP / CC
CON	-6.57 (**)	1991	1998	LM / CC
OTH	- 3.68	1980	1998	LP / CC

In the case of this paper, if the analysis had terminated with the implementation of panel unit root tests, we would have concluded that long-term differences in energy intensities among industrials sectors were stochastic. However, after allowing for structural breaks in the series, the opposite conclusion can be reached.

The policy implications of this study are related to the timing and frequency of policy interventions and to the type of instruments. As long-term differences among industrial sectors are mainly deterministic, short-term random difference dissipates across time. For this reason, policy makers with an interest in the series following a certain time pattern do not need to intervene at any single minor departure of the series from the desired trend. In relation to the type of instruments which can be used by policy makers, we can conclude that policies taking into account the evolution of the series and the peculiarity of the economic actors cannot be justified on the basis of the analysis discussed here. In fact, the presence of long-term deterministic differences among the series supports the adoption of policy instruments which are applied across productive sectors without frequent changes. Among the policies in the energy sector, energy and CO_2 taxes applied to all industrial subsectors would be an example of this type of instruments. A final advice to policy-markers is related to the caution needed when drawing conclusions from the use of a narrow set of statistical methodologies. As discussed above, the results from tests allowing for breaks are opposite to univariate and panel tests assuming structural stability. Policy advice from the latter would be completely different from the conclusions drawn from this article and would ignore the importance of structural breaks. The robustness of the results discussed here could be assessed by re-running the tests on the 1978-2004 time period. Over this timespan data are available for less aggregated sector than those used in this study.

ACKNOWLEDGEMENTS

The authors would like to thank Ron Smith of Birkbeck College, University of London for the encouragement and support received during the preparation of this paper and Kirsten Wiebe of the Institute of Economic Structures Research, Osnabrueck for feedback on a previous draft of this paper. Funding from the Anglo-German Foundation under the programme Resource Productivity, Environmental Tax Reform and Sustainable Growth in Europe is also gratefully acknowledged.

REFERENCES

Bai J. and S. Ng, (2002). Determining the Number of Factors in Approximate Factor Models *Econometrica* 70 pp.191-221

Bai J., and S. Ng (2004) A PANIC attack on unit roots and cointegration *Econometrica* 72 p.1127-1177

Banerjee A., Marcellino M., and C Osbat (2005) Testing for PPP: Should we use panel methods? *Empirical Economics* 30 pp.77–91

Bhargava, A. (1986) On the Theory of Testing for Unit Roots in Observed Time Series, Review of Economic Studies 53 pp.369-384

Breuer J. B., McNown R. and M. Wallace (2002) Series-specific Unit Root Tests with Panel Data *Oxford Bulletin of Economics and Statistics* 64 5 pp.527-546

Breitung J. (2000) The Local Power of Some Unit Root Test for Panel Data in B. Baltagi (eds.) *Advances in Econometrics, Vol. 15: Nonstationary Panels, Panel Cointegration, and Dynamic Panels*, Amsterdam: JAI Press, p. 161–178

Breitung, J. and M.H. Pesaran (2005) Unit Roots and Cointegration in Panels in L. Matyas, and P. Sevestre (eds) (forthcoming) The Econometrics of Panel Data (Third Edition), Kluwer Academic Publishers

Campbell J.Y. and P. Perron (1991) Pitfall and Opportunities: What Macroeconomists Should Know About Unit Roots in Blanchard O.J. and S. Fischer (1991) NBER Macroeconomics Annual 1991 144-201 Cambridge, Mass; London: MIT Press

Carlino G. A. and L. O. Mills (1993) Are U.S. Regional Incomes Converging? *Journal of Monetary Economics* 32 pp.335-346

Carrion-i-Silviestre J., del Barrio-Castro T. and E. Lopez-Bazo (2005) Breaking the Panels: An Application to the GDP per Capita Econometrics Journal 8 pp.159-175

Carrion-i-Silvestre and Sansó (2005) The KPSS Test with Two Structural Breaks Universitat de les Illes Balears, Departament d'Economía Aplicada DEA Working Papers 13

Chang T., Lee K., Nieh C. and C. Wei (2005) An empirical note on testing hysteresis in unemployment for ten European countries: panel SURADF approach Applied *Economics Letters* 12 pp.881–886

Chen P. F. and C. C. Lee (2007) Is energy consumption per capita broken stationary? New evidence from regional-based panels *Energy Policy*, in press

Choi I. (2001) Unit Root Tests for Panel Data *Journal of International Money and Finance* 20 pp.249–272

Choi I. (2002) *Combination Unit Root Tests for Cross-Sectionally Correlated Panels* Hong Kong: Hong Kong University of Science and Technology

DeJong D. N. Nankervis J. C. Savin N. E. and C. H. Whiteman (1992) Integration versus Trend Stationary in Time Series *Econometrica* 60 (2) pp.423-433

Department of Trade and Industry (DTI) (2006) *Energy: its impact on the environment and society – 2006* London: DTI

Elliott G., Rothenberg T. J. and J. H. Stock (ERS) (1996) Efficient tests for an autoregressive unit root *Econometrica* 64 (4) pp813-836

Fisher R.A. (1932) *Statistical Methods for Research Workers* London: Oliver and Boyd

Gaffeo E., Gallegati M. and M. Gallegati (2005) Requiem for the unit root in per capita real GDP? Additional evidence from historical data *Empirical Economics* 30 p.37–63

Gegenbach C., Palm F.C. and J. P. Urbain (2004) *Panel Unit Root Tests in the Presence of Cross-Sectional Dependencies: Comparison and Implications for Modelling* METEOR Research Memorandum 04039 Universiteit Maastricht

Gutierrez L. (2005) Panel Unit Roots Test for Cross-Sectionally Correlated Panels: A Monte Carlo Comparison *Oxford Bulletin of Economics and Statistics* forthcoming

Hadri K. (2000) Testing for Stationarity in Heterogeneous Panel Data *Econometrics Journal* 3 148-161

Hadri K. and R. Larsson (2005) Testing for Stationarity in Heterogeneous Panel Data where the Time Dimension is Finite *Econometrics Journal* 8 55-69

Harris D., Leybourne S., and B. McGabe (2005) Panel Stationarity Tests for Purchasing Power Parity with Cross-Sectional Dependence *Journal of Business and Economic Statistics* 23 pp.395-409

Harris D., McCabe B. P.M. and S. J. Leybourne (2003) Some Limit Theory for Infinite-Order Autocovariances *Econometric Theory* 19 p.829–864

Hlouskova J. and M. Wagner (2006) The Performance of Panel Unit Root and Stationarity Tests - Results from a Large Scale Simulation Study *Econometrics Reviews* 25 pp85-116

International Energy Agency (IEA) (2005) Energy Balances Paris: OECD/IEA

Im K. S., J. Lee and M. Tieslau (2005). Panel LM unit root test with level shits *Oxford Bulletin of Economics and Statistics* 67 pp.393–419

Im K. S., Pesaran M. P. and Y. Shin (2003) Testing for Unit Roots in Heterogeneous Panels *Journal of Econometrics* 115 pp.53 – 74

Lee C. C. (2005) Energy consumption and GDP in developing countries: A cointegrated panel analysis *Energy Economics* 27 p. 415-427

Lee C. and C. Chang (2005) Structural breaks, energy consumption, and economic growth revisited: Evidence from Taiwan *Energy Economics* 27 p.857-872

Lee K., Pesaran M. H. and R. Smith (1997) Growth and Convergence in a Multi-country Empirical Stochastic Solow Model *Journal of Applied Econometrics* 12 pp. 357-392

Lee J. S. and M. Strazicich (2004) *Minimum LM Unit Root Test with One Structural Break* Boone, NC: Department of Economics, Appalachian State University

Lee J. S. and M. Strazicich (2003) Minimum Lagrange Multiplier Unit Root Tests With Two Structural Breaks *Review of Economics and Statistics* 85(4) p.1082-1089

Lee J. and M. C. Strazicich (2001a) Break point estimation and spurious rejections with endogenous unit root tests *Oxford Bulletin of Economics and Statistics* 63(5) pp.535-558

Lee J. S. and M. Strazicich (2001b) Testing the null of stationarity in the presence of a structural break *Applied Economic Letters* 8(6) pp.377-382

Levin A., Lin, C. and C. J. Chu (2002) Unit Root Tests in Panel Data: Asymptotic and Finite-Sample Properties *Journal of Econometrics* 108 pp. 1–24

Leybourne S. J. (1995) Testing for Unit Roots Using Forward and Reverse Dickey-Fuller Regressions *Oxford Bulletin of Economics and Statistics* 57(4) pp.559-571

Leybourne S.J., Kim T.H. And P. Newbold (2005) More Powerful Modifications of Unit Root Tests Allowing Structural Change *Journal of Statistical Computation and Simulation* 75(11) pp.869-888

Lopez C. (2006) A Panel Unit Root Test with Good Power in Small Samples University of Cincinnati, Economics Working Papers Series downloadable at http://www.artsci.ucch/Wppdf/2005-01.pdf

Lumsdaine R. L. and D. H. Papell (1997) Multiple trend breaks and the unit-root hypothesis *Review of Economics and Statistics* 79(2) pp.212-218

Kwiatkowski D. Phillips P. C. B., Schmidt P. and Y. Shin (1992) Testing the Null Hypothesis of Stationarity against the Alternative of a Unit-Root - How Sure Are We that Economic Time-Series Have a Unit-Root *Journal of Econometrics* 54(1-3) pp159-178

Maddala G.S. and S. Wu (1999) A Comparative Study of Unit Root Tests with Panel Data and a New Simple Test *Oxford Bulletin of Economics and Statistics* 61 Special Issue pp.631- 652

Mehrara M. (2006) Energy consumption and economic growth: The case of oil exporting countries *Energy Policy* in Press

Moon H.R. and B. Perron (2004a) Testing for a Unit Root in Panels with Dynamic Factors *Journal of Econometrics* 122: 81-126

Moon H.R. and B. Perron (2004b) *Asymptotic Local Power of pooled t-ratio Tests for Unit Roots in Panels with Fixed Effects* USC working paper

Moon, H.R. and B. Perron (2005) An Empirical Analysis of Nonstationatity in Panels of Interest Rates with Factors *Journal of Econometrics* forthcoming

Moon H.R., Perron B. and P.C.B Phillips (2005) *Incidental trends and the Power of Panel Unit Root Test* Cowles Foundation Discussion Paper No 1435

Murray C. J. and E. Zivot (1998) *Inference on Unit Roots and Trend Breaks in Macroeconomic Time Series* Houston: Department of Economics, University of Houston

Narayan P. K. and R. Smyth (2007) Are shocks to energy consumption permanent or temporary? Evidence from 182 countries *Energy Policy* 35 p.333-341

Ng S. and P. Perron (2001) Lag length selection and the construction of unit root tests with good size and power *Econometrica* 69(6) pp.1519-1554

Newey W. K. and K. D. West (1994) Automatic Lag Selection in Covariance Matrix Estimation *Review of Economic Studies* 61 pp.631-653

O'Connell P. G. J. (1998) The Overvaluation of Purchasing Power Parity *Journal of International Economics* 44 pp1-19

Office for National Statistics (ONS) (2006a) *Detailed Index of Production* London: ONS Office for National Statistics (ONS) (2006b) *Economic Trends* London: ONS

Office for National Statistics (ONS) (2006c) *Input-Output - United Kingdom National Accounts* London: ONS

Perron P. (1989) The Great Crash, the Oil Price Shock, and The Unit-Root Hypothesis *Econometrica* 57(6) pp.1361-1401

Perron P. (1997) Further Evidence on Breaking Trend Functions in Macroeconomic Variables *Journal of Econometrics* 80 pp355-385

Pesaran M. (2006) A Simple Panel Unit Root Test in the Presence of Cross Section Dependence *Journal of Applied Econometrics* forthcoming

Pesaran M. H., Smith R. P., Yamagata T. and L. Hvozdyk (2006) Pairwise Tests of Purchasing Power Parity Using Aggregate and Disaggregate Price Measures Cambridge: Department of Economics University of Cambridge, downloadable at http://www.econ.cam.ac.uk/faculty/pesaran/wp.htm#ptpp

Phillips P. C. B. and P. Perron (1988) Testing for a Unit Root in Time Series Regression *Biometrika* 75(2) pp.335-346

Phillips P. C. B and D. Sul (2003) Dynamic panel estimation and homogeneity testing under cross section dependence *Econometrics Journal* 6 pp.217–259

Schmidt P. and P. C. B. Phillips (1997) LM Tests for a Unit Root in the Presence of Deterministic Trends *Oxford Bulletin of Economics and Statistics* 54(3) p.257-287

Schwert G.W. (1987) Effects of model specification on tests for unit roots in macroeconomic data *Journal of Monetary Economics* 20 pp.73-103

Sen A. (2003) On Unit Root Tests when Alternative is a Trend-break Stationary Process *Journal of Business and Economic Statistics 21*(1) p.174-184

Shin D. W. and B. S. So (2001) Recursive Mean Adjustment for Unit Root Tests *Journal of Time Series Analysis* 22(5) pp595 612

Stouffer S.A., Suchman E.A., DeVinney L.C., Star S.A. and R.M. Williams Jr (1949) *The American Soldier, I.* Princeton University Press, Princeton, NJ.

Tauchmann H. (2006) Cost of electric power sector carbon mitigation in India: international implications *Energy Policy* 34 p.1619-1629

World Energy Council (2004) Energy Efficiency: A Worldwide Review London: World Energy Council

Zivot E. and D. W. K. Andrews (1992) Further Evidence On The Great Crash, The Oil-Price Shock, And The Unit-Root Hypothesis *Journal Of Business and Economic Statistics* 10(3) pp.251-270

In: Energy in Europe: Economics, Policy and Strategy
Editors: F. L. Magnusson and O. W. Bengtsson
ISBN: 978-1-60456-829-5

Chapter 3

THE PUBLIC PROMOTION OF RENEWABLE ELECTRICITY IN EUROPE: THE CASE FOR TRADING GUARANTEES OF ORIGIN

Pablo del Río-González[1]
Facultad de Ciencias Jurídicas y Sociales de Toledo.
Universidad de Castilla-La Mancha. C/ Cobertizo de S. Pedro Mártir s/n.
Toledo-45071, Spain

ABSTRACT

The promotion of electricity from renewable energy sources (i.e., renewable electricity) is a general public policy priority for, both, the Member States and the European Commission. A Directive on the promotion of renewable electricity (Directive 77/2001/EC) has been in force since 2001, setting indicative targets for the share of electricity coming from renewables in 2010 in the Member States, and allowing these to apply the instrument they deem best in order to reach those targets, in line with the subsidiarity principle. An ambitious overall target for renewable energy of 20% of energy consumption has been set for the year 2020. This means that great efforts will have to be made by individual countries to reach this target, renewable electricity being the main element of this obligation (the others are heat and renewable fuels for transport).

The European Commission is about to propose a new Renewable Electricity Directive, which is likely to include flexibility provisions allowing the cost-effective attainment of those targets. One of the possibilities that are being considered is to allow countries to reach their individual targets by buying their renewable electricity deployment deficit to other countries which have a surplus (i.e., a renewable electricity deployment which is above their targets). This trade is likely to take the form of an exchange in guarantees of origin (GOs). Guarantees of origin are currently implemented in Member States to fulfil the Renewable Electricity Directive requirement that each country have a system that allows the tracing of the source of each kWh of renewable electricity and informs on this source.

[1] Corresponding author: Pablo del Río-González, T el: +34 91 7768027. E-mail: pablo.rio@uclm.es.

This paper analyses the advantages and drawbacks of a system of trading renewable electricity attributes based on the GO concept and discusses several design alternatives and options to make such idea workable, effective and efficient.

1. Introduction

The promotion of renewable energy sources (RES) in general and electricity from renewable energy sources (RES-E) in particular is a key policy priority at the EU level. RES-E has traditionally been supported by Member States (MSs) in Europe. An ambitious target for renewable energy of 20% of energy consumption has been set for 2020 in Europe.

Justification for granting support to renewable electricity at both EU and MS levels has been, at least, twofold. On the one hand, RES-E pollutes much less than conventional electricity and, therefore, avoids its negative environmental externalities. More specifically, RES-E contributes to the achievement of the climate change mitigation goals of the Kyoto Protocol (KP). In addition, emissions of local pollutants are reduced as well.

Apart from the environmental benefits, RES-E provides other socioeconomic advantages such as development, employment and investment opportunities. Finally, by having renewable energy deployment in its territory, Europe can also reduce its fossil fuel dependency and mitigate the risks related to the security of energy supply, which is certainly a major policy concern nowadays.

However, those benefits stemming from renewable electricity would not be enough by themselves to justify the implementation of policy support schemes, because if the market valued such advantages (positive externalities), then RES-E would penetrate the market without any need for public support. Unfortunately, renewable electricity competes on an unequal playing field with thermal-based conventional electricity (del Río and Gual 2004).

The European Commission is about to propose a new Directive, which is likely to include flexibility provisions allowing the cost-effective attainment of the aforementioned 20% ambitious target. One of the flexibility provisions currently being considered is to allow countries to reach their individual targets by buying their renewable electricity generation shortfall from other countries exceeding their targets. This EU-wide trade will probably take the form of an exchange in guarantees of origin (GOs), which are issued by MS to ensure that electricity has been produced from renewable energy sources.

A major discussion on the design elements of a GO trading scheme concerns how this trade should be organised. So far, the recent and scarce literature on the analysis of GO trading has focused on trade between firms, whereas the exchange of GOs between governments has not received a comparable attention.

This paper analyses the situation of the adoption of a system of GOs with respect to a situation where GOs have not been implemented. In addition, considering the current debate spurred by the European Commission and that the implementation of a system of GOs is highly likely in Europe, the paper compares the advantages and drawbacks of GO trading options. We adopt a qualitative policy-analysis approach considering several assessment criteria (i.e., multicriteria).

Accordingly, the paper is structured as follows. The following section provides a brief overview of renewable electricity policy in the EU and its Member States. It then analyses whether a system of GOs could improve the current situation according to several criteria (section 3). Section 4 identifies several options for GO trade which are currently being discussed in policy implementation and evaluates the GO-trading alternatives according to these criteria. The final section concludes.

2. Renewable Electricity Policy in the EU: An Overview

2.1. The EU Lvel

The first major milestone in renewable electricity policy in the EU was the White Paper on Renewable Energy Sources, published in 1997, which set a target of 12% of gross inland energy consumption coming from RES in 2012.

Later on, in 2001, the renewable electricity Directive[2], established a target of electricity from RES (21% of electricity consumption in 2012, 12.5% if large hydro is excluded), in order to achieve the aforementioned 12% target. There are also differentiated indicative targets per MS. The range of these targets is wide (from 5.7% for Luxembourg to 78% for Austria)(Figure 1). Target setting depended on the share of RES-E in 1997, the available renewable potential in each country and the resulting political negotiation. It should be taken into account that these targets are indicative, i.e., not mandatory. This means that non-compliance with them does not lead to financial sanctions[3].

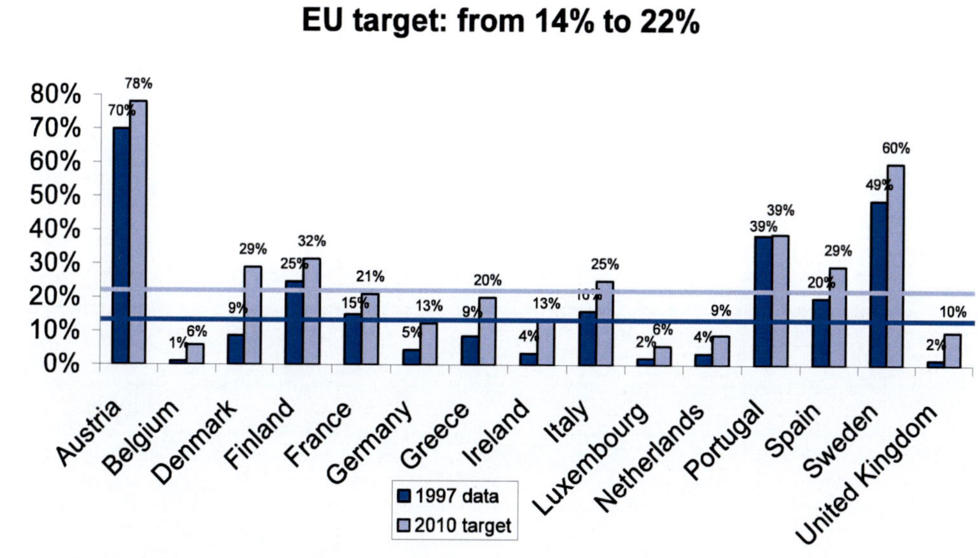

Source: Uyterlinde et al. (2003).

Figure 1. Indicative targets for RES-E per Member State

---

[2] Directive 2001/77/EC on the promotion of electricity produced from renewable sources of energy in the internal market (OJ L 283, 27.10.2001, p. 33).

[3] Current projections indicate that the 12% target for 2010 will not be met (at most, 10% is likely). In addition, with current policies and efforts in place, and unless current trends change, the EU will probably not reach the 21% RES-E Directive indicative targets in 2010 (19% is likely in this regard)(European Commission 2007a).

On the other hand, the Directive does not establish a common support scheme for the promotion of RES-E in Europe, although it envisages the possibility that a harmonised framework for support schemes will be implemented in the future. It allows each MS to choose the support scheme that "corresponds best to its particular situation" (Subsidiarity Principle). Therefore, national RES-E support schemes will continue to be used by MSs (del Río 2005). The following table provides the main elements of the 2001 RES-E Directive.

Table 1. Summarising the RES-E Directive

Issue	Comment
Indicative Targets for RES-E	An indicative, non-mandatory target for the deployment of RES-E in Europe is specified in the Directive, which also sets indicative targets per MS.
National Support Schemes and EU Community Framework	The directive states that it is too early to set a Community wide framework regarding support schemes. By 27 October 2005 the Commission should present a report on the experience gained with the application and coexistence of different support schemes in the MSs. This report may be accompanied by a proposal for a Community framework for RES support schemes (Article 4.2). The directive also stipulates that such a proposal for a harmonized support framework should allow a transition period of at least seven years (thereafter) in order to maintain investors' confidence and avoid stranded costs.
Mandatory Guarantees of Origin (GOs)	By 27 October 2003 at the latest MSs shall ensure that the origin of RES-E can be guaranteed as such according to objective, transparent and non-discriminatory criteria laid down by each MS. They will issue a GO containing data about the energy source from which the electricity was produced (including dates and places of production). They will enable RES-E producers to demonstrate that the electricity they sell has been indeed produced from RES. GOs have to be accepted by other Member States and do not have to be tradable, but if they are made to be tradable, then they will work as a Tradable Green Certificate (TGC) scheme.
Guarantee for grid access.	MSs should adopt the necessary measures to ensure that the system operators guarantee grid access to RES-E in their territory.
Mandatory reports for European Commission and MSs.	FOR THE COMMISSION: -To publish a report which assesses whether MSs are on their way to achieve their national indicative targets (10/27/2004 and thereafter every five years) -To present a report on the application of the Directive (12/31/2005 and thereafter every five years).

Table 1. (Continued)

Issue	Comment
	FOR THE MSs: -Not later than 27 October 2002 and every five years thereafter, Member States shall adopt and publish a report setting national indicative targets for future consumption of RES-E in terms of a percentage of electricity consumption for the next 10 years. -Member States shall publish, no later than 27 October 2003 and thereafter every two years, a report which includes an analysis of success in meeting the national indicative targets taking account, in particular, of climatic factors likely to affect the achievement of those targets and which indicates to what extent the measures taken are consistent with the national climate change commitment. -MS shall issue GOs for RES-E (10/27/2003). -MS shall publish a report assessing the procedures for authorisation of RES-E plants (10/27/2003).

Source: Own elaboration.

In a couple of communications after the RES-E Directive, the European Commission (European Commission 2005, 2007a) reaffirms the RES-E Directive statement that it is too early to set a harmonised framework for RES-E. In particular, the European Commission (2005, p.16) stresses that "it is too early to compare the advantages and disadvantages of well-established support mechanisms with systems with a rather short history. Therefore, and considering all the analyses in this Communication, the Commission does not regard it appropriate to present at this stage a harmonised European system". Furthermore, this report states that "the Commission will closely monitor the state of play in EU renewable energy policy and, not later than December 2007, make a report of the level of Member States systems for promoting renewable electricity in the context of the on-going assessment related to 2020 targets and a policy framework for renewable energy beyond 2010" (op.cit., p.18).

In the Renewable Energy Road Map, published in 2007, the European Commission already proposed a binding target of increasing the level of renewable energy in the EU's overall mix from less than 7% in 2007 to 20% by 2020 (European Commission 2007a, 2007b). In this Road Map, the Commission provides its strategic vision for the energy future of Europe.

A draft of a new RES-E Directive has recently seen the light on 23rd January 2008. Among other aspects, this draft incorporates the 20% target, making it binding for the EU as a whole as well as for Member State, and lays down rules relating to guarantees of origin, administrative procedures and electricity grid connections in relation to RES-E[4].

[4] A key aspect in the context of this chapter is that the draft Directive allows the transferability of GOs between Member States, subject to some restrictions.

2.2. The Member State Level

In order to comply with the RES-E Directive targets, RES-E is currently promoted by the MSs through a wide array of support schemes. There are four main types of these schemes (del Río 2007; del Río and Gual 2004):

- *Feed-in tariffs*. These are subsidies on output (i.e. per kWh generated) paid in the form of guaranteed premium prices and combined with a purchase obligation by the utilities. This system can take two forms. Either a total payment per kWh of electricity of renewable origin is given or a payment per kWh on top of the electricity 100 wholesale-market price is granted. In this later case RES-E producers have to sell 101 the power themselves. In both cases, the costs are usually borne by consumers.
- *Tendering/bidding systems*. The government invites RES-E generators to compete for either a certain financial budget or a certain capacity of RES-E generation. Within each technology band, the cheapest bids per kWh receive the subsidy (Schaeffer et al. 1999). The operator pays the bid price per kWh. A fund financed by a levy on electricity consumers covers the difference between this bid price and the market price of electricity.
- *Quota with Tradable Green Certificates (TGCs).* TGCs are certificates that can be sold in the market, allowing RES-E generators to obtain revenue. This is additional to the revenue from the sales of the electricity they produce and feed into the grid. Therefore, RES-E generators benefit from two streams of revenue: the market price of electricity plus the market price of TGCs. The issuing (supply) of TGCs takes place for every specified amount of RES-E (i.e., MWh), while their demand generally originates from an obligation (usually termed "quota").
- *Fiscal incentives* work via the tax system. They can be exemptions or rebates on (energy, corporate or income) taxes, tax refunds, lower VAT rates or attractive depreciation schemes. They might affect old and recent installations (generation-based incentives) or only the new ones (capacity-based incentives) (Uyterlinde et al. 2003).

The most widespread system in Europe is feed-in tariffs. Nineteen Member States use this system, whereas seven use quotas with TGCs, three use bidding schemes and two countries use fiscal incentives (see Ragwitz et al. 2007 for further details)[5]. An analysis of their effectiveness and cost-effectiveness shows that feed-in tariffs tend to be highly effective in promoting RES-E at moderate costs. Bidding schemes have shown to be ineffective, whereas the jury is still out on TGC schemes, given that they have been recently implemented and it is probably too early to conclude on their success or failure, although the initial evidence shows that they are behaving poorly both in terms of the effectiveness and the cost-effectiveness criteria (see European Commission 2007a for further details).

---

[5] These main instruments are usually complemented with other "secondary" schemes, whose use is widespread in Europe, including investment subsidies and green pricing schemes.

3. The Advantages and Drawbacks of an EU-Wide GO Trading Scheme fr Renewable Electricity

GOs are currently implemented in Member States to fulfil the Renewable Electricity Directive requirement that each country have a system that allows the tracing of the source of each kWh of renewable electricity and informs on this source. GOs, which are issued by Member States (MS), ensure that electricity has been produced from RES sources. They specify the energy source from which the electricity was produced and serve to enable RES-E producers to demonstrate that the electricity they sell is produced from RES (Directive 77/2001/EC, art. 5). GOs would be recorded in a centralised electronic registry (European Commission 2007c).

Therefore, GOs represent a proof of renewable energy production and are based on the idea of "physical detachment", i.e., they could be traded independently of physically exchanged energy. In other words, GOs would be a vehicle for trade which can be used by governments to show compliance with the aforementioned 2020 target (RECs 2007)[6]. An EU-wide GO system would establish uniform and clear accounting rules, thus ensuring that trade in GOs is as reliable as trade in the actual product. Comparability ensures that GOs can be used by a MS to demonstrate compliance with the RES target regardless of the MS in which the RES has been produced (European Commission 2007c).

If GOs are made tradable, they could be regarded as a TGC scheme. This section discusses whether this trading of GOs is preferable to no trading (i.e., to continuation of the MS domestic schemes). First, we discuss several assessment criteria according to which this analysis should be made.

3.1. Assessment Criteria

The implementation of GO trading, as well as the alternatives for GO-trading (see next section) can be assessed according to several criteria, including:

- -*Effectiveness*. This refers to the achievement of RES-E policy targets, which is related to an effective increase in the amount of RES-E generation fed into the grid (both at the MS and Europe-wide levels). Reaching the target depends on the level of support as well as on the security (continuity) of the support scheme. The later contributes to investment risks for investors being kept at a low level.
- -*Cost-effectiveness* involves reaching the target at the lowest possible social costs. More specifically, this can be interpreted as low costs for electricity consumers who

6 This role of GOs as a vehicle for trade is stressed by Directive 77/2001/EC itself when stating in its introduction that "This Directive does not require Member States to recognise the purchase of a guarantee of origin from other Member States or the corresponding purchase of electricity as a contribution to the fulfilment of a national quota obligation. However, to facilitate trade in electricity produced from renewable energy sources and to increase transparency for the consumer's choice between electricity produced from non-renewable and electricity produced from renewable energy sources, the GO of such electricity is necessary. Schemes for the guarantee of origin do not by themselves imply a right to benefit from national support mechanisms established in different Member States. It is important that all forms of electricity produced from renewable energy sources are covered by such GO".

are at the end the ones paying for the policy in their electricity bills. Cost-effectiveness includes several aspects and particularly the equimarginality rule and transaction costs. In principle, the costs for the consumers would be kept low if countries with high marginal RES-E generation costs had a lower generation level, compensated by a greater RES-E generation in countries with low levels of RES-E generation costs. Ideally, the targets should be attained with the same marginal generation costs across countries (equimarginality) with a single EU support price at such equimarginality. However, a single support price might lead to windfall profits, i.e., too much support for the low-cost technologies.

- -Transaction costs. Although part of the cost-effectiveness criteria, its relevance deserves a particular mention. The transaction costs related to the implementation and functioning of a RES-E support scheme should be considered. An instrument satisfying the equimarginality rule may not be cost-effective if it involves high transaction costs.
- We should distinguish between system installation, system operation and system adjustment (Madlener and Stagl 2005). Transaction costs may fall on public administration (administrative) or on companies (company transaction costs).
- -*Dynamic efficiency* refers to the promotion of technological diversity, i.e., encouraging the development and adoption of a basket of technological alternatives, including the currently more expensive ones. If they are not supported in the short term, low-cost technologies, which will be necessary to reach the future targets cost-effectively will not be available, and target attainment will be more expensive than it would otherwise be.
- -*Political feasibility*. The implementation of a system which meets all the aforementioned criteria may still not be politically feasible, either due to the rejection of governments or other actors such as firms and consumers. In turn, political feasibility depends on the distribution of the costs of reaching the targets and on awareness on potential local benefits.
- -*Equity*. This refers to cost and benefit distribution. Do they fall on consumers or on society in general? How do alternative designs influence distribution of benefits and costs of RES-E generation across socioeconomic actors in a country? Would some countries be more affected than others?
- -*Local benefits*. The exploitation of renewable energy sources and technologies contributes not only to achieve environmental goals (climate change mitigation and reduction of local pollutions) but also to security of energy supply, to the development of a knowledge-based industry creating jobs, economic growth, competitiveness, and rural development as well as regional and local development opportunities, export prospects, social cohesion and employment opportunities (European Commission 2008). Therefore, countries may want to make local generation of RES-E a policy priority, because of its local benefits (socioeconomic and environmental) bring and not care so much about reaching the RES-E targets cost-effectively by resorting to the purchase of GOs, which involves encouraging the

generation of RES-E abroad[7]. Indeed, the willingness of citizens to pay for RES-E generation when the local benefits are not enjoyed by those citizens may be very low. A system would thus be considered superior if it stimulated the local generation of renewable electricity. Clearly, there is a trade-off between this and the cost-effectiveness criteria.

3.2. The Advantages

Theoretically, an EU-wide unrestricted system of GO trading could lead to the same price for GOs in the EU. This means that the equimarginality principle would be accomplished, i.e., that RES-E would be developed where it is cheaper and that a cost-effective deployment of RES-E would be achieved. Countries/companies with a shortfall would buy the needed GOs to other companies/countries instead of developing RES-E domestically, which would be more expensive. In theory, this would be more efficient that the current system, based on the existence of different support schemes across countries in which countries achieve their own RES-E targets by relying on their own support schemes. The following discussion, based on del Río (2005) shows why.

Figure 2 shows the RES-E marginal cost curves for two countries, A and B, which represent the increase in RES-E generation costs when an additional kWh of RES-E is produced (MCA and MCB, respectively). We can assume that both countries have a RES-E target (Q_A and Q_B respectively). Assuming an EU-wide GO market, it is shown that achieving Q_A leads to a higher marginal and total costs (area below the MCA curve between 0 and Q_A) for country A while, for the same target, marginal costs (and, thus, total costs) for country B would be much lower (MCB and the area below this curve up to Q_B). The total costs of reaching the target for country A are given by the sum of areas a, b, c and d while the costs for country B are given by the sum of the areas c and d: The total costs of achieving the targets without GO trading would be the sum of all those areas.

When looking at Figure 2, a policy instrument could be implemented so that the production of RES-E in the higher-cost country would be reduced (reducing its total costs accordingly) and production of RES-E in the low-cost country was increased, so that the aggregate compliance costs would be reduced. This could be possible if a harmonisation of support schemes in Europe together with an EU-wide GO trading system was implemented. Trade would then be driven by the possible gains for both countries with different marginal cost curves (and targets).

Theoretically, assuming an overall target of Q, a unitary GO price in Europe would result from the intersection of the aggregate marginal cost curve and the total quota Q. Figure 3 depicts the aggregate marginal cost curve for both countries (constructed by adding the domestic marginal cost curves of each country). An EU-wide GO price would result. The domestic support level in country A is higher than P_{GO}; meaning that country A has no potential to produce RES-E at that price if a harmonised market with a single GO scheme for both countries was implemented (and trade allowed). On the contrary, the support level in

[7] The local benefits of RES-E would be especially valuable in the case of countries depending to a large extent on primary energy imports. In addition, RES-E deployment would be very attractive as a development alternative in rural regions, with few options in this regard apart from the traditional and declining agricultural activity.

country B would be below P_{GO}; meaning there is a large RES-E potential in country B. Note that the more ambitious the targets for a country, the more likely his (internal) support level should be high.

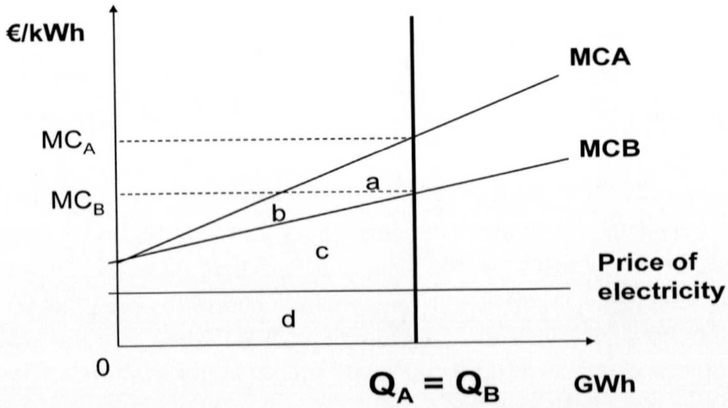

Source: Adapted from del Río (2005).

Figure 2. The costs of reaching the RES-E targets with national promotion schemes and no GO trading.

In harmonisation with GO trading, for an overall GO price of P_{GO}; it would be attractive for country A to deploy RES-E up to Q'_A. It will not be profitable to go beyond that amount, because then it would be cheaper to buy GOs in the market than to produce RES-E by itself. Therefore, in order to reach its own target (Q_A), country A will buy GOs at a price of P_{GO}. By following this strategy, country A would save area 1 in Figure 3 compared to the previous scenario (Figure 2). The picture is the opposite for country B, which given its low-cost RES-E generation, would increase its deployment with harmonisation and GO trading. It would produce up to Q'_B because the additional costs of generating RES-E are more than compensated by the revenues B would obtain from selling its surplus certificates to country A at a price of P_{GO}. This additional revenue is given by area 2 in Figure 3.

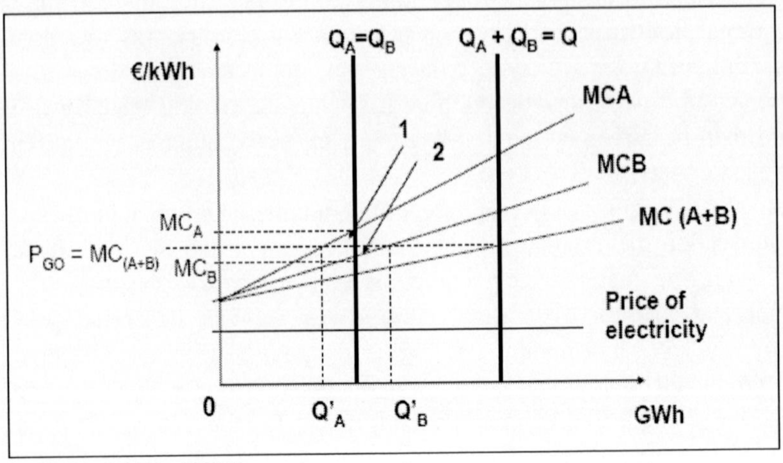

Source: Adapted from del Río (2005).

Figure 3. The costs of reaching the RES-E targets with GO trading

This analysis shows that both countries individually would be better off if harmonisation of support schemes combined with a GO system was implemented. Country B would obtain additional revenues while country A would save costs. Not only would individual countries be better off, but also the EU-wide target would be achieved at lower costs.

3.3. The Drawbacks

However, the aforementioned cost-effectiveness would come at a high price concerning the rest of criteria.

- *Effectiveness*. Although a GO trading scheme is a quantity-based RES-E promotion scheme which, at least in theory, sets the quantity of RES-E to be attained and allows this quantity to be reached in the most cost-effective manner, in the reality it is doubtful and to be seen whether a European-wide GO trading system would be more effective than the current domestic schemes, which have shown a great amount of effectiveness in some countries (notably, Germany and Spain). The risk is that we replace promotion schemes which have worked reasonably well in this regard for a new system with uncertain effectiveness.
- *Transaction costs*. A Europe-wide GO trading scheme involves a radical institutional change with respect to the current schemes. It would probably bring more transaction costs for the administration (Member States as well as for the European Commission) and for firms. Some are one-off. For example, new administrative structures would have to be created in the Member States and the European Commission. Others transaction costs are related to the functioning of the scheme. The problem of a Europe-wide GO trading alternative is that, compared to feed-in tariffs, it involves controlling for compliance and imposing sanctions, which certainly entails additional administrative costs compared to the existing schemes. Transaction costs for companies would also tend to be higher. These are related to getting familiar with the new system (information costs), finding trading partners, entering transactions and demonstrating compliance, activities that they currently do not have to carry out under feed-in tariff schemes. Since transaction costs are part of the cost-effectiveness criteria, they would tend to reduce the cost-effectiveness benefits mentioned in the previous subsection.
- *Dynamic efficiency*. GO trading would clearly score worse than the no GO trading option regarding this criteria. The reason is that, compared to national support systems, EU GO trading does not encourage technological diversity and the most expensive technologies would not be promoted. The reason Is that the implementation of a Europe-wide harmonisation of support schemes with a GO scheme including all RES-E technologies would allow the achievement of the RES-E Directive targets with the most competitive and mature technologies (i.e., wind and biomass), while other technologies would not have a chance and would not penetrate the market (i.e., PV). It would be ensured that the most competitive RES-E technologies are deployed first, but currently high-cost renewable energy technologies would not be deployed.The problem is that non-mature technologies cannot compete with those technologies that have benefited in the past from

economies of scale and learning effects. A vicious circle would prevent them from penetrating the market: their high costs (but high potential for further improvements) make them relatively unattractive and are a barrier to adoption. But if they are not being adopted, they will not benefit from economies of scale and their unitary costs will not be reduced. In this situation, harmonisation with GO trading would only complicate things for these technologies and it would not contribute to the breaking of such a perverse and vicious circle. The non-reduction in their costs would be a problem especially in the long-term, since currently expensive technologies could be necessary in the future to achieve RES-E targets. If they are expensive then (because they have not diffused), future RES-E targets will be reached at higher costs than could have been the case (del Río 2005). This is certainly not desirable. In contrast, by allowing the creation of niches for all types of renewable energy technologies, current support schemes (particularly the German, Danish and Spanish ones) have shown great level of dynamic efficiency. There is no reason to think that this dynamic efficiency of domestic systems could not continue in the future.

- *Local benefits*. GO trading is clearly worse than no trading in this regard. Following del Río (2005, p.1248), the following lines discuss why. If EU-wide GO trading is set up in order to cost-effectively support RES-E, the national (local) benefits of RES-E (socioeconomic and environmental) can no longer be stimulated directly. This would probably be cost efficient, but, for example, by not deploying RES-E in Spain and having the RES-E plants located somewhere else (i.e., UK), also the local benefits would be gone to this latter country. This would have very negative effects on a country such as Spain, where employment creation in the RES-E sector has been very significant[8]. Taking into account these negative effects, harmonisation of support schemes combined with a GO-trading system might not be very attractive for this latter country. Not only socioeconomic, but environmental benefits would be foregone as well. RES-E provides both global and local environmental benefits. From a climate change point of view, a RES-E plant located in Denmark would lead to the same CO2 emissions reductions as a plant located in Spain. If it is more cost efficient to install a wind turbine in Spain than in Denmark, then, in order to reach cost efficiency, the plant should be installed in Spain. But if we take into account other (local) environmental benefits stemming from the installation of the RES-E plant (SO2, NOx and VOC emissions), then harmonisation and GO trading would not have such beneficial consequences because the Danish would like to see the plant located in their territory in order to replace dirtier generation plants and experience a reduction in local pollution levels.To sum up, if two countries with different RES-E marginal generation costs participate in a joint GO-trading scheme, then the higher-cost country will import GOs from the lower-cost country. This would have both positive and negative effects for the importing country compared to a continuation of the domestic schemes. Costs would be saved but some of the national socioeconomic and environmental policy objectives of the high-cost country would not be met.
- *Equity*. The assessment in the context of these criteria is uncertain. Of course, a priori it could be expected that some countries would be affected differently with respect to others by GO trading. Some countries would be worse-off and others better off.

[8] Currently, around 200,000 people work in this sector (of which half are direct jobs).

However, it is quite difficult at this stage to tell which types of countries would be more affected than others. This would require a difficult balance between some of the criteria mentioned before, which would be certainly a too speculative exercise. It would also depend on the costs of complying with the RES-E targets and on the perceived national benefits of RES-E deployment. Empirical analysis in the form of in-depth case studies would certainly be useful in this context. In theory, a system that links the efforts in renewable energy support to a greater local RES-E deployment would probably lead to a more balanced distribution of the costs and benefits of RES-E deployment. However, if RES-E targets were binding per country, some Member States may prefer to have an additional flexibility element to reach those targets. The GO trading system provides this element.

- *Political feasibility*. This is related to the previous two criteria. Accordingly, local benefits would remain in the supporting country under a continuation of the present schemes, which would make this option highly politically feasible. A GO trading scheme is likely to be rejected by some governments.

Table 2 summarises the above discussion.

Table 2. Assessing GO trading with respect to no-trading

	No trading of GOs	GO trading
Effectiveness	?	+
Cost-effectiveness	+	?
Transaction costs	-	+
Dynamic efficiency	- -	+
Political feasibility	-	+
Equity	??	??
Local benefits	-	+

Source: Own elaboration. Note: (-) means a negative score and (+) a positive score.

4. Discussing the Options for GO Trade

4.1. The Alternatives for GO Trade

Two main alternatives clearly stand out in this context: either companies or governments are allowed to trade GOs:

(I) Company-Based Trading

RES-E targets fall on MS, which can comply with them by surrendering GOs at the end of the compliance period. Not complying with the target would involve the payment of a penalty at a level which should be set either in absolute terms (as in the EU ETS for CO2 allowances) or in relative terms (% above the market price of GOs). MS delegate target-attainment to their RES-E generators (supply-side) and electricity suppliers (demand side). The later have to surrender a number of GOs per kWh of electricity sold.

Countries maintain their own RES-E support scheme. RES-E generators receive a GO per each kWh of electricity from renewable origin and may apply for the support granted by a different country.

RES-E generators would sell their GOs to the country offering the highest support level, receiving its support[9]. Eligibility to receive the RES-E support of the importing country would be contingent upon handing out the GOs, which avoids double counting[10].

The country with high support levels would have a surplus of GOs, i.e., GOs exceeding its target, which would be greater the higher its support with respect to RES-E generation costs in this country, the less stringent its objectives and the lower its restrictions to imports (if any). GO exporting countries would be those with low support levels. They will have a GO deficit and problems to comply with their quota since their RES-E generators will sell GOs abroad to obtain the higher support level. In order to avoid this overflow of GOs from the exporting to the importing country, countries may put a limit on trade i.e., restrictions on the amount of GOs imported or exported. A secondary market for GOs would be created. Country "A" (or their suppliers, in case they have delegated the obligation to comply on them) will have an excess of GOs to sell to suppliers in B. This would finally lead to a single EU GO price[11]. The import restrictions of the government in "A" would contribute to avoid an overflow of GOs to this country. Export restrictions by "B" would have a similar effect, increasing the opportunities to local trade of GOs.

(II) Government-Based Trading

In this system the responsibility to comply with a target rests on countries, which are also the ones liable to trade. In this case, targets also fall on MS but these are not delegated to subnational entities (i.e., companies). Each kWh of RES-E leads to the issuing of GOs and countries have to surrender as many GOs as the level of their targets. Countries with an ambitious RES-E policy would pay their own (but not foreign) RES-E generators for their RES-E production, generating an amount of GOs which they could sell to countries with a GO deficit. The revenues thus obtained can be recycled back to electricity consumers to reduce their costs for supporting RES-E, be used to finance RandD activities in renewable energy or go to the general public budget. A market for GOs would be created at the government level, with a supply side represented by GO-seller countries and a demand-side resulting from GO-purchasing countries.

---

[9] In the case governments have delegated their responsibility on electricity suppliers, which would be the obligated parties, these would buy GOs. They would give the support level to the purchaser and afterwards require their government to reimburse them in exchange for the GO bought to the foreign RES-E generator. The GOs would finally end up with the governments, which would hold them to show compliance with their RES-E targets.

[10] i.e., the RES-E generators is not allowed to receive the support of its country of origin plus the support given by the foreign country.

[11] The final price of GOs would be within a range defined by the support level of "A" (which is the amount paid by the government of "A" to purchase those GOs), assuming this level would be below the non-compliance penalty and the support level of "B". The final price would be the result of the negotiation between the parties (the government of "A" and utilities in B).

4.2. Assessing the Options for GO Trade

The starting point of the analysis is related to the fact that company-based trading is a radical innovation in RES-E promotion, which in effect leads to an EU-wide TGC scheme, with the problems that this would have: windfall profits, no technological diversification, great uncertainty on future support levels (and thus, greater risks for investors), small generators burdened with transaction costs and distortions created by different electricity prices across Member States (see del Río 2005 for further details).

In contrast, government-based trading would not involve such a drastic institutional change and would make fully compatible the implementation of existing support schemes (including the currently very effective feed-in tariffs) with GO-trading. An additional flexibility element would be introduced in existing support schemes allowing those countries with a RES-E deficit (in relation to their targets) to buy GOs from other countries. The GO price would work like a safety valve, i.e., a cap on the costs of RES-E promotion.

4.3. Effectiveness

As mentioned above, this depends on the level of support and the continuity of the support scheme, leading to low risks for investors.

Both systems introduce the necessary flexibility to allow those with a RES-E generation deficit (either governments or electricity suppliers) with respect to their targets to cover it by buying GOs, while simultaneously encouraging those willing to have a greater generation than required to receive a reward for the surplus. However, in the case of company-based, this flexibility is at the expense of an increase in risks and transaction costs for investors whereas with government-based trading the currently effective domestic schemes are maintained and there are no distortions as a result of the combination of these schemes with the trading element. Company-based trading in effect creates an EU-wide tradable green certificate (TGC) system. The drawbacks of this system have been mentioned elsewhere (see del Río 2005). Furthermore, currently implemented domestic TGC schemes have not proven to be particularly effective in RES-E promotion.

Thus, a government-based trading would probably involve less risk for investors, because it would entail a continuation of the currently effective support schemes. In contrast, under company-based trading, further investments in RES-E capacity and generation would depend on the price of GOs, which is unknown ex-ante and can be assumed to be more volatile than the support levels granted in currently effective FIT schemes.

In addition, government-based would involve lower transaction costs for small RES-E generators than a fully new system like company-based. This might be a key aspect constraining RES-E investments (see 4.3).

4.4. Cost-Effectiveness

When calculating the (consumer) costs of reaching the RES-E targets, we should consider several aspects: whether the different options comply with the equimarginality principle, whether they lead to windfall profits, if they lead to higher risk premiums for investors, if they allow a revenue for the government which can be recycled and what are the transaction costs (administrative and firm) of implementing and operating the system. This later aspect will be dealt with separately in 4.3.

First, none of the three cases leads to an "ideal" cost-effective achievement of the RES-E targets because the equimarginality principle (equal marginal costs of RES-E generation) is not attained. Figure 4 shows why. For illustrative purposes we have assumed that there are only two countries (A and B), with different marginal cost curves for RES-E generation (MC_A and MC_B), and that the overall target would be the sum of both country targets (A+B). It is shown that, for the same level of RES-E generation, the marginal (and the overall) costs are lowest in an ideal case of equimarginality and highest in the case of no trade[12]. The cases in between are represented by GO trading coexisting with national promotion schemes. The reason is that neither of both GO-trading systems (government or company-based) ensure that the equimarginality rule is fully accomplished, given the coexistence of a GO price with different support levels in MS. Still, these two cases have lower costs than the "no-trade" option because some flexibility in meeting targets is allowed through trading.

Notwithstanding, government-based trading could probably have more difficulties in attaining the equimarginality rule. Being more aware about the costs of RES-E generation than governments are, companies are arguably more capable of exploiting the RES-E cost differences across countries and technologies.

On the other hand, another aspect affecting cost-effectiveness in attaining RES-E targets is the issue of windfall profits. These are more likely to occur under a company-based GO scheme than under a government-based one.

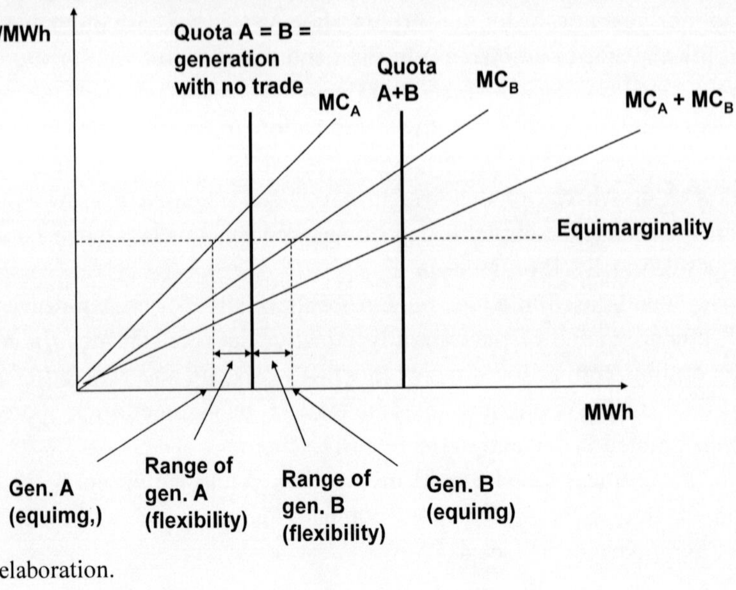

Source: Own elaboration.

Figure 4. Alternatives for GO-trade and the equimarginality principle.

[12] The total costs of reaching the joint targets of A and B in the equimarginality case would be calculated as the area below the combined marginal cost curve (MCA + MCB) up to the level of the combined quota (quota A+B). In the case of no trade, thc total costs would be the sum of the areas below MCA and MCB , up to the level of their respective quotas (i.e., quota A = B). It can be seen that the area would be lower in the equimarginality case (see del Río 2005 for a detailed comparison of both scenarios).

The reason is that the first system tends to lead to a single GO price, especially if there is unrestricted trade[13]. This means that the cheapest technologies will receive too much support with respect to their costs. Under a government-based GO scheme, FIT schemes can be used, which allow governments to adapt the level of support to the costs of the technologies[14].

Furthermore, the aforementioned greater uncertainty and risks of company-based trading for investors does not only have a negative influence on the effectiveness, but probably also on the cost-effectiveness of RES-E promotion. Greater costs related to the higher risk premiums can be expected, leading to a higher interest rate on bank loans.

Finally, it should be taken into account that government-based trading allows the government of the selling country to obtain revenue, which can be recycled back to consumers or be used to reduce pre-existing distortionary taxes, much in the vein of the "double dividend" hypothesis put forward by environmental economists. This recycling of revenues would therefore have an impact on the overall economic efficiency of the scheme[15]. In the case of GOs between firms, governments are not able to obtain such revenues. The financial flows are between firms (which may aggravate the problem of windfall profits).

4.5. Transaction costs

We should distinguish here between the administrative costs and the transaction costs for firms. Regarding the former, a further relevant distinction is between the administrative costs of implementing the system and its operating costs.

Concerning the implementation costs for the MS, this mostly depends on the novelty of the system and its compatibility with existing administrative structures and support schemes. Given that government-based trading is an incremental institutional innovation compared to the company-based alternative, and that it fully preserves the continuation of current promotion schemes, lower implementation costs can be expected. Compared to the current situation, the additional costs would be related to the (probably small in volume) trading of GOs between governments.

Regarding the operation of the system, efficient monitoring is necessary in order to avoid double counting of renewables across MS (European Commission 2007c). This means that a European GO registry should be put in place in both trading cases. However, greater controls would probably be necessary in the case of company-based trading to comply with the targets, because the registry has to descend to the company level in addition to the MS level, which is not the case with government trading. The reason is that in company-trading the risk of double accounting and double support should be avoided at all costs, i.e., a RES-E generator receiving the national support should not be allowed to sell its GOs abroad. This means that an effective administrative system (monitoring, verification) should be put in place, whereas this is not necessary with government-trading of GOs, for which only a central EU registry of GOs would be needed.

---

13 Still, in the company trade case, and even if a single price of GOs does not result, windfall profits are highly likely to occur because RES-E producers would sell to suppliers in countries with the highest support levels.

14 Indeed, current FIT schemes applied in the EU show that technology differentiation of support is a common feature (see, for example, del Rio and Gual 2004).

15 The money would also be earmarked for climate and energy policy actions.

Regarding transaction costs for firms, they would probably be higher under the company-based trading alternative. The costs associated to this trading (information costs, negotiation and search costs) can make trade difficult for small actors. Transaction costs have proved very low in currently successful FIT schemes (see del Río and Gual 2007), which would coexist with the government trading scheme, while they can be significant in schemes with a trading component such as TGC schemes (Wiser et al. 2005, Sawin 2004), which would be the main element of a company-based scheme.

4.6. Dynamic Efficiency

Company-based trading is likely to lead to a single EU price for GOs. This price which would be above the costs of low-cost technologies (leading to windfall profits), would be below the one needed to encourage the adoption of the most expensive renewable energy technologies, which will be needed in the future when more stringent targets are set. This means that the cheaper technologies would be promoted with too much support, whereas the expensive technologies would not have a chance to penetrate the market. This lack of technological differentiation would contrast with the case of government-based trading because, with their domestic national support schemes, governments may decide to provide different support levels to different technologies, as they currently do. It is well known that dynamic efficiency properties of a FIT are superior to other systems (see Menanteau et al. 2003, among others) and government-trading would be fully compatible with a FIT system.

It could be argued that, in a company-based system, some sort of "technology banding" could be adopted to increase the technological diversity of the scheme. However, this would be highly complicated. Two options may exist in this regard: either it is required that a part of the quota be met with GOs from expensive renewable technologies or no differentiation is made in the quota, but in the amount of GOs granted per kWh of renewable electricity generated, giving more GOs to expensive electricity generation from RES. Both solutions would be problematic. The first one because it would narrow the market for GOs and, thus, its liquidity (in addition to the difficult decision on what should be the part of the quota (%) to be met with GOs from expensive RES-E projects). The second option would run up against the difficulty of the exchange rate to be set, i.e., the amount of GOs to be granted per kWh of electricity from these high-cost projects. This problem would be avoided in a government-based system because FITs allow differentiation of the level of support per technology and there is no need to calculate quota percentages of GOs from expensive RES-E or "exchange rates".

4.7. Political Feasibility

A government-based GO trading scheme would be politically more feasible. Some countries are not keen on the idea of abandoning their current support system. The alternative of government-based trading allows countries to keep their scheme and the trading component does not involve a new system, but only provides some flexibility for the existing systems. MS retain the capacity to decide on the level of RES-E support they deem appropriate, according to their national priorities and the expected local benefits stemming from RES-E deployment.

In the case of company-based trading, a redistribution of RES-E deployment across MS would occur, depending on differences in RES-E generation costs and trade between firms. This means that national deployment (generation) might be lower in some countries with respect to their national target (which might not be a desired result for the country, if it attaches a high value to RES-E deployment, given its local benefits) or it might be higher, (which might also be controversial, if it leads to an excessive concentration of RES-E deployment in certain locations, leading to undesired environmental impacts or rejection by the local actors). In the case of government trading (and even no trade) the government might avoid these problems.

Furthermore, as already mentioned above, the fact that governments could obtain some revenue in the government-trading case would make this system more attractive for governments than the company-based alternative.

Finally, and taking into account political-economy arguments, company-based trading is likely to be rejected by existing companies on the grounds of stranded costs, i.e., they could argue that they took investment decisions under a system and rules of the game that are now changing. Government-based is only an incremental institutional innovation which would leave current systems untouched and it is unlikely to generate such rejection. Furthermore, the fact that under government-based the level of support would be certain whereas it would be volatile in the case of GO-trading is an additional reason for companies, and particularly RES-E generators, to prefer government-based.

4.8. Local Benefits

The government-based trading alternative allows governments to support RES-E deployment in case they attach great value to the local benefits of RES-E. In contrast, with company-based trading, countries lose the control of national support for RES-E deployment and, thus, the benefits of an ambitious RES-E promotion policy are reduced. If a country has a high RES-E generation cost but the government attaches great value to the contribution of RES-E to the local sustainability of the country, under a company-based RES-E would tend to be deployed abroad, and so the highly valued local benefits would be lost. In other words, it would partly lose the control of promoting RES-E deployment with high associated local benefits. It is doubtful that a country would be willing to pay for the local RES-E benefits accruing to third countries only for reasons of target compliance.

Table 3 summarises the above discussion on the comparison of the three alternatives according to the considered criteria, with a situation with no trading. It is shown that government trading is clearly preferable.

5. Conclusion

This paper has focused on two major issues regarding trading in GOs. On the one hand, it has analysed the advantages and drawbacks of GO trading in the EU with respect to a situation where GOs are not implemented. In addition, considering the current debate spurred by the European Commission and that the implementation of a system of GOs is highly likely, the paper has compared the advantages and drawbacks of GO trading options.

Table 3. Comparing alternatives for GO-trading

	Company-based trading	Government-based trading
Effectiveness.	?	+
Cost-effectiveness.	+	+
Administrative costs (adjustments)	-	+
Dynamic efficiency.	-	+
Political feasibility.	-	+
Local benefits	-	++

Source: Own elaboration. Note: (-) means negative change with respect to a "no-trading" scenario, (+) positive change.

Regarding the former analysis, the paper has shown that, using a multicriteria assessment, GO trading is not unambiguously preferable to no trading. While, according to the cost-effectiveness criteria, GO trading can be deemed positive, other criteria (mainly dynamic efficiency and local benefits) seems to favour no trading. Therefore, the jury is still out on this system and whether we should favour it would depend on the weight we attach to the different assessment criteria.

On the other hand, if a GO system is finally implemented in the EU, the preferable option according to the criteria considered in this paper seems to be government-based trading. This is so because government mitigates the trade-off between different criteria and, particularly, between static (cost-effectiveness) and dynamic efficiency and between static efficiency and local benefits. Compared to company-based trading, government-based trading allows taking advantage of the flexibility provided by introducing a market component (trading) while not incurring the major drawbacks and negative impacts of company-based trading, particularly regarding technological diversity. The draft renewable energy directive seems to go in this direction.

The analysis in this paper has been theoretical and qualitative, taking into account several assessment criteria (i.e., multicriteria). Future empirical research, either in the form of case studies or simulations, should be carried out. Furthermore, the advantages and drawbacks for specific countries of GO trading versus no trading and, in case of GO trading, the impact of different GO trading designs on distinct MS should be analysed.

REFERENCES

European Commission (2007a): An energy policy for Europe. Communication from the Commission to the European Council and the European Parliament. Brussels, 10.1.2007 COM(2007) 1 final.

European Commission (2007b): Renewable Energy Road Map. Renewable energies in the 21st century: building a more sustainable future. Communication from the Commission to the European Council and the European Parliament. Brussels, 10.1.2007 COM(2006) 848 final

European Commission (2007c): Towards a coherent renewables and climate change policy implementation. Brussels.

European Commission (2008): Proposal for a Directive of the European Parliament and of the Council on the promotion of the use of energy from renewable sources. COM(2008) 19 final. 2008/0016 (COD) Brussels, 23.1.2008

Madlener, R. and S. Stagl (2005): "Sustainability-guided promotion of renewable electricity generation". *Ecological Economics.*

Menanteau, P., Finon, D., Lamy, M. (2003): ”Prices versus quantities: choosing policies for promoting the development of renewable energy”. *Energy Policy,* 31, 799–812.

Ragwitz, M., Held, A., Resch, G., Faber, T., Haas, R., Huber, C., Morthorst, P.E., Jensen, S.G., Coenraads, R., Voogt, M., Reece, G., Konstantinaviciute, I., Heyder, B. (2007): OPTRES: Assessment and optimisation of renewable energy support measures in the European electricity market; Final report of the research project OPTRES of the European Commission, DGTREN, Intelligent Energy for Europe Programme. Karlsruhe, Germany.

RECs (2007): Memo – Concept for an Internal Market RES-E. RECS International, 12 July 2007. Available at: *http://www.recs.org*

Del Río, P. (2005): “A European-wide harmonised tradable green certificate scheme for renewable electricity: Is it really so beneficial?” *Energy Policy*, 33 (10), 1239-1251.

Del Río, P. (2007): “The interaction between emissions trading and renewable electricity support schemes. an overview of the literature”. Mitigation and Adaptation Strategies for Global Change, 12(8), 1363-1390.

Del Río, P. and Gual, M.A. (2004): “The promotion of green electricity in Europe: Present and future”. *European Environment Journal*, 14, 219-234.

Del Río, P. and Gual, M.A. (2007): “An integrated assessment of the feed-in tariff system in Spain”. *Energy Policy*, 35, 994-1012.

Sawin, J. (2004): National Policy Instruments. International Conference for Renewable Energies, Bonn.

Uyterlinde et al. (2003): Renewable electricity market developments in the EU. Final Report of the ADMIRE-REBUS project. ECN-C—03-082. 2003, Petten (Holanda).

Wiser, R.; Porter, K. and Grace, R. 2005. Evaluating experience with renewable portfolio standards in the United States. Mitigation and Adaptation Strategies for Global Change, 10, 237-263.

In: Energy in Europe: Economics, Policy and Strategy
Editors: F. L. Magnusson and O. W. Bengtsson
ISBN: 978-1-60456-829-5

Chapter 4

SOCIAL BENEFITS AND PUBLIC POLICY TO SUPPORT GREEN ELECTRICITY IN SPAIN: AN APPLICATION TO BIOMASS

Mario Soliño*[1]*, Albino Prada and María Xosé Vázquez
Department of Applied Economics.
Faculty of Economics. University of Vigo, Lagoas Marcosende, 36310 Vigo, Spain

ABSTRACT

The social costs of electricity generation may be classified into private costs and external costs or externalities. Private costs of electricity production range from the initial investment of capital in the power plant through purchase and transport of the primary energy source, infrastructures, etc. These are costs with known market prices. Related to the externalities component, environmental costs include the effect on atmosphere pollution, climate change and the impacts on landscape and the environment, while non-environmental external costs relate to socio-economic effects such as price and supply instability, energy dependence, etc. All these impacts have no market price, resulting in market failure and inefficient allocation of resources. In this chapter we will show how the existing data on external costs could be used to define a public intervention to re-structure the energy pricing in Spain. We will begin, therefore, by analysing private costs in electricity generation processes based on both renewable (wind power, hydro, and biomass) and non-renewable energy sources. We will also describe how private costs of electricity generated from biomass (biomass-e) are higher than with other sources, mainly due to the high supply costs of the biomass. We will then analyze the price system of electricity generated from renewable energy sources (RES-e) in Spain and we will explain how the market price does not always cover the private costs of production. We will show how public intervention in the electricity market, through the design and implementation of price-premiums to RES-e production has encouraged the growth of wind power. This system has been highly ineffective in increasing the production of electricity from forest biomass, and in achieving regional and national objectives stated on energy plans. The chapter will finish with an estimation of the relative social benefits

---

[1] E-mail: mario@uvigo.es.

of biomass-e and concludes that the internalization of social benefits is a strong argument for a public policy to reinforce a higher penetration of this energy source in the Spanish electricity market.

Keywords: *renewable energy, wind power, biomass, electricity pricing.*

1. Introduction

The social costs of electricity generation can be divided into private and external costs (Figure 1). External costs may be environmental and non-environmental. The former include atmosphere pollution, climate change, effects on landscape, etc. Non-environmental costs include costs of a socio-economic nature such as price and supply instability, energy dependence, employment, rural development, etc. In both cases these are impacts that have no market price, and there is no compensation to affected population. These effects are known as (negative) externalities. As they are not considered by the price system (or internalised) they result in market failings and inefficient allocation of resources.

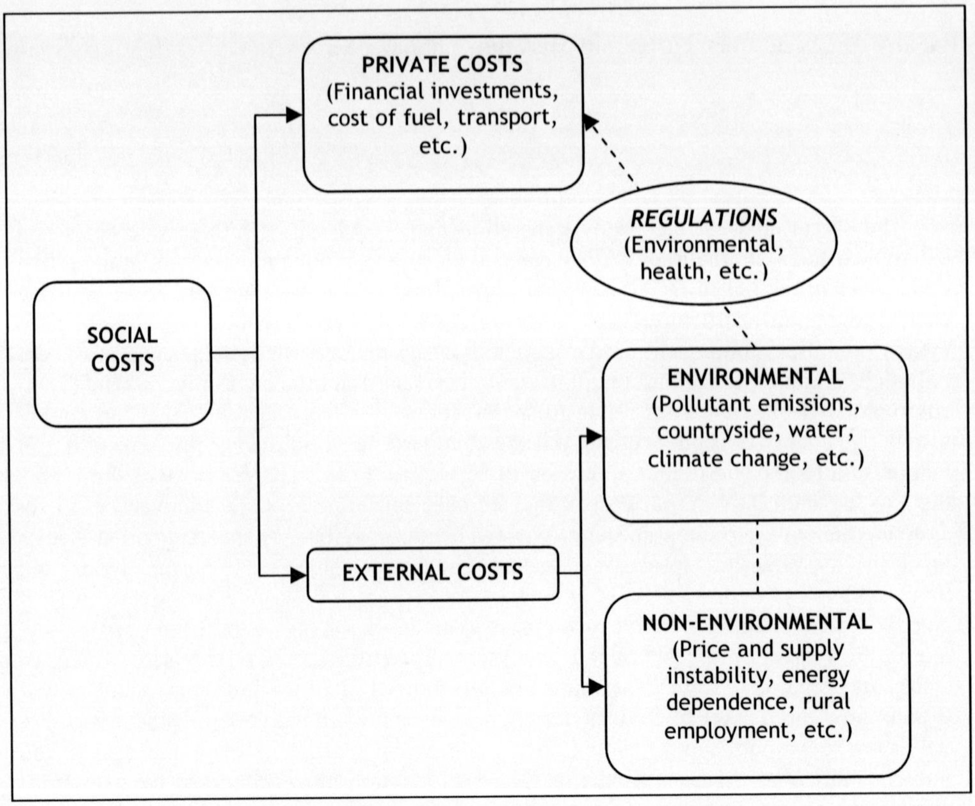

Source: own figure based on International Energy Agency (1995).

Figure 1. Types of energy-related social costs.

For their part, private costs of electricity generation range from the initial investment of capital in the power plant through to various inputs needed to the generation of energy. This is a process involving various stages such as the purchase of the primary energy source (oil-derived, gas, biomass, etc.), its transport and the construction of infrastructures. These are, therefore, costs with market price. In this chapter, we will show how the existing data on external costs could be used to re-structure the energy pricing in Spain.

We will begin, therefore, by analysing private costs in electricity generation processes based on both primary renewable energy sources (wind power, hydro, primary and secondary biomass) and non-renewable. We will also explain how the costs of electricity generated from biomass (biomass-e) are higher than with other technologies, both renewable and non-renewable, mainly due to the high fuel supply costs that producers have to face.

We will then analyse the payment system for electricity generated from renewable energy sources (RES-e) in Spain and we shall see how the sale price of green electricity does not always cover the private costs of producers. We will explain how public intervention in the electricity market, implemented through a system of incentives for producing RES-e paid by the collection of taxes linked to electricity consumption, has encouraged the growth of wind power. We will show, however, that this system has proved ineffective in increasing the amount of electricity generated from forest biomass, the stated objective of energy plans in Spain. Consequently, the last section of the chapter details the social benefits of this technology and shows that the internalisation of these avoided costs or social benefits encourages its penetration in the Spanish electricity market.

2. The Private Costs of Electricity Generation

Estimates of the private costs of electricity generation from a range of primary energy sources are available. In general terms, the SAFIRE database details the prices and costs of RES-e in Europe (European Commission, 2000b). In Spain, based on data collated from a number of studies (European Commission 2000a; Menéndez 1997, Linares 1997, Spanish Finance Ministry 2003), we may summarise the private costs of eight electricity generation technologies, three of them renewable (Figure 2), and their price intervals.

In general, the average private costs of renewable sources are higher than those of conventional (non-renewable) energy sources. The Spanish Finance Ministry provides information on average private costs and states the electricity schemes the various technologies are subject to (Spanish Finance Ministry, 2003). The private costs of technologies subject to the standard scheme, namely traditional generators and new combined cycles, are EUR 0.0374 (3.74 cents) per kWh. Of this total price, production costs account for 3.13 cents/kWh, capacity costs 0.48 cents/kWh, and complementary service costs 0.12 cents/kWh. The private costs of special scheme technologies (renewable and cogeneration technologies) amount to 6.22 cents/kWh, i.e., 66.57% higher than conventional technologies.

In the case of renewable, the private costs of biomass are greater than those of wind power. The minimum private cost of a kWh produced from biomass is 1.3 cents higher than a kWh of wind power. This gap perhaps explains the different implementation and use of these two energy sources, both said to play a greater role in the Spanish and European energy systems (European Commission, 1997; IDAE-MINER, 1999).

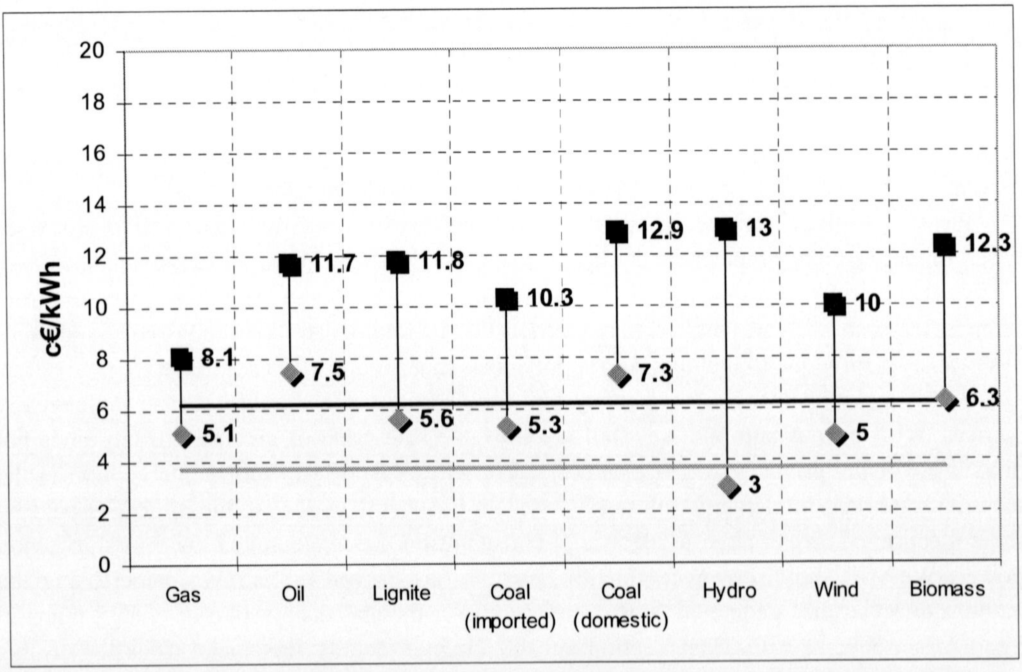

Source: own figure based on European Commission (2000a), Menéndez (1997), Linares (1997) and Spanish Finance Ministry (2003).

Figure 2. Private costs of electricity generation in Spain.

However, for biomass, it is also very important to specify and profile the scale of the private costs interval, which ranges from 6.3 cents/kWh to 12.3 cents/kWh. In addition, given that we are focusing on forest biomass and its use in electricity generation, the private costs associated with it need to be analyzed. The aforementioned minimum costs thus correspond to primary biomass or energy crops (6.3 cents/kWh). These costs increase when forest biomass is used, because it is a fuel whose cost is situated at the top of the price interval.

Given the extent to which estimates of private costs vary and the fact that biomass costs have not been sufficiently disaggregated, we have analyzed direct data sources that have enabled us to observe the components of the private cost. At the Allarluz[2] biomass plant (Spain) we were given useful information that helped us ascertain the private costs of generating electricity from biomass for each type of fuel used. We conducted several interviews to obtain information on the plant's current private costs: those associated with the use of waste from the wood processing industry as a main fuel, and those resulting from the use of waste (undergrowth and waste wood from logging) taken directly from scrubland as an alternative fuel source.

As Figure 3 shows, if the supply of fuel is based on waste taken directly from scrubland[3] the total private cost increases to 12.54 cents/kWh (scrubland scenario). Fuel costs account

[2] Allarluz is Spain's first operational biomass plant (it opened in 1994) and all the energy it generates goes to the electricity market. It has a production capacity of 2.35 MW and its main fuel source is waste from the wood processing industry.

[3] Increasing the amount of land devoted to biomass will raise the price of land and the private costs. Nevertheless, scrubland does not present this problem. The resource has gone unmanaged across the last decades and it has increased the risk and severity of forest fires.

for 68% of the costs, whereas the remaining 32% is accounted for, in equal proportions, by personnel, operating and maintenance costs[4]. On the other hand, if the fuel is made up of wood-processing waste, the total private cost falls to 7.37 cents/kWh, with fuel costs accounting for 45% of that total, and personnel, operating and maintenance costs accounting for just over 27% each.

With regard to fuel costs (the main component of private costs), we have taken the current costs incurred at Allarluz as the model for a supply based on wood-processing waste. As for a supply based on biomass taken directly from scrubland (scrubland scenario), after confirming that output at Allarluz is below efficient output levels[5] (2.34 kg/kWh as opposed to 0.49 kg/kWh), we estimate the average fuel cost in that scenario to be 8.52 cents/kWh. We noted that in a situation of technological efficiency[6] (technological efficiency scenario), fuel costs fell to 42% (2.96 cents/kWh) of the total private cost, equating to 6.98 cents/kWh.

As detailed in Figure 3 above, the private cost of production from biomass ranged between 6.3 and 12.3 cents/kWh. The private cost using industry forest waste falls within this interval therefore, whereas the exclusive use of undergrowth (scrubland scenario) leads to a slightly higher figure (12.54 cents/kWh)[7]. This reveals that as far as forest biomass is concerned, it is essential to consider the type of fuel used.

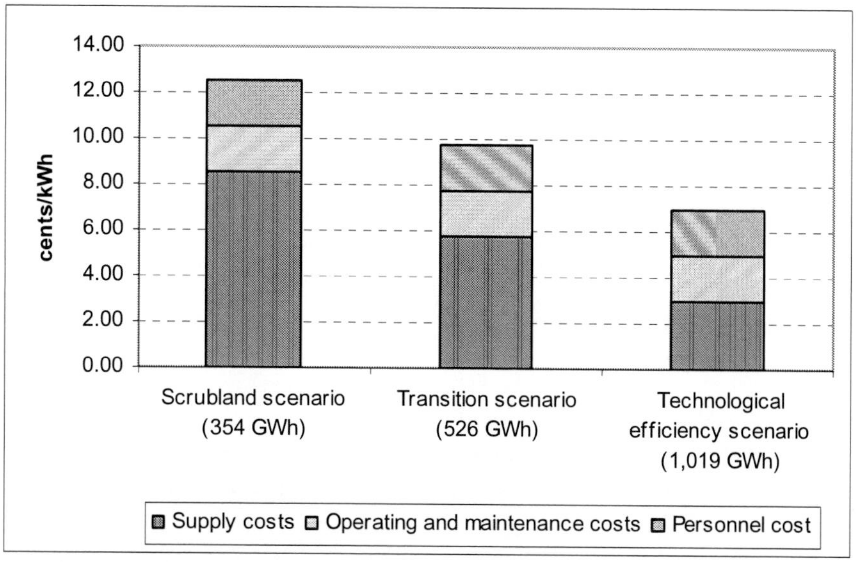

Source: own figure.

Figure 3. Private costs from forest biomass.

---

[4] Personnel, operating and maintenance costs make up the generating cost (4.02 cents/kWh), which also includes the depreciation of the initial investment (Spanish Ministry of Industry-IDAE, 2005). Diakoulaki et al. (2000) calculate a similar generating cost of 3.65 cents/kWh.

[5] Calculations based on Núñez-Regueira et al. (2004).

[6] Carrasco (2002) points out that one of the obstacles preventing biomass from breaking into the market concerns the lack of proven production and collection technology, in addition to the need to improve combustion boiler technology.

[7] In the technological efficiency scenario (assuming that technological efficiency is attained), the private cost falls within the cost interval detailed in Figure 2.

As we have seen, private costs vary considerably depending on whether it is industrial waste or biomass taken directly from scrubland that is used to generate energy. These costs are always higher than the average private costs of special-scheme technologies (wind, hydro, solar, etc.) and the highest band of biomass energies (above the primary level).

3. Energy Subsidies: The Price System and Incentives in the Spanish Electricity Market

At a European level, subsidies currently awarded to conventional energy sources are much higher than those designed to promote renewable sources or energy saving/efficiency measures (European Environment Agency, 2004). The price-induced incorporation of externalities in the social costs of energy also operates in the same way as a subsidy in that consumers do not directly pay for the environmental, social and security-related impacts of the energy options available. Although not considered subsidies, externalities represent unpaid costs that society as a whole has to assume and which, in the same way as explicit subsidies, have a distorting impact (Worldwatch Institute, 2004) and a negative effect on the environment.

Current energy production in the EU is affected, therefore, by a system of explicit subsidies in the form of grants[8], and other implicit ones in the form of the non-internalisation of external costs. Both types of subsidies reward the least environmentally friendly energy sources.

In such a context, the European Commission aims to achieve more ambitious energy taxation targets by internalising external costs and creating indirect measures contributing to the development of RES (European Commission, 2001b). In this scenario so heavily weighted in favour of conventional energy sources (CES), both explicitly in the form of grants, and implicitly due to the internalisation of external costs, the European Commission has suggested the following: "One way which could be explored is that profitable energies could finance the development of renewable energies, which unlike traditional energy sources, have not benefited from substantial support." (European Commission, 2000a).

Consequently, the European Commission recently drafted a series of energy-related directives on public support for environmental protection (European Commission, 2001a). In view of the technical difficulties (variations in unitary costs) preventing RES from competing effectively with their conventional counterparts, points 54 and 55 of these directives envisage the creation of specific provisions for promoting RES. The European Commission also proposes that this support be considered as based on environmental protection reasons. Such support would help bridge the existing gap between the private costs of power generation costs of RES and the market price at which the energy produced is sold (European Commission, 2001a: Directive 56) to the electrical grid.

In order to internalise the social costs it is necessary to base public aid on the external costs (or external costs avoided) of all alternative technologies. One possible channel would be the introduction of taxes on the consumption of conventional electricity as a means of

---

8 Grants for renewables are implemented in the form of increased payments on top of the sale price of green energy, securities and certificates, capital subsidies, tax exemptions, etc. (Stenzel et al., 2003; Jäger-Waldau, 2004; Reiche and Bechberger, 2004).

financing RES. Such action would provide the basis for an effective mechanism for internalising[9] external costs adapted to the "polluter pays principle"[10], which is also a crucial aspect in amending energy consumption guidelines (Márquez, 2002)[11].

This system is already operating in the Spanish electricity market, where the revenue generated by green electricity producers derives from the sum of the market price and an incentive set by the government every year. In addition to the sale price of electricity[12], consumers' electricity bills also include an energy-environmental surcharge (for diversification and security of supply) on all electricity consumption. This surcharge feeds a fund to support incentives for electricity production under special schemes.

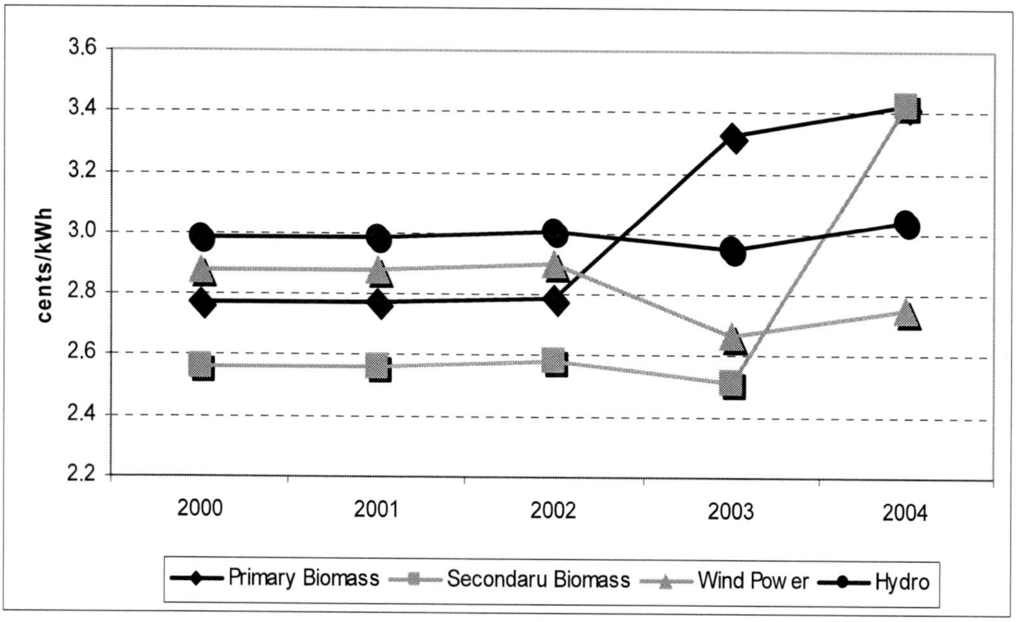

Source: own figure.

Figure 4. Incentives for producing electricity under special schemes.

---

9 The Rome Club Report of 1972 proposed that the price of fossil fuels should cover the damage they cause and that renewable energies receive grants for the benefits they lead to (Meadows et al., 1972).

10 The "Treaty establishing a Constitution for Europe" refers explicitly to energy (article III-256) and to the preservation and improvement of the environment (article III-233) and states the objective to "promote energy efficiency and energy saving and the development of new and renewable forms of energy" (article III-256, section 1.c.). The promotion and development of renewable energies can be implemented by mechanisms designed to internalise externalities based on the "polluter pays" principle, as set out in the Environmental Liability Directive (Official Journal of the European Union, 2004).

11 The internalisation of social costs or the application of the Polluter Pays Principle can also be achieved through tax incentives in the form of environmental taxes. Nevertheless, given the inflexibility of energy demand in relation to prices, a "price signal" needs to be sent out; in other words, a sufficiently high level of fiscal pressure has to be exerted (European Commission, 2000b).

12 The market sees the interaction of the producers' supply with the demand of eligible consumers, distributors and vendors, with the final kWh market price being identified as that corresponding to the last unit of production required to meet the demand for electrical energy. The calculation of the sale price of electricity generated under a special scheme is based on the market price, an incentive, a bonus for participating in the market and an additional reactive energy incentive payment. In order to simplify our analysis, we have assumed that the sale price is made up of the sale price in the energy pool in addition to the production technology incentive.

Table 1. Revenue and costs of electricity produced using RES (cents/kWh) (2004)

cents/kWh	Wind power	Secondary biomass	Primary biomass
Private cost	5 – 10	12.54	6.3 – 12.3
Minimum private costs	5	12.54	6.3
Electricity output bonus (2004)	2.75	3.42	3.42
Average sale price (1999-2003)	3.85	3.79	3.91
Current revenue	6.60	7.21	7.33
REVENUE – COSTS	1.60	-5.33	1.03

Source: own table.

Up until 2003 incentives for the production of electricity from secondary or forest biomass were below those paid out for other technologies such as wind power or primary biomass (Figure 4). Furthermore, the gap between incentives for using primary and secondary biomass has continued to grow in favour of the former. Laws of 2004[13] removed the existing distinction between primary and secondary biomass and placed both in the same renewable category. As a result of this legislation, producers now receive the same subsidies for using energy crops as their main fuel as they do for using forest biomass (3.42 cents/kWh). For the first time, the secondary biomass incentive is the highest of all those paid for power generated using renewable energies.

We have calculated above a private cost of 12.54 cents/kWh (Figure 3) for forest biomass, with the average purchase price for biomass-e produced in Spain between 1999 and 2003 standing at 3.79 cents/kWh[14]. Furthermore, as stated in the energy-related Community directives on state aid for environmental protection (European Commission, 2001a), there is justification for providing public subsidies for the generation of electricity from forest biomass in order to compensate for the shortfall between cost and revenue.

By incorporating the incentive defined in 2004, the revenue generated by biomass-e producers would amount to 7.21 cents/kWh (Table 1), a figure that would not cover private costs which, in the case of electricity generation using waste wood from logging and undergrowth, would stand at 12.54 cents/kWh. In this particular case an increase of the electricity production incentive would be needed as, with regard to the supply structure, there would still be a shortfall of 5.33 cents/kWh between revenue and cost.

4. Internalisation of Social Costs: The Economic Viability of Forest Biomass

Rather than taking into consideration externalities resulting from the replacing of CES-e with RES-e, the calculation of current incentives to renewable technologies has followed criteria based on the economic viability of the various renewable technologies. As we shall

---

13 Royal Decree 436/2004, dated 12 March, establishing the methodology for updating and systemising the legal and financial arrangements pertaining to the production of electrical power under special schemes (Official State Journal no. 75, dated 27 March 2004) and amendments published in Official State Journal 85, dated 8 April 2004.

14 This figure was calculated using data provided by the Spanish National Energy Commission (CNE, 2004).

see, modulating incentives in line with a criterion of internalising external costs and benefits (as proposed by the European Commission) would imply increasing incentives for wind power and secondary biomass.

We will look now at the social costs if fossil fuels were replaced by RES in electricity generation. In doing so we will estimate the benefits of replacing fossil fuels (oil and coal) with wind energy and biomass, both from primary (energy crops) and secondary (forest waste) to generate 1,019 GWh of electricity in Galicia[15].

The ExternE Project (European Commission, 1995; European Commission, 1999) analyses (bottom-up life cycle assessment[16]) the external costs resulting from electricity generation from eight different energy sources (coal, lignite, oil, gas, nuclear, biomass, hydro and wind). This cycle can generate over 200 impacts. These can be classified as follows: *i)* atmospheric pollution (effects on health, crops, materials, woodland, fishing grounds), *ii)* health and safety in the workplace, *iii)* noise, *iv)* global warming, *v)* effects on water (freshwater, seawater, underground sources, etc.) and *vi)* effects on the marine environment. However, the impact function does not consider a number of well-documented effects. This is due to a variety of reasons such as lack of information on probabilities, lack of data, uncertainty, etc.[17]

Figure 5 reveals that of all technologies oil and coal create the most negative impacts[18]. The differences in external costs greatly favour RES-e. The coal cycle generates external costs of 5.06 cents/KWh and the oil cycle 4 cents/KWh. However, external costs rise to 0.14 and 1.08 cents/KWh for wind and biomass energy respectively, resulting in a net difference in costs avoided in relation to the two main fossil fuels of 3.86 – 4.92 cents/KWh and 2.92 - 3.98 cents/KWh respectively.

In the proposed substitution scenario (1,019 GWh), the external costs avoided in the electricity generation life cycle would be, therefore, €39.33m – €50.13m in the case of wind power and €29.76m - €40.56m for secondary and primary biomass.

As we have already mentioned, the external cost estimates calculated in ExternE overlook a number of environmental impacts. Furthermore, and perhaps more importantly for our study, they do not take into consideration the socio-environmental benefits of replacing CES-e with RES-e. It is for this reason that, in addition to the external costs of the fuel-electricity life cycle, we have also considered other components in the social costs of electricity (external costs avoided and external benefits generated).

With regard to external costs avoided, Leiby et al. (1997) estimate that the social costs of importing oil (military/defence, spills, cartels, sharp market fluctuations, etc.) would vary between €2 and €5 a barrel.

---

[15] Galicia is the most Atlantic Region of Spain and it represents the more suitable area for biomass-e. The sustainable exploitation of the scrubland biomass is situated around of 502,542 dry tons / year which, applying a conversion factor of 0.17 toe / dry ton (Núñez-Regueira et al., 2004), would give us a potential of 1,019 GWh for generation of electricity.

[16] The analysis involves six different stages: exploration, construction of plants, extraction, power generation, residues storage and the various transport processes.

[17] Navrud (1994) argues that the transference of results, using financial estimates from other studies, leads to impacts being underestimated. Sundqvist (2002) confirms that the results provided by ExternE are an underestimation.

[18] Calculations are based on the estimated external costs of electricity generation (European Commission, 2003) in Denmark, Finland, France, Greece, Germany, the Netherlands, Portugal, Sweden and the UK.

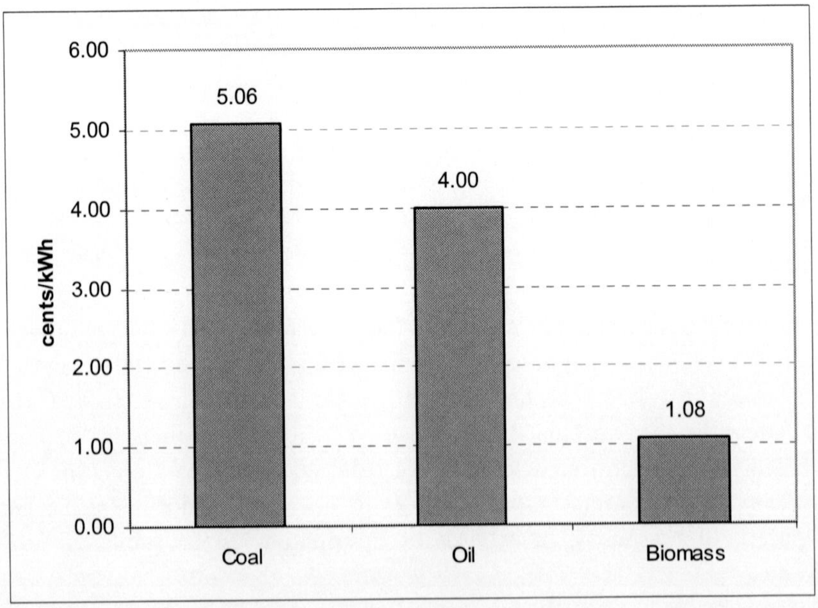

Source: own figure based on European Commission (2003).

Figure 5. External costs of CES-e versus RES-e.

Consequently, in terms of imports security only, over €3m could be saved in security costs by reducing imports of crude oil discharged in Galicia by 1.65%, with this €3m as only part of the indirect costs of fuel dependence.

These two components are factors common to the replacement of CES-e by RES-e in general. However, unlike other renewable technologies, certain specific external costs/benefits are associated with secondary biomass. Subsequently, in contrast to other renewable energies analysed herein, electricity generation from forest biomass would be beneficial in terms of controlling forest fires (Ministry of Industry-IDAE, 2005), as the ecological removal of the undergrowth needed to produce 1,019 GWh of electricity could reduce the risk of fires in Galicia by half. This could, in turn, help cut annual public expenditure on preventing and fighting fires by over €23m.

Another external benefit associated with electricity generation from forest biomass which, like the three aforementioned components, forms part of the total social costs, concerns the creation of rural employment (Spanish Ministry of Industry-IDAE, 2005). Biomass production provides jobs in specific geographical areas which keep communities from declining (as Galician rural areas). In the proposed scenario, 4,000 jobs a year could be created, 3,200 of them in rural areas. Taking into account the unemployment subsidies that Spanish Government assigns to these areas, the creation of these jobs could lead to annual savings of €26.4m in unemployment payments, not to mention the benefits these new jobs would have for social and territorial cohesion.

The net difference in benefits[19] (Table 2) as a result of the partial replacement (1,019 GWh) of fossil fuels (oil and coal respectively) with wind in the generation of electricity

[19] The external costs of electricity generation (European Commission, 1995; European Commission, 1999), the social costs of importing oil (Leiby et al., 1997), unemployment benefits avoided and costs saved by reducing the risk of fires were all taken into consideration in calculating benefits, all within the scenario of replacing

would, therefore, be 4.17–4.92 cents/kWh. This difference would provide grounds for a minimum increase in the current incentive of 1.42 cents/kWh (Table 3).

As regards secondary biomass, a greater difference was found to exist between the benefits and the incentive; the former standing at a minimum of 8.08 cents/kWh and the latter at 3.42 cents/kWh (Table 3). A minimum increase in the forest biomass incentive of at least 4.66 cents/kWh would be justified in these circumstances. The same is not true for primary biomass, however, as this is the only case in which the incentive would already fall within the estimated benefits range.

In the case of secondary biomass, however, it is clear that even raising bonuses to the level of our externalities estimate[20] (8.08 cents/kWh) would not generate sufficient revenue to cover the private costs of this technology, which would continue to be financially unviable (Figure 6). As the European Commission argues, more ambitious targets in relation to energy taxation must be set as a means of internalising all external costs (not just part of them) and creating in direct measures contributing to the development of RES (European Commission, 2001b).

Table 2. Net difference in costs avoided by the partial replacement of fossil fuels with wind power and primary and secondary biomass in Galicia

Benefits (millions of euros)	Wind	Secondary biomass	Primary biomass
Impacts on the electricity production life cycle	39.33 - 50.13	29.76 - 40.56	29.76 - 40.56
Forest fires	-	23	-
Social costs of importing oil	3.13	3.13	3.13
Unemployment benefit	-	26.44	-
Total benefits(millions of euros)	42.46 - 50.13	82.33 - 90	32.89 - 40.56
Total benefits (cents/kWh)	*4.17 – 4.92*	*8.08 – 8.83*	*3.23 – 3.98*

Source: own table.

Table 3. Difference between benefits and incentives for production involving the partial replacement of fossil fuels by renewables

cents/kWh	Wind	Secondary biomass	Primary biomass
Electricity production bonus (2004)	2.75	3.42	3.42
Benefits resulting from partial replacement of oil	4.17	8.08	3.23
Benefits resulting from partial replacement of coal	4.92	8.83	3.98
Minimum benefits – Bonus	1.42	4.66	-0.19

Source: own table.

fossil fuels (oil and coal) with RES to generate 1,019 GWh of electricity. See Soliño and Prada (2004) for a detailed explanation.

20 This figure can be categorised as a minimum estimate given that it does not consider the impacts on the industrial fabric, sustainable development in energy use, the advantages of supply diversification, and the progress made in preserving our forests, etc.

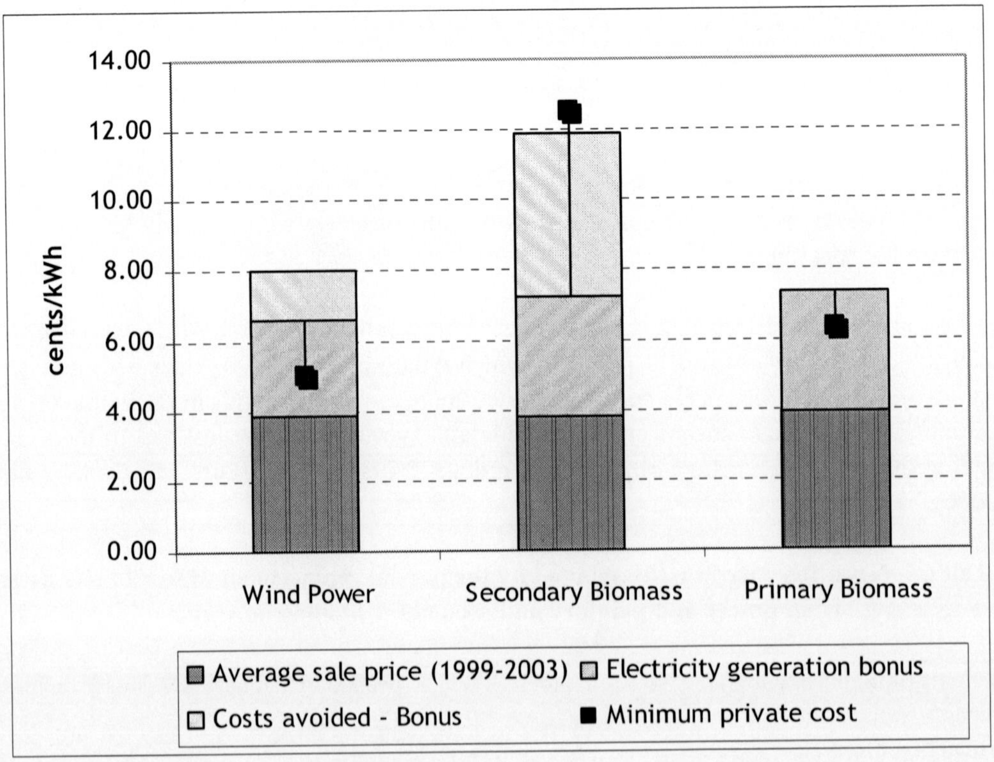

Source: own figure.

Figure 6. Financial viability of renewables.

Far from this being achieved, however, current revenue exceeds the minimum private costs of wind and primary biomass power although it is in no way sufficient to cover the high costs of forest biomass, which involve a negative private costs differential of 5.33 cents/kWh (Table 1 and Figure 6).

With regard to RES, the existing system of incentives for electricity generation under special schemes is the correct one in terms of covering the higher private costs of wind power. In fact, this system has led to wind farming becoming the most profitable of renewable technologies, with a spectacular growth (Figure 7) of electricity generation from wind technology in Spain, a growth that Galicia has also played an important part in.

The dynamic nature of the wind power sector is largely attributable to its ability to absorb most of the funds generated as a result of energy-environmental surcharges on electricity consumption. Thus, in analysing the distribution of subsidies (solely through production incentives) to technologies under special schemes, we have found that over 50% of funding currently goes to the wind sector, something that did not figure among the objectives set beforehand in the Renewable Energies Promotion Plan for Spain (IDAE-MINER, 1999).

In fact, it has been noted that other energy sources that figure prominently in Spanish energy strategy, such as forest biomass, receive very little funding as the incentives set aside for them are insufficient to encourage their use in electricity generation or, what is the same, to reduce the private costs to the same level of profitability as the price for kWh generated using fossil fuels offers.

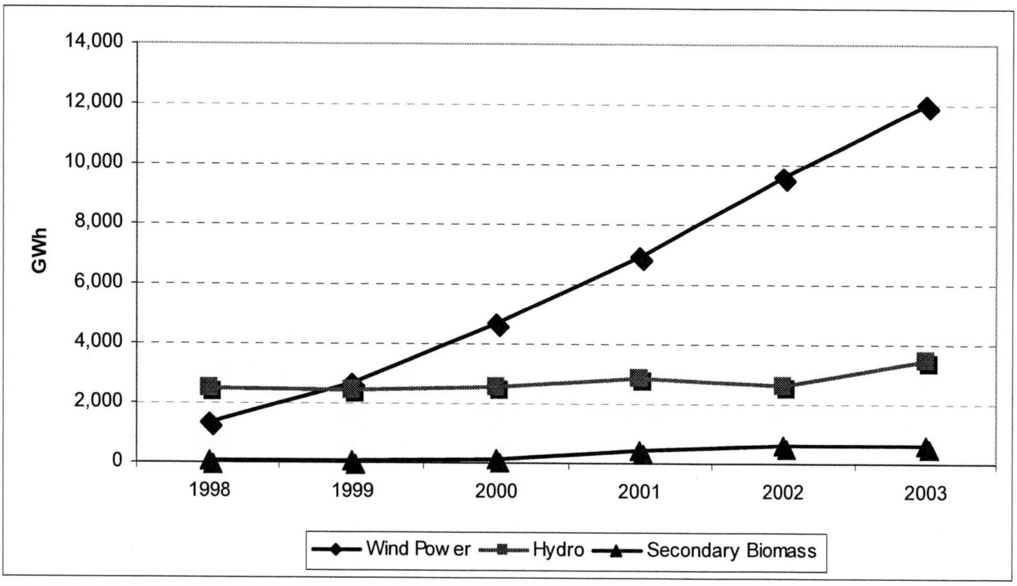

Source: own figure.

Figure 7. Growth of wind-power production in Spain.

With the current levels of RES-e output and public intervention, there is a clear weakness in the design of incentives for electricity production under special schemes, as they do not meet market competitiveness criteria or (and more appropriately in our opinion) criteria on the internalisation of external costs. We believe that planners ought to consider adapting incentives to the economic viability threshold (Figure 6), as they are currently weighed heavily in favour of wind power and primary biomass, whereas secondary biomass is under-funded.

Bearing in mind that 60% of the objectives of the Spanish Renewable Energies Promotion Plan are targeted at biomass, and that the technological development of biomass is beginning, it is necessary to re-define and implement the support policies. If this is not done, it is likely that these objectives will remain largely unfulfilled (IDAE, 2003). The European Commission thus states that *"the level of support is not sufficient to make them into credible alternatives for wide use"* (European Commission, 2000a). In addition, the Spanish Renewable Energies Promotion Plan states that the development of biomass for generating electricity is not financially viable and that public support is necessary for investing in, exploiting (incentives) and producing fuels in order to make the price competitive (IDAE-MINER, 1999).

Broadly speaking, the higher private cost of generating electricity from RES can be attributed to investment and not to the price of the fuel[21] (Mirasgedis et al., 2000), which is usually the most important factor in the private cost of CES. However, we have seen how forest biomass is an exception among RES as the related fuel costs have an important impact on the total private cost, being much higher than for CES and constituting over half the total private cost. This is a key factor that makes private costs of biomass-e twice as high, on average, as the costs of other electricity-producing technologies (Diakoulaki et al., 2000).

[21] Direct fuel costs for RES such as wind, solar or hydro are zero.

Furthermore, these operating and supply costs help prevent significant environmental damages and generate a number of benefits for society. Therefore, forest biomass should be treated in a specific way among RES. Related fuel costs and supply logistics represent an obstacle to its entering the electricity market, an obstacle which, critically, applies solely to this renewable technology. Therefore, electricity prices should reflect total social costs. In the specific case of biomass, they should also take into account the type of fuel used[22] (forest biomass, agricultural biomass, energy crops), as fuel costs can account for over half the production costs in this case.

The aim should be to reward external costs avoided and the benefits resulting from the substitution of "dirty electricity" by biomass-e so that, at the very least, higher operating costs can be covered, thereby making the collection of biomass from forests (areas that are currently prone to a catastrophic cycle of fires) viable. In order to make this a reality, indirect support mechanisms complementing electricity production incentives must be designed, the most suitable of these being the creation of fuel supply grants, which can be adapted to our external costs estimate (by a minimum of 8.08 cents/kWh).

The price of biomass-e could thus be calculated by adding to the total market price a bonus from the incentives system and a public subsidy based on the internalisation of externalities (Table 4).

In this new scenario, producers' revenue could increase to 15.29 cents/kWh, i.e., more than double their current revenue (Table 1). In this way, even if externalities were internalised for a figure below our estimate (8.08 cents/kWh) in order to compensate for the current losses detailed in Table 1 (5.33 cents/kWh), a situation of economic viability could be brought about (Figure 8) as well as, therefore, the utilisation of scrubland to generate electricity.

Consequently, the subsidy resulting from the internalisation of externalities (5.33 cents/kWh) plus the market price (3.79 cents/kWh) would enable the removal of the existing incentive. This incentive would only be maintained to bring about the transition to a situation of technological efficiency (Figure 8). Even then, the capacity of the programme could be extended beyond 1,019 GWh.

5. Conclusion

We have explained how the private costs of generating electricity from renewable energy sources[23] are higher than for conventional ones –the dominant energy source that has such a negative impact on society and the environment. The dominance of CES can be partly explained by the fact that, up until now, they have benefited more from both explicit subsidies in the form of grants and implicit ones through the non-internalisation of the external costs they generate. Such support mechanisms distort the electricity system and hamper the sustainable use of RES. Electricity generation from certain RES remains economically unviable as the market price of the electricity generated by the producers is too low to cover their private costs. A financial support system has been set up in Spain to encourage the use of renewable sources and involves, in particular, the levying of a surcharge on the price of

[22] The price of the resource for uses other than energy generation should also be considered.

[23] Hydro is the exception.

electricity. The revenue generated feeds a fund to finance incentives for electricity generation under special schemes.

Table 4. Revenue accrued by generating electricity from forest biomass

	CRITERIA	COMMENTS	REVENUE
Market sale price	Market	Clash of supply and demand in the energy pool. Average sale price in 1999-2003 was 3.79 cents/kWh.	3.79 cents/kWh
Electricity production incentive	State intervention: support to help renewables penetrate the market	Incentives are set in accordance with financial viability criteria, and are weighed in favour of less competitive technologies and, therefore, their penetration into the electricity market. In line with Community directives on state aid for environmental protection(European Commission, 2001a), the maximum incentive is 5 cents/kWh	Current: 3.42 cents/kWh
Fuel subsidy	State intervention: internalisation of social costs	Fuel subsidies would be based on the criteria of internalising the external costs and benefits of biomass-e. Subsidies for kWh generated from biomass should, therefore, equate to a minimum amount set in line with the estimates currently available.	Minimum: 8.08 cents/kWh
TOTAL 15.29 cents/kWh			

Source: own table

Although this incentive system has proved adequate for certain technologies such as primary biomass (energy crops) and wind power, it is clearly inadequate for secondary biomass (undergrowth and waste wood from logging). The total private costs of generating electricity from waste taken directly from scrubland currently stand at 12.54 cents/kWh, more than double the revenue (sale price plus electricity production incentive) producers receive for this energy source. These incentives, therefore, fail to cover the high private costs incurred by biomass-e producers (resulting, in the main, from fuel supply costs). As a result, proper support enabling this technology to break into the market is lacking (one of the stated priority objectives in Spanish and EU energy strategies).

Biomass is a notoriously specific case. Unlike other renewable sources, the generation of electricity from biomass gives rise to externalities not only in its generation but also in the fuel supply processes. Whereas fuel costs for most RES are low, in the case of forest biomass they account for up to 68% of the total private cost. This fact is particularly important in that the removal of waste and undergrowth from woodland generates a number of significant social benefits that are also deserving of internalisation.

We have put forward a reasoned proposal for redesigning state aid for the generation of electricity from forest biomass. This proposal is based on two lines of support that would permit the use of this RES as the main fuel in electricity generation processes:

- The first, implemented through the current *production incentive,* would help forest biomass penetrate the electricity market.
- The second, through *fuel subsidies* internalising the social costs (associated benefits and costs avoided) incurred as a result of clearing and then using this renewable energy source to generate electricity.

Reform of the programmes supporting the environmental and energetic resource analysed in this chapter (unmanaged scrubland that has increased the probability and severity of fires), as incentives compensating for all related social impacts, would eliminate the market distortions that cause its inefficient allocation and prevent the development of this type of RES. The internalisation of all costs (negative, positive and costs avoided) on a gradual basis will provide an opportunity for modifying this energy allocation in line with all the environmental, financial and social assets in play, thereby giving forest biomass and RES in general a more prominent role in relation to fossil fuels, a key stated objective in both Spain's and the European Union's future energy strategies.

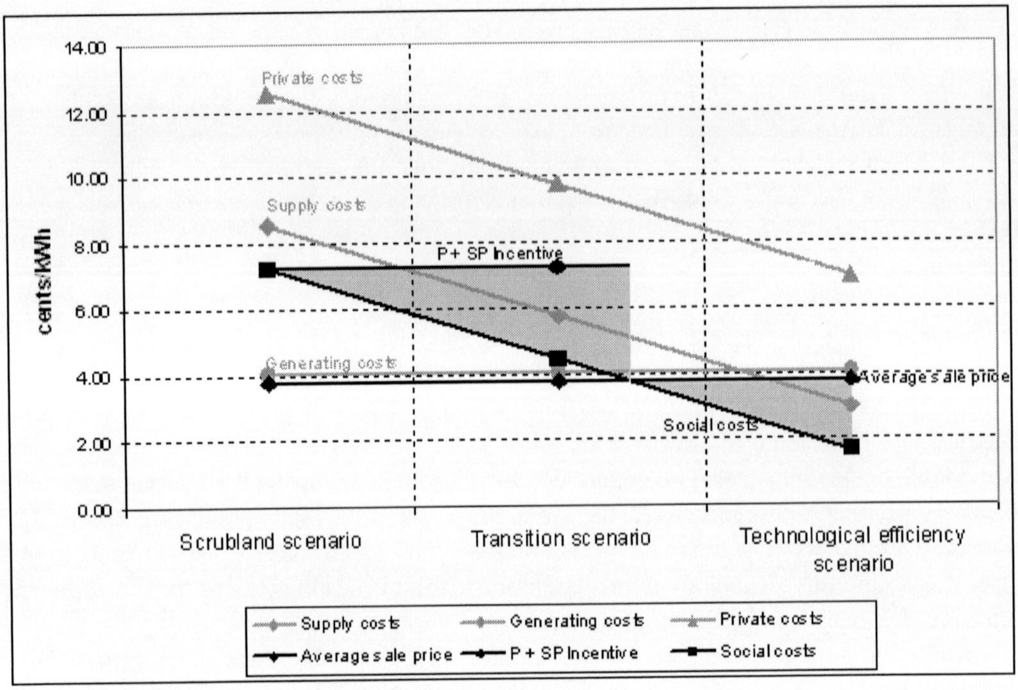

Source: own figure.

Figure 8. Forest biomass: social costs, price, incentives and subsidies.

ACKNOWLEDGEMENTS

This paper forms a part of the project "Economic and environmental evaluation of the energy valorisation of biomass in the forests" financed by the Spanish Ministry of Science and Technology (AGL2002-04753) and the Caixa-Galicia Foundation (CO-104-2002).

References

Camps, M. and Marcos, F. (2001). *Los Biocombustibles*. Madrid, Spain: Mundi-Prensa.

Carrasco, J. E. (2002). Principales Barreras y Retos para el Desarrollo Comercial de la Biomasa. *CIEMAT Curso Situación actual y futuro de la biomasa como recurso energético*, Madrid, 4-8 November.

CNE (2004). *Régimen Especial en España: Información Estadística sobre las Ventas de Energía del Régimen Especial*. Madrid, Spain: Comisión Nacional de Energía.

Diakoulaki, D., Mirasgedis, S. and Tziantzi, M. (2000). Environmental Externalities and the Development of Renewable Energy Sources. *EAERE Annual Conference*, Rethymnon (Greece), 30 June-2 July.

European Commission (1997). Energy for the Future: Renewable Sources of Energy, White Paper for a Community Strategy and Action Plan. COM(97) 599 final.

European Commission (2000a). Towards a European strategy for the security of energy supply. COM (2000) 769 final.

European Commission (2000b). Proposal for a European Parliament and Council Directive on the promotion of electricity from renewable energy sources in the internal electricity market. COM (2000) 884 final.

European Commission (2001a). Community guidelines on state aid for environmental protection. *Official Journal of the European Communities* C 37, 3-15.

European Commission (2001b). A Sustainable Europe for a Better World: A European Union Strategy for Sustainable Development. COM (2001) 264 final.

Diakoulaki, D., Mirasgedis, S. and Tziantzi, M. (2000). Environmental Externalities and the Development of Renewable Energy Sources. *EAERE Annual Conference*, Rethymnon, 30 June-2 July.

Official Journal of the European Union (2004). Directive 2004/35/EC of the European Parliament and of the Council of 21 April 2004 on environmental liability with regard to the prevention and remedying of environmental damage. L 143/56.

European Commission (1995). *ExternE, Externalities of Energy Vol. 1-6*. Luxembourg: Office for Official Publications of the European Communities.

European Commission (1999). *ExternE: Externalities of Energy Vol. 7-10*. Luxembourg: Office for Official Publications of the European Communities.

European Commission (2003). *External Cost: Research Results on Socio-Environmental Damages Due to Electricity and Transport*. Luxembourg: Office for Official Publications of the European Communities.

European Environment Agency (2004). *Energy Subsidies in the European Union: A Brief Overview*. Luxembourg: Office for Official Publications of the European Communities.

IDAE (2003). *Eficiencia Energética y Energías Renovables. Boletín nº 5*. Madrid, Spain: Instituto para la Diversificación y Ahorro de la Energía.

IDAE-MINER (1999). *Plan de Fomento de las Energías Renovables en España (2000-2010)*. Madrid, Spain: Instituto para la Diversificación y Ahorro de la Energía.

International Energy Agency (1995). *Global Warming Damage and the Benefits of Mitigation*. IEA Greenhouse Gas RandD Programme, Cheltenham.

Jäger-Waldau (Ed.) (2004). *Energy End-Use Efficiency and Electricity from Biomass, Wind and Photovoltaics in the European Union*. Luxembourg: Office for Official Publications of the European Communities.

Leiby, P. N., Jones, D. W., Curlee, T. R. and Lee, R. (1997). *Oil Imports: An Assessment of Benefits and Costs*. Tennessee, USA: Oak Ridge National Laboratory.

Linares, P. (1997). Una Aplicación de la Programación Multiobjetivo a la Planificación Eléctrica. *Energía,* 03/04, 71-75.

Márquez, R. (2002). La Fiscalidad de la Energía en la Unión Europea: El Estado de la Cuestión. *Fundación Biodiversidad-Papeles de Trabajo sobre Medio Ambiente y Economía,* 2.

Meadows, D. H., Meadows, D. L., Randers, J. and Behrens III, W. W. (1972). New York, USA: *The Limits to Growth*, The American Library.

Menéndez, E. (1997). *Las Energías Renovables*. Madrid, Spain: Los libros de la Catarata.

Mirasgedis, S., Diakoulaki, D., Papagiannakis, L. and Zervos, A. (2000). Impact of Social Costing on the Competitiveness of Renewable Energies: The Case of Crete. *Energy Policy,* 28, 65-73.

Navrud, S. (1994). Economic Valuation of External Costs of Fuel Cycles. Testing the Benefit Transfer Approach. In: A. de Almeida, A. Rosenfeld, J. Roturier, and J. Norgard, *Integrated Electricity Resource Planning* (50-66). The Netherlands: NATO ASI Series.

Núñez-Regueira, L., Proupín-Castiñerias, J. and Rodríguez-Añón, J. A. (2004). Energy Evaluation of Forest Residues Originated from Shrub Species in Galicia. *Bioresource Technology*, 91, 215-22.

Reiche, D. and Bechberger, M. (2004). Policy Differences in the Promotion of Renewable Energies in the EU Member States. *Energy Policy*, 32, 843-849.

Soliño, M. and Prada, A. (2004). Environmental Externalities of Biomass Power Plants in an Atlantic European Region. In: W. P. M. Van Swaaij, T. Fjällström, P. Helm and A. Grassi (Eds.), *Second World Biomass Conference: Biomass for Energy, Industry and Climate Protection* (2370-2373). Florence, Italy: ETA-Florence/WIP-Munich.

Spanish Finance Ministry (2003). *La Tarifa Eléctrica para el Año 2004*. Madrid, Spain: Ministerio de Economía.

Spanish Ministry of Industry-IDAE (2005). *Plan de Energías Renovables en España 2005-2010*. Madrid, Spain: Instituto para la Diversificación y Ahorro de la Energía.

Stenzel, T., Foxon, T. and Gross, R. (2003). *Review of Renewable Energy Development in Europe and the US*. London, UK: Imperial College London. Centre for Energy Policy and Technology.

Sundqvist, T. (2002). *Power Generation Choice in the Presence of Environmental Externalities*. Doctoral Thesis. Department of Business Administration and Social Sciences – Division of Economics, Luleå University of Technology.

WorldWatch Institute (2004). *La Situación del Mundo*. Barcelona, Spain: Fuhem-Icaria.

In: Energy in Europe: Economics, Policy and Strategy ISBN: 978-1-60456-829-5
Editors: F. L. Magnusson and O. W. Bengtsson

Chapter 5

REGIONAL PRICE DIFFERENCES IN THE SPANISH PETROL MARKET. WHAT DO THEY REFLECT?

Alejandro Bello Pintado and Ignacio Contín-Pilart[1]
Departamento de Gestión de Empresas, Universidad Pública de Navarra

ABSTRACT

This chapter deals with competition in the Spanish petrol retail market. We analyze whether regional price differences have been recently observed in the pricing of retail petrol. Retail petrol prices were regulated by a ceiling price system from June 1990 to October 1998. Such a price regulation imposed some degree of regional price uniformity nationwide. With the abolition of the ceiling price regulation, one would not expect uniform prices since there are no longer legal impediments to the existence of regional petrol price differences. Our empirical findings show a more competitive scenario than the early years after the abolition of the ceiling price regulation, when political interference in price setting was documented. On the one hand, pre-tax petrol prices and retail margins in Spain have converged recently with European levels. On the other hand, retail petrol price differences exist among the Spanish regions. These differences are due both to different regional taxation regimes, and to the regional demand conditions which the service stations operate under.

1. INTRODUCTION

This chapter deals with competition in the Spanish retail petrol market. Specifically, it analyzes whether regional price differences have recently been observed in the pricing of retail petrol. These retail petrol prices were regulated by a ceiling price system from June 1990 to October 1998, to avoid large price differences between the rich coastal and the poor inland regions (Contín et al., 1999). In fact, the price formula was too inadequate to cover the higher transport cost of supplies in the central part of Spain (Ministerio de Industria, 1995).

[1] The authors gratefully acknowledge the financial support from the Ministerio de Educación y Ciencia (SEJ-67895-C04-01/ECON; SEC2007-67737-C03-02/ECON).

This price regional uniformity implied a cross-subsidization between the coast and the inland areas.

With the abolition of the ceiling price regulation, there are now no legal impediments to the existence of regional petrol price differences. Therefore, uniform prices would not be expected, rather prices that vary over distances in response to local conditions of demand and the degree of local competition; in other words, a non-cooperative equilibrium (Eckert and West, 2004). Non regional petrol price differences would suggest a tendency for parallel pricing behaviour among the major oil operators, given their dominant position in the industry.

Retail petrol prices were regulated within the context of the restructuring and liberalization of the Spanish oil industry, which took place during the 1980s and 1990s, and culminated with the abolition of the state oil monopoly in 1993 (Contín, 1999; Correljé, 1994). Between 1927 and 1992, consumers of oil products in Spain had been supplied by a state-monopoly, operated by the private concessionary CAMPSA. The gradual dismantling of the Spanish oil monopoly, in which the industry's refining, distribution and retail trade segments were opened up step –by step to new domestic and foreign operators, was agreed to by the refiners, the Spanish government and the European Commission. It aimed to address the adverse consequences of the prevailing system of regulation, such as excess capacity, inadequate product yield and the failure to recover costs. Furthermore, it set out to "prepare" the Spanish-based refiners to operate in a competitive market, and to create a "national champion" in the oil sector.

Table 1. Evolution of the Number of Service Stations in Spain by Brand.1995-2006

OPERATOR	1995		2001		2006*	
	Number of Outlets	Market Share	Number of Outlets	Market Share	Number of Outlets	Market Share
Repsol-Ypf	3,500	55.32	3,704	47.50	3,556	42.94
Cepsa	1,500	23.71	1,437	18.43	1,523	18.39
Bp	400	6.32	582	7.46	592	7.15
Shell	120	1.90	278	3.57	-	
Total	127	2.01	187	2.40	-	
Galp	120	1.90	187	2.40	209	2.52
Meroil	70	1.11	184	2.36	200	2.41
Disa	100	1.58	140	1.80	470	5.67
C.Oil	25	0.40	0	0.00	-	
Agip	75	1.19	127	1.63	321	3.88
Petrocat	50	0.79	69	0.88	65	0.78
Avanti	45	0.71	61	0.78	-	
Esso	25	0.40	67	0.86	89	1.07
Esergui	30	0.47	65	0.83	100	1.21
Tamoil	20	0.32	43	0.55	43	0.52
Kuwait P.	20	0.32	17	0.22	40	0.48
Texaco	15	0.24	50	0.64	-	
Independent	85	1.34	600	7.69	1082	13.06
TOTAL	6,327	100	7,798	100	8,282	100
C1		0.55		0.47		0.43
C3		0.86		0.73		0.61
HHI		3,653		2,859		2,254

Source: Enciclopedia Nacional del Petróleo, Petroquímica y Gas, 1995-2006.
*Number of service stations in June 2006.

In addition, the restructuring of the Spanish oil industry has allowed the Spanish refiners to create a very efficient distribution system for oil products, a pipeline system that has rendered all alternative means of petrol transport uncompetitive. Furthermore, the system, which is owned and operated by CLH, has a near-monopoly position in the post-refinery storage capacity of oil products (Contín, 2004).

As a result of such a liberalization and restructuring process, a highly concentrated oligopoly emerged, in the Spanish petrol market. By 1993, the "national champion", Repsol-Ypf, and the other two refiners, Cepsa, and BP, controlled together more than 80% of the Spanish service stations. Since then, the new domestic and foreign operators that have entered the market have eroded the refiners' market share, which may have enhanced competition.

In June 1990, when the three Spanish-based refiners virtually controlled the Spanish automotive fuel market, a ceiling price regulation system replaced the system of state administered prices. As stated, the price ceilings remained in force until October 1998 for unleaded Euro 95. However, since the abolition of the price regulation there have been frequent allegations of political interference in price setting, in particular during the period mid-1999 to 2003. As a result, the regional price differences that emerged following the abolition of the price regulation have almost disappeared (Contín et al., 2006). Since 2004, however, Spanish and European pre-tax retail petrol prices seem to have moved in parallel. In addition, a common pattern of competition in all the Spanish regions has been recently identified. Large brand name petrol companies compete with independent unbranded service stations and hypermarkets whose strategy is to set relatively low prices (Bello and Cavero, 2007; 2008).

The chapter sheds new light on competition in the Spanish petrol market, after the abolition of the ceiling price regulation, by analyzing regional price level differences. In doing so, we use a unique random sample of 565 service stations that accurately represents all stations both at national and regional level. We have collected the Euro 95 petrol prices for each sampled station over the period January 1st to October 30th 2007. In addition, we use several variables that allow us to know the specific local conditions under which each service station operates, as well as the supply and demand conditions at a regional level.

The chapter is structured as follows. Section 2 describes the market structure of the downstream Spanish oil market. Section 3 examines the evolution of retail petrol prices in Spain, as compared to Europe, since the abolition of the ceiling price regulation. Section 4 analyzes petrol price differences across the Spanish regions. Section 5 concludes the chapter.

2. The Spanish Retail Petrol Market

The Spanish oil industry has experienced important changes in the last two decades. In only a few years the industry has shifted from being a state monopoly to competing in a free market. At the retail level, after dismantling the state oil monopoly, the opening and liberalization process was directed at leaving the national firms with a good platform to compete with newcomers[2]. Undoubtedly, these actions noticeably affected the competitive structure of the industry and thus the pricing setting process[3].

However, several changes, which have taken place in the last few years at a retail level, have furthered a new competitive environment within the industry. As such, the number of service stations has risen considerably, from 4,800 in 1992 to 8,600 in 2006. The elimination of administrative and strategic entry barriers has promoted a gradual entry of newcomers, among others Esso, Galp, Avanti and Tamoil. Furthermore, since the end of the 1990s the number of independent unbranded service stations has increased considerably, and in terms of numbers is currently the third most important group of retail outlets in Spain. As a result, the retail market concentration has gradually reduced (Table 1). In 1995, the three main operators controlled 86% of the market share, but in 2006 the C3 index indicated that these operators controlled only 61% of the final market[4]. Nevertheless, concentration in the retail market is still high.

Table 2. Concentration indexes and distances between service stations by region, in 2007

Spanish Regions	C2 Concentration Index (%)	HHI (10,000)	Average Distance Between Service Station (Km)
Andalucia	53.26	1,584.47	2.19
Aragón	68.65	2,755.95	5.63
Asturias	52.91	1,922.59	2.06
Cantabria	44.22	1,375.82	1.59
Castilla-La Mancha	62.33	2,326.83	4.77
Castilla y León	72.26	3,157.04	5.44
Cataluña	62.28	2,515.66	2.54
Comunidad Valenciana	68.18	2,968.38	2.32
Extremadura	60.30	2,261.34	3.20
Galicia	69.26	2,981.19	3.75
Islas Baleares	82.54	3,798.05	1.80
La Rioja	81.15	4,398.24	1.50
Madrid	65.73	2,776.92	1.47
Murcia	52.20	1,772.15	2.11
Navarra	50.70	1,725.59	2.45
País Vasco	58.41	2,464.12	1.69
Average	62.77	2,426.35	2.78

Source: Enciclopedia Nacional del Petróleo, Petroquímica y Gas, 2007; Catalist, 2007

2 On the one hand, the Spanish government favoured the national refinery industry and basically positioned Repsol-Ypf as the industry leader. On the other hand, the government implemented various measures to foster competition (Bello and Cavero, 2008).

3 Repsol-Ypf currently has 4 refineries with an approximate nominal capacity of 37 MT/year, which is 57.5% of Spain's refining capacity; Cepsa has 3 refineries or 33.3%, and BP España has one refinery or 9.2%. Furthermore, the Spanish-based refiners control 45% of CLH's capital shares; CLH being the Spanish oil industry's "essential facility" (Contín et al., 2001).

4 The three firm concentration index is the sum of the three highest market shares. The Herfindahl index is the sum of the squares of the market shares of all operators. It can take values between 0 and 10,000. The average Herfindhal Index is above the 3,000 points mark. This is higher than the threshold, established by the United States competition authorities, for categorizing the market as very concentrated; a similar threshold has been established by the European Union.

Table 2 describes the concentration in the retail market, and the average distance in kilometres between each service station and the next, for 2007. Although the market concentration varies among regions, the two main operators, Repsol-Ypf and Cepsa have a joint market share of over 50% in practically all regions, which renders competition difficult. More recently, the increase in hypermarket and independent unbranded service station numbers may have enhanced competition (Oilgas, 2007). This could be particularly important in those regions with the lowest concentration indexes, Navarra, the País Vasco and Cantabria. The average distance between service stations varies among the regions; Madrid, the País Vasco and Cantabria have, on average, the shortest distances between service stations.

3. Evolution of Retail Petrol Prices in Spain and in Europe

Petrol taxation has historically been lower in Spain than in the main European countries. As a result, the average Spanish retail price for petrol has been lower than the average European retail price for petrol since the ceiling price regulation was abolished (Figure 1). Between October 1998 and December 2002 Spain as a whole had the same taxation policy for petrol. However, since January 2003 the Spanish regions have been able to add excise taxes of up to 2.4 cents (Euro cents) per litre to the national excise duty, in order to finance their public health systems. Consequently, different levels of taxation have emerged among the Spanish regions; in particular, the Asturias, Castilla-La Mancha, Cataluña, the Comunidad Valenciana, Galicia and Madrid are the regions that are currently charging the regional excise tax.

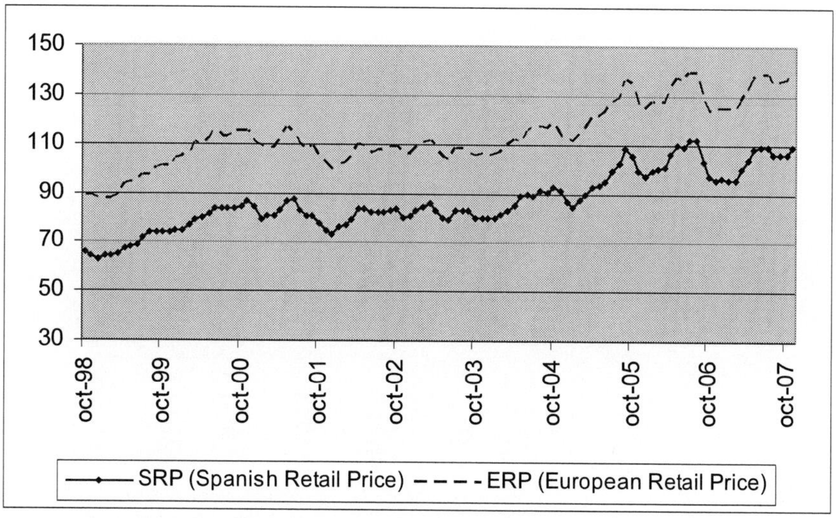

Note: ERP is the average retail price in six European countries: The Netherlands, United Kingdom, Germany, Belgium, Italy and France.

Source: Ministerio de Industria.

Figure 1. Average Monthly Petrol Prices After Taxes (Cents per Litre).

During the period of ceiling prices, January 1993 to September 1998, the Spanish government directly influenced the price dynamics of the Spanish petrol market by using this regulation. Ceiling prices were set weekly, according to a mathematical formula that took spot and pre-tax retail prices in other European countries into consideration.

Between January 1993 and September 1996 the ceiling price was binding, as the prices were in fact at about the same level. As a result, the average Spanish pre-tax price for petrol (SPRP) was higher than the average pre-tax retail petrol price of the six European countries (EPRP). This EPRP was considered by the Spanish government as the "competitive benchmark" for the Spanish petrol market (Contín et. al., 1999).

Since September 1996 Repsol-Ypf and subsequently Cepsa and the other operators have responded to the Spanish government's pressure to lower Spanish pre-tax petrol prices by setting retail prices, on average, 1.2 cents below the prevailing maximum prices (El Pais, 15/09/1996 and 17/09/1996). Yet, it was only in the service stations close to supermarkets and to other newly opened sites, sometimes in association with promotional campaigns, where the refiners reduced their prices below the maximum price. As result, the Spanish average pre-tax retail price for unleaded petrol (SPRP) converged towards the European average pre-tax price used in the ceiling price formula (EPRP). This induced the government to remove the ceiling price regulation in October 1998 (Contín et al., 1999) (Figure 2).

Between October 1998 and August 1999, the SPRT and the EPRP moved closely in parallel (Figure 2). By January 1999 the spot price of petrol began to rise. In September 1999 the government called upon the oil companies to cooperate in controlling the inflation rate. Between 1996 and 1998 the government successfully controlled the Retail Price Index (RPI). The stated objective for 1999 was an increase of 1.8%. However, in the first quarter of that year already half of the annual objective had been reached, and this gave rise to serious political concern (Banco de España, 1999).

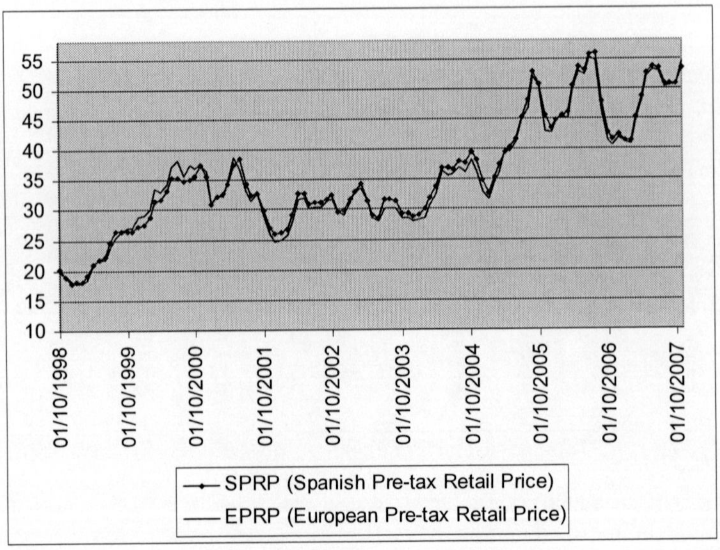

Note: EPRT is the average pre-tax retail price in six European countries: The Netherlands, United Kingdom, Germany, Belgium, Italy and France.

Source: Ministerio de Industria.

Figure 2. Monthly average petrol prices before taxes (Cents per litre).

The Spanish based refiners agreed to pass on only part of the spot price increase on to posted petrol prices, and other less powerful operators had to follow suit. The promised compensation would result from allowing a slower rate of decrease of petrol retail prices, when the crude oil and petrol spot prices went down again (Contín and Huerta, 2000).

In the second part of 1999 and during all 2000, a period of increasing oil and petrol spot prices, the average pre-tax petrol prices increased less in Spain (SPRP) than in Europe (EPRP); whereas between 2001 and 2003, a period of falling oil and petrol spot prices, the Spanish petrol retail prices decreased at a slower rate than that in Europe (Figure 2). As a result, the Spanish retail margin, per litre of petrol sold, was below that in Europe in 2000; whereas in the years 2001, 2002 and 2003 it was higher (Figure 3).

Additionally, in 2000 the European Commission expressed it's "surprise" at the lack of variation in the petrol prices among the Spanish regions and at the different service stations (El Pais, 25/05/2000). In 1999, the price difference for Euro 95 in the most and the least expensive Spanish regions was 2 cents per litre, whereas in 2000 it was only 0.3 cents per litre (Oilgas, February 2001). The retail operators explained this by pointing out that the oil companies did not pass on the full spot price increases.

These facts suggest that the major operators and the government cooperated in price setting during the first years after abolishing the ceiling price regulation, in order to maintain overall price stability in an attempt to serve general interests. This pricing behaviour was made possible because of the specific structure of the Spanish oil industry, facilitated by Repsol-Ypf's market dominance and the relationship between the firms and Spain's political elite.

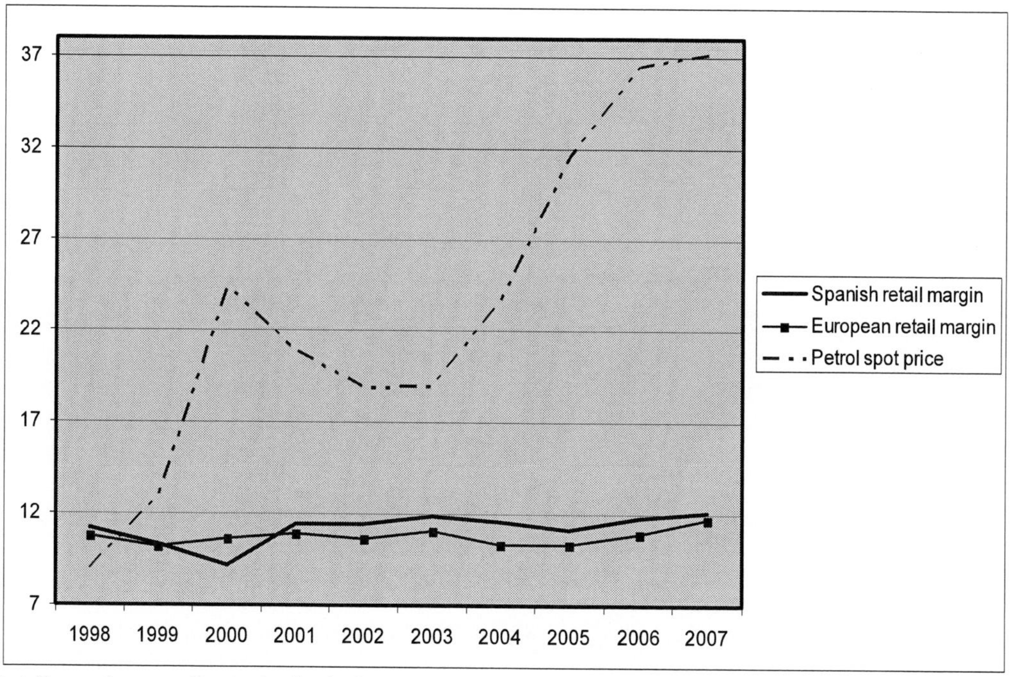

Retail margin = retail petrol price before taxes – petrol spot price.
Source: Ministerio de Industria.

Figure 3. Average Annual Retail Margins and Spot Prices for Petrol (Cents per litre)

However, Figure 2 shows how since the beginning of a new period of increasing petrol spot prices in 2004, the average Spanish and the European pre-tax petrol prices have moved closely together. As a result, Spain's and Europe's retail margins have moved together, although from 2004 to 2006 the Spanish retail margin was higher than in Europe (Figure 3). However, in 2007 both the pre-tax petrol price and the retail margin in Spain have converged with European levels (Figures 2 and 3). As previously stated, this chapter analyzes regional petrol price differences for 2007, which in turn sheds new light on how competition works in the Spanish retail petrol market.

4. Regional Price Differences in the Spanish Retail Petrol Market

This section focuses on regional petrol price differences in Spain. Specifically, we analyze whether the pre-tax and after-tax petrol price differences across the regions are statistically significant. We will also see which regions have the highest, and which have the lowest prices. Furthermore, we study the supply and demand factors that could explain those differences.

To do so, we use a unique sample of 565 randomly selected service stations (Table 3). Our sample is based on the regional market shares of the different operators, so it accurately represents all of the services stations, both nationally and regionally. For each of the sampled stations we have collected Euro 95 petrol prices, over the period January 1st to October 30th 2007 from the Ministry of Industry (http://www.mityc.es/es-ES/index.htm). In addition, we know the distance in kilometres between one station and the next.

Table 3. Descriptive Statistics– 565 Spanish Service Stations, in 2007

Spanish Regions	Number of Outlets	Income €/Inhabitant	Population Density	Vehicles per Inhabitant	Refinery
Andalucia	102	10,798	91.06	0.58	Yes
Aragón	23	14,563	26.78	0.58	No
Asturias	15	13,376	101.56	0.54	No
Cantabria	10	14,012	108.15	0.60	No
Castilla-La Mancha	50	11,142	24.33	0.61	Yes
Castilla y León	52	13,470	26.89	0.59	No
Cataluña	74	15,269	222.33	0.64	Yes
Extremadura	20	10,327	26.09	0.59	No
Galicia	39	11,944	93.58	0.61	Yes
Islas Baleares	13	14,743	200.53	0.82	No
La Rioja	5	14,532	60.93	0.56	No
Madrid	40	16,074	748.96	0.66	No
Murcia	22	11,048	121.13	0.63	No
Navarra	11	16,858	61.41	0.64	No
País Vasco	20	17,066	300.99	0.55	Yes
Comunidad Valenciana	70	12,306	206.71	0.64	No
Total	565	13,384	88.59	0.62	

Source: Catalist 2007. Centro de Estudios Económicos de la Caixa, 2007. Spanish Statistic National Institute (INE).

Our sample also includes supply and demand side variables at a regional level. On the demand side, they are: average income per capita, the population density, and the number of registered vehicles per capita for every Spanish region; see Table 3. These variables have been largely used in other studies (Plummer et al. 1998; Barron et al. 2000; Hastings, 2004). On the supply side, the variables are: the regional concentration indexes, C2 and HHI, and a dichotomy variable that takes on value 1 when the region has a refinery and 0 if not. All of this allows us to know the regional supply and demand conditions under which each station operates.

Our view is that, in a competitive environment retail prices would reflect regional supply and demand conditions, and uniform prices among regions would not be expected. On the one hand, those regions with high market concentration and more isolated stations would have less competitive markets and therefore higher prices (Marvel, 1978; Slade, 1986; Borenstein, 1991; Shepard, 1991). On the other hand, those regions with high population densities, high vehicle ownership and high per capita income would have relatively higher prices, given that petrol is a normal good. Non-regional prices differences would suggest a tendency for parallel pricing behaviour among the major oil operators.

Tables 4 and 5 summarize the pre-tax and after-tax average monthly petrol price for all regions, respectively, over the January to October 2007 period. Firstly, we have performed an analysis of variance (ANOVA), whose results show significant pre and after tax price differences among the 565 services stations for every month. Secondly, we have carried out a Ryan-Einot-Gabriel-Welsch- Multiple Step-Down Procedure (REGW), based upon the *F*-test, which identifies the monthly petrol prices regional homogeneous groups according to differences in pricing after and before taxes; this allows us to identify those regions with the highest and the lowest petrol prices.

With regard to after-tax petrol prices, the REGW test shows that the number of regional homogenous groups varies from 2 to 4 over the 10 months analyzed (Table 4). The results also show that those regions charging regional excise tax, Castilla-La Mancha, the Asturias, Cataluña, the Comunidad Valenciana, Galicia and Madrid, generally have the highest retail prices. The lowest prices are in the País Vasco, Navarra, La Rioja and Cantabria, where the regional excise tax is not charged. These results suggest that retail price differences are mainly due to different regional taxation policies.

With regard to pre-tax petrol prices, the REGW test only identifies regional homogenous subgroups, according to their petrol prices for the January to June period but not for the July to October period (Table 5). This means that during the second part of our period of analysis, the regional petrol price differences are not statistically significant. During this latter period, spot prices were higher, at €0.387 per litre, compared to the first period the cost was €0.346 per litre. In periods of increased oil and petrol spot prices, variations in regional prices tend to decrease (Contín et al., 2006), which could explain our empirical results.

Over the January to June period, the REGW test indicates that the number of regional homogeneous groups varies from 2 to 5. Again, prices are the lowest in the País Vasco, Navarra, La Rioja and Cantabria, and the highest in the Islas Baleares, Cataluña and Madrid. As stated, regional prices differences are not statistically different over the July to October period, although the Islas Baleares and Madrid continue to have the highest prices.

Table 4. Analysis of Variance – Average Monthly Petrol Prices at Retail Outlets in 2007 (€ per litre)

Jan 07		Feb 07		Mar 07		Apr 07		May 07	
Analysis of Variance (ANOVA) for average monthly petrol prices of the 565 Spanish service stations.									
F=72.04 Significance = 0.00		F=67.46 Significance = 0.00		F=37.84 Significance = 0.00		F=17.18 Significance = 0.00		F=10.439 Significance = 0.00	
Ryan-Einot-Gabriel-Welsch (REGW) Multiple Stepdown Procedures. Monthly average petrol prices.									
Spanish regions	Average Prices (2)	Spanish regions	Average Prices (3)	Autonomous Region	Average Prices (4)	Spanish regions	Average Prices (2)	Spanish regions	Average Prices (4)
Cantabria	0.9373*	Cantabria	0.9333*	País Vasco	0.9764*	País Vasco	1.0093*	País Vasco	1.0420*
La Rioja	0.9375*	Castilla y León	0.9408	Cantabria	0.9770*	Navarra	1.0110*	Navarra	1.0430*
Aragón	0.9390*	Extremadura	0.9411	Extremadura	0.9776	Extremadura	1.0113*	Extremadura	1.0456*
Castilla y León	0.9391*	La Rioja	0.9420	Castilla y León	0.9788	Castilla y León	1.0158*	Castilla y León	1.0524*
Murcia	0.9398*	Navarra	0.9432	Navarra	0.9803	Cantabria	1.0193*	Cantabria	1.0584*
Navarra	0.9405*	Murcia	0.9432	Aragón	0.9814	Andalucia	1.0208*	Andalucia	1.0593*
Extremadura	0.9405*	Aragón	0.9432	Andalucia	0.9853	Aragón	1.0216*	Aragón	1.0610
Andalucia	0.9431*	Andalucia	0.9446	Murcia	0.9863	Murcia	1.0237*	Murcia	1.0632
País Vasco	0.9438*	País Vasco	0.9456	La Rioja	0.9885	La Rioja	1.0289*	La Rioja	1.0700
Islas Baleares	0.9452*	Islas Baleares	0.9500**	Islas Baleares	0.9911	Islas Baleares	1.0343**	Islas Baleares	1.0757
Madrid	0.9660**	Asturias	0.9656**	Asturias	1.0072	Castilla-La Mancha	1.0449**	Castilla-La Mancha	1.0839
Asturias	0.9668**	Galicia	0.9703**	Castilla-La Mancha	1.0098**	Asturias	1.0466**	Asturias	1.0847
Galicia	0.9691**	Castilla-La Mancha	0.9705**	Galicia	1.0100**	Comunidad Valenciana	1.0485**	Comunidad Valenciana	1.0908
Comunidad Valenciana	0.9693**	Cataluña	0.9706**	Cataluña	1.0142**	Galicia	1.0509**	Galicia	1.0937
Castilla-La Mancha	0.9698**	Comunidad Valenciana	0.9711**	Comunidad Valenciana	1.0143**	Cataluña	1.0542**	Cataluña	1.0941**
Cataluña	0.9701**	Madrid	0.9726**	Madrid	1.0175**	Madrid	1.0550**	Madrid	1.0968**

Table 4. (Continued)

Jun 07		Jul 07		Aug 07		Sep 07		Oct 07	
Analysis of Variance (ANOVA) for average monthly petrol prices of the 565 Spanish service stations.									
F=10.03 Significance = 0.00		F=22.49 Significance = 0.00		F=25.79 Significance = 0.00		F=29.51 Significance = 0.00		F=29.35 Significance = 0.00	
Ryan-Einot-Gabriel-Welsch (REGW) Multiple Stepdown Procedures. Monthly average petrol prices.									
Spanish regions	Average Prices (4)	Spanish regions	Average Prices (3)	Spanish regions	Average Prices (2)	Spanish regions	Average Prices (2)	Spanish regions	Average Prices (2)
País Vasco	1.0475*	País Vasco	1.0483*	País Vasco	1.0445*	País Vasco	1.0451*	País Vasco	1.0447*
Navarra	1.0490*	Navarra	1.0510*	La Rioja	1.0457*	Castilla y León	1.0459*	Extremadura	1.0459*
Extremadura	1.0571*	Extremadura	1.0589*	Castilla y León	1.0470*	Extremadura	1.0477*	Castilla y León	1.0472*
Castilla y León	1.0623*	Castilla y León	1.0626*	País Vasco	1.0477*	Cantabria	1.0485*	Cantabria	1.0489*
Aragón	1.0671*	Aragón	1.0663*	Extremadura	1.0487*	La Rioja	1.0496*	Murcia	1.0490*
Andalucia	1.0708*	Andalucia	1.0690*	Cantabria	1.0487*	Navarra	1.0503*	Navarra	1.0494*
Murcia	1.0722	Murcia	1.0705	Navarra	1.0499*	Murcia	1.0509*	La Rioja	1.0497*
Cantabria	1.0735	Cantabria	1.0737	Aragón	1.0504*	País Vasco	1.0516*	País Vasco	1.0503*
La Rioja	1.0780	La Rioja	1.0765	Murcia	1.0522*	Aragón	1.0525*	Andalucia	1.0505*
Islas Baleares	1.0851	Islas Baleares	1.0838	Andalucia	1.0529*	Andalucia	1.0557**	Aragón	1.0557**
Comunidad Valenciana	1.0962	Comunidad Valenciana	1.0955	Islas Baleares	1.0741**	Islas Baleares	1.0769**	Islas Baleares	1.0767**
Castilla-La Mancha	1.0977	Castilla-La Mancha	1.0961**	Asturias	1.0742**	Galicia	1.0776**	Asturias	1.0767**
Galicia	1.1019**	Galicia	1.0983**	Galicia	1.0748**	Asturias	1.0776**	Galicia	1.0782**
Asturias	1.1031**	Asturias	1.1000**	Madrid	1.0786**	Madrid	1.0789**	Madrid	1.0784**
Madrid	1.1046**	Madrid	1.1018**	Cataluña	1.0788**	Castilla-La Mancha	1.0791**	Comunidad Valenciana	1.0790**
Cataluña	1.1073**	Cataluña	1.1038**	Comunidad Valenciana	1.0802**	Cataluña	1.0793**	Castilla-La Mancha	1.0791**

( ). The number of homogeneous groups are identified using the Ryan–Einot–Gabriel–Welsch F test for α= 0.05.
*Included in the group of regions with the lowest petrol prices.
** Included in the group of regions with the highest petrol prices.

Table 5. Analysis of Variance – Average Monthly Pre-Tax Petrol Prices at Retail Outlets in 2007 petrol (€ per litre)

Jan 07		Feb 07		Mar 07		Apr 07		May 07	
Analysis of Variance (ANOVA) for monthly average pre-tax petrol prices of the 565 Spanish service stations.									
F=2.922 Significance = 0.00		F=4.753 Significance = 0.00		F=4.062 Significance = 0.00		F=2.650 Significance = 0.00		F=2.128 Significance = 0.00	
Ryan-Einot-Gabriel-Welsch (REGW) Multiple Stepdown Procedures. Monthly average pre-tax petrol prices.									
Spanish regions	Average Prices (3)	Spanish regions	Average Prices (5)	Spanish regions	Average Prices (3)	Spanish regions	Average Prices (3)	Spanish regions	Average Prices (3)
La Rioja	0.4111*	Cantabria	0.4080*	País Vasco	0.4469*	País Vasco	0.4731*	País Vasco	0.5028*
Cantabria	0.4113*	Castilla y León	0.4133*	Cantabria	0.4472	Navarra	0.4736*	Navarra	0.5037*
Aragón	0.4125*	Extremadura	0.4138	Extremadura	0.4482	Extremadura	0.4751*	Extremadura	0.5064
Castilla y León	0.4128	La Rioja	0.4146	Asturias	0.4484	Castilla y León	0.4786	Castilla y León	0.5124
Navarra	0.4130	Murcia	0.4148	Castilla y León	0.4493	Castilla-La Mancha	0.4813	Castilla-La Mancha	0.5160
Murcia	0.4131	Asturias	0.4149	Navarra	0.4503	Cantabria	0.4820	Comunidad Valenciana	0.5162
Extremadura	0.4139	Navarra	0.4150	Aragón	0.4516	Asturias	0.4826	Aragón	0.5171
Asturias	0.4159	Aragón	0.4158	Castilla-La Mancha	0.4516	Andalucia	0.4828	Andalucia	0.5187
Andalucia	0.4161	Andalucia	0.4165	Galicia	0.4519	Comunidad Valenciana	0.4832	Cantabria	0.5201
País Vasco	0.4163	País Vasco	0.4172	Andalucia	0.4548	Aragón	0.4833	Galicia	0.5216
Comunidad Valenciana	0.4171	Comunidad Valenciana	0.4196	Cataluña	0.4551	Murcia	0.4851	Murcia	0.5219
Islas Baleares	0.4179	Cataluña	0.4198	Comunidad Valenciana	0.4553	Galicia	0.4857	Asturias	0.5237
Galicia	0.4185	Castilla-La Mancha	0.4199**	Murcia	0.4561	Cataluña	0.4885	Cataluña	0.5244
Madrid	0.4185**	Galicia	0.4200**	La Rioja	0.4585	La Rioja	0.4890	La Rioja	0.5278
Cataluña	0.4186**	Islas Baleares	0.4210**	Islas Baleares	0.4596	Islas Baleares	0.4943**	Madrid	0.5333**
Castilla-La Mancha	0.4190**	Madrid	0.4237**	Madrid	0.4665**	Madrid	0.4965**	Islas Baleares	0.5334**

Table 5. (Continued)

Jun 07		Jul 07		Aug 07		Sep 07		Oct 07	
Analysis of Variance (ANOVA) for monthly average pre-tax petrol prices of the 565 Spanish service stations.									
F=1.870 Significance = 0.024		F=1.600 Significance = 0.069		F=1.065 Significance = 0.387		F=1.438 Significance = 0.125		F=1.471 Significance = 0.111	
Ryan-Einot-Gabriel-Welsch (REGW) Multiple Stepdown Procedures. Monthly average pre-tax petrol prices.									
Spanish regions	Average Prices (2)	Spanish regions	Average Prices (1)	Spanish regions	Average Prices (1)	Spanish regions	Average Prices (1)	Spanish regions	Average Prices (1)
País Vasco	0.5094*	País Vasco	0.5089	La Rioja	0.5043	Castilla y León	0.5055	Extremadura	0.5057
Navarra	0.5106*	Navarra	0.5101	Asturias	0.5044	Extremadura	0.5069	Castilla y León	0.5063
Extremadura	0.5172*	Extremadura	0.5190	Galicia	0.5045	Galicia	0.5071	Galicia	0.5070
Castilla y León	0.5214*	Castilla y León	0.5220	Castilla y León	0.5062	Asturias	0.5076	Asturias	0.5072
Aragón	0.5260*	Aragón	0.5249	País Vasco	0.5070	Cantabria	0.5077	Cantabria	0.5072
Comunidad Valenciana	0.5274*	Comunidad Valenciana	0.5265	Extremadura	0.5087	La Rioja	0.5085	Murcia	0.5089
Andalucia	0.5286*	Andalucia	0.5271	Cantabria	0.5087	Castilla-La Mancha	0.5090	Navarra	0.5090
Castilla-La Mancha	0.5291*	Castilla-La Mancha	0.5277	Navarra	0.5087	Cataluña	0.5092	Castilla-La Mancha	0.5090
Murcia	0.5310*	Galicia	0.5290	Cataluña	0.5088	Navarra	0.5096	Comunidad Valenciana	0.5092
Galicia	0.5327*	Murcia	0.5291	Comunidad Valenciana	0.5096	Comunidad Valenciana	0.5101	Cataluña	0.5092
Asturias	0.5331*	Asturias	0.5300	Aragón	0.5099	Murcia	0.5103	La Rioja	0.5094
Cantabria	0.5333*	Cantabria	0.5329	Castilla-La Mancha	0.5102	País Vasco	0.5109	País Vasco	0.5097
La Rioja	0.5370*	La Rioja	0.5330	Murcia	0.5104	Aragón	0.5116	Andalucia	0.5103
Cataluña	0.5372*	Cataluña	0.5336	Andalucia	0.5118	Andalucia	0.5123	Aragón	0.5105
Madrid	0.5391**	Madrid	0.5375	Islas Baleares	0.5129	Islas Baleares	0.5157	Islas Baleares	0.5157
Islas Baleares	0.5403**	Islas Baleares	0.5381	Madrid	0.5151	Madrid	0.5175	Madrid	0.5180

( ). The number of homogeneous groups are identified using the Ryan–Einot–Gabriel–Welsch F test for $\alpha = 0.05$.
*Included in the group of regions with the lowest pre-tax petrol prices.
** Included in the group of regions with the highest pre-tax petrol prices.

Table 6. Matrix of Correlations

	1	2	3	4	5	6	7	8
1-Average Pre-tax Retail Price (P_i)	1							
2- Average Income per Inhabitant.	0.087*	1						
3- Population Density	0.222**	0.573**	1					
4-Vehicles per Inhabitant	0.145**	0.299**	0.418**	1				
5-C2 Concentration Index	0.023	0.269**	0.114**	0.461**	1			
6-Herfindahl Index	0.035	0.411**	0.232**	0.343**	0.499**	1		
7-Distance between one service station and the next (D_i)	-0.016	-0.054	- 0.182**	-0.080	0.134**	0.026	1	
8-Refinery	0.002	-0.189	-0.095*	-0.161**	-0.198**	-0.306**	0.066	1

*$p < 0.05$, ** $p < 0.01$

In order to asses the reasons that could explain the regional pre-tax petrol price differences observed in Spain, first, for each service station in our sample we have calculated the average pre-tax petrol price over the period January 1st to October 30th 2007 (P_i). As stated, we know the distance in kilometres between one station and the next (D_i). Such a distance is a proxy of a station's market power, since the more isolated it is, the more market power it has. Next, we conduct a bivariate correlation analysis, both between these two variables P_i and D_i and between P_i and the variables that indicate the regional supply and demand conditions under which each station operates (Table 6).

There is a positive and statistically significant correlation between P_i and the regional demand variables; i.e., average income per capita, population density and density of vehicles. However, the correlations between the average pre-tax price for each service station (P_i) and the regional supply conditions under which each service station operates, the C2 and HHI indexes, are not statistically significant. Neither the correlations between P_i and D_i nor between P_i and the dichotomy variable, which takes on value 1 when the region has a refinery and 0 if not, are statistically significant.

These results suggest that different conditions of regional demand could cause the differences observed in regional pre-tax petrol prices in Spain. However, neither the regional supply conditions nor the distance between one station and the next seem to affect pre-tax retail prices. Therefore, regional pre-tax petrol price differences in Spain seem to be caused mainly by the specific conditions of demand under which the service stations operate.

5. Conclusions

Although the Spanish Oil Monopoly was abolished in January 1993, retail petrol prices were regulated by a ceiling price system until October 1998. The weekly ceiling prices were set according to a formula that to some degree imposed regional price uniformity throughout the country. However, with the abolition of the price regulation there are no longer legal impediments to the existence of regional petrol price differences.

In a competitive environment, retail prices would reflect regional supply and demand conditions; therefore, uniform prices among regions would not be expected. However, in 2000 the European Commission expressed its "surprise" at the lack of variation in the petrol prices among the Spanish regions and at the different service stations. Political interference in price setting would explain the regional price uniformity observed just after the abolition of the ceiling price regulation.

This chapter seeks to shed light on how retail petrol prices are currently set in the Spanish petrol retail market. Specifically, it analyzes whether regional price differences have recently been observed in pricing retail petrol. In doing so, we use a unique sample of 565 randomly selected service stations, which accurately represents all service stations both at national and regional level. For each sampled station we have collected Euro 95 petrol prices over the first ten months of 2007. We have also several variables that reflect the regional supply and demand conditions under which the service stations operate.

Our empirical results show that average monthly retail petrol prices vary among the regions. These differences in the regional petrol retail prices are mainly due to different regional taxation regimes. Specifically, the Spanish regions can add excise taxes up to 2.4 cents per litre to the national excise duty. The average difference between the most expensive and the cheapest regions was 4.05 cents per litre over the first ten months of 2007.

Although there are pre-tax differences in regional petrol price, they are only statistically significant from January to June 2007. During this period, the average difference between the most expensive and the cheapest regions was 1.62 cents per litre. We have also conducted a bivariate correlation analysis, and the results suggest that demand not supply explains the regional pre-tax petrol price differences that have recently been observed in Spain.

This chapter also sheds new light on the competitive behaviour within the Spanish petrol market. Overall, our results show a more competitive scenario than the years just after the ceiling price regulation was abolished. On the one hand, pre-tax petrol prices and retail margins in Spain have recently converged with European levels. On the other hand, petrol retail price differences exist among the Spanish regions. These differences are due both to different regional taxation regimes, and to the regional demand conditions that the service stations operate under. The retail market structure does not vary greatly among the Spanish regions, and this fact could explain the lack of significant correlations between the regional supply conditions and the pre-tax prices at service stations.

REFERENCES

Banco de España (1999). Boletín Económico. Diciembre, 1999. Madrid, Spain.

Barron, J., Taylor, B., and Umbeck, J.,(2000). A theory of quality-related differences in retail margins: why there is a "premium" on premium gasoline. *Economic Inquiry* 38, 550–569.

Bello, A. and Cavero, S.(2007). Competencia Estratégica en la Distribución Minorista de Carburantes y Combustibles de Automoción en España.", Revista de Economía Aplicada, 45,125-54.

Bello, A. and Cavero, S. (2008). The Spanish Retail Gasoline Market. New patterns of competition since the liberalization of the industry. *Energy Policy*, 36, 612-626.

Borenstein, S.(1991). Selling Costs and Switching Costs: Explaining Retail Gasoline Margins. *Rand Journal of Economics*, 22, 354-369.

Contín, I, Correljé, A. and Palacios, M.B. (2006). Competition, regulation and pricing behaviour in the Spanish gasoline market. Working Paper, 2, 2006, Dpto.: Economía Aplicada III-Universidad del País Vasco.

Contín, I. and Correljé, A. Lessons from restructuring the Spanish distribution system of oil products. In Editor Frank Columbus, European Economic and Political Issues, (edition 8, 155-170). New York .Nova Science Publisher, Inc.2004

Contín, I., Correljé, A. and Huerta, E. (2001). The Spanish distribution system of oil products: An obstacle to competition?. *Energy Policy*, 29 , 103-112

Contín, I. and Huerta, E., (2000). El mercado español del gasolinas: ¿falta competencia?. El País, on Tuesday 22 nd of February.

Contín, I., Correljé, A., and Huerta, E. (1999). The Spanish gasoline market: From ceiling regulation to open market pricing. *The Energy Journal*, 20, 9-16.

Correljé, A. (1994). The Spanish oil industry: Structural change and modernization. Amsterdam: Tinbergen Institute Research Series, 84. Thesis Publishers.

Eckert, A. and West, D.S.(2004). A tales of two cities: Price uniformity and price volatility in gasoline retailing. *The Annals of Regional Science*, 38, 25-46

Eckert, A. and D. West, (2005). Price uniformity and competition in a retail gasoline market", *Journal of Economic Behavior and Organization*, 56, 219–237.

Hastings, J (2004). Vertical Relationships and Competition in Retail Gasoline Markets: Empirical Evidence from Contract Changes in Southern California. *The American Economics Review,* 94, 317-328.

Marvel, Howard P.,(1978). Competition and Price levels in the Retail Gasoline Market. *Review of Economics and Statistics*, 6, 252-258.

Ministerio de Industria (1995). Informe sobre la formación de precios en el sector petrolero, Mimeo.

Plummer, P., Haining, R., and Sheppard, E., (1998). Spatial pricing in interdependent markets: testing assumptions and modeling price variation.Acase study of gasoline retailing in St. Cloud, Minnesota. *Environment and Planning*, 30, 67–84.

Slade, M. E. (1986). Conjectures, Firms Characteristics, and Market Structure: An Empirical Analysis. *International Journal of Industrial Organization*, 4,347-370.

Shepard, A. (1991). Price Discrimination and Retail Configuration. *Journal of Political Economy*, 99, 30-54.

In: Energy in Europe: Economics, Policy and Strategy ISBN: 978-1-60456-829-5
Editors: F. L. Magnusson and O. W. Bengtsson

Chapter 6

Regulatory Reform and Competition in the Turkish Electrcity Market

Tamer Cetin
Department of Economics, Zonguldak Karaelmas University
Zonguldak/Turkey

Abstract

By enacting the Electricity Market Law in 2001, Turkey reformed the electricity market and initiated to unbundle the state-owned monopolistic structure of the market by separating vertically. Thus, the regulatory endowment and the market structure before the reform fully changed. The regulatory reform has aimed to introduce competition to the market. As an independent regulatory agency with full regulatory discretions on the market, Energy Market Regulatory Agency (EMRA) was established. The aim of this chapter is to review the transition process to the reform, the effects of the regulatory reform on the new market structure and evaluate the current economic, political and legal institutional problems of the market.

Introduction

Turkey attempted unbundling the monopolistic market structure and reforming the regulatory endowment surrounding the electricity market by the Electricity Market Law enacted in 2001. The regulatory reform has aimed to direct the market to a more competitive environment. As an independent regulatory agency with full regulatory discretions on the market, Energy Market Regulatory Agency (EMRA) was established. The aim of this chapter is to review the transition process to the reform, the effects of the regulatory reform on the new market structure and evaluate the current economic, political and legal institutional problems of the market.

The chapter roughly consists of five sections. The first section introduces a historical background to review the pre-reform structure of the electricity market and the main reasons for the reform. The analysis of the pre-reform monopolistic structure and the troublesome legacy of the past-term are pioneer topics to understand the difference between the pre-reform and post-reform structure of the market. On the other hand, the section also includes the main reasons for the reform like economic, political, and the external factors.

The second section analyzes main reasons for the reform of the Turkish electricity market. The section scrutinizes rationales for the reform, as the economic reasons, the political and legal reasons, and accession to the EU and external factors.

The third section presents an outlook of the new market structure. The outlook comprises the crucial properties and components of the reform like the 2001 Law, EMRA, and disintegration of the monopolistic structure.

The fourth section reviews the main economic issues in the current market structure. These issues are problems related to existing contracts, lack of excess supply, stranded costs in generation and wholesale segments, retail tariffs and cross subsidies, vertical integration and disintegration, the "Strategy Paper" of March 2004, and delay in the privatization process of distribution assets and the equalization mechanism.

Finally, the fifth section suggests some policy recommendations for future of the Turkish electricity market. In the end, the paper expires with a conclusion section.

1. Historical Background

A historical development of the structure of the Turkish electricity market is introduced in Figure 1, including the structure after the Electricity Market Law of 2001 as well.

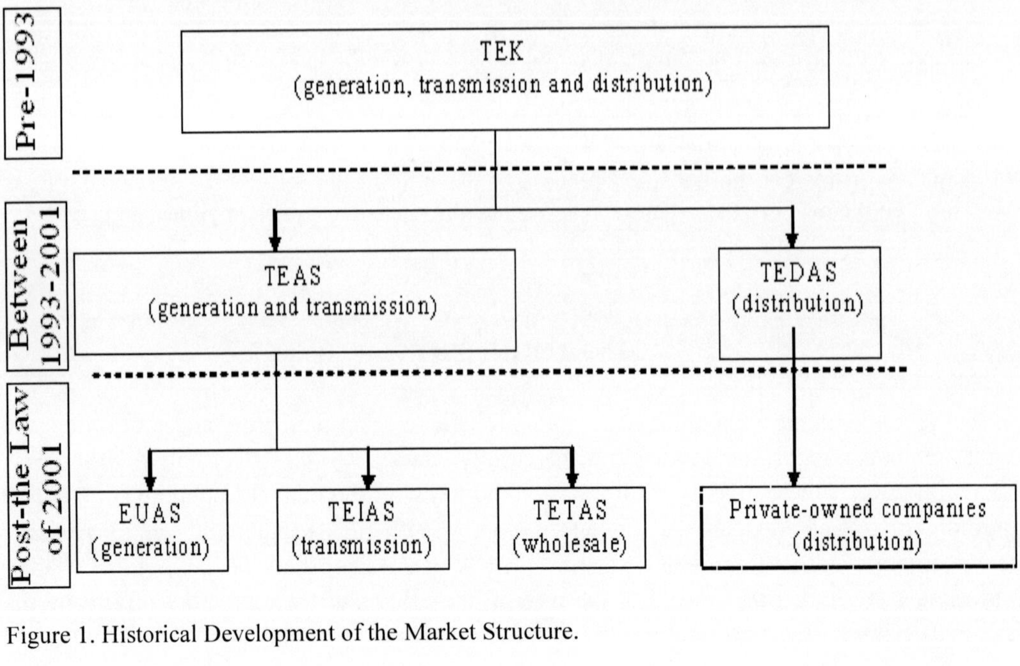

Figure 1. Historical Development of the Market Structure.

1.1. The Structure before 1980

Electricity in Turkey was firstly generated in 1902 in Tarsus by a foreign group. This first enterprise was properly a private one and installed capacity of electricity generation was 2 kWh. Afterwards, a consortium consisting of the Ottoman Company (Osmanlı Anonim Şirketi), GANZ Company, Banque de Bruxellese and Banque Generale de Credit begun to generate electricity in Silahtarağa Plant in Istanbul as of 1914. This privileged consortium was expropriated in 1938 by the new Republic of Turkey. The term 1930-1950 was an interventionist period to the Turkish electricity market [Kulali, 1997].

Municipality Law enacted in 1933 gave right municipalities to establish and to operate electricity plants. In order to plan and to regulate electricity utilities by the state, Electricity Office of Studies (Elektrik İşleri Etüt İdaresi - EİEİ) was established in 1935. By the period 1938-1944, all of foreign and domestic private enterprises in the country, with the exception of Kayseri Electricity Company (Kayseri ve Civarı Elektrik Türk A.Ş.), had been expropriated. Electricity production in this term was carried out by the plants municipalities and different state-owned enterprises had. In 1950, the installed capacity reached 407.8 MW, electricity production reached 789.5 GWh, per capita electricity production 38 kWh, and per capita annul electricity consumption reached 32 kWh. The ratio of the population able to use electricity in this term had reached 23%. 97 of average installed capacity in the term 1930-1950 were by thermal power plants. In this period, growth rate of electricity production was 10.7% and growth rate of installed capacity was 9%, as growth rate of national product decreased by 2.9%.

During the period 1950-1960, large and regional coal-fired and hydroelectric plants rather than small and domestic diesel-fired power plants were established. Tuncbilek and Soma coal-fired power plants were constructed and begun to generate electricity in 1956 and 1957, respectively. General Directory of State Hydraulic Works (Devlet Su Isleri - DSI) in 1953 and Turkish Coal Enterprises (Turkiye Komur Isletmeleri - TKI) in 1957 was established. In addition, an important feature of the term is the realization of the participation of private sector to electricity facilities. This participation does not include foreign capital. Thermal installed capacity reached from 389.9 MW in 1950 to 860.5 MW in 1960 and hydraulic installed capacity from 17.9 MW in 1950 to 411.9 MW in 1960. In 1960, electricity generation was 2815.1 GWh and per capita annual consumption was 86 kWh, as total installed capacity was 1272.4 MW.

The term 1960-1980 was years of planned economy for electricity policies in Turkey. In particular, the third development plan (the period 1973-1977) adopted an energy policy controlled by the state. However, although Seyitomer, Aliaga ve Hopa thermal and Keban hydroelectric power plants were in operation in the 1970s, electricity demand could not be met in the third planning period. Besides, during the planned term, growth rate of energy generation decreased, as energy consumption increased. Towards the 1980s, dependency of electricity to imports reached 50 %. In the term before 1980, high and sunk-cost energy investments were realized by the state resources. Although there is the share of the private enterprise in these investments, they are rather limited. Energy policies dominated by the development plans impeded almost all the participation of the private sector to energy investments.

In 1980, thermal installed capacity reached 2987.9 MW, as hydraulic installed capacity reached 2130.8 MW. The share of thermal power plants in electricity generation decreased to

62%. Per capita annual electricity consumption reached from 32 kWh in 1950 to 459 kWh in 1980. The ratio of the population able to use electricity in 1980 had reached 79.9%. In the term before 1980, growth rate of electricity generation became 11.2%, as growth rate of installed capacity was 7.4% [Yilmaz and Uslu, 2007].

1.2. The Structure after 1980

Until the middle of 1980s, the Turkiye Elektrik Kurumu (TEK), a vertically integrated monopoly owned by the state, dominated all production, transmission and distribution segments of the Turkish electricity industry. In 1993, this institution was vertically unbundled two separate state institutions in parallel with deregulation and liberalization policies of the term after 1980. One of them is Turkey Electricity Generation and Transmission Company (Turkiye Elektrik Üretim İletim Anonim Sirketi - TEAS) and the other is Turkey Electricity Distribution Company (Türkiye Elektrik Dağıtım Anonim Şirketi - TEDAS). The both companies began operation in 1994 [Ozkivrak, 2005].

In the term after 1980, the state-controlled economy policies in Turkey replaced with more liberal policies in accordance with the global economic change. This change also influenced energy policies of the governments and Turkey initiated to bring into force privatizations in electricity market after 1980. During the 1980s and 1990s, various laws in relation to the electricity market were passed to encourage the initiatives of private sector to electricity generation and distribution facilities. With these legal regulations, it was designed some concession agreements for the participation of private-sector actors into the electricity generation and distribution investments, the so-called Build Operate Transfer (BOT), Build Operate (BO) and Transfer of Operating Rights (TOR) contracts. However, as discussed below, these concession contracts brought about some issues and could not be showed the expected impact over energy investments, in spite of the aim of protecting investors from financial risks.

The Law no. 3096 enacted in 1984 began the first deregulation initiatives and introduced the implementations of privatization to the Turkish electricity market by unbundling the vertical structure of the market. The Law encouraged the private participations to private in the all segments of the market. Two private companies, Kepez Elektrik and CEAS, granted a concession to operate in transmission facilities in Antalya and Adana, Mersin, Hatay, and Osmaniye, respectively. Afterwards, with the Law passed in 1984, the private sector initiatives were only allowed to operate in generation and distribution segments of the market. The first privileged contract could be in force in 1996, because of some constitutional and bureaucratic issues taking place while passing the Law no. 3096.

The Law no. 3096 enjoyed heavily two methods to transfer the state owned public utilities in generation and distribution segments of the electricity market to private sector participations. As expressed above, these methods are BOT and TOOR, including franchise rights for the owners. With an additional law, (the Law no. 4283) enacted in 1997, it was introduced a new contract method the so-called BOO to encourage private investments especially in operation of new thermal energy plants. All these franchise contracts have guaranties obtained by Treasury and tax exemptions.

However, the Constitutional Court and Danistay (an administrative high court) adjudicated that all electricity utilities are public utilities rather than a private ones and must be subject to public law by taking into consideration the Turkish administrative law and the Turkish Constitution. In spite of subsequent some legal amendments, the Constitutional Court voided the struggles to convert the franchise contracts from public law to private law[1]. Accordingly, the Constitutional Court and Danistay annulled the franchise contracts and privatization of electricity utilities again and again until the beginning of the 2000s [Ozkivrak, 2005; Ulusoy and Oguz, 2007].

In the end, the decisions of the Danistay and the Constitutional Court over the franchise contracts in the electricity and the lack of international arbitration for international contractors in case of any legal dispute impeded the implementation of the expected investments and privatizations by causing an ambiguity in terms of foreign investors. The Law no. 3096 could not provide a favorable legal environment for successful privatizations. Therefore, the government passed a constitutional amendment[2] related to these issues in 1999. Thus, the constitutional amendment of 1999 enabled;

- Parliament to provide public utilities via private law contracts,
- International arbitration for international investors in case of any dispute in the franchise contracts,
- A restriction in the Danistay's advisory role.

Law no. 4493 and Law no. 4501, which were enacted after the constitutional amendment, altered the legal endowment of the franchise contracts in the electricity market. The Law no. 4493 converted BOT contracts to private law contract status and the Law no. 4501 obtained the opportunity to include directly the international arbitration in the contracts. As a result, although the Law no 3096 allowed the implementation of privatizations through BOT and TOR contracts by private participations in the electricity market and some private entrepreneurs signed electricity generation agreements with the Ministry of Energy and Natural Resources (MENR), most of them have not been implemented. In additional, the franchise contracts of KEPEZ, CEAS and AKTAS (Istanbul Anatolian region) in distribution market were recognized as TOR contracts. However, the franchise contracts of KEPEZ and CEAS were cancelled unilaterally by MENR. In another judicial decree[3] as well, Danistay repealed the franchise contract of Aktas due to faulty operations of the company [Ulusoy and Oguz, 2007].

1.3. The Market Structure before the 2001 Reform

As mentioned above, the TEK was vertically separated into two different companies the so-called TEAS and TEDAS, generation-transmission company and distribution company, respectively. By doing so, it was aimed to introduce competition to the market gradually. The distribution segment of the market was split into 29 distribution regions in 1996 and the TOR

---

[1] See: CC Decisions, 9.12.1994, E.1994/43, K.1994/42-2; 28.06.1995, E.1994/71, K.1995/23.

[2] Law no. 4446 of August 14, 1999.

[3] See Danistay 10th Chamber Decision no. 1993/1752 of April 29, 1993.

concessions for distribution utilities were sold at an auction to the highest bidder in the subsequent year[4].

However, the controversial structure of the franchise contracts, fiscal constrains, the constitutional restrictions over the legal regulations and the interventions of the judiciary to the initiatives of privatization in generation and distribution segments of the market had continuously impeded a participation of private sector with full competition. The typical BOT, BOO and TOOR[5] contracts, signed between the private initiatives and TEAS or TEDAS and included privileged "take or pay" obligations with fixed guaranties and a fair revenue over capital costs of the privileged company at least for 15-39 years, failed to ensure a structure for the aimed competition *in the market* [Atiyas and Dutz, 2005; 191].

2. Main Reasons for The Reform of Turkish Electricity Market

The problems of the previous market structure triggered transition to the reform in the Turkish electricity industry. It is possible to find some reasons for the reform of the Turkish electricity market as follows. We can align economic, political, legal and external reasons for the reform, respectively.

2.1. The Economic Reasons

The structure of electricity prices in Turkey was a main economic reason for the reform of the market. Since, electricity prices in Turkey are subject to various taxes and levies that are not directly concerning the market activities or the cost structure. For example, a 3.5% levy for Turkish Radio and Television Corporation (TRT) was imposed on the prices for end-users of electricity[6]. In addition, electricity prices are subject to the municipality consumption tax that is 1% for industry users and 4% household consumers. Furthermore, the VAT rate for electricity in Turkey is 18% then and now [OECD, 2005; 30].

Another economic reason for the reform is the increasing demand in electricity demand and the lack of supply from the indigenous sources. As of 1998, electricity generation from the indigenous sources covered 39% of electricity demand. Turkey was dependent on imports in electricity consumption with 61% [Yigitguden, 1999].

A reason that triggers the reform was fiscal constraints. The government decelerated that budget deficits do not allow to meet future energy demand and the financing by private sector is needed to meet it [Erdogdu, 2007]. On the other hands, financial structures of public entities, TEDAS and TEAS, which dominated the market, were rather fragile. For example,

---

4 With the cabinet decisions, the franchise rights of the TOR contracts were granted to private distribution companies by MENR. However, the Danistay repealed the validity of the cabinet decrees. Thereupon, MENR annulled unilaterally these contracts by taking the Danistay decree into consideration (Ulusoy and Oguz, 2007; 5026).

5 In Turkey, one electricity distribution region and two power plants are operating under the TOOR (OECD, 2005; 131).

6 Although the new Electricity Law of 2001 forbids and EMRA does not prefer it, this levy on electricity prices is still in force and applies a 2%.

only the duty losses of TEAS were 656 million dollars as of 2000. This partly originated from high cost BOT contracts. In addition, the earlier privatization attempts were largely unsuccessful.

The national tariff structure of electricity prices involving significant cross-subsidies from industry users to household users and from a region to another region were another reason for the reform [Çetin and Oguz, 2007b; OECD, 2002].

2.2. The Political and Legal Reasons

As in any other developing country, the state-owned monopolies in Turkey as well have been economically inefficient and politicians have manipulated this inefficiency for their own political ends [Erdogdu, 2007]. For that reason, the main reason behind the poor performance and then the financial weakness was political interventions and pressures on the public electricity utilities. In such an environment, the managements who do not know the distinctive properties of electricity industry were frequently employed to management of these companies. Thus, the instable governments have intervened in all regulatory policies of the market like price, investment and employment. Political considerations were generally preferred more in determining these policies rather than economic considerations. As in the other public utilities, the political interventions and instability brought about crises in the electricity industry as well[7] [Ozkivrak, 2005].

Another reason for the reform was the failure of the Law 3096 and the previous legal structure. The legal and bureaucratic endowment of the market and the political domination on the market caused the liberalization goal to deviate from the aimed point and discouraged privatizations. For example, anyone who wants to operate a generation plant was needed;

- to obtain a permission from the State Planning Organization,
- to obtain a guarantee from the Treasury,
- to negotiate the Ministry,
- to obtain a recommendation from the Ministry,
- to obtain a decree from the Cabinet and, in the end,
- to sign an electricity sale contract with TEDAS.

These procedural conditions were also in force for the transfer of operating rights in any distribution region. This institutional structure of the market clearly constituted barriers to entry [Ulusoy and Oguz, 2007].

2.3. Accession to the EU and External Factors

Turkey is a candidate country for full accession to the European Union (EU). For that reason, Turkey needs to fulfill its obligations during the membership process into the EU. In this context Turkey requires to reshape the structure of the electricity market according to the electricity directive issued by the European Parliament in 1996. The main goal of the electricity directive of 1996 is to introduce competition to the competitive generation and supply to end users segments of the electricity market and to regulate the monopolistic transmission and distribution segments in the internal market of the EU. In this sense, it is possible to mention four main components for the reform. These are market opening, vertical separation, third party access, and public service obligations and regulation.

By the directive, all member states should open a determined and later, increasingly expanding percentage of its market within the next six years. Accordingly, each member state needs to designate its own set of eligible consumers. However, according to the directive, consumers who use electricity more than 100 GWh will be recognized as eligible consumers in all member states.

On the other hand, the directive stipulates the member states to unbundled vertically their own markets. Accordingly, it is needed the vertically integrated monopolies in each member state to separate the management of the generation, transmission, distribution, and non-electricity activities. In order to impede discrimination during the vertical separation, each member state is obligated to establish an independent transmission system operator (TSO). The main responsibility of a TSO is to ensure dispatch of any generator under fair, impartial and transparent conditions by means of this system.

Another goal of the directive is to enable access of third parties to transmission network again under fair conditions. Accordingly, consumers and generators directly sign a contract with each other by negotiating transmission and distribution companies for access the transmission and distribution networks. In order to ensure a fair pricing for all parties, access prices to the network is regulated by the independent regulator agencies. In case of a single buyer model, this single buyer pays the generator the regulated price minus network charges. Thus, the prices of end users equal to the contract prices due to the generator can compensate the consumer.

As a result, from a social point of view, the directive stipulates the member states to implement some obligations for their electricity markets. These obligations are security of supply, regularity on service, quality performances, price stability, and environmental protection. However, the feature of the obligations would be determined by the member states [Atiyas and Dutz, 2005; 189-190].

As a candidate country to the EU, Turkey was coerced and is coerced to take into considerations the obligations of the electricity directive of 1996 and reformed its electricity market. However, although the reform is a precondition for Turkey's membership into the EU, the only effective external pressure was not from the EU. In addition, the pressures of some other international institutions like IMF, World Bank and OECD were effective in triggering the reform of the electricity market [Ozkivrak, 2005; Erdogdu, 2007].

[7] In Turkey, it is found a positive correlation between economic crises and political instability. See Kibritcioglu (2001) for more information about this issue.

3. The Outlook of the New Market Structure

3.1. Properties of the Reform

In the end, the Turkish government reformed the electricity market by passing a law in 2001. With the law, the market was again unbundled an independent regulatory agency was established. This sub-section observes the features of the reform. The government aimed to meet the following objectives by reforming the market. These are:

1) to diversify of primary energy sources,
2) to increase use of domestic energy resources, including increasing the share of renewable energy sources in electricity generation,
3) to establish a liberal and competitive electricity market compatible with the EU directives,
4) to enable a good investment environment for new generating capacity as well as transmission and distribution networks,
5) to unbundle and privatize the state-owned companies,
6) to establish an environment-friendly power system, and in this context, investing in environmental retrofitting projects for the existing power plants,
7) to develop regional interconnections and participating in regional markets,
8) to increase efficiency in electricity generation and consumption,
9) to decrease the cost and end-user prices of electricity [OECD, 2005; 131].

3.2. The Electricity Law of 2001

The Electricity Market Law enacted in February 2001 reformed the Turkish electricity market. The Electricity Market Law is the main law regulating the electricity market. The new Law enacted in February 2001 was in force on 3 March 2001, but implementation of the Law was subject to a two-year transition period. The main aim of the Law of 2001 is to introduce competition to the market, protect consumers, ensure an efficient market regulation, and encourage private sector participation without treasury guarantees.

Regulations concerning all segments of the market are subject to the new Law. However, we can express that the Law has two main components. These are general provisions related to regulation of the market and the legal structure governing EMRA. The first component of the Law constitutes the operation of the market, including market activities and licensing, as the second part contains the duties and regulatory discretion of EMRA. Besides, the Law of 2001 is fully compatible with the EU Electricity Directive of 2003 [OECD, 2005; 143].

Furthermore, by following Ozkivrak [2005], we can define the key features of the new Law as in Figure 2. Firstly, the new Law vertically unbundled the market as three different public utilities. These are a generation company (the Turkiye Elektrik Uretim A. S. / TEUAS), a transmission company (the Turkiye Elektrik Iletim A. S. / TEIAS) and a state-owned wholesale company (the Turkiye Elektrik Taahhut ve Ticaret A. S. / TETAS). The Law stipulates to obtain a different license for each market activity.

The Law aims to introduce competition to competitive segments of the market, generation and retail sale. However, there are some restrictions over competition in generation for the transition period to full competition. For example, the total market share of a private generation company can not exceed 20% of the published figure for the total installed capacity in Turkey in the previous year. Any transmission company cannot engage in another market activity and distribution companies cannot generate electricity more than 20% of consumption in their own regions, given the preceding year [OECD, 2002; 114].

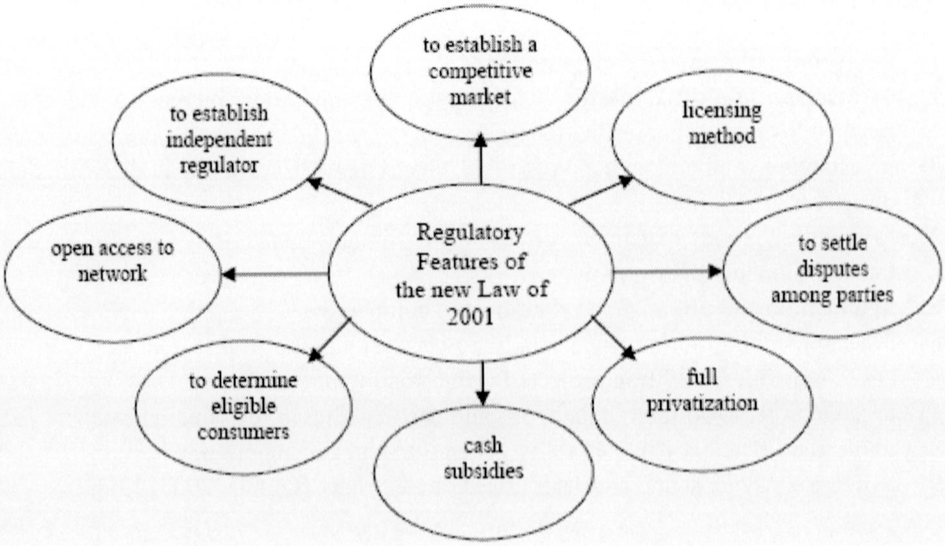

Source: Ozkivrak [2005].

Figure 2. Regulatory Features of the new Law.

All companies are needed to obtain a license for each market activity. The obligations concerning the contract, including how prices are to be regulated, how long license periods are, what conditions related to cancellation of a license are, how disputes are to be solved by the regulator, are clearly set on licenses.

In addition, the Law aims:

- to establish an independent regulatory agency,
- to give place to eligible consumers
- to support cash subsidies to consumers,
- to ensure open access to transmission and distribution networks,
- to establish a national competitive electricity market ,
- and in the end, to fulfill full privatization.

3.3. The Independent Regulator: EMRA

The Electricity Law of 2001 established the Electricity (now Energy) Market Regulatory Agency as an independent regulatory agency. EMRA has a board with 9 members. The main features of EMRA are as follows:

- to set up a new licensing structure,
- to grant licenses related to the market activities,
- to determine secondary legislation,
- to ensure third party access (TPA) under fair conditions,
- to apply a new transmission and distribution code,
- to regulate prices and tariffs for transmission, distribution and wholesale activities,
- to control the operations of all market participants,
- to protect consumer rights,
- to apply sanctions to faulty parties,
- to solve disputes between the parties in the market.

EMRA is an independent institution in the sense of administrative and financial from political and private interest groups. It collects its own incomes from licensing fees and from a surcharge on TPA tariffs. When EMRA's revenues exceed its expenditures, the surplus is transferred to the state budget. The regulatory discretions of EMRA relation to energy markets has been expanded with the subsequent some laws, including the Natural Gas Market Law enacted in May 2001 and the Petroleum Law enacted in 2003.

3.4. Unbundling and the New Structure of the Market

As mentioned above, the new Law *de facto* split the vertically integrated structure of the market into three state-owned public utilities and *de jure* enabled new entries into all segments of the market, with the exception of transmission. The current structure of the market with the historical development is depicted in Figure 1 above. TEIAS is a state-owned company, which responsible for the implementation of transmission activities, including the balancing market and financial reconciliation. The main responsibility of TETAS is to facilitate the financing of the costs of stranded contracts of the preceding period.

Under the new market structure, by the new Law, all market activities are only subject to the Electricity Law of 2001 and in responsibility of EMRA. Generation, distribution, wholesale, and retail activities can only be fulfilled within the electricity market and under EMRA's audit [Atiyas, 2006; 12]. In this case, according to the new Law, the market activities in all segments are as follows.

Generation

With the enactment of the new Law in 2001, the Turkish government *de jure* introduced competition to generation segment of the Turkish electricity market. According to the new Law, the total share of a generation company cannot exceed 20% of Turkey's total installed capacity. Although companies performing the market activities in generation segment may get in touch with distribution ones, this affiliation must not to bring about a control over them [Ozkivrak, 2005].

However, the expected investments in generation from the regulatory reform have not yet performed. The electricity indicators have not gave a good signal. Electricity generation capacity of Turkey is not adequate to meet its own energy demand. Electricity interruptions are still locally seen. Big shopping centers and some households have their own co-generators fuelled by diesel [OECD, 2005; 134].

In 2005, as shown in Figure 3, the net electricity generation is 155.5 TWh compared to the total final electricity consumption of 128.6 TWh.

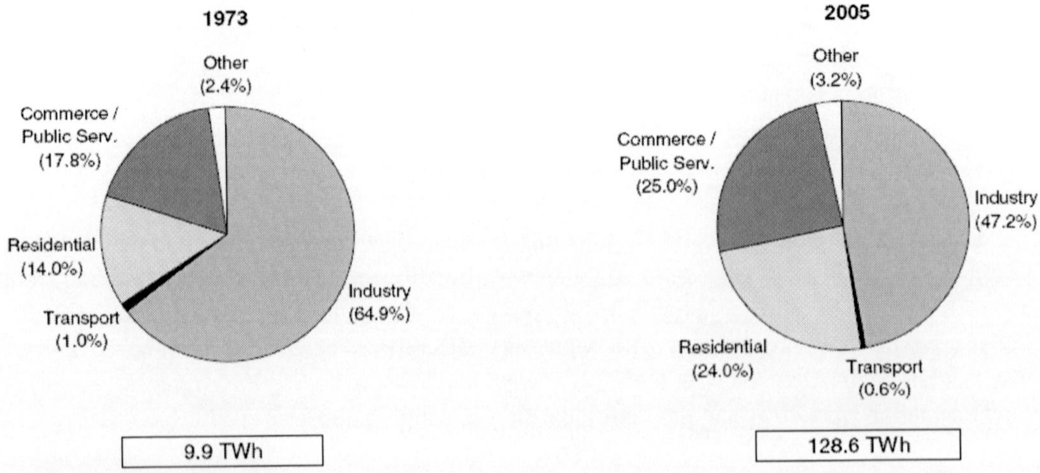

Source: IEA/OECD, 2007; 571.

Figure 3. Total Final Electricity Consumption by Sector.

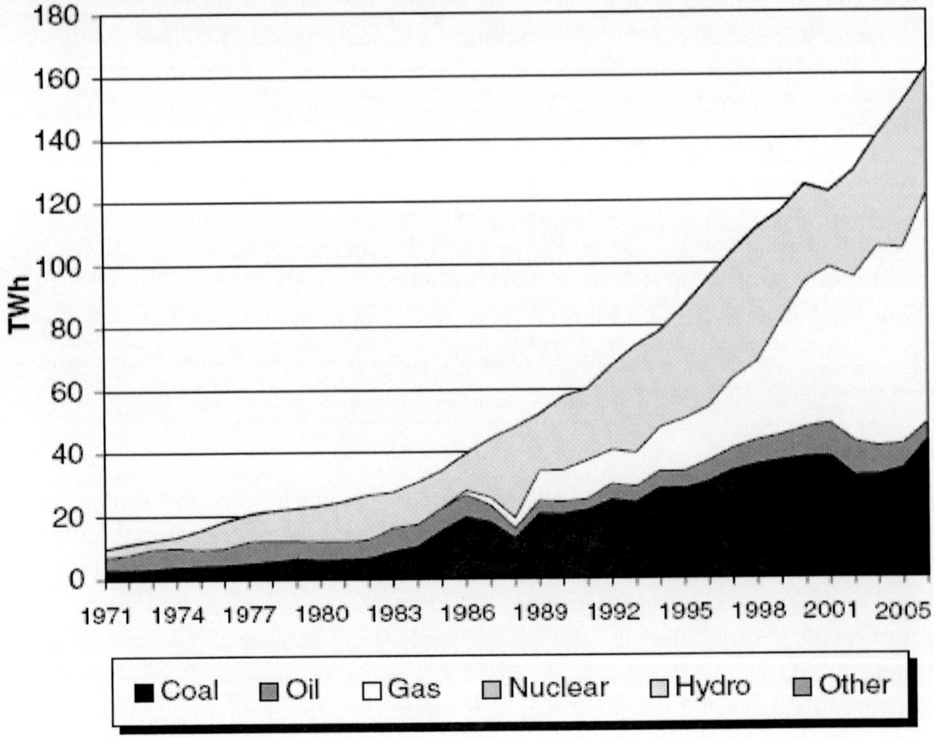

Source: IEA/OECD, 2007; 571.

Figure 4. Electricity Generation by Fuel.

On the other hand, as shown in Figure 4, more than 45% of electricity generation in Turkey is provided from natural gas. The main reason for the share of natural gas in electricity generation is the 2001 economic crisis that reduced gas demand in other sectors [Çetin and Oguz, 2007a].

However, as seen in Table 1, the net maximum electricity generating capacity in 2005 is 34.78 GW. Hydropower has the highest share in electricity generation with 35%. Natural gas, lignite, hard coal and oil, following hydropower have 33%, 18%, 5%, and 9% shares, respectively. Although non-hydro renewable resources have been increasingly bigger share in electricity generation, they have still a negligible share at 34 MW as of 2003 [OECD, 2005; 134].

Table 1. Net Maximum Electricity Generating Capacity (GW)

	1974	1980	1990	1995	2001	2002	2003	2004	2005
MAIN ACTIVITY PRODUCERS									
Classification by Fuel									
Total Capacity	3.31	4.52	15.12	19.61	24.96	28.11	31.05	32.44	34.78
Nuclear	-	-	-	-	-	-	-	-	-
Hydro	1.44	2.12	6.75	9.85	11.62	12.14	11.94	11.99	12.34
of which: Pumped Storage	-	-	-	-	-	-	-	-	-
Geothermal	-	-	0.02	0.02	0.02	0.02	0.02	0.02	0.02
Solar	-	-	-	-	-	-	-	-	-
Tide, Wave, Ocean	-	-	-	-	-	-	-	-	-
Wind	-	-	-	-	0.02	0.02	0.02	0.02	0.02
Other (e.g. Fuel cells)	-	-	-	-	-	-	-	-	-
Combustible Fuels	1.87	2.40	8.35	9.74	13.30	15.93	19.07	20.42	22.41

Source: IEA/OECD, 2007; 585.

Table 2. Sectoral Electricity Demand by Years (TW)

	2003	2005	2010	2020
Industry	53.2	64.7	101.1	227.8
Residential and services	52.2	60.5	94.1	195.3
Agriculture	3.7	4.0	5.0	7.4
Transport	0.9	1.1	1.7	4.0
Net demand	110.0	130.2	201.9	434.6
Gross demand	141.2	163.2	242.0	499.5
Gross consumption per capita (kWh)	1 994	2 232	3 085	5 692

Source: OECD, 2005; 133.

On the other hand, Table 2 gives the current and future figures for sectoral electricity demand by years. The figures show that electricity demand in all of the sectors will also increase in the following years until 2020.

Transmission and Third Party Access

According to the Law of 2001, although transmission facilities will be fulfilled by a state owned monopoly, TEIAS, private sector-owned generation companies would construct private direct transmission lines. The only restriction on this case is that EMRA's granting of generation licenses is conditional on no any congestion in the transmission and distribution lines linking the new plant to the network or directly to consumers[8]. TEIAS is a responsible company from transmission assets, system operation and maintenance, planning and building of new investments for transmission facilities, balancing the market activities physically and financially.

The new Law stipulates a third party access regime for access to national transmission and distribution networks. According to the TPA regime, the third parties are needed to be enabled under fair and transparent conditions an open access to transmission facilities performed by the state owned TEIAS. The probable disputes between parties will be solved by EMRA. For those reasons, in the new model, TEIAS is a system operator compatible with the EU electricity directives and administrator of needed settlements for access of third parties to transmission facilities [Atiyas and Dutz, 2005; 194].

With the Law of 2001, legal and functional separation of transmission facilities from generation has strengthened the vertically unbundling, TPA to transmission, and hence, introduction competition to the market. EMRA has to approve tariffs and conditions for access to transmission network *ex ante* and to monitor and present a report to the Commission on the evaluations related to especially supply-demand balance of the market. As a result, ownership rights of all transmission facilities owned and operated by other companies are needed to be transferred to TEIAS. Likewise, the franchise rights of Kepez Elektrik and CEAS were annulled by the Ministry and their property rights were transferred to TEIAS in June 2003, because of both companies had failed to operate efficiently them and it is legally a necessity [Atiyas and Dutz, 2005].

Distribution

A main aim of the Electricity Market Law was the implementation of privatizations in distribution as soon as possible, although it has not yet been possible. Nevertheless, the government and EMRA have tried to implement privatization of distribution assets owing to the provisions of the Law as well. In the Turkish electricity market, distribution facilities have been dominated by TEDAS and seven regional distribution companies. As shown in Figure 5, distribution network is divided into 21 regions based on energy demand, geographical nearness, administrative structure, and some technical and financial components. As one of them has been operated presently under a type TOOR contract, the government has aimed to privatize other 20 distribution regions owned by TEDAS [OECD, 2005].

According to the Law, obtaining a generation license, any company engaging in distribution facilities has right to construct and to operate generation facilities in the region specified in its license. However, electricity generation of distribution companies cannot exceed 20% of consumption in this region in the previous year.

---

[8] A congestion in the grid is most likely to be solved via some type of auctions between the companies that would benefit from the transmission investments (Atiyas and Dutz, 2005; 193)

Although TEDAS will heavily perform distribution and retail facilities of electricity until competition will be fully introduced in the market, TEDAS and private companies having a retail sale license are needed to segregate both facilities each other [Ozkivrak, 2005; 1344].

There have been some serious issues delaying the implementation of privatizations in the distribution market. A major issue is losses and illegal use of electricity [OECD, 2002; 107]. The rate of electricity losses in the distribution networks is four times as high as the average for other IEA member countries. The rate reached 21.5 in 2000, but declined to 19.5% in 2003 and 18% in 2004. Approximately half of the losses in electricity results from illegal use. The government aims to reduce the losses and to facilitate needed investments and efficiency considerations by applying policies to privatize distribution assets [OECD, 2005; 139].

The new Law stipulates TEDAS to issue distribution and retail sale tariffs and MENR to approve them, but not EMRA. Tariffs have been sustained at uniform levels regardless of the differences between east-west regions because of the national pricing rather than the cost-based regional pricing. EMRA and the new Law intend to convert this price structure to a new one, including the regional pricing. In addition, the government deals with to develop an equalization mechanism. The equalization mechanism includes the cash subsidies between east and west regions under the audit of the council of Ministers [OECD, 2002].

Market Opening and Eligible Consumers in Wholesale and Retail Segments

The Electricity Law enacted in 2001 gave place to the components of market opening and eligible consumers for a competitive electricity market. The Law enabled eligible and non-eligible consumers to connect directly with their own suppliers. Eligible consumers can *de jure* buy electricity from any generation, distribution, wholesale and retail company directly by signing a contract with them, as non-eligible ones can purchase only distribution or retail companies at rates regulated by EMRA. In addition, they can *de jure* directly access to the transmission grid. Open access or market opening is a crucial factor to ensure a competitive electricity market. By ensuring market opening, it is provided third parties to access to distribution and transmission networks in a competitive environment. However, access to the grid is implemented via regulation by EMRA and prices and the conditions are regulated by EMRA [OECD, 2002; Ozkivrak, 2005]. Besides, eligible consumers have the right to sign a contract to purchase electricity with TETAS under the regulated price and tariff conditions [OECD, 2005].

According to the Law, the initial eligibility threshold for eligible consumers is 9 GWh for each year. The Law gave the right EMRA to re-determine the eligibility threshold every year up to full market opening aim has been reached. Thus, eligible consumers started to alter their own suppliers as of 3 March 2003. In January 2004, EMRA re-determined the eligibility threshold as 7.8 GWh. Accordingly, market opening is 29%[9]. In October 2004, the figures of eligible consumers who signed a bilateral contract with a new supplier reached approximately to 270 [OECD, 2005; 146-147].

---

[9] In 1999, this rate was 20% (OECD, 2002; 22).

1. Dicle Elektrik Dagitim A.S
2. Vangolu Elektrik Dagitim A.S
3. Aras Elektrik Dagitim A.S
4. Coruk Elektrik Dagitim A.S
5. Firat Elektrik Dagitim A.S
6. Camlibel Elektirk Dagitim A.S
7. Toroslar Elektric Dagitim A.S
8. Meram Elektrik Dagitim A.S
9. Baskent Elektrik Dagitim A.S
10. Akdeniz Elektrik Dagitim A.S
11. Gediz Elektrik A.S
12. Uludag Elektrik Dagitim A.S
13. Trakaya Elektrik Dagitim A.S
14. İstanbul Anadolu Yakasi Elektrik Dagitim A.S
15. Sakarya Elektrik Dagitim A.S
16. Osmangazi Elektrik Dagitim A.S
17. Bogazici Elektrik Dagitim A.S
18. Kayseri ve Civari Elektrik T.A.S
19. Menderes Elektrik Dagitim A.S
20. Goksu Elektrik Dagitim A.S
21. Yesilirmak Elektrik Dagitim A.S

Figure 5. Turkey's Electricity Distribution Companies.

Although wholesale and retail markets are opened to competition, market activities in these segments have been subject to regulation. For example, share of wholesale companies in the market cannot exceed 10% of total consumption. On the other hand, although wholesale companies can purchase electricity from the state-owned generation companies, eligible consumers have not the right to do so [OECD, 2005; 143].

There are some limitations on competition in retail segment as well. For example, although retail companies can engage in retail sale without any restrictions relation to regions, distribution companies obtaining a license for retail activities can sell electricity to eligible consumers only in the region where a distinct distribution company fulfills the market activity, including provisions concerning retail activities in its retail license [Ozkivrak, 2005].

With the new Law, TETAS has been a dominant wholesale company in the market. Since the enactment of the new Law, all energy sale and purchase contracts of TEDAS and TEAS have been took over by TETAS. This case means all BOT, BOO and TOOR projects and export and import contracts to be executed by TETAS. The Law stipulated EUAS to sell all electricity generated by it to TETAS. But, the dominant role of TETAS in the electricity market has still continued, because of this case did not expire in 2007 [OECD, 2002; 113].

4. Main Issues in the Current Market Structure

4.1. The Problem of Stranded Costs and Disintegration

A major problem of the electricity market is stranded costs stemming from investments implemented by the long-termed BOT, BOO and TOOR contracts of the previous regulatory model. Some contracted BOT projects are given in Table 3. Because such costs in the electricity markets cannot be recovered in a competitive electricity market some non-competition ways are engaged in to solve the problem[1]. For example, the government subsidies, a charge on consumers, and bundling with stranded benefits are some of these ways. Turkey preferred the latter approach among them. For that reason, the new Law temporarily allowed affiliation between TEAS in generation and TETAS in trade to recover stranded costs within this temporary time. By choosing this way, it was aimed prices not to rise. On the other hand, the bundling way is technically easer than recovery methods.

Table 3. Contracted BOT Plants

NAME	TYPE	CAPACITY (MW)	CONTRACTED GENERATION(GWh)
Mazı I	Wind	39.2	131.9
Kapıdağ		35.3	87.7
Bandırma		15.0	49.6
Datça		28.8	84.3
Kocadağ II		26.3	80.5

[1] In such cases, less competitive applications can easily ensure the recovery of stranded costs or higher privatization revenues for the state budget.

Table 3. Contracted BOT Plants

NAME	TYPE	CAPACITY (MW)	CONTRACTED GENERATION(GWh)
Çeşme	Geothermal	12.8	36.0
Germencik		25.0	187.2
Yukarı Akçay	Hydro	2.0	15.0
Arıt		9.0	46.0
Pamuk		19.8	81.4
Keklicek		17.0	42.2
Suşehri Projesi		11.4	59.2
Lamas III-IV		41.0	198.0
Aralık		16.0	50.8
Kavşak Bendi		140.0	652.3
Çamlıhemşin		42.0	195.0
Akköy I		65.0	259.0
Boyabat		510.0	1290.0
Yedigöze		250.0	785.6
Akköy II		190.0	679.0
Andırın Enerji Grubu		41.0	147.4
Keban Deresi		5.0	31.5
Fımıs		7.3	36.0
Çaldere		8.0	35.0
Kovada III		2.8	56.7
Kdz. Ereğli	Gas	206	1413
Yalova		307	2297
Kirklareli		75	597
Eskisehir		199	1391
Konya-ilgin	Coal	425	2835

Source: MENR.

Finally, an important disadvantage of the method used to recover stranded costs is obstruction of competition up to recovery of stranded costs. Because, the dominant role of TEAS in generation and TETAS in trade has still went on. For that reason, competition in generation and wholesale segments is dependent on how long the dominant role of TEAS and TETAS continue. True competition in generation and later wholesale would only materialize, as generation assets are privatized, as the dominant role of TETAS was removed by the competitive entries to wholesale activities, hydropower was released in the market again, and as the excess of expensive BOT projects is met [OECD, 2002; 114-115].

Although the new Law disintegrated industry into generation, transmission, distribution, wholesale and retail segments, the transition to competition has brought about some problems. TEUAS in generation, TEDAS in distribution and TETAS in trading are currently state owned monopolies. It appears rather difficult these public utilities to privatize and competition to introduce to these segments in a near future.

4.2. Lack of Excess Supply

A crucial problem of the new structure is the lack of excess supply. As seen in Table 4, in Turkey, the excess in the supply of electricity has not been provided up to now and the future projections regarding this issue are not optimistic. Since, it does not seem possible that the supply can the demand as well, while the demand is estimated to increase until 2010. This case impedes competition in the market and erodes EMRA's competitive policies.

On the other hand, Turkey has been dependent on imports in energy, because of insufficiency of the supply. In addition, the ratio of consumption/generation is not expected to increase but decrease in the future. It appears that the state owned assets or the governments will go on to dominate the market, because of the excess supply would not be provided cheaply in the future. Furthermore, the secondary legislation has not enough incentive firms to implement their activities in plants on time. Although there are 118 applications for wind plants in 2005, most of them did not correct even the feasibility reports needed to give sufficiency to firms.

Consequently, while excess supply would facilitate transition to competition, the existing lack of supply increases the dominant role of the state in the electricity market. This also gives rise to the government to intervene in the market and hence, postpones competition to introduce all segments of the electricity market [Oguz and Çetin, 2005].

Table 4. Energy Production-Consumption and Import Expectations

YEARS	2000	2005	2010	2020
Consumption	79.671	129.625	171.339	298.448
Production	27.593	34.116	47.329	70.238
Import	52.078	95.509	124.010	228.210
Consumption/Production Rate (%)	34.6	26.3	27.6	23.5

Source: The Undersecretariat of Foreign Trade, Turkey, 2002.

4.3. Retail Tariffs and Cross Subsidies

Electricity prices in Turkey contained before the reform and at present go on to contain cross-subsidies between consumer groups, specifically from industrial users to household users, between different geographical regions, and between various public enterprises, in spite of the government's initiative to remove from the market them during the competitive market reform of 2001. However, it is not expected that at least regional cross-subsidies will go on up to 2010.

As the case of another cross-subsidy, various taxes and levies have continued to cover the electricity prices, although the new Law prevents inclusion of such costs on electricity prices[2]. For example, as mentioned above, a 2.5% levy for TRT, and a 1% the municipality tax means a cross-subsidy from electricity users to TRT and the municipalities by means of the electricity prices. Additionally, an 18% the VAT rate for electricity in Turkey are rather

[2] The surcharges of EMRA on electricity TPA tariffs are an exception to this application.

high especially for the industrial consumers compared to other OECD countries, but mid or lower levels for household users, as shown in Figure 6 and 7. Furthermore, when electricity prices for industry users are compared with ones of other OECD countries in Figure 6, it is seen that Turkey is at the highest third level. This case has weakened the rivalry power of the industrialists in Turkey against the international competitors.

On the other hand, it is seen in Figure 8 that electricity prices for both industry and household consumers have been stable since the mid-1990s. However, when it is compared electricity prices for industry and house users in Figure 8, it appears that they are nearly same. This case means a cross-subsidy from the industrial users to the residential consumers of electricity.

Another problem concerning electricity prices in Turkey is that wholesale prices are currently high, because of the ongoing existence of prices determined according to the previous BOT and BOO contracts. At present, the wholesale prices are approximately $ 0.052 per kWh. However, it is estimated prices to decline gradually, if 40% share of the previous contracts in total generation decreases [OECD, 2005].

A vital problem relation to electricity prices in the current structure is pricing method. In Turkey, it is engaged in uniform pricing method throughout the country. However, both the new Law and EMRA have preferred a cost-based regional pricing, due to the cost of supply varies from region to region. EMRA will regulate prices for the end users by using the so-called "price equalization mechanism" during transition to the cost-based regional pricing. With this mechanism, all supply costs will be taken into account; the price will be determined according to these tariffs; and cash subsidies will be made for the distribution companies confronting a supply cost exceeding the uniform national price [OECD, 2005; 141-142].

4.4. The "Strategy Paper" of March 2004

A Strategy Paper was enacted in March 2004 by High Planning Council at the initiative of government. The Paper introduces a detailed strategy to implement privatizations in the electricity market.

The main idea of the Paper is to give the regulatory power to the government to implement privatizations at the expense of reduction in the regulatory discretion of EMRA over the market. The Paper is focused on distribution rather than other market activities and postponed the transition from the national pricing to the regional pricing, in spite of EMRA's insistence on the regional tariffs. It has impeded competition almost in all segments of the market and revealed the divergence between EMRA and the government over the regulatory issues [Ulusoy and Oguz, 2007]. We can summarize the differences of opinion between EMRA and the government arising in the Strategy Paper as follows.

The Paper gave priority involvement of distribution companies to generation and also retail activities. Accordingly, TEDAS will go on to dominate the electricity market by remaining a public monopoly and retaining property of the distribution network. Private companies will be needed to follow both all the related regulations of MENR via TEDAS and EMRA because they are needed to sign TOR contracts with TEDAS and to obtain a license for distribution activities from EMRA. This case will at least cause overregulation in the market instead of competition. Thus, the Paper impeded disintegration in the market by constructing a barrier to the unbundling end against EMRA's opinions [Ulusoy and Oguz, 2007; 5031].

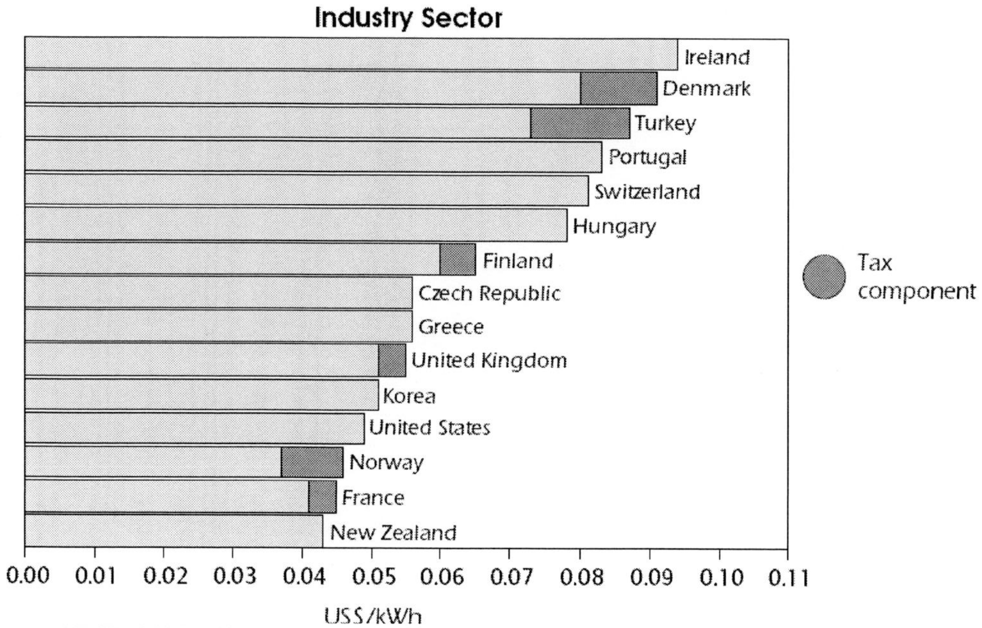

Source: OECD, 2005; 140.

Figure 6. Electricity Prices and Tax Components for Industry Consumers in IEA Countries.

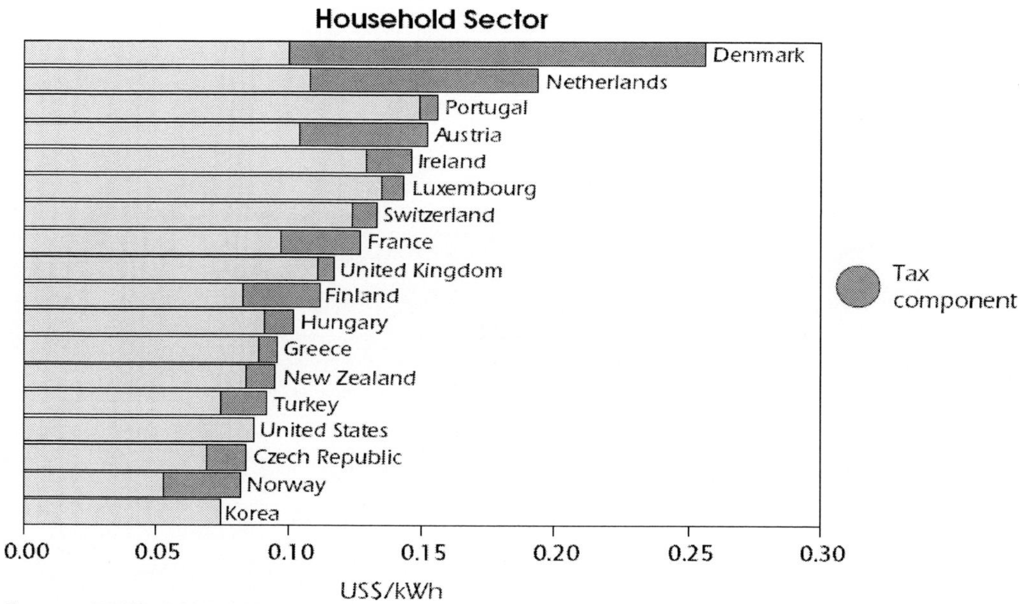

Source: OECD, 2005; 140.

Figure 7. Electricity Prices and Tax Components for Household Consumers in IEA Countries.

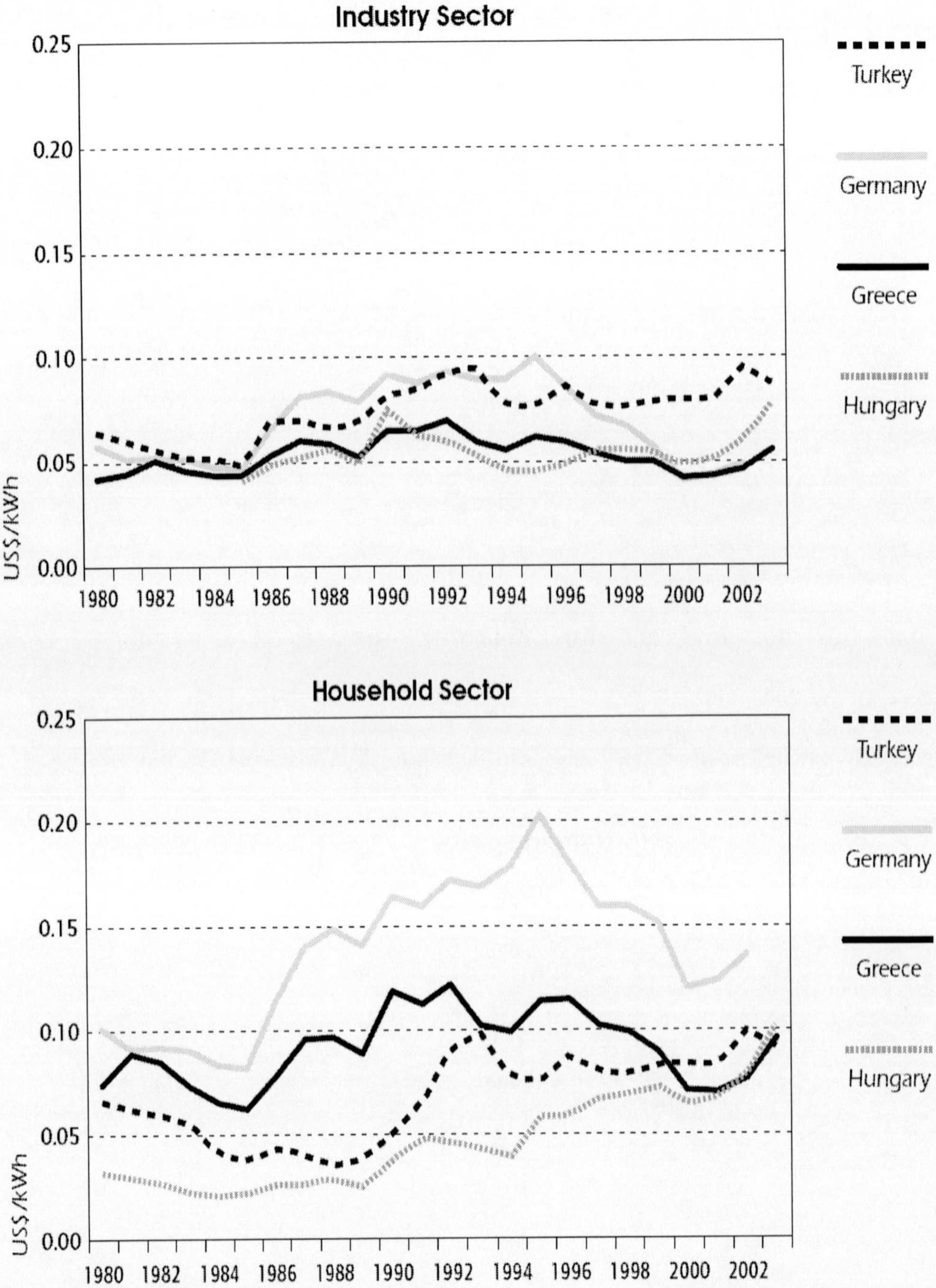

Source: OECD, 2005; 141.

Figure 8. A Comparison for Electricity Prices between Turkey and Some Other Selected IEA Countries (1980-2003)

Although the Law gave the right EMRA to re-determine the eligibility threshold every year up to full market opening aim has been reached, the Paper fixes the threshold at 7.8 GWh up to the beginning of 2009 and envisages the implementation of full market opening by the beginning of 2011 [OECD, 2005; 147]. For that reason, competition in the retail market is *de facto* seriously restricted until 2009, in spite of *de jure* status of the market. Because, according to the Strategy Paper, thresholds for eligible consumers will remain fixed and non-eligible consumers cannot purchase electricity from suppliers other than the distribution company in their regions. However, while it is expected that welfare losses from delays in the implementation of competition in the retail market are not significant ones, it is anticipated that welfare losses arising in the wholesale activities are more important. Such welfare losses can be prevented by reducing eligibility threshold to rates that would cover most consumers than residential and small commercial users [Atiyas, 2006; 19].

4.5. Privatization of Distribution Assets and the Equalization Mechanism

In given in Table 5, one of the most serious problems in the Turkish electricity market is the highness and interregional variability of illegal/inefficient use rates of electricity in the distribution regions. Illegal/inefficient use of electricity stems from the technical losses and theft in the electricity distribution. While the losses and theft rates can reach to 75.8% in Mardin and 66,2 in Van, in the Southern and Eastern Anatolia Regions, in turn, they can reduce 6.8% in İzmir and 7% in Bilecik, in the Ege and Marmara Regions, respectively. Accordingly, the losses and theft uses for the distribution companies in the Eastern and Southern Anatolia Regions are the highest ones. However, the rates in the West Regions are lower than the East Regions.

It is another problem that the government strives to privatize the distribution assets before any initiative is taken against the illegal and inefficient use. It means that the struggle obligation against the technical losses and theft will be left to the private distribution companies after privatization of the distribution assets.

Table 5. Illegal-inefficient use rates in some regions (2005)

Regions	Cities	Illegal-Inefficient use rate (2005)
East and Southeastern Anatolia (The highest 10 cross-subsidized cities)	Mardin	75.8
	Van	66.2
	Hakkari	65.6
	Batman	65.6
	Diyarbakir	65.2
	Sanliurfa	60.2
	Sirnak	59.9
	Bitlis	55.1
	Mus	53.3
	Agri	50.9

Table 5. (Continued)

Some other cities in Akdeniz, Marmara and Ege (Cross-subsidizing cities)	Mersin (Akdeniz)	19.9
	Istanbul (Marmara)	13.3
	Kocaeli (Marmara)	12.5
	Mugla (Ege)	11.3
	Tekirdag (Marmara)	10.4
	Edirne (Marmara)	9.8
	Antalya (Akdeniz)	9.5
	Kırklareli (Marmara)	9.0
	Bilecik (Marmara)	7.0
	Izmir (Ege)	6.8

Source: www.tedas.gov.tr and www.belgenet.com.

This policy is controversial. The loss and theft rates should be reduced by a public authority before privatization. Thus, the delay problem of privatization in distribution would also be solved.

On the other hand, the cost of the current system is cross-subsidized from regions with the low level to the high level regions due to the national pricing in electricity. In order to solve this matter, the government aims to transfer $2.5 billon to these regions as subsidized electricity until 2011 via the new price equalization mechanism. However, the equalization mechanism is not enough to overcome the problem by oneself. Additionally, it should be passed to the regional pricing as well, instead of the national pricing. Privatization of the distribution assets should be implemented after these policies [Atiyas, 2006].

5. Policy Recommendations for Future

Although the reform started with the good intentions regarding competition of the market, some problems have occurred during the following process. In order to surmount such problems, the institutions whose are responsible for regulation of the electricity market should be taken initiative. Otherwise, the reform process can deviate from the aim and fail at the very beginning of the reform. For that reason, this section of the chapter includes some policy suggestions for future of the market to the responsible institutions. Accordingly, the government, EMRA and the state-owned electricity companies should:

- insist on to encourage the private participants for investments in the market under fair and competitive conditions.
- take into account the independence of EMRA to ensure a credible commitment for the investors and to introduce more competition to the market.
- ensure independence of TEIAS, the transmission grid and the market operator, from the political pressures.
- go on the attempts to be compatible of the Turkish power system with the European grid.

- encourage the source diversification in the electricity generation, such as nuclear, thermal and wind power.
- note the security of supply and environmental issues.
- reduce the illegal and inefficient use rates in distribution and later pass to privatization of the distribution assets.
- define clearly duties and responsibilities of EMRA and the government over all the market.
- comply with especially independence of the independent regulator.
- take measures to minimize the discretional interventions of the judiciary to the market activities, privatization and independence of EMRA.
- follow the roadmap the Electricity Law of 2001 has shaped [OECD, 2002; 2005].

CONCLUSION

Enacting the Electricity Law of 2001, Turkey has reformed the electricity market. The aim is to introduce competition to competitive segments of the market and to regulate the natural monopoly elements. For that reason, EMRA is established as an independent regulator and the market has been opened to competition gradually. According to the Law, EMRA is responsible for all regulations of the market and independence from the political authority and private interest groups in the market.

However, the reform process has brought about some problems. Timetable relation to competition and privatizations determined by EMRA and the Law of 2001 has often been interrupted by the government. The reform process has deviated from the beginning aim. Privatizations have not been able to implement on time. Some state-owned assets whose are needed to be privatized according to the Law and EMRA have remained as a monopoly in the market. EMRA's insistence on competitive pricing methods and the nullification of cross-subsidies over the tariffs has been ignored by the political authority. In the end, this case has impeded competition in the market.

To solve the problems, the government should take measures. It should comply with the Electricity Law and the EU Electricity Directives. It should be passed from the national tariff to the cost-based regional tariff in pricing of electricity. In particular, privatizations in distribution have to be implemented immediately. The legislature and the judiciary are needed not to intervene in the regulatory discretions of EMRA related the market.

REFERENCES

Atiyas, Izak [2006]; Elektrik Sektorunde Serbestlesme ve Duzenleyici Reform (Liberalization and Regulatory Reform in the Electricity Sector), TESEV Publications, Istanbul, (in Turkish).

Atiyas, Izak and Mark Dutz [2005]; "Competition and Regulatory Reform in Turkey's Electricity Industry“, in (eds.) Hoekman, Bernard M. and Subidey Togan, Turkey:

Economic Reform and Accession to the European Union. Herndon, VA, USA: World Bank, The, 187-208.

Çetin, Tamer and Fuat Oguz [2007a]; "Reform in the Turkish Natural Gas Market: A Critical Evaluation" *Energy Policy,* Vol.35, No.7, July 2007, 3856-3867.

Çetin, Tamer and Fuat Oguz [2007b]; "The Politics of Regulation in the Turkish Electricity Market", *Energy Policy,* Vol.35, No.3, March 2007, 1761-1770.

Erdogdu, Erkan [2007]; "Regulatory Reform in Turkish Energy Industry: An analysis", *Energy Policy,*

IEA/OECD [2007]; Electricity Information 2007, IEA/OECD Paris, 2007.

Kibritcioglu, Aykut [2001]; "Economic Crisis and the Governments in Turkey, 1969-2001", *Yeni Türkiye Dergisi,* Vol.1, No.7, 174-182 (in Turkish).

Kulali, Ihsan [1997]; Elektrik Sektöründe Özelleştirme ve Türkiye Uygulaması (Privatization and Turkey Application in the Electricity Sector), DPT Publications, Publication No: 2479, (in Turkish).

OECD [2005]; Energy Policies of IEA Countries, Turkey-2005 Review, IEA/OECD Paris, 2005.

OECD [2002]; Regulatory Reform in Turkey: Regulatory Reform in Electricity, Gas and Road Freight Transport. OECD Reviews of Regulatory Reform, 2002, OECD, Paris France.

Oguz, Fuat and Tamer Çetin [2005]; "The Regulatory Dilemma in the Turkish Electricity Market: Who has the Power?", *International Journal of Business, Management and Economics,* Vol.1, No.2, 5-21.

Ozkivrak, Ozlem [2005]; "Electricity Restructuring in Turkey", *Energy Policy,* Vol.33, No.10, 1339-1350.

Ulusoy, Ali and Fuat Oguz [2007]; "The Privatization of Electricity Distribution in Turkey: A Legal and Economic Analysis", *Energy Policy,* Vol.35, 2007, 5021-5034.

Yigitguden, H. Y. [1999]; "Privatization and Regulation in the Electricity Sector (Elektrik Sektöründe Özelleştirme ve Regülasyon)", Devletin Düzenleyici Rolü: Türkiye'de Elektrik ve Telekomünikasyon Sektörlerinde Özelleştirme ve Rekabet, TESEV, Istanbul (in Turkish).

Yilmaz, A.O. and T. Uslu, [2006]; "Energy Policies of Turkey during the Period 1923–2003", *Energy Policy*, in press, doi:10.1016/j.enpol.2006.02.008.

In: Energy in Europe: Economics, Policy and Strategy
Editors: F. L. Magnusson and O. W. Bengtsson
ISBN: 978-1-60456-829-5

Chapter 7

RESEARCH AND DEVELOPMENT STRATEGIES FOR SUSTAINABLE SECURITY OF ENERGY SUPPLY IN THE FUTURE

Filomena Pinto[1], Rui Neto André, I. Gulyurtlu, and I. Cabrita
INETI, Estrada do Paço do Lumiar, 22, 1649-038 Lisboa, Portugal

ABSTRACT

Due to the constant growth of world population and the increasing needs in life quality, it is predictable that energy demand will have a great increase in the near future. Hence, it is foreseen that in the next decades, fossil fuels will continue to have a great contribution to energy production, particularly to power generation. World's economy depends very much on oil, not only because of its great contribution to the worldwide energy production, but also because it is the natural raw material for the production of many indispensable materials. Such a large dependency on oil has become threatening to economic stability, due to the limited lifetime of known oil reserves and to the constant rise of oil prices, which may compromise the sustainability of economic development. On the other hand, due to the threat of global warming and climate change, CO_2 and other greenhouse gases emissions have to be reduced drastically. Therefore, all the European Union countries have encouraged the use of renewable energies and the use of biomass and wastes for energy production, to reduce the dependency on imported fossil fuels and simultaneously to decreasing CO_2 emissions. To attain these goals, EU regulations, governmental political measures and research and development (R&D) strategies supported by research financing programs are needed.

As in years to come fossil fuels will continue to have a great contribution mainly to power generation, the new power plants should use advanced coal technologies like supercritical steam cycles, IGCC, natural gas based combined cycles, etc. However, to ensure sustainable security of energy supply in the future, several areas still need further scientific and technological developments, such as: 1) the use of oxygen for combustion and gasification, instead of air, 2) the production of H_2 from coal and biomass through gasification, 3), H_2 separation using membranes (dry processes), 4) production of liquids fuels, through methanol synthesis or Fischer-Tropsch synthesis, 5) CO_2 removal and

---

[1] E-mail: filomena.pinto@ineti.pt Tel: 351 21 092 4786 Fax: 351 21 716 6569.

storage at large scale, 6), pyrolysis of wastes to produce fuels or raw materials, 7) nuclear power, 8) biofuels production and 9) renewable energy. The main R&D needs of each process or technologies analyzed are underlined. The role of science is very important and crucial for the development of new robust technologies to ensure sustainable security of energy, but these goals can not be achieved without governmental political support, at national level and integrated in international wide policies.

Keywords: *energy sector overview;* R&D *strategies; new energy technologies; sustainable and security energy supply.*

1. Introduction

The continuous growth of world population, together with increasing needs in life quality has led to a substantial increase in energy demand. It is predictable that energy use will have a great increase in years to come by demand from large emerging economies of Asian countries like India and China.

Oil is the fossil fuel with the greatest contribution to the worldwide energy production and is also the natural raw material for the production of many indispensable materials like plastics and tyres. The large dependency on oil has become threatening to economic stability, due to the constantly oil prices increase, which may compromise the sustainability of economic development. As all fossil fuels reserves have limited lifetime, especially oil, it is advisable to have a better management of the remaining resources, by increasing the efficiency of energy production technologies from conventional resources and by diversifying sources of energy. All the European Union (EU) countries have encouraged the use of renewable energies and the use of biomass and wastes for energy production, with the aim of reducing the dependency on imported fossil fuels and at the same time of decreasing carbon dioxide (CO_2) emissions to prevent global warming, since increasing biomass availability on a sustainable basis absorbs most of CO_2 released. As a consequence of the reduction of greenhouse gases emissions imposed by Kyoto protocol, EU is committed to increase biomass contribution to primary energy in Europe to levels of 10% by year 2010. In the EU, rising concerns about security of energy supply, environmental impacts in general and about global warming in particular have resulted in environmental policies at the level of each Member State that obliges to reduce CO_2 levels.

Most renewable energy technology development has not responded to the presented challenge as fast as expected and rapidly enough to meet needs to diversify energy resources, to substitute fossil fuels, at competitive prices and simultaneously to ensure an effective reduction of CO_2 emissions. Nano-sciences, genomics, superconductivity, electrochemistry and electrophysics, catalysis, biophysics and biochemistry, advanced computation and complex, adaptive systems, were the areas identified as opportunities for basic research to provide the required breakthroughs for emerging energy technologies.

However, in the near future it is expected that fossil fuels will continue to have a great contribution, particularly to power generation. The new power plants would use advanced coal conversion technologies like supercritical steam cycles, IGCC, etc. Some of the new technologies for fossil fuels can already meet the Kyoto targets, like clean coal utilization for power generation or natural gas based combined cycles, but some others still need further scientific and technological developments, such as: 1) the use of oxygen for combustion instead of air, 2) gasification with steam and/or O_2, 3) production of H_2 from coal and biomass through gasification, 4) CO_2 removal and storage at large scale, 5) H_2 separation using membranes (dry processes), 6) steam reforming of natural gas alone or in gasifiers.

Oil still continues to be the predominant fuel source in the transport sector, but in future, it is predictable that both biofuels and electric cars will have an important contribution. The use of electricity produced from renewable sources in electric cars will allow reducing CO_2 emissions and other pollutants associated with oil derived fuels. Prior to biofuels large utilization, further research is still needed in the following areas: 1) biotechnology application to biofuels (biodiesel and bio-ethanol), 2) biomass thermochemical conversion technologies (gasification, Fischer-Tropsch process), 3) pyrolysis of waste materials, namely plastic wastes to recover hydrocarbons, 4) nuclear power and 5) renewable energy. By-products from these processes could have various applications, thus giving these processes an added value.

In the domestic sector, apart from the full generalization of natural gas, solar energy use should be widespread for heating and cooling, especially in southern European countries. However, some technology improvements are still needed, which include the integration of solar systems in buildings, new materials for solar collectors with better efficiencies and lower costs, and coupling of solar thermal systems with absorption cycles.

In this essay it is first presented an overview of the actual energy sector and the predictions for the years to come. Next the energy supply security and environmental consequences of energy production will be analyzed, together with main EU directives and measures to decrease energy insecurity. And last, the necessary R&D strategies to ensure the security of energy supply needed to guarantee sustainable growth and economic development are analyzed. It is emphasized the need for governmental regulations to promote public and private research in energy technologies, to encourage their adoption by private sector and their dissemination in the market. The role of science is very important and imperative to provide the basis and the development of new robust technologies for the future and to lower the costs of already clean and efficient existing solutions.

2. World Energy Position

2.1. Overview of the Actual Energy Sector

Nowadays world is facing two main threats related to energy consumption; one because of not having sufficient and safe supplies of energy at reasonable prices and the other due to the environmental damage caused by consuming too much energy, which may cause changes in global climate. The aims of getting energy security and environmental protection can only be achieved by strong governmental actions that need to have public support.

As may be observed in Figure 1, oil is the fuel with the highest contribution to energy production, being followed by coal and natural gas. Coal share in global energy production is around one quarter and probably will remain constant in the next years, according to the predictions of IEA (IEA, 2007[a]).

Coal reserves are distributed by many countries, but more than 80% of the proven reserves are located in only six counties, as shown in Figure 2. China, USA and India have about half of the global reserves, but these countries are also the largest consumers. Russia, Australia and South Africa account for around 31% of the total coal reserves. According to the last IEA previsions, these reserves will last about 155 years at current production rates (IEA, 2007[a]).

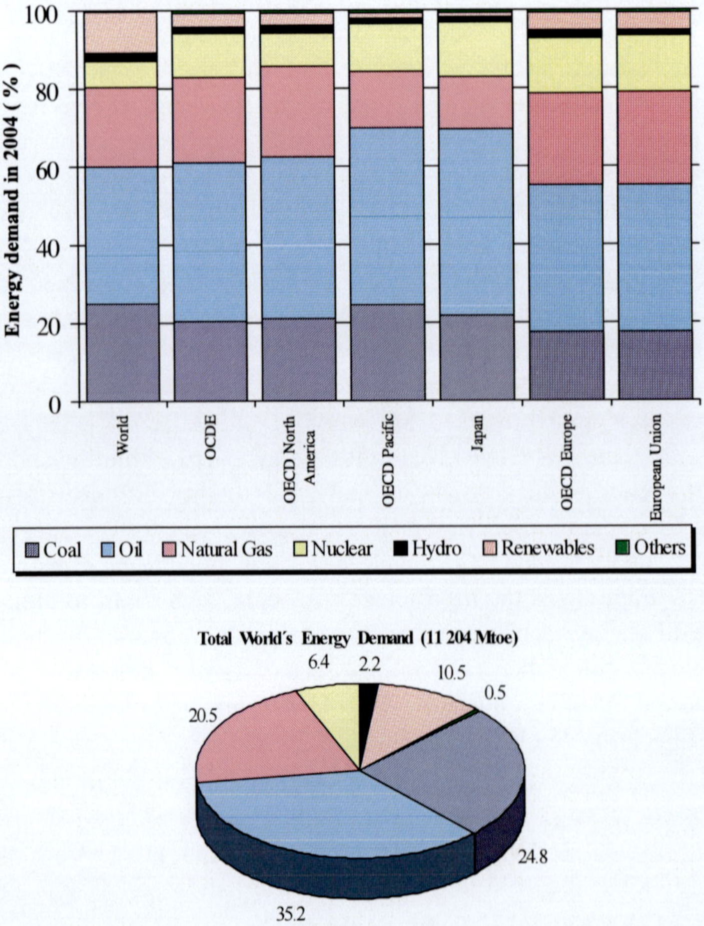

Figure 1 Energy demand in 2004 (IEA, 2007a).

Oil reserves are concentrated in the Middle East and North Africa, these regions account to 62% of the world total known reserves, considering crude oil, natural gas liquids, condensates and non-conventional oil. Saudi Arabia is the country with the largest reserves. Canada has the least developed reserves, which are able to sustain current production for more than 200 years, as presented in Figure 3. However, the world oil proven reserves, counting with non-conventional oil, could only sustain current production levels for 42 years (IEA, 2006[a]). Since 1986, the reserves/production ratio has varied between 39 and 43 years, this variation is not because of new important discoveries, but mainly due to revisions of the estimates of ongoing production fields. The use of new technologies has also improved reserves management and enhanced recovery rates.

However, the predictions for oil reserves duration depend on several factors, which may modify the actual figures, for instance there are some doubts about the ability and motivation of oil companies to increase the installed capacity. OPEC counties plan to launch new major projects to add an estimate of 11.4mb (million barrels) per day of gross crude oil (IEA, 2007a), but much of this new capacity will be needed to replace the lost capacity of declined fields. On the other hand, global oil demand, especially in China may increase more that what is expected and the price of oil barrel will also have a great effect on the balance between offer and demand.

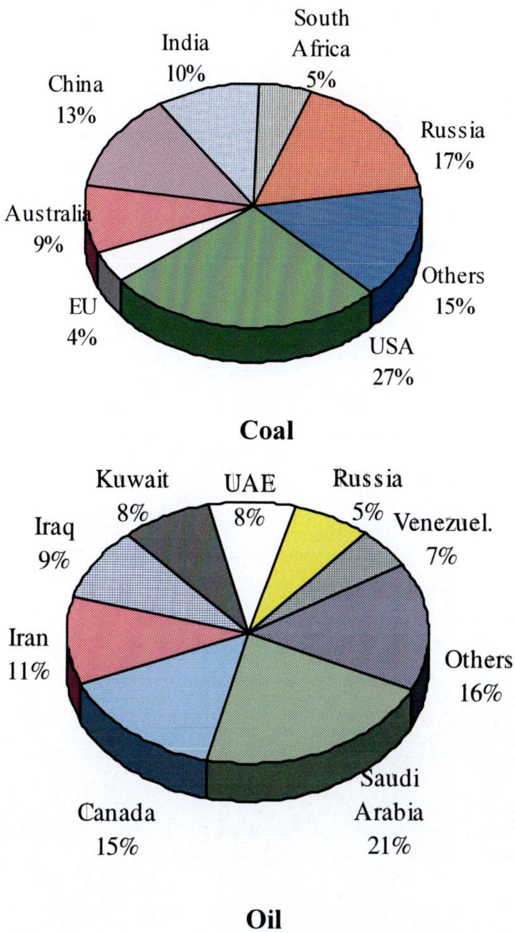

Figure 2. Distributions of coal and oil reserves overall the world (IEA, 2007a).

About 56% of the known gas reserves were found in three countries: Russia, Iran and Qatar. Proven gas reserves have increased 80% in the last two decades, but in recent years, the increase observed in known reserves have come from revisions to fields already discovered that are undergoing development (IEA, 2006[a]). Most of the gas reserves have been discovered during oil exploitation.

There is a great discrepancy between fossil fuels reserves and utilization, as shown in Figure 4. Known oil reserves will last around 42 years, however, this fossil fuel has the largest utilization. While, coal has the greatest known reserves, nowadays, it is less used than oil.

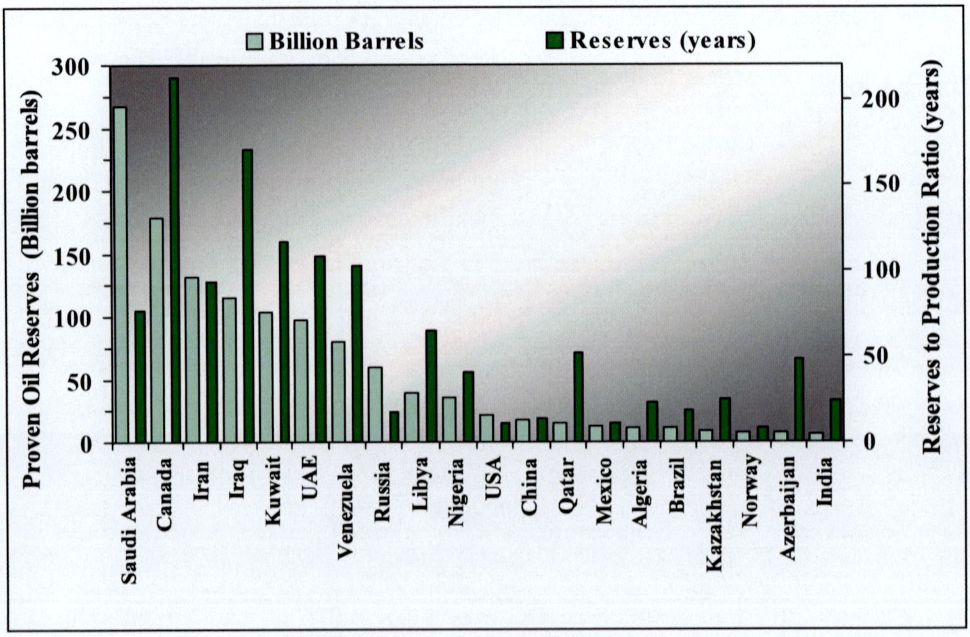

Figure 3. Proven oil reserves at the end of 2005 (IEA, 2006a).

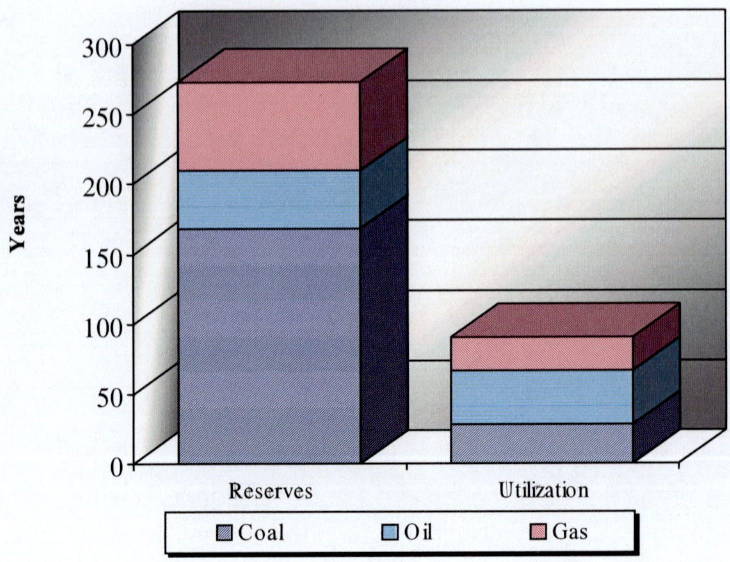

Figure 4. Relationship between energy reserves and utilization (IEA, 2007a).

Proven gas reserves will last around 64 years at current rates, but if the production is going to grow about 2% a year, these reserves will last only 40 years (IEA, 2006[a]).

2.2 Overview of the Energy Sector in the Future

According to the previsions presented in IEA, 2007[a], oil will continue to have the greatest share in energy production, as shown in Figure 5. Apart from oil, both coal and gas will also have great contributions to energy production, being the contribution of non-fossil fuels relatively lower. Two scenarios were considered in IEA predictions: Reference and Alternative Policy Scenario. Both scenarios are based on traditional assumptions about economic growth and the predictions of both scenarios are affected if the evolution of crude oil import prices is going against the expected evolutions.

The Reference Scenario takes account the government policies and measures that were ratified or adopted during 2006. It is also assumed that population growth is in average about 1% per year. Crude oil import price is assumed to be slightly above $60 in 2007, decreasing to about $47 in 2012 and rising again to $55 in 2030. Coal import prices are assumed to stabilize at about $55 per tone in the next few years, increasing to $60 in 2030. In this scenario it is also assumed that energy supply and end use technologies become more efficient. However, crude oil prices have shown great increases during 2007 and especially 2008, reaching values that more than doubled the ones considered in these prediction models, and as crude oil prices are not expected to decrease to the values considered by the models, the predictions done according to them will probably be greatly affected.

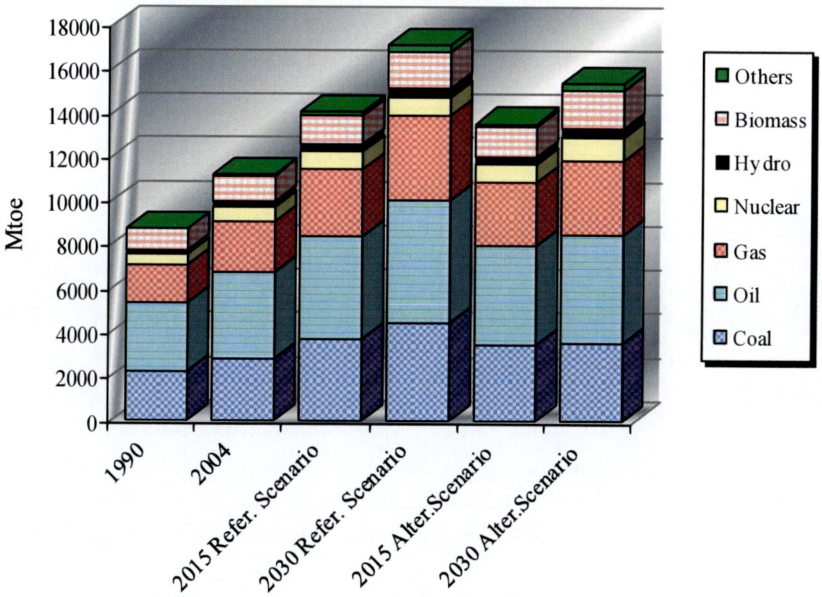

Figure 5 Prevision for energy demand overall the world, considering two scenarios: reference (Refer.) and alternative policy (Alter.) scenarios (IEA, 2007a).

In the Alternative Policy Scenario are considered that governments around the world are going to implement all the policies currently considered to limit the growth in fossil-energy demand and related emissions and to assure energy security. Therefore, the previsions of Alternative Policy Scenario for energy demand all over the world are lower than those of Reference Scenario. Though, in the Alternative Policy Scenario, the world's energy demand and related emissions are considerably reduced, to accomplish greater reduction in both energy demand and emissions, extraordinary policy actions and technological transformations would be needed (IEA, 2007[a]).

Globally, fossil fuels are expected to be dominant in energy production. Oil will stay the single largest fuel, though its share in global demand will probably decrease from 37% (in 2006) to 32%. It is expected that oil demand will grow continuously at an average annual rate of 1.3%, reaching 99 mb/d (mb/d = million barrels per day) in 2015 and 116 mb/d in 2030, while in 2006 this value was 32 mb/d (IEA, 2007[a]). Most of this increase will be due to economic growth of developing countries. China and India together will probably account for about 46% of this increase scenario. Until the 1990s, these countries were minor importers, but their imports have increased, while their own production has remained relatively stable. Nowadays, these countries are major net importers, as are most of IEA member countries. However, these predictions are affected by crude oil import price.

OPEC's contribution to world oil will increase from 42% now to 52% in 2030. Non-OPEC production will probably increase slowly to 2030, coming most of this rise from no conventional sources, mainly Canadian oil sands (IEA, 2007[a]).

The transport sector will probably be responsible for about 63% of the increase in total oil demand between 2004 and 2030 (IEA, 2007[a]). In OECD countries oil use in other sectors apart from transportation will hardly increase, in non-transport sector the increase in energy demand will be of the responsibility of gas, coal, renewable sources and electricity. In non-OECD countries the transport sector is also the biggest contributor to oil demand increase, but it is expected that industry also show significant increase.

The volume of oil trade will probably enlarge from 41 mb/d in 2006 to 51 mb/d in 2015 and 65 mb/d in 2030 (IEA, 2007[a]). The Middle East, the transition economies, Africa and Latin America will probably export more oil. All other regions, including China and India will have to import more oil. As refining capacity for export increases, a growing share of trade in oil is expected to be in the form of refined products, notably from refineries in the Middle East and India.

It is expected that gas demand grows by an average of 2% per year from 2004 till 2030, below the value observed between 1980 and 2004. Therefore, the share of gas in the global primary energy production is expected to grow more modestly than other fossil fuels, from 21% in 2004 to 23% in 2030 (IEA, 2006[a]). Gas demand will increase faster in Africa, Middle East and developing Asian countries, mainly China. Half of the increase in primary gas demand will probably be due to power sector, being this increase, possibly, 2.5% per year from 2004 to 2030. In some regions, gas is preferred to other fossil fuels, because of its cost, competitiveness and environmental advantages. It is predictable that the use of gas to produce liquids will increase in the future. In 2006, a new plant called Oryx, with the capacity of 34 kb/d was built by Qatar Petroleum and Sasol in Qatar. Several other plants are under construction or planned (IEA, 2006[a]).

Coal use is predicted to grow about 1.8% a year from 2004 to 2030. This increase is lower than that observed in the last five years, 5%, mainly due to the great increase in China and India. However, it is expected that coal will have the biggest increase in demand in absolute terms, reaching 73% between 2005 and 2030 and raising its share of total energy demand from 25% to 28%. China and India, which already account for 45% of world coal use, will take around 80% of the increase to 2030 in the Reference Scenario (IEA, 2006[a]). Higher oil and gas prices make coal more competitive as a fuel. However, coal use depends not only on relative fuel prices, but also on government policies on fuel diversification, climate change and air pollution, and developments in clean coal technology in power generation.

Power generation is the sector with the largest utilization of coal, although, the importance of coal for power generation will depend much on the region considered, it is expected that power generation accounts for 81% of the increase in coal demand (IEA, 2006[a]). Even if the deployment of more efficient power-generation technology, would decrease the amount of coal needed to produce electricity, it would probably increase the request of coal over other fuels, therefore, leading to higher coal demand. The use of coal for liquids production is expected to remain marginal, probably because of the high cost of these processes.

In global terms, IEA predicts that the world's energy needs would increase around 55% between 2005 and 2030, at an average annual rate of 1.8% per year. It is expected that energy demand will reach 17.7 billion tones of oil equivalent, while in 2005 this value is 11.4 billion tones. Fossil fuels will remain the main source of primary energy, accounting for 84% of the overall increase in demand from 2005 to 2030 (IEA, 2006 [a]).

In the Alternative Policy Scenario, global primary energy demand grows by 1.3% per year between 2005 and 2030, which means 0.5 (in percentage) less than in the Reference Scenario. Global oil demand is predicted to be 14 mb/d lower in 2030, which is equal to the entire current amount of the United States, Canada and Mexico together (IEA, 2007 [a]).

Developing countries will probably account for 47% of the increase in global energy market in 2015 and more than half in 2030, in comparison with only 41% of nowadays. China and India together will probably account for 45% of this increase scenario (IEA, 2007 [a]). Quality of life of China and India citizens is improving and consequently they are using more and more energy to run their offices and factories and they are buying more electrical equipments and cars. This legitimate desire has to be accepted and supported by the rest of the world. About half of the increase in global energy demand goes to power generation, while around one-fifth goes to transport sector, most of it in the form of petroleum based fuels (IEA, 2007 [a]).

Although considerable progress is expected in enlarging the use of modern household energy services in developing countries, many people will still depend on traditional biomass in 2030. Nowadays, around 2.5 billion people use fuel wood, charcoal, agricultural waste and animal dung to achieve their daily energy needs for cooking and heating. Therefore, around 1.3 million people die prematurely every year because of exposure to indoor air pollution from biomass. According to Reference Scenario, the number of people using biomass will rise to 2.7 billion by 2030, mainly due to the rise of population, which means that one-third of the world's population will still be relying on these fuels, a share scarcely smaller than today. There are still 1.6 billion people in the world without electricity, but this value should decrease to less than one billion by 2015 (IEA, 2006 [a]).

3. Energy Supply Security

The supply of common energy sources at current prices is threatened in the near future, due to different aspects: decreasing of available resources, rising global energy demand and instability in main supplier countries of most important energy resources. It is expected that global energy demand will rise due to the increase of life quality of developed countries, but mainly of under developing ones, like China and India. Rising global energy demand causes a real and growing threat to the world's energy security. China's and India's combined oil imports were around 5.4 mb/d in 2006 and are expected to increase till 19.1 mb/d in 2030, which is more than the currently combined imports of Japan and the United States (IEA, 2007[a]).

By 2030, the OECD as a whole, will probably import two-thirds of its oil needs according to the Reference Scenario, in comparison with the 56% of these days (IEA, 2007 [a]). As non-OPEC production of conventional crude oil and natural gas liquids is expected to attain a peak within a decade, much of the additional imports will come from the Middle East, through vulnerable maritime routes. The concentration of oil production in a small group of countries with large reserves, especially Middle East OPEC members and Russia, will increase their market domination and their ability to impose higher prices. These regions are particularly political and social instable, which may contribute to a further rise of prices. An increasing use of gas is also expected to happen, being the gas supplied by pipelines or in the form of liquefied natural gas from distant suppliers.

Increasing import dependence could enlarge the risk of increasing the energy insecurity for all consuming countries, since main oil reserves are concentrated in a few countries. Much of the added oil imports are expected to come from the Middle East, the region responsible for most disturbances in past supplies, and will travel through vulnerable maritime routes to both eastern and western markets. In case of supply interruption oil prices will increase and the impact in world's economy will be huge.

The more the energy prices rise, the greater the threat to economic growth in importing countries. Uncontrolled rise of oil and gas demand would increase the weakness of consuming countries to a severe supply disturbance and could lead to a price shock. Such a shock caused by a sudden and severe supply disruption would be harmful for all countries, but especially for heavily indebted poor countries, causing strikes in vital economy sectors and revolts and manifestations, as observed in June 2008 in several European countries. As long as the energy demand increases in all regions, the world would have to face higher energy prices, unless high investments are made to increase the supply of low quality fuels or stronger policy actions to control demand growth in all countries are adopted by the governments.

4. Environmental Consequences of Energy Production

These days, the most used energy production technologies are responsible for pollutants emissions that put in danger life in earth as we know it and pose threats to many animal and vegetal species. There are already available new technologies to reduce the emissions of these pollutants, namely particulates, sulfur and nitrogen compounds. Despite their cost, they are being used in developed countries and old coal fired power stations, have been substituted by new ones with more environmental friendly technologies. However, carbon dioxide (CO_2) emissions are still a threat, as they are an important cause in climate change in the last few years and its effect may be worse in the future if there are no changes on energy conversion technologies. Therefore, government actions must focus on limiting the rapid growth in CO_2 emissions from coal fired power stations and from the transport sector. Clean coal technology, particularly CCS (carbon capture and sequestration), is one of the most promising routes for mitigating emissions in the longer term, especially in countries where coal use is growing fastest, like: China, India and the United States. CCS could merge the need of increasing coal burning for energy production with the need of decreasing emissions in the longer term, although to achieve this goal the technology needs to be demonstrated on a large scale and adequate incentives to investment are need with the support of political authorities.

Energy efficiency and conservation will need to play a central role in reducing the increase of electricity demand and decreasing inputs to generation. Nuclear power and renewable energies can also make an important contribution to lowering CO_2 emissions.

As fossil fuels are expected to have a great contribution to energy production, it is expected a continuous growth in energy related emissions of CO_2 and an increased dependence of consuming countries on imports of oil and gas, most of them from the Middle East and Russia. Both trends would intensify concerns about climate change and energy security. According to IEA predictions, CO_2 emissions will rise by 55% between 2004 and 2030, or 1.7% per year, in the Reference Scenario. However, by 2030, they are expected to be 19% lower in the Alternative Policy Scenario than in the Reference Scenario. CO_2 emissions are expected to reach 40 Gt (gigatonnes) in 2030, an increase of 14 Gt in relation to the values of 2004 (IEA, 2006[a]).

If the actions considered in the Alternative Policy are to be taken, emissions in the OECD and in the transition economies will stabilize and then decline before 2030. CO_2 emissions due to energy production are expected to be cut by 1.7 Gt, or 5%, in 2015 and by 6.3 Gt, or 16%, in 2030, in relation to the Reference Scenario. CO_2 emissions in 2030 will be still slightly higher than in 2004, but well below the Reference Scenario level (IEA, 2006[a] and IEA, 2007[c]), because European Union and Japan are expected to take the right measures to decrease current levels. On the other hand, CO_2 emissions in developing countries are expected to keep on increasing, but the rate of growth will slow substantially over the considered period, in comparison with the predictions of Reference Scenario.

Developing countries will be responsible for over three-quarters of the increase in overall CO_2 emissions between 2004 and 2030. The contribution of developing countries to the world emissions are expected to rise from 39% in 2004 to over one-half in 2030. This increase is faster than that of their share in energy demand, because developing countries use proportionately more coal and less gas (IEA, 2006 [a]). For instance, China's emissions will probably more than double between 2004 and 2030, due to strong economic growth and heavy dependence on coal in power generation and industry. Therefore, China will be responsible for about 39% of the rise of global emissions. Before 2010, China will be world's biggest CO_2 emitter, overtaking the United States. Other Asian countries, like India, will also contribute heavily to the increase in global emissions. Nevertheless, the per-capita emissions of non-OECD countries will stay beneath those of the OECD countries (IEA, 2006 [a]).

In 2003, CO_2 emissions due to coal utilization overtook those due to oil and these positions are expected to remain till 2030. Emissions are estimated to grow slightly faster than primary energy demand, overturning the tendency of the last two decades, because the average carbon content of primary energy sources will increase.

All countries should put in motion a transition to a more secure, lower-carbon energy system, without compromising economic and social development. All governments must face the challenge to implement vigorous, immediate and collective policy actions to ensure a more sustainable energy pathway. The cheapest and fastest way to limit demand and emissions growth in the near future is the implementation of measures to improve energy efficiency. But even in these circumstances, in 2030, CO_2 emissions are expected to be still around 25% above current levels (IEA, 2006[a]). To attain bigger reductions in CO_2 emissions, immediate policy actions are required, together with technological transformation on a great scale. Next years will be essential, as the rate of expansion in energy supply infrastructure is expected to be particularly fast. China's and India's energy challenges are most important, as they are major energy consumers. The adoption and implementation of policies by developed countries to deal with energy security and climate change concerns are crucial, but not sufficient, as the contribution of all countries is essential and there can be no successful long-term solutions to the threat of climate change, unless all major energy consumers participate.

5. Policies Available to Decrease Energy Insecurity

In most countries, especially in the developed ones, the public is becoming familiar with the need of guaranteeing the security of energy supply and of protecting the environment by taking measures to encourage more efficient energy use and to enhance the role of renewable energy sources. However, politicians need to explain clearly the need of these measures and their benefits to the economy and to the society, as some of them are likely to find resistance from some industry and consumer interests. For more severe government policy measures, the support of private sector and the international cooperation is fundamental.

Although most energy related investments will have to come from the private sector, the right politic measures are essential to create, implement and support these investments. As the most advanced technologies, to achieve more efficiency and lower the negative impact on the environment, usually are initially promoted by industrialized countries, they will have to help developing countries in adopting these new technologies. International programs to promote technology transfer, capacity of building new plants and collaborative research and development are most needed. A great cooperation between countries and between industry and government is required immediately. IEA has made a lot of effort in establishing this kind of collaboration between more industrialized countries and others like China and India. This may be even more important for small developing countries which, unlike China and India, will probably have more difficulty in attracting investment.

Cooperation between IEA countries and developing countries, including China and India, may increase the deployment of the new mechanisms needed to make easy and encourage the financing of such technologies in developing countries. It is fundamental to attain substantial increase in public and private funding for energy technology research, development and demonstration. The financial burden of supporting research efforts will continue to fall largely on more industrialized countries.

A great quantity of energy and emissions savings could be achieved by stronger policy measures to improve energy efficiency, by enhancing nuclear power and renewables based electricity generation and by supporting CO_2 capture and storage technologies. Some of these technologies are already available or close to commercialization. However, even if these technologies are technically feasible, they still need more efforts to increase their deployment.

Many of the procedures to decrease energy insecurity can also help to diminish local pollution and climate change and they may also contribute to lowering energy costs. These policies depend on the balance of costs and benefits, which differ among countries. Regulatory measures, like standards and mandates, will also be required, together with government support for long-term research, development and demonstration of new technologies.

The implementation of the right political measures to decrease energy insecurity and to reduce emissions is urgent, as a delay of some years would have a great effect on the near future. According to IEA previsions, if the implementation of political measures is delayed till 2015, by 2030, the cumulative avoided emissions predicted by the Reference Scenario would be 2%, while those predicted by Alternative Policy Scenario would be 8% (IEA, 2006[a]), because in Alternative Policy Scenario it is considered that governments around the world are going to implement more policies to limit the growth in fossil-energy demand than those considered in the Reference Scenario. On the other hand, delays in implementing energy related research and development efforts about CO_2 capture and storage, would hold back the possibility of decreasing emissions. Even if governments implement all the policies to control energy imports and emissions, both would still increase through to 2030 (IEA, 2006[a]).

Even keeping global CO_2 emissions at current levels would require much stronger policies, as the normal trend is to increase present values. Technological advances that change deeply the way energy is produced and consumed are most needed. However, the difficulty of implementing these measures is well known, but the faster they start, the quicker a new generation of more efficient and low or zero carbon energy systems can be established, otherwise the environmental cost and the long-term economic consequences will have a very high price to pay.

To decrease energy insecurity, massive investments in energy supply infrastructure are required. More than half of this enormous investment is needed in developing countries, where demand and production increase most rapidly. According to IEA predictions, the cumulative investment is over $20 trillion (in year-2005 dollars) between 2005 and 2030. China alone will need to invest about $3.7 trillion, 18% of the world total (IEA, 2007[a]). However, there are no guarantees that all of the investment needed will be helpful. On the other hand, government policies, geopolitical factors, unexpected changes in unit costs and prices, and new technologies could change the mentioned previsions and may also affect the opportunities and incentives for private and publicly companies to invest in different parts of the various energy supply chains.

Investments on refining are expected to rise to achieve a production capacity of almost 8 mb/d by 2010. Further than the current decade, higher investment will be needed to maintain the growth of capacity. If investments are postponed and OPEC crude oil production is lowered, non-OPEC production is expected to increase and to rise oil prices by one-third, in this case and according to Reference Scenario global oil demand will be about 7 mb/d, in 2030 (IEA, 2007[a]). The investment decisions of the main oil and gas producing countries are of great importance, as they will affect the volume and cost of imports in the consuming countries.

A great increase in reducing CO_2 emissions can be achieved by policies that encourage more efficient production and use of energy. The change to low and/or zero carbon fuels will further decrease CO_2 emissions. More efficient use of electricity in a wide range of applications, including lighting, air-conditioning, appliances and industrial motors, could lead to 30% reductions in CO_2 emissions. More efficient cars and trucks could account for almost 36% of the emissions saved. More efficient energy production contributes to decreases of about 13%, while renewables and biofuels together contribution could be another 12% and the remaining 10% could be achieved by nuclear energy (IEA, 2006[a] and IEA, 2007[c]). In general, the most effective policies in reducing emissions are also those that lead to the biggest reductions in oil and gas imports.

EU have encouraged R&D activities to ensure sustainable security of energy supply by financing research projects in key areas through research programs, like for instance Framework Programs (now the 7th FP is ongoing) and FRCS (Research Program for Coal and Steal). The later finances research about production, preparation, conversion or utilization of coal. 7th Framework Program have supported research projects about the following main areas: hydrogen and fuel cells, CO_2 capture and storage technologies for zero emission power generation, smart energy networks, renewable electricity generation, renewable fuel production, renewables for heating and cooling, clean coal technologies, energy efficiency and savings and knowledge for energy policy making.

Intelligent Energy - Europe Programme main objective is to contribute to secure, sustainable and competitively priced energy for Europe and covers action in the following areas:

- Energy efficiency and rational use of resources (SAVE), including: improving energy efficiency and the rational use of energy, in particular in the building and industry sectors and supporting the preparation and application of legislative measures.
- New and renewable energy resources (ALTENER), including: promoting new and renewable energy sources for centralized and decentralized production of electricity, heat and cooling, and biofuels, thus supporting the diversification of energy sources; integrating new and renewable energy sources into the local environment and the energy systems and supporting the preparation and application of legislative measures.
- Energy in transport (STEER) to promote energy efficiency and the use of new and renewable energies sources in transport, including: supporting initiatives relating to all energy aspects of transport and the diversification of fuels; promoting renewable fuels and energy efficiency in transport and supporting the preparation and application of legislative measures.

European Union is actively promoting several energy related programs focusing mainly on energy efficiency and alternative energy sources. The main EU directives published in last years are summarized in Table 1. The objectives of these EU directives is the promotion of alternative fuels for road transportation, including biofuels, enlarge biomass utilization for energy production, in creasing energy efficiency, promotion of innovative technologies and of renewable energy.

Table 1. Main EU directives focusing mainly on energy efficiency and alternative energy sources

<table>
<tr><th>Date and Reference</th><th>EU Directives and Commission Communications</th></tr>
<tr><td>15 October 1997
COM(97) 514</td><td>A Community Strategy to Promote Combined Heat and Power (CHP) and to Dismantle Barriers to Its Development</td></tr>
<tr><td colspan="2">In preparation to the Kyoto conference it was already predicted that by 2010 it was expected a growth of CO_2 emissions of 10%, compared with 1990 levels. As a negotiation position, the major industrialized countries proposed a reduction of 15% of CO_2 emissions, corresponding to a combined required reduction of 23% on these emissions. It was considered that Combined Heat and Power (CHP) plants are helpful in the reduction of CO_2 emissions with an average amount of 500kg of CO_2 saved per MWh. Including also the then current production potential of around 900-1000 TWh for CHP, corresponding to a level of 40% of the total electricity production of the 15 member countries in 1994. The main goal of the strategy is to increase the level of CHP installed capacity on electricity production from 9% in 1994 to over 18% in 2010. This would imply a reduction of 4% on CO_2 emissions.
The strategy also includes the Member States policies on CHP, which must have the necessary incentives for the penetration of CHP in the internal energy markets. The main barriers to overcome are mainly: economic, related to energy and fuel prices; regulatory, related with licenses and planning authorizations and institutional barriers related with the restrictions and costs associated with the connection to existing electrical grids. There was also the need to promote CHP technology with tax incentives, internalization of external costs and benefits and setting new emission levels for power plants.
This strategy increases European funding on R&D on CHP related technologies, including the improvement of existing solutions and the development of mini and micro CHP plants, with capacities below 30 kWe and 0.5 kWe, respectively. A special attention was made to the use of renewable fuels like biomass and other low heating value fuels and their mixtures. It was also proposed the need of agreements with the industry sectors where there was the highest potential of energy savings by the use of CHP, where several industrial partners could improve their processes and promote the dissemination of the developed technology to third-party companies.</td></tr>
<tr><td>29 November 2000
COM(2000) 769</td><td>Towards a European Strategy for Energy Supply Security - Green Paper</td></tr>
<tr><td colspan="2">Based on the European Union perceived structural weakness concerning a growing dependence on oil as energy source, several choices have to be made in Europe regarding new investments in energy for the growing market, with the due constraints from the related environmental concerns. The Green Paper emphasizes the need for new taxation measures that can change the consumer behavior for the solutions with a lesser environmental impact. On the supply side, energy producers must fight against global warming by developing new technologies using renewable fuels that can increase their share on the supply quota from 6% in 1999 to 12% in 2010, corresponding to raising its share on electricity production from 14% to 22%.
This Green Paper emphasizes the current profiles of energy supply in Europe, stressing the main evolution trends. The fossil fuels being a major part of energy consumption and imports, they impose large restrictions on future increases of their share as the market has been presenting large</td></tr>
</table>

Date and Reference	EU Directives and Commission Communications
variation both in the available amounts and on fuels costs The reference framework for energy presented includes also the restraints related with the reduction of CO_2 emissions from the industrial use of fossil fuels and from the transport sector. It is proposed the harmonization on tax disparities of the energy sector between EU Member States, both on industrial and domestic markets. These changes should improve the renewable sector; to avoid existing trends that instead of promoting national renewable based energy production sometimes only promote imports of fossil fuel based energy from neighboring countries where the taxation profiles are different. State aids in the energy sector must also be evaluated to ensure higher priorities to renewable energy sources. The ultimate goal is the formation of integrated and open energy markets for each energy sector, which can reduce the existing limitations on the establishment of energy networks that can supply its Member States with a maximum contribution from internal and renewable energy sources.	
27 October 2001 Directive 2001/77/EC	Promotion of electricity produced from renewable energy sources in the internal electricity market
This directive promotes an increase in electricity production from renewable sources. These sources include non-fossils energies like, wind, solar, geothermal, wave, tidal, hydropower, biomass, landfill gas, sewage treatment plant gas and biogases. Based on each country energy production profile individual targets are presented, to reach a global indicative target of the contribution from renewable sources in 2010 to 12 % of national energy consumption and 22.1% of the produced electricity. The Member States and the European Commission must ensure the production of electricity from renewable sources, designating regulating bodies to certify and verify the renewable origin of every electricity producer. Regulatory systems must also promote a free and priority access to the electricity grid system for electricity produced from renewable sources. All connecting transmission and distributions fees in the internal electricity market should be clear and non-discriminating against electricity produced from renewable sources.	
7 November 2001 COM (2001) 547	Communication on the promotion of alternative fuels for road transportation and on a set of measures to promote the use of biofuels for transport alternative fuels for road transportation
This communication considers the needs to decrease both the dependence on oil and greenhouse gas emissions to comply with the Kyoto agreements for the industrialized countries. It also meets the current expectations on personal transportation of the population, regarding an easy and comprehensive supply of fuel and with very low restrictions on circulation or parking of their vehicles. These objectives can be achieved by replacing a fraction of the existing diesel and gasoline by biodiesel or ethanol. The European Commission seeks to attain an increase of the potential use of alternative fuels for the total energy needs of the automotive sector by 2020, to a combined level of up to 23%. This includes the increase of the use of biofuels, natural gas and hydrogen of up to 8%, 10% and 5% of total energy needs. Additional benefits could be reached by improvements on the basic vehicle technologies like the promotion of hybrid cars that can achieve similar levels of fuel saving in comparison with the use of alternative fuels by combining combustion engines and electric drives. The individual goals to accomplish include:	

Date and Reference	EU Directives and Commission Communications
Increase fuel economy and reduce CO_2 emissions by 2010, to reach values of 120g CO_2/km and fuel efficiencies of 5.8 litre/100 km for gasoline and 5.3 litre/100 km for diesel. This goal includes commitments on technological improvements of the car manufacturers sector from Europe, Japan and Korea. Increase the use of biofuels as alternative motor vehicle fuels by imposing minimum amounts of biofuels to include in transport fuels. These include direct conversion of plant oils (like colza, soybean or sunflower oils) to a diesel substitute to be used in different amounts as mixtures with conventional diesel. Gasoline can also be also improved by mixtures of bio-ethanol (obtained from the fermentation of sugar beets, cereals and other crops, or even wood or straw) that can be converted to ETBE. Organic wastes can also be used as energy sources by converting waste cooking oil into biodiesel, or agriculture and household wastes to bio-gas or bio-ethanol. Considering the limitations associated with the predictable costs associated with the use of biofuels, distribution and use of biofuels that would be reached by an active support of the non-food agriculture sector and increasing the competitiveness of biofuels in the energy market by tax differentiation. The promotion of natural gas is considered desirable as an alternative fuel. The auxiliary supply network necessary to obtain, compress and distribute natural gas must be optimized to decrease methane losses as this is also a potent greenhouse gas. The existing natural gas network could be used as the infrastructure to supply the transport sector, without major investments. Hydrogen can also be considered as an alternative energy source for transportation, but only if there are accomplished different technological limitations related with storage and fuel cell development. Deployment of a supply network will also need a large investment, which will only be profitable after the necessary improvements on the car related technologies.	
17 May 2003 Directive 2003/30/EC	<u>Promotion of the use of biofuels or other renewable fuels for transport</u>
This directive promotes the replacement of diesel and petrol by diesel and other renewable fuels, for transport purposes. It defines minimum amounts of biofuels incorporation with limits of 2% by 2005 and 5.75%, in terms of energy content. Member States should actively promote this replacement through monitoring, while they consider the environmental balance of different biofuels and renewable fuels. Higher priority should go to solutions with a better environmental balance, competitiveness and security of supply. Information on biofuels should be available to the public, with specific labeling when the limits of 5% of biofuels or bioethanol are exceeded. Starting on 2006 biennial reports should be presented to the European Commission of the economical and environmental impacts levels resulting from the measures taken by each member state.	
21 February 2004 Directive 2004/8/EC	<u>Promotion of cogeneration based on a useful heat demand in the internal energy market</u>
This directive includes all the previous recommendations of the European Commission and aims to create a framework for the promotion of high efficiency cogeneration of heat and power to increase both energy security of supply and energy efficiency. All the Member States should evaluate the potential for application of high-efficiency cogeneration, identifying both the potential heating and cooling demands and consider all the prices and costs barriers associated	

<table>
<tr><th>Date and Reference</th><th>EU Directives and Commission Communications</th></tr>
<tr><td colspan="2">with the relevant fuel markets. Member States must actively support cogeneration units, based on heat demand and energy savings and considering other economically viable or environmental beneficial measures. Access to the existing electricity grid system should be facilitated for cogeneration units, even if they are not eligible under national legislation.</td></tr>
<tr><td>22 June 2005
COM(2005) 265</td><td><u>Doing More With Less – Green Paper on Energy Efficiency</u></td></tr>
<tr><td colspan="2">The Green Paper on Energy Efficiency presents non-binding targets for energy efficiency across the European Union. The plan focuses on efficiency in the transport and housing sectors with the aim of returning EU energy consumption to 1990 levels. It proposes actions at community level on promoting best practices on energy efficiency and renewable energy sources. It includes support on research and technology development on a few key priorities including renewables for power generation and fuel production, clean coal technologies, smart energy networks and energy efficiency improvements. National action plans for growth and employment should also integrate the focus of improving energy efficiency. A special tax harmonization should be promoted, to rationalize all the production and consumption sectors. Specific policy guidelines are applied to increase efficiency on the building sector, on domestic appliances and on fuel consumption of private cars, together with public awareness campaigns for the energy consumers.
At a national level, energy networks should be improved with a better regulation of the supply sector and improvements on energy generation, ensuring the use of the best fuel-efficient technologies and promoting distributed production and cogeneration. The industry sectors should be involved on voluntary agreements to increase efficiency on every industrial process. The transport sector can be improved through the promotion of clean vehicles through taxation and minimum levels of acquisition by public administration. Some benefits could be found from air-traffic reorganization, and change individual behavior by optimizing car traffic management, promote better tyres use and even introducing new tariffs on urban areas to discourage the use of private cars.</td></tr>
<tr><td>7 December 2005
COM(2005) 628</td><td><u>EU Biomass Action Plan</u> (BAP)</td></tr>
<tr><td colspan="2">The EU Biomass Action Plan identified 32 key activities for increasing the bioenergy market and encourages Member States to establish national biomass action plans (nBAPs). Only 4% of EU energy needs came from biomass. If EU made full use of its potential, it would more than double biomass use by 2010 (from 69 mtoe in 2003 to 185 mtoe in 2010), through fulfilling with good agricultural practice, safeguarding sustainable biomass production and without major effects on domestic food production.
According to scientific and economic studies this increase in biomass use could result on the following benefits by 2010:
• Raise the share of renewable energy by 5%, to diversify Europe's energy supply and reduce energy imports from 48 to 42%;
• Decrease greenhouse gas emissions by 209 million tonnes CO_{2eq} a year;
• Increase the employment in rural areas for the range of 250-300 000 people;
• Lower the needs of oil imports, resulting on a reduced trend of oil price increases;
• Extend the European leader leadership in the sectors of energy efficiency and biomass use;
• Reduce the additional pollution and environmental damage resulting from conventional energy production technologies.</td></tr>
</table>

Date and Reference	EU Directives and Commission Communications
8 February 2006 COM(2006) 34	EU Strategy for Biofuels
The Strategy for Biofuels dealt with the greenhouse gas emissions of fuel combustion. To reduce the fuel cycle's emissions, the EU seeks to decarbonise transport fuels, diversify fuel supply sources, offer new income opportunities in rural areas and develop long-term replacements for fossil fuel. EU Biofuels Strategy clarifies that biofuels may be particularly useful in the transport sector, which was responsible in 2004 for around 21% of the EU's greenhouse gas emissions. Biofuels are produced from biomass and being a direct substitute for traditional petrol and diesel can readily be integrated into existing fuel supply systems and also prepare the way for other advanced transport fuel alternatives. The EU Strategy for Biofuels contemplates seven policy axes: (1) Stimulating demand for biofuels (2) Capitalizing upon environmental benefits (3) Developing production and distribution of biofuels (4) Expanding feedstock supplies (5) Enhancing trade opportunities (6) Supporting developing countries (7) Supporting research and development. This strategy intends to help Member States to achieve the 2010 target of 5.75% market share for biofuels in the overall transport fuel supply and to accomplish this objective, several possible measures to encourage biofuels are described including: revision of the Biofuels Directive in 2006 that might include mandatory biofuel targets for the Member States; active promotion of second-generation biofuels (e.g. from wood); inclusion of biofuel use on the evaluation of the CO_2 emission reduction targets for the transport sector; development of the "appropriate environmental standards" ensuring the sustainability of biofuel feedstock cultivation in and outside the EU; promote an "ad hoc" group to consider the production of biomass and biofuels in national rural development programmes; possible inclusion on a forestry action plan to be developed in 2006 of special measures for the production of material for second-generation biofuels; development of a code for trade of biofuels; development of a biofuel assistance package for developing countries and promotion of research into the "bio-refinery" concept with further developments of an industry-led "biofuel technology platform".	
8 March 2006 COM(2006) 105	Green Paper on a European Strategy for Sustainable, Competitive and Secure Energy
This Green Paper provides a strategic framework for imminent improvements on common energy policy. The European Commission identified several preliminary goals to strengthen and secure the European energy market. The Green Paper also lists energy efficiency and renewable energy production in the context of sustainable, secure sources of energy: Improving network security, possibly by securing the cooperation of network operators and through a new formal European grouping, which could be develop into a European Centre for Energy Networks: A Community wide debate on the different energy sources, including costs and contributions to climate change. This could ensure that the EU's energy mix pursues the objectives of security of supply, competitiveness and sustainable development. An action plan on energy efficiency with commonly agreed objectives. In addition, the European Community could propose an international agreement on energy efficiency. A road-map for renewable energy sources, to ensure a predictable framework for the long-term development of renewable energy sources in the EU. It could encompass a proposal for renewable targets for 2020 and a clear short, medium and long term plan to stabilize and gradually reduce the EU's dependence on imported oil.	

Date and Reference	EU Directives and Commission Communications
A strategic energy technology plan, building on European technology platforms and possibly leading to joint technology initiatives to develop leading markets for energy innovation. A strategic EU Energy Review, for presentation to the European Council and European Parliament in Spring 2007. This would cover external energy policy and the possibility of limiting dependence on imports of oil and gas, the future of the internal gas and electricity markets, diversity in the EU energy mix and a Road Map for renewable energy. A European energy observatory, enhancing transparency on energy supply and demand issues within the EU.	
27 April 2006 Directive 2006/32/EC	Directive on energy end-use efficiency and energy services
This Directive promotes the cost-effective and efficient end-use of energy in Member States. It provides all the required targets, mechanisms, incentives and institutional, financial and legal frameworks to remove existing market barriers and imperfections for the efficient end use of energy. As each Member State must transpose the directive into national law by 17 May 2008, with the exception of three sub-articles to be incorporated by 17 May 2006, each national action plan on existing energy saving practices must be supplied shortly to the European Commission that will use these reports in developing harmonized European energy efficiency indicators and benchmarks by the end of June 2008. According to the Directive, the Member States shall adopt and aim to achieve an overall national indicative energy savings target of 9 % for the ninth year of application of the Directive, to be reached by way of energy services and other energy efficiency improvement measures. The Member States must then draw up three energy efficiency action plans. The first action plan must be presented to the European Commission by 30 June 2007, the remaining two by 30 June 2011 and 30 June 2014. The Member States must describe their energy efficiency measures in all three plans. The Directive also creates the conditions for developing the market in energy services, and requires the public sector to lead the energy efficiency improvements by its example.	
6 October 2006 COM(2006) 583	Global Energy Efficiency and Renewable Energy Fund (GEEREF)
GEEREF proposes the creation of a global risk capital fund to assemble private investment in energy efficiency and renewable energy projects in developing countries and economies in transition. GEEREF will accelerate the transfer, development and deployment of environmentally sound technologies and help to bring secure energy supplies to the poorer regions of the world. These projects will also combat climate change and air pollution. The European Commission intends to kick-start the fund with a contribution of up to € 80 million over the next four years, and expects that financing from other public and private sources will take funding to at least € 100 million. This means that it will contribute to the financing of investment projects of a value up to € 1 billion. GEEREF will stimulate the creation of regional sub-funds tailored to regional needs and conditions, rather than investing in projects directly. Sub-funds are envisaged for the African, Caribbean and Pacific (ACP) region, North Africa, non-EU Eastern Europe, Latin America and Asia. The focus will be on investments below € 10 million as these are mostly ignored by commercial investors and international finance institutions. Corporate finance will be offered to support small and medium-sized enterprises as well as project finance.	

Date and Reference	EU Directives and Commission Communications
12 October 2006 13855/06 (Presse 284)	Competitiveness and Innovation Programme (CIP)
The Competitiveness and Innovation Programme (CIP) was approved by the EU Council and Parliament to be applied in the period 2007–2013, with a budget of € 3.2 billion. Its main objectives are the promotion of innovative technologies, the increase energy efficiency and new and renewable energy sources in all sectors including transport. These objectives will be pursued through the implementation of three specific sub-programmes, with a total budget envelope that amounts to € 3.2 billion for the seven-year period: The entrepreneurship and innovation programme, which provides new types of financial instruments aimed at supporting eco-innovation in SMEs, will have around 60% of the budget; The information and communications technology policy support programme, with 20% of the total budget, will focus on the development of coordinated actions, the sharing of good practices and the deployment of interoperable solutions within Europe's information society; The intelligent energy-Europe programme, also with 20% of the total budget, aims at encouragement of sustainable development, security of supply and competitiveness as it relates to the energy sector.	
19 October 2006 COM(2006) 545	Action Plan for Energy Efficiency "Saving 20% by 2020"
The Action Plan for Energy Efficiency was published under the title "Saving 20% by 2020" including a complete framework of policies and measures to reduce annual primary energy consumption by 2020 and it also proposes a selection of cost-effective energy efficiency improvement initiatives to be implemented before 2012, attending to energy efficiency improvements across all sectors to transform the internal energy market, with a special focus on products, services and buildings. The plan includes 10 energy conservation priorities and more than 70 planned actions, including new appliance efficiency standards under the 2005 Energy-using Products (EuP) Directive, a revision of energy labeling rules and more-stringent energy performance standards for buildings - including new EU minimum standards and the inclusion of more buildings. Twenty of these actions were implemented in 2007. The plan sets forth future binding efficiency requirements for new electricity, heating and cooling installations lower than 20 megawatts, and a renewed effort to strengthen taxation policies encouraging energy efficiency, including in 2008 a review of the EU's minimum energy tax levels. The European Commission reiterates plans to propose legislation if car makers fail to meet a voluntary target to reduce average new vehicle emissions.	
10 January 2007 COM(2006) 848	Renewable Energy Road Map – Renewable energies in the 21st century: building a more sustainable future
Despite the previously proposed targets of 12% of renewable forces on the global energy consumption by 2010, this communication assumes that this target could not be met; being 10% the more likely level of renewable sources. Different progress levels have been reached for every energy sector. On electricity generation renewable energy sources (non-hydro) have increased their share by 50%, with a higher than predicted growth from wind energy. Biofuels use on the transport sector has not reached the proposed 2% levels, due mainly to market difficulties and	

Date and Reference	EU Directives and Commission Communications
suppliers resistance to the use of bioethanol. On the heating and cooling sector insufficient legislation and promotion has not increased the renewable energy sources, with few positive developments on higher efficiency equipments. To reach the level of 20% of renewable energy sources by 2020, the European Commission proposes a framework where the biofuels should correspond to 10% of the global diesel and petrol consumption in the transport sector. Each Member State circumstances should be considered to evaluate its contribution to the global goals. Each National Action Plan should include targets for renewables inclusion on each energy sector, removing all existing barriers and promoting every aspects related to production, distribution and consumption.	
10 January 2007 COM(2007) 1	An Energy Policy for Europe
Energy supply in Europe must meet the challenges of sustainability, security of supply and competitiveness. Environmental impacts should be decreased, to reach a level of 20% of greenhouse gas emissions by 2020. The proposed energy policy includes an Action Plan to transform EU economy with high efficiency, low CO_2 energy sector while maintaining its competitiveness. A set of measures are proposed to create a real Internal Energy Market, allowing freedom of choice to both citizens and businesses. This includes an unbundling of the energy networks between energy producers and carriers, while creating an effective and independent regulating system at national and European level. Solidarity and cooperation between Member States must be encouraged to assure security of supply, greenhouse gases reduction, higher efficiency and higher integration of renewable sources.	
22 November 2007 COM(2007) 723	A European Strategic Energy Technology Plan (SET-Plan)–"Towards a low carbon future"
This SET-Plan aims to meet the 2020 targets of Energy in Europe with some key technology challenges, while a more ambitious set of goals is also presented to be achieved by 2050. For the 2020 goals, industrial initiatives must be implemented from 2008, on several key areas. These include second generation biofuels, commercial use for CO_2 capture, transport and storage, increases on the capacity of large-scale wind and solar powered electricity generators and commercialization of energy conversion and end-use devices with high efficiency, like poly-generators and fuel cells for the industry, buildings and transport sectors. It must also be developed a single integrated European electricity grid to integrate this diverse set assembly of renewable and decentralized energy sources. The technology plan must combine the effort from the Member States, the Community, and industrial and research organizations. There must be a coherent implementation at all levels, with coordinated activities based on an open-access information and knowledge management system, which must include all technology roadmaps on the energy area to allow regular progress reports of the SET-Plant. On the research sector a European Energy Research Alliance will be implemented to improve current models on project collaboration between National Research Institutes.	

Date and Reference	EU Directives and Commission Communications
23 January 2008 COM(2008) 19	Promotion of the use of energy from renewable sources
Use of energy from renewable sources on 2020 should reach a level of 20% global energy share. Each country can choose its own mix between the sectors of electricity, heating and cooling and transport, with a recommended level of 10% for biofuels use on the transport sector. Until 2010, each member state must adopt and apply a national action plan with regular reports to the European Commission and make all the adjustments if it does not fulfill its own goals. Accounting of energy share from renewable sources must be considered as the sum of contributions from the three main energy sectors. Electricity imports from third countries can also be considered for each country goal if the installations become operational after the entry of force of the directive and it can obey to the renewable guidelines of the directive. Member States should implement Guarantees of Origin to each producer of renewable energy, which will be needed for accounting and tracking of each energy system. These Guarantees of Origin must be recognized by every Member State. Other steps must be taken, including the free access to the electricity grid system and the definition of environmental sustainability criteria for biofuels and other bioliquids, including a calculation of their greenhouse gas impact.	

6. Alternative Technologies to Ensure Energy Supply Security

6.1. Coal Based New Processes

Power generation from fossil fuel is responsible for over one-third of global annual CO_2 emissions and the production and use of fossil fuels contribute to around 64% of the greenhouse gas emissions produced by mankind all over the world (McKee, 2002). The increase of greenhouse gas emissions (methane (CH_4), nitrous oxide (N_2O), halogens and CO_2) is a threat to the planet's natural systems. CO_2 molecules are the least powerful greenhouse ones, but due to its large quantity and increasing emissions levels, it is urgent to reduce their emissions.

Although, nowadays Europe uses more natural gas and renewable energy sources than 20 years ago, the latest significant increase in world oil and natural gas prices forces Europe to look back at coal, a cheaper fossil fuel. Europe has done substantial R&D effort to develop efficient processes of coal utilization mixed with biomass or other wastes, like co-gasification to produce syngas (synthesis gas) and hydrogen rich syngas for an efficient application in fuel cells or storage. The syngas can also be used in motors or turbines for electricity generation or in chemical sysmthesis for the production of liquid fuels. Moreover, European Union has also contributed to the development of oxy-combustion, oxy-gasification and carbon sequestration technologies, through the financing of R&D projects in these areas.

It is of most importance to carry on the research and development of new technologies that allow the use of fossil fuels and wastes for energy production in a more environmentally friendly way.

6.1.1. Oxy-Combustion

Regarding that it is necessary to develop new technologies that can promote the production of CO_2 in concentrations that can enable its further capture or storage, its been proposed since the 1980s a method of promoting the combustion of coal with pure oxygen instead of air, designed as oxy-combustion. This method has been applied in several installations and with the suitable adaptations can be used in existing coal fired furnaces (Buhre et al., 2005 and McDonald et al., 2007).

The idea of oxyfuel firing plants is being developed to enable CO_2 to be captured during fuel combustion. Air is separated into O_2 and N_2 prior to combustion and CO_2 is recycled to replace N_2, as the temperature moderator in the furnace. The flue gas consists only of CO_2, steam and minor species. The CO_2 can readily be isolated for sequestration.

As a first approach, it is clear that by using oxy-combustion, the removal of nitrogen (about 79% vol.) from the air supplied to the furnace will enable the formation of flue gases with much higher CO_2 concentrations. This fact will avoid the need to use more complex or expensive methods for the separation of CO_2 and nitrogen, as the flue gases will contain mainly mixtures of CO_2 and water (Buhre et al., 2005).

However, the retrofitting of existing installations to oxy-combustion will need to be carefully adapted to accommodate the different energy and mass flows that result from this change. The construction materials and the radiation profiles inside the furnace are optimized for air combustion, so some compromises must be reached to modify existing installations (McDonald et al., 2007).

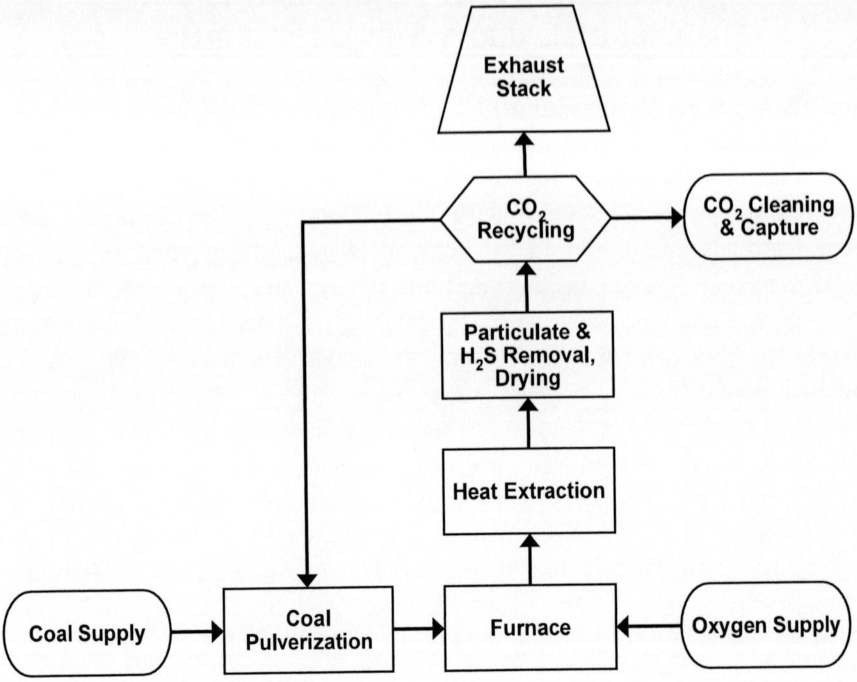

Figure 6. Diagram of oxy-combustion process.

Therefore, in installations with oxy-combustion, it is necessary to supply oxygen with a high purity. As the amount of flue gases will have a strong reduction, it is necessary to promote a partial recycling of the flue gas to increase the volume of gases and help to control the temperature inside the furnace. A flow diagram of the oxy-combustion process, adapted and simplified from (McDonald et al., 2007) is presented in Figure 6.

Some of the main differences of using oxy-combustion, instead of air in conventional processes include (Buhre et al., 2005 and McDonald et al., 2007):

- A large reduction on the amount of flue gas leaving the stack after the recycling, up to 80%.
- The size of the exhaust gas cleaning equipment can be reduced accordingly, with lower requirements on its dimensions.
- There are lower heat losses, due to the exhaust gas reduced flow rate, as the heat of condensation can be captured and reused in the process, instead of being lost in the exhaust gas.
- There are lower requirements for the characteristics of the gas.
- The separation of pollutants (including sulfur gases) from the exhaust gases is easier due to their higher concentrations.
- The larger component of the exhaust gas is CO_2 instead of nitrogen making it particularly suitable for sequestration.
- The density of the flue gas is also higher, as the molecular mass of CO_2 is 44 instead of 28 for N_2.
- Separation by compression is easier, as a large fraction of the flue gases is condensable.
- It has a major impact on nitrogen oxides production, which is very low, as nitrogen from air is not present in the system. The cost associated with the removal of these pollutants is severely reduced.

As a first proof of this concept, several existing air based coal combustion facilities have been retrofitted to oxy-combustion (McDonald, 2007). As a method of reducing the risk of operation, the start-up of the installations is still made using the conventional air promoted combustion at a partial load. After the unit is stabilized the exiting flue gases are partially recycled to the input of the power plant. This step occurs gradually while air input is being closed, being replaced by oxygen. When the power plant shows a stable operation, using only oxygen, the power plant load is raised to the desired values (McDonald, 2007).

These retrofitting technologies have shown that existing power plants can be successfully converted to oxy-combustion, requiring only a proper choice of the recycle ratio to compensate the nitrogen reduction from the gaseous mixtures. Some negative impact arising from the higher CO_2 concentrations is only noticeable on fuel efficiency, but it is largely compensated by the benefits associated with the ease of CO_2 capture due to its higher concentrations. For new installations the beneficial impact of oxy-combustion is even higher as several components can be designed with reduced dimensions with corresponding lower investment and operational costs (McDonald, 2007).

6.1.2. Oxy-Gasification

Gasification is preferable to combustion processes, because of the smaller plant size and the easier control, apart from this, emissions are lower and more controllable, and gasification products have higher added value, above all hydrogen. Though gasification is a complex process with many chemical reactions occurring simultaneously, the main ones may be summarized by equations (1) to (7).

Oxidation $C + O_2 \leftrightharpoons CO_2$ (1)

$C + ½\, O_2 \leftrightharpoons CO$ (2)

Boudouard $C + CO_2 \leftrightharpoons 2\, CO$ (3)

Water-gas:

primary $C + H_2O \leftrightharpoons CO + H_2$ (4)

secondary $C + 2\, H_2O \leftrightharpoons CO_2 + 2\, H_2$ (5)

Methanation $C + 2\, H_2 \leftrightharpoons CH_4$ (6)

Water-gas shift $CO + H_2O \leftrightharpoons CO_2 + H_2$ (7)

Gasification processes tended to be air blown, as the heat necessary to the process could be obtained through partial combustion of the fuels used, through reactions (1) and (2). However, air blown processes have an important disadvantage, as the presence of great amounts of N_2 in the fuel gases decreases syngas heating value and makes CO_2 capture and storage more difficult. Oxygen blown processes are currently seen as more environmentally acceptable coal gasification options, as they can be adapted for the production of H_2 with CO_2 capture. Oxygen blown processes have also the advantage of producing a syngas with higher energetic value, which is advantageous if the gas is used for energy production, due to the absence of the diluting effect of nitrogen. An important disadvantage of this process could be the higher cost of oxygen to be fed to the gasifier. However, the need to captured CO_2 for subsequent storage, obliges to remove N_2 from either the fuel or exhaust gases, which are complex, onerous and expensive processes.

The same idea of oxy-combustion can be applied to gasification. Oxy-gasification gas does not contain N_2, but CO_2, CO, H_2, CH_4, steam, S compounds, N compounds, dust and tars. The flue gas after cleaning only contains CO_2 and steam, as shown in Figure 7. Afterwards, the syngas can be burnt for energy production and after combustion of the fuel gas in O_2, it only contains CO_2 and steam. The steam could be condensed and part of the CO_2 can be recycled to gasifier, the remaining could be transported to storage. The gasification medium is a mixture of O_2 and recycled CO_2. The partial pressure of CO_2 will be much higher than in a conventional air blown system and this should enhance the gasification performance.

Another option for the syngas could be its use in a water gas shift converter to form a hydrogen rich gas for other applications, like in fuel cells.

By this process, a CO_2 rich stream is produced, which is suitable for sequestration, by piping this gas to the storage site. Therefore, this process seams to be more suited for larger scale installations. However, application at a smaller scale could be feasible if the CO_2 could be captured in a solid form, such as mineral carbonates.

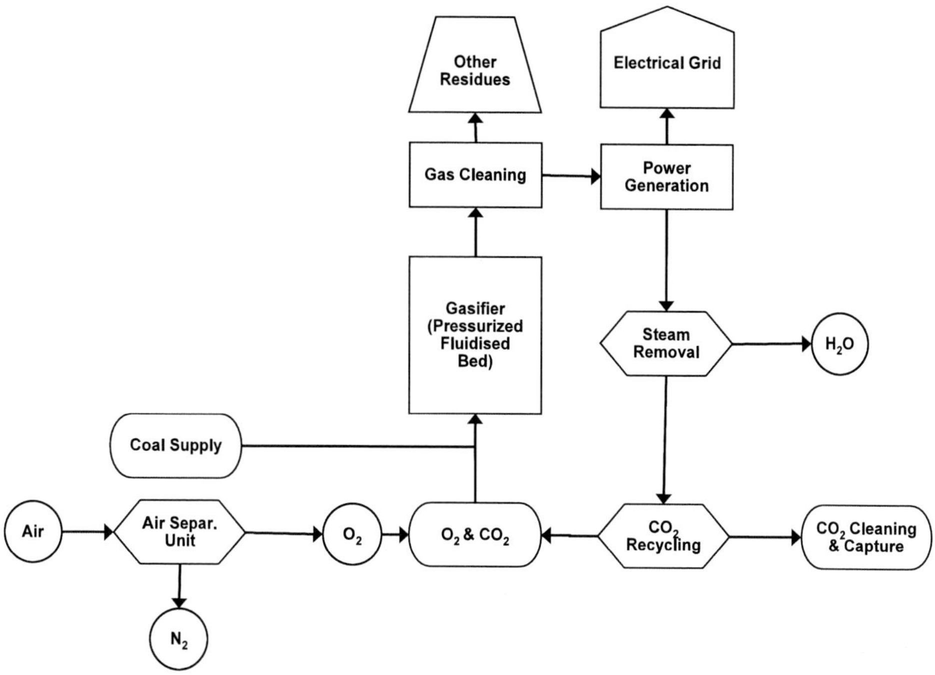

Figure 7. Diagram for the concept of oxy-fuel gasification.

6.1.3 Production of H_2 From Coal through Gasification

Around 95% of the used H_2 is obtained by steam reforming of light hydrocarbons ranging from natural gas to naphtha (Loiacono 2001). Though the cost of H_2 produced is relatively low (Gray et al., 2002) this process depends on price and availability of natural gas. The production of H_2 from coal gasification may be a viable alternative, due to coal availability and relatively stable costs. The syngas produced by gasification may be treated by novel cleaning technologies in integrated process units to make the gas suitable for gas turbines (like in IGCC plants) or for fuel cells in a near future. The gasification process is called 'Integrated Gasification Combined Cycle' (IGCC) when it is combined with a power generation system to recover the chemical energy contained in the purified syngas, through gas turbine/generator equipment and/or fuel cell, followed by waste heat boiler/steam turbine/generator equipment.

Coal IGCC plants have several challenges mainly due to the lower gas pressure and to the low H_2 content of the clean gas (Kubek et al., 2002). More research work is still needed to increase H_2 content in syngas, to improved the efficiency of CO_2 separation processes and to enhance the integration of the H_2 production unit in the overall IGCC process, in order to diminish costs, minimize auxiliary power consumption and heat losses.

The main components of gasification gas are CO, H_2, CH_4, CO_2, but small amount of others undesirable compounds are also formed like tars, NH_3, HCN, H_2S, COS, CS_2, HCl, HBr, or HF. Heavy metals are also formed or released as bottom ash and as fine particulates. The main components to be removed from gasification gas, either physically or chemically are: tars, NH_3, HCN and NO_x, halogens (Cl, F, Br), H_2S, COS and CS_2, heavy metals (mainly Be, Cr, Ni, As, Cd, Pb, Hg), CH_4, CO_2, fine particulates, solid residues (bottom and fly ash).

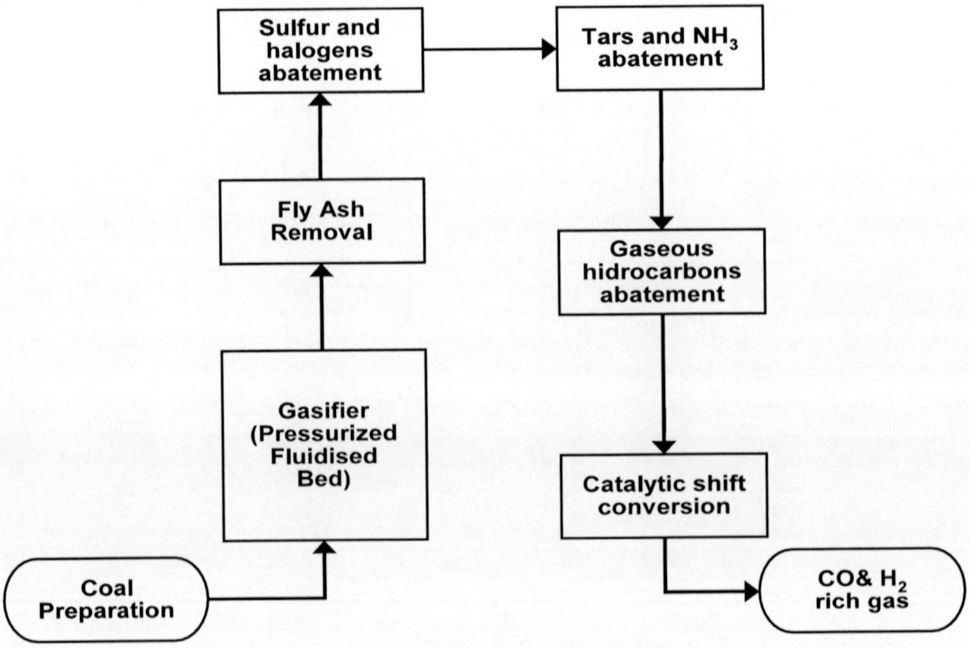

Figure 8. Possible diagram for gasification gas cleaning process.

The main abatement operations should include: catalytic cracking, catalytic steam reforming, catalytic shift conversion, physical or reactive absorption, physical or reactive adsorption, vapor condensation, gas separation, high and low temperature filtration. In Figure 8 is shown a possible diagram for gasification gas cleaning process. The hot syngas that leaves the gasifier contains ash particulates and fine soot when steam, CO_2 and sub-stoichiometric oxygen gasification are used at temperatures usually higher than 800°C. These solids are removed by cyclone and hot ceramic candle filters.Cyclone filters are used as an initial gas cleanup step in most gasifier systems, because they are effective and relatively inexpensive to build and operate and are available commercially from many suppliers, but they are useless with sub-micron particles. Ceramic candle filters are usually more efficient in particulates removal, but they are very expensive and not appropriate for smaller scale installations. Therefore, R&D efforts are still needed in this area.

Tars and NH_3 Abatement

Next to the removal of particulates, the gas should be submitted to catalytic cracking and/or catalytic steam reforming to remove tars and NH_3 present in the gasifier raw gas. There are several catalysts that have proven to be effective for this purpose (Corella et al., 2004 and 2005), nickel based catalysts are usually a good choice, concerning effectiveness and cost (Pinto et al., 2007). Some of these catalysts may be poisoned by the presence of halogens and sulfur compounds in the raw gas, which need to be removed prior to the use of tars abatement catalysts.

The retention of sulfur and halogens in the solid phases produced during gasification process can be increased through the addition of specific sorbents that contain Fe, Ca and other elements that favor the formation of sulfides and halogenides (Pinto et al., 2008). When, even in presence of specific sorbents inside the gasifier, the contents of H_2S, HCl, HF, HB in the raw gas are high enough to compromise the long life of catalysts for tar and NH_3 abatement, the raw gas should be prior treated with catalysts or sorbents.

On the other hand, the presence of H_2S is also undesirable, because it leads to the formation of SO_2 after combustion in gas turbines or to fuel cells poisoning. H_2S can be separated from the syngas, either in physical scrubbing plus stripping units in the presence of organic solvents, or in physico-chemical absorption plus desorption units using promoted selective polyamine aqueous solutions. However, these options are not suitable, because the gas is first cooled down to be heated again during catalytic cracking processes.

R&D work is still needed in this area, in order to find efficient and low cost catalysts that may guarantee effective abatement of tars, S, N and halogens compounds. It is also necessary to define effective operation conditions and to analyze the economic viability of such processes.

Gaseous Hydrocarbons Abatement by Catalytic Steam Reforming

Methane and other hydrocarbons produced during gasification process can be transformed into CO, CO_2 and H_2 by catalytic steam reforming by reactions (8) to (11). However, as syngas is rich in CO_2, hydrocarbons conversion may also occur through reactions (12) and (14).

Steam reforming

$$CH_4 + H_2O \leftrightarrows CO + 3\,H_2 \quad (8)$$

$$CH_4 + 2\,H_2O \leftrightarrows CO_2 + 4\,H_2 \quad (9)$$

$$C_nH_m + n\,H_2O \leftrightarrows n\,CO + (n + m/2)\,H_2 \quad (10)$$

$$C_nH_m + n/2\,H_2O \leftrightarrows n/2\,CO + (m\text{-}n)\,H_2 + n/2\,CH_4 \quad (11)$$

Dry reforming

$$CH_4 + CO_2 \leftrightarrows 2\,CO + 2\,H_2 \quad (12)$$

$$C_nH_m + n\,CO_2 \leftrightarrows 2n\,CO + m/2\,H_2 \quad (13)$$

$$C_nH_m + n/4\,CO_2 \leftrightarrows n/2\,CO + (m\text{-}3n/2)\,H_2 + (3n/4)\,CH_4 \quad (14)$$

The heat necessary for these endothermic reactions is supplied by different ways: 1) by external heat (from electrical power), or 2) by exothermic reactions or 3) by combination of the two last options to reach an isothermal running. In option 2) the heat necessary to the process may be obtained by partial oxidation of part of the gas by O_2 or by carbonation of CaO into $CaCO_3$, which also allows the CO_2 recovery in a separate reactor from $CaCO_3$ regeneration. These different options are under study and development and the selection of the best option still needs more R&D.

Hydrocarbons steam reforming is a process that may be largely improved by the presence of suitable catalysts. However, these catalysts are usually expensive and may compromise the economic viability of the process. Therefore, the research of new and cheaper catalysts that may lead to higher conversions is fundamental to ensure the success of these processes.

Catalytic Shift Conversion

After the removal or reduction of tars, S, N and halogens compounds and also after the reduction of syngas hydrocarbons contents, the syngas can react with steam through the water gas shift reaction (7) to produce additional hydrogen and a concentrated stream of CO_2.

Higher conversions of CO into CO_2 and H_2 are usually achieved in presence of suitable catalysts that have the disadvantage of increasing the operational costs, though a reduction in temperature may be achieved due to catalyst presence. The research challenge in this area is to develop new and cheaper catalysts that allow achieving higher conversions of CO. It is also of great importance that these catalysts may resist to some of the undesirable compounds of syngas like S and halogens compounds that usually poison this kind of catalysts. Therefore, and depending on syngas content of this type of compounds some of the gas treatments mentioned previously could be avoided or done with cheaper less efficient catalysts. Some research work has been done in order to develop water gas shift catalysts that may also convert tars and hydrocarbons, but also that may retain sulfur compounds. Another challenge is the catalyst regeneration and its technical, economical and environmental viability.

CO_2 Separation

CO_2 may be removed from syngas produced at coal IGCC plants by physical absorption solvents, using the Rectisol and Selexol processes. However, these processes have large energy consumption and require extensive equipment for solvent circulation and heat exchange which leads to a reduction in efficiency. Therefore, new processes for CO_2 separation in gasification technology need to be developed to ensure efficiency improvements and cost reductions. R&D activities should focus on developing:

1) Lower cost and more efficient membranes for H_2 separation, which could be combined with water-gas shift reactors.
2) More efficient adsorbents for CO_2 removal that could operate at higher partial pressures than those currently used for pressure swing adsorption (PSA). These adsorbents need to be regenerated, by temperature swing adsorption (TSA) and by PSA, thus producing CO_2 for sequestration. By this technique, it is expected to remove CO_2 more easily and with lower energy consumption than that needed for traditional absorption/scrubbing processes.

Both alternatives still require further investigation. This theme will be further analyzed in 6.1.5.

H_2 Membranes

H_2 membranes can be either organic or inorganic, but organic membranes are usually sensitive to process conditions and trace contaminants and because of that they have limited applications for coal-based hydrogen production. Inorganic membranes appear to be more suitable for gasification gas applications, they can be either porous or dense and the latter can be metallic or solid electrolytes (ceramic). Several authors have developed different types of membranes (Stiegel et al., 2006, Dolan et al., 2006, Vasileiadis et al., 2004). Stiegel et al., 2006 developed dedicated metallic membranes, most of which are based on Palladium (Pd). However, these membranes have been found to be susceptible to degradation due to the presence of sulfur and more work needs to be done to improve its performance.

Bounaceur et al., 2006 stated that membranes can provide significant improvements in relation to the best available technology, amine absorption. Membrane processes based on already existing materials can purify flue gas with about 0.5–1 GJ/tonne CO_2 recovered, while the energetic costs of amine absorption range from 4 to 6 GJ/tonne CO_2 recovered.

Kaldis et al., 2004 studied two alternative separation options: (a) low temperature separation, using polymer membranes, and (b) high temperature separation, by ceramic membranes, whose cost analysis show that CO_2 removal was technically feasible. Incorporation of shift reaction and membrane separators results in an energy penalty of 8–14%. However, this technology would permit reduction of CO_2 emissions exceeding 50%, compared to conventional IGCC (Kaldis et al., 2004).

Instead of developing new sophisticated H_2 membranes of high price and of uncertain performance, another option is the use of cascade membrane systems similar to the distillation column. Therefore, sophisticated cascade schemes, either in counter-current recycle cascade or in continuous membrane column, have been considered to meet industrial requirements for high purity products for air separation, methane recovery from biogas, and argon ultra-purification from propane impurities. However, more research work is still needed to prove the technical and economical viability of these processes.

Bottom-Ash Treatment

Another important aspect, though not directly related to H_2 production from coal through gasification is the gasifier bottom-ash treatment. Depending on their basicity bottom-ashes can be recycled to the top of a moving bed reactor for sulfur and halogens counter-current abatement, before final disposal.

On the other hand, some heavy metals can be trapped in bottom-ashes, either externally or internally depending on their porosity and pores size. Therefore, gasification solid residues, both bottom-ashes and particulates collected in the cyclone need to be fully characterized before deciding its disposal in landfills or its reuse, for instance, in building materials.

Before H_2 production from coal or biomass gasification is largely spread, cleanup technologies for gases leaving the gasifier need to be fully demonstrated. Further research work on the areas of reaction kinetics, reactor designs, and feeding of fuel to pressure vessels are still needed to reduce capital and operating costs and to ensure low environment impact processes.

6.1.4 Production of Liquid Fuels through Methanol Synthesis or Fischer-Tropsch Synthesis

The recent interest in producing liquid fuels through methanol synthesis or Fischer-Tropsch synthesis is a consequence of environmental demands, technological developments and changes in fossil energy reserves. In either methanol synthesis or Fischer-Tropsch synthesis a syngas with predominantly CO and H_2 is used to obtain liquid mixtures of multi-component hydrocarbons to be used as fuels or for other chemicals synthesis. Syngas can be obtained from coal and wastes gasification, as described previously, or it can be produced from natural gas. The global process to obtain liquid fuels is an integrated one, which involves different important processes: i) Gas cleaning; ii) Catalytic reforming; iii) Fischer-Tropsch Synthesis or methanol synthesis.

Gas cleaning is especially important if syngas is produced from coal and/or wastes and aims of removing the contaminants from the gasification gas. The main contaminants are essentially particulates, condensable tars, alkali compounds, sulfur, nitrogen and halogens compounds and its removal may be done as described in 6.1.3. These contaminants can lower efficiency of Fischer-Tropsch synthesis or of methanol synthesis by poisoning and deactivating catalysts, by sticking to the existing active site. Deactivation of catalyst affects the activity and selectivity for the desired hydrocarbon products.

In the second step, the catalytic reforming is used for producing a synthesis gas, which contains mainly H_2 and CO. This step involves catalytic steam reforming, as mentioned in 6.1.3, with the proper catalyst to obtain a gas with the suitable H_2/CO ratio to be used in Fischer-Tropsch or methanol synthesis. The increase of H_2 content in the syngas can be achieved through the water gas shift reaction (7) applied to part of the gas, which afterwards is joined to the main stream. The CO_2 present in the obtained syngas needs to be removed, by the processes described in 6.1.5, prior to its utilization in Fischer-Tropsch Synthesis or methanol synthesis.

Methanol Synthesis

In methanol synthesis, methanol is produced from a syngas with the suitable H_2/CO ratio, through reactions (15) and (16) in presence of a suitable Cu-Zn catalyst and at 250 to 300 °C and 5 to 25 MPa. The challenge is to develop new and cheaper catalysts that allow decreasing operating conditions without losing efficiency and selectivity.

$$2\,H_2 + CO \leftrightarrows CH_3OH \tag{15}$$

$$3\,H_2 + CO_2 \leftrightarrows CH_3OH + H_2O \tag{16}$$

The methanol obtained is purified and converted into gasoline by Mobil MTG process, in which occur reactions (17) to (20).

$$2\,CH_3OH \leftrightarrows (CH_3)_2\,O + H_2O \tag{17}$$

$$2\,CH_3OH + (CH_3)_2\,O \leftrightarrows \text{light olefins} + H_2O \tag{18}$$

$$\text{light olefins} \leftrightarrows \text{heavy olefins} \tag{19}$$

$$\text{heavy olefins} \leftrightarrows \text{aromatics, paraffins, cycleparaffins} \tag{20}$$

Another process, also developed by Mobil, is Mobil Olefin to Gasoline/Distillate (MOGD) process (Radtke et al., 2006). Different types of compounds may be formed according to operating conditions and catalysts used, usually zeolites. Hydrocarbons produced, usually present a distillation range similar to those of gasolines and a high octane number, due to its high content of aromatic compounds, in opposition to what normally happens in Fischer-Tropsch synthesis, whose products usually have higher contents of paraffins and olefins.

Fischer-Tropsch Synthesis

Products obtained by Fischer-Tropsch are hydrocarbons of different molecular weight and type (paraffins and olefins), but also oxygenated compounds like alcohols, as shown by reactions (21) to (23). Process selectivity is influenced by a number of factors catalyst dependent and non-catalyst dependent. The first ones are influenced by the type of metal, support, preparation, pre-conditioning and age of catalyst. The last ones are temperature, pressure, H_2/CO ratio in feed syngas and reactor type.

$$(2n + 1)\, H_2 + n\, CO \leftrightarrows C_nH_{2n+2} + n\, H_2O \qquad (21)$$

$$2n\, H_2 + n\, CO \leftrightarrows C_nH_{2n} + n\, H_2O \qquad (22)$$

$$2n\, H_2 + n\, CO \leftrightarrows C_nH_{2n+1}\, OH + (n-1)\, H_2O \qquad (23)$$

The Fischer-Tropsch process generally operates at pressures ranging from 20 to 40bar. The temperature, depending on operating mode, ranges from 180 to 250 °C (low temperature Fischer-Tropsch) and from 300 to 350 °C (high temperature Fischer-Tropsch). High temperatures are usually more favorable to produce little wax and more gasoline, linear hydrocarbons (liquid fuels), however, with lower selectivity. Also higher pressure leads to higher selectivity. Generally, the catalysts used in Ficher-Tropsch synthesis are metals of VIII group of periodic table, like Co, Ru and Fe. The type of catalyst used affects products selectivity and operating conditions. When a wide range of different types of compounds are formed, Ficher-Tropsch products need to be separated in different fractions and up-graded, before its utilization as fuels.

Fischer Tropsch plants are in operation in South Africa, using syngas produced from coal gasification. The challenge is to use Fischer Tropsch synthesis applied to syngas produced from co-gasification of coal and wastes. Another challenge is to develop selective catalysts towards gasolines production at milder temperature and pressure conditions and with lower production of waste products, as this will improve the technical feasibility and the economic viability of these processes.

6.1.5. CO_2 Removal and Storage at Large Scale

6.1.5.1. CO_2 Removal

CO_2 large production sources are mainly industrial exhausts from cement, iron, steel, and chemicals production, CO_2 release from oil and natural gas operations, CO_2 by-product from H_2 production and CO_2 produced for power generation in different types of installations, which include: a) pulverized coal-fired plant equipped with flue gas desulfurization facilities, b) natural gas combined cycle power plant (NGCC), c) Integrated coal gasification combined cycle (IGCC) d) pulverized coal-fired plant that uses O_2/CO_2 flue gas recycle combustion. These technologies are at different level of development and demonstration.

At present, there are three main CO_2 capture approaches (McKee, 2002):

a) Post combustion capture - CO_2 is captured from combustion products in power plant flue gas or industrial exhaust.
b) Oxygen combustion approach - O_2 and flue gas are recycled to increase CO_2 concentrations in flue gas combustion installation prior to capture.
c) Hydrogen/syngas approach - coal is gasified or natural gas is reformed to produce syngas with CO and H_2, which are converted into H_2 and CO_2 for capture.

a) Post Combustion Capture

Prior to CO_2 capture, contaminants like particulates, sulfur dioxide (SO_2), and nitrous oxides (NO_x) need to be removed by gas pre-treatments. Post combustion capture methods may include: absorption after contact with solvents, adsorption on activated carbon or other materials, cryogenic separation, and membrane separation.

Absorption with solvents

The most common solvents used for CO_2 chemical absorption are: diethanolamine (DEA), monoethanolamine (MEA) and methyldiethanolamine (MDEA). The amine absorbs the CO_2 and afterwards the CO_2-rich amine goes into the desorber tower to release the CO_2, by reducing the pressure and/or by the increasing the temperature till around 120°C. The regenerated amine is recycled into the absorption column. If the gas is produced at higher pressure, as it happens in IGCC, physical absorption is probably a better option. Commonly used solvents include: dimethylether of polyethylene glycol (Selexol process), cold methanol (Rectisol process), propylene carbonate (Fluor solvent process) and sulfolane. Another option is the use of hybrid solvents that take profit of the best characteristics of chemical and physical solvents. Common hybrid solvents include: Sulfinol (an organic solvent mixture made up of sulfolane (tetrahydrothiophene) and alkanolamines in an aqueous mixture) (Schäfer, 2002), aMDEA (activated Methyldiethanolamine) (Kanniche et al., 2007) and UCARSOL (Jackson et al., 2003).

Though absorption technology is a well known one readily available to be used in new plants or in existing power plants, it has important disadvantage such as: high degradation rates, mainly due to the presence of impurities like SO_x and NO_x, corrosion in the presence of O_2 and the most important, the large amount of energy requirements for regeneration. Therefore, new absorbents need to be found to guarantee: high degradation resistance, low corrosivity in the presence of O_2, higher capacities and faster CO_2 absorption rates and low vapor pressure solvents. Other R&D needs include the development of high efficiency gas-liquid contactors to eliminate the need for large scrubbers and thus reducing the regeneration costs.

Adsorption

CO_2 can be captured by materials with high surface areas, like activated carbons and zeolites, being the process dependent on temperature, partial pressures, surface forces, and adsorbent pore sizes. Different technologies can be used: pressure swing adsorption (PSA), temperature swing adsorption (TSA) and electrical swing adsorption (ESA).

Pressure swing adsorption (PSA)

PSA processes are based on preferential adsorption of CO_2 on porous sorbent materials at a high pressure. The adsorbent bed is then isolated and depressurized to release the adsorbed CO_2 and then the sorbent material can be used again for successive CO_2 adsorption. PSA technology has relatively low energy requirements and low capital investment costs, though the equipment used needs to be able to support the high pressures suitable for the process. Therefore, PSA is a technology with potentialities to be applied for removal of CO_2 from high pressure gas streams, such as those produced in IGCC processes. However, CO_2 captured in gasification processes needs to be compressed to around 70 bar before being piped to storage sites, which is an additional cost. This energy penalty could be decreased by the suitable design of efficient PSA processes to ensure CO_2 desortion at high CO_2 partial pressures.

However, PSA technology also needs further research to develop regenerable adsorbents with high adsorption capacities. Both activated carbons and molecular sieves have been studied, due to theirs high sorbent capacity at high pressures. Activated carbons seem to be a better choice than molecular sieves, because of their better strengths of adsorption for gases, which makes easy the desorption process and their better resistance to chemical attack (Sircar et al., 1996 and Siriwardane et al., 2001). These adsorbents need to be prepared in a form that ensures great flexibility and avoids pressure drop, for instance in monolithic or in granular form.

Temperature Swing Adsorption (TSA)

This process is similar to the previous one, but regeneration takes place by increasing the temperature.

Electrical Swing Adsorption (ESA)

This process uses a new adsorption material highly conductive, so the regeneration of the material is done by a low-voltage electrical current, without significant variations in pressure and temperature.

While, PSA and TSA processes are available at commercial scale, ESA is not yet accessible. However, all these processes do not seem to be suitable for large-scale installations, because they work at low temperature, the energy needed for regeneration is high and their capacity and CO_2 selectivity is low (McKee, 2002). Therefore, more research is still needed to develop adsorbents with more suitable characteristics.

Cryogenic Separation

If the flue gas is cooled below CO_2 boiling point, CO_2 condense into a liquid state and may be separated from other gas components. By this process, particulates and other gaseous components may be also removed from the flue gas. However, the separation into pure components depends on flue gas composition, as it affects boiling points. Though, liquid CO_2 could be easily transported with compression requirements, the use of this process at large scale units is not advisable, due to economic reasons, even if efficient refrigeration processes would be developed.

Membrane Separation

CO_2 can be separated from the flue gas by membranes, however, a high degree of separation can not be achieved, unless multiple stages or recycling of one of the streams is done. On the other hand, to obtain CO_2 with high purity and due to the several components of the flue gas, several membranes with different characteristics will be needed.

Two different kinds of membranes may be used for CO_2 separation: gas separation membranes and gas absorption membranes. In gas separation membranes, the separation depends on the solubility or diffusivity of the gas molecules in the membrane and on the differences in partial pressure from one side of the membrane to the other. In gas absorption membranes, CO_2 that diffuses through the membrane is removed from the gas stream.

Membranes technology for CO_2 separation is still under development and membranes have not yet been optimized for large scale installations. On the other hand, membranes are usually expensive and as the gas needs to be compressed the overall cost of the process is high. Therefore, the developing of new membranes more efficient and cheaper is still needed.

B) Oxygen Combustion Approach

In normal combustion processes air is used and therefore the major component of flue gas is N_2, which dilutes CO_2 concentrations in the flue gas. If O_2 is used instead of air and if O_2 and flue gas is recycled, CO_2 concentrations in flue gas increases, as already described in 6.1.1. By this procedure and with a relatively pure O_2 feed, a flue gas with around 90 percent vol. dry of CO_2 can be obtained (McKee, 2002). However, prior to CO_2 storage, the removal of O_2, N_2 and other components from the flue gas may still be needed. The main disadvantage of this process is the high cost of oxygen production. Therefore, scientific developments of less expensive ways of producing O_2 are of most importance, either by improving air separation technologies or by developing new membranes for O_2 production.

C) Hydrogen/Syngas Approach

As described in 6.1.3 fossil fuels, usually coal and natural gas, but also biomass and other wastes may be converted into H_2 and CO_2, through water gas shift reaction. This gas can be used for electricity production in a gas turbine and the stream highly concentrated in CO_2 can be captured and stored. If the objective is the production of H_2, to be used in a fuel cell to produce electricity or for transportation or to be used in chemical synthesis, CO_2 has to be separated from H_2 by the methods described previously. However, further research work is still required. Fuel cells need further development for application in large power plants and the possibility of using H_2 in existing gas turbine with little modification, still needs to be demonstrated (McKee, 2002).

D) Other Options

Although the methods just described still need further development or demonstration at larger scales, they are known approaches. However, there are some other options that have not yet been proven, but as stated by McKee, 2002, they provide new ideas that may offer step forwards. These new options are described by McKee, 2002 and may be listed as follows: new power cycles, chemical looping, dry ice co-generation, biological CO_2 fixation with algae, hydrate formation, direct capture of CO_2 from air, and zero emissions from coal.

6.1.5.2 .CO_2 Transportation

CO_2 transportation after capture is also a challenge. It can be transported by high pressure pipeline or by tanker, being tanker capacity and distance till the storage site the determining issues. CO_2 has been transported in ships in North Western, using tankers similar to that for LPG, the constructions of larger vessels has been considered for gases with only CO_2. North America has transported CO_2 in pipelines since the 1980s. GHG R&D Program supported by IEA has put some effort and money to develop and improve CO_2 transportation, either by developing models to assess CO_2 transport and energy transmission options or by collecting current CO_2 pipeline design data (McKee, 2002).

6.1.5.3 CO_2 Storage

CO_2 may be used in different applications, for instance, as a chemical feedstock, in food production, fish farms and agricultural greenhouses, to make solids, like carbonate minerals and in oil recovery operations. However, CO_2 utilization is much smaller than the amount emitted per year. Therefore, CO_2 needs to be stored and different alternatives can be considered: in geologic formations, in the deep ocean and in natural sinks, either ocean or terrestrial ecosystems (McKee, 2002).

Another option for CO_2 utilization is in coal bed methane exploitation, which may be done through CO_2 injection to recover a gas purified for sale. However, research work is still needed to prove the stability and permanence of CO_2 in coal bed reservoirs and to optimize CO_2 sequestration and methane production,

Though, geologic formations and the ocean have great CO_2 storage capacity, the environmental impacts and safety aspects need to be further studied and understood. Some R&D projects are going on, mainly in USA, Canada and Norway with the aim of understanding geologic storage, maximizing CO_2 sequestration potential and monitoring and risk evaluation. GHG R&D Program from IEA has developed some activities in studying CO_2 injection into deep saline aquifers (McKee, 2002).

On the other hand, CO_2 sequestration in terrestrial ecosystems has no great health, safety or environmental problems, but its capacity is limited. Therefore, to maximize its storage potential, to measure storage capacity and to verify storage rates, further basic and applied research is still needed. R&D activities should focus on developing carbon measurement equipments, on methods to raise organic content, water retention and protection of organic matter and to improve soil texture to increase carbon impregnation.

6.2. Nuclear Power

Nuclear energy has the advantage of being a source free of carbon and other gaseous pollutants. The operation of a 1-GW nuclear power plant can avoid the emission of 6 to 7 million tonnes of CO_2 per year in relation to an equivalent coal-fired power plant. The increase of fossil fuels energy prices is becoming nuclear power more competitive and raising the interest in building nuclear reactors. However, if coal prices are less than $70 per tonne, nuclear power would still be more expensive than conventional coal-fired plants, unless CO_2 emissions taxes are introduced (IEA, 2007[e]).

Nuclear energy fuel is natural uranium (U) or enriched U with ^{235}U isotope. U fission reactions release heat that is received by a coolant fluid that directly impels the turbine or provide heat to a secondary fluid that drives a turbine to power an electricity generator. There are different types of nuclear reactors, depending on neutron energy level (thermal or fast reactors), on coolant (water, gas, liquid metal) and on neutron moderator (water, heavy water and graphite).

Most of the existing nuclear power plants were built in the 1970s and have second generation technology. In the 1990s the third generation reactors were developed with longer lifetime and modular design to decrease costs, licensing and manufacture time, and high fuel burn-up to minimize waste and optimize fuel utilization and safer conditions, like reactor shutdown in case of accident. According to IEA, 2007[e], by 2030, the 4th generation reactors will be developed and are expected to be available in the market with several different concepts: fast reactors and thermal reactors. But all of them with improved safety conditions, reduced costs and lower production of long life radioactive waste. Thermal reactors use high fuel burn-up to extract more energy from uranium, while fast reactors run on closed fuel cycles to burn actinides, recycle Pu and use ^{238}U. All models have high cooling temperature to attain high efficiency (40% to 50%), but while thermal reactors may be cooled by high temperature helium (VHTR), molten-salt (MSR) or supercritical water (SCWR), fast reactor are cooled by liquid lead (LFR), sodium (SFR) or gas (GFR) (IEA, 2007[e]).

Present reactors already have the safety levels defined internationally. However, the proliferation of new reactors still depends on many factors, including: public approval, investment risk and waste management.

One important barrier to nuclear reactors expansion is waste management. Usually, a 1-GW power plant produces per year 200 to 350m^3 of low-level wastes (with short life radioactivity) and medium-level waste and 10 to 20m^3 of high-level waste (with long life radioactivity). After recycling, high-level waste may be transformed into less than 3m^3 of vitrified waste that have to be stored for thousands of years (IEA, 2007[e]). The new challenges in this area are the development of new ways of recycling and reprocessing wastes in order to decrease the amounts of high-level wastes. Partitioning and transmutation processes could transfer long-lived radio-nuclides into short-lived elements, consequently reducing high-level waste volume. However, the need of geological disposal will not be eliminated and further R&D will be required. Unless, all these barriers are overtaken, nuclear power will not spread and its contribution to electricity generation might even decrease in the future.

As uranium resources are plentiful and widely distributed around the globe and nuclear energy production costs are less susceptible to fuel price changes than coal or gas fired generation, nuclear power may be a potentially attractive alternative for increasing the security of electricity supply. However, plant safety, nuclear waste disposal and the risk of proliferation needs to be solved to guarantee public acceptance. On the other hand, governments of countries where nuclear power is acceptable have to actively promote private investment and liberalize the market.

6.3. Wastes Pyrolysis for Energy Production

6.3.1. Pyrolysis of Polymers

Pyrolysis of waste polymers, like plastics, rubber and bio-polymers has been studied with the aim of producing liquid fuels or raw materials, by inverting the polymerization process used in their production. The importance of waste polymers pyrolysis is increasing, as land filling and combustion become more expensive, due to the need of decreasing the negative consequences of these processes to the environment. On the other hand, the public acceptance of these methods is decreasing and they do not allow taking profit completely of its energetic content. Polymers pyrolysis occurs in an inert atmosphere at a temperature between 400 to 700°C. Polymers chemical bonds are cracked into small fractions and a complex mixture of hydrocarbons is produced, which being similar to oil derived products, can be processed by the well known technologies applied to oil derived products, to produce fuels or different types of raw materials. Pyrolysis products are gaseous and liquid hydrocarbons (oils) and a solid residue containing fillers and impurities such as heavy metals.

Pyrolysis of waste polymers has been studied by many research groups all over the world. Mixtures of the more abundant plastics in municipal solid wastes were studied by Pinto et al., 1999: PS (polystyrene), PP (polypropylene) and PE (polyethylene). In all tests, total conversion values were above 90%, most products were liquid, while gas yields were lower than 10%. The gaseous compounds formed in higher concentrations were: methane, ethane and propane, being alkenes formation favored by the presence of PP. Liquids composition depended on the plastic waste used. The presence of PE increased alkanes content, whilst the presence of PP favored alkenes formation. The highest aromatic content in the end product was obtained with PS. Liquid alkanes from C5 to C20 were detected. Hexene and heptene were the alkenes produced in higher amounts. The aromatics mainly consisted of toluene, benzene and) xylene.

Kaminsky has developed a very important research work about plastics pyrolysis. Through pyrolysis of plastics benzene, toluene and xylene were produce at temperatures of about 700°C (Kaminsky et al., 2000). In other activities the objective was the recovery of monomers at lower temperatures of 450 to 600°C. Polymethylmethacrylate was depolymerised at 500°C to produce more than 98% of the monomer and similar results were also obtained with polystyrene (Kaminsky et al., 2001).

Pyrolysis technologies have been developed in Europe, Fuji Tech in Japan and Conrad in Canada. The main processes developed in Europe for plastics pyrolysis are: a) Hamburg University Pyrolysis Process, using a fluid bed reactor that operates at high temperature with mixed plastics, but the demonstration plant at Ebenhausen had different operating problems, b) BP low temperature pyrolysis of mixed plastics, using a fluid bed reactor that produces a more uniform product with less unwanted side reactions, c) BASF low-temperature pyrolysis of mixed plastics, using a series of stirred tank reactors for liquefaction, d) Veba Oil hydrogenation technologies (Buekens, 2006).

Plastics pyrolysis has been studied in academic circles and still continues to encourage industrial R&D and demonstration projects. However, the economic viability of pyrolysis of mixed plastics still needs to be proven, as it is dependent on the price of crude oil and also because of the huge costs of collecting, cleaning and grading considerable amounts of plastics. Only few polymers can be recycled by pyrolysis under economically favorable conditions, as is the case of PMMA and PA-6 pyrolysis, because of the high value of the monomers produced (Buekens, 2006).

As some of the plastics pyrolysis processes are still in development, their economic viability has not yet been established and is very much dependent on crude oil price. Present situation is expected to be changed in the future, due to technical progress and to the increasing amounts of available plastics wastes. In the meanwhile, further research is still needed to optimize the amounts of monomers and oil from different polymers and to demonstrate this technology at commercial scale, through the construction of industrial pilot plants.

6.3.2. Biomass Pyrolysis

Pyrolysis of different types of biomass, in the form of wood and agro-industrial residues, sewage sludge, lignins and bark has been carried out by different researchers with the principal aim of producing liquid fuels. Different routes may be followed to achieve this objective: a) indirect route, through biomass gasification, gas conditioning and liquid synthesis by a Fischer-Tropsch, as described in 6.1.4, b) hydrous pyrolysis, c) hydro-pyrolysis and d) flash pyrolysis.

For hydrous pyrolysis biomass conversion process, it is more suitable to use biomass and sludge with moisture content higher than 50%. This process occurs in presence of steam at 10 to 200 bar and 340 to 360 °C, close to steam supercritical conditions, with a residence time of 6 to 72 hours and a homogeneous catalyst like sodium hydroxide.

Hydrous pyrolysis converts biomass into a hydrogen rich gas, at pressures around 100 bar with a residence time in the range of 1 to 30 seconds. The product has low oxygen content and in opposition of flash pyrolysis products is miscible with petroleum products.

Biomass flash pyrolysis occurs at 500–1000°C with an extremely short residence time, normally below 1 second and the products are cooled down very fast of to prevent its further reactions (Bridgwater, 1999). The polymeric constituents of biomass, cellulose, hemicelluloses and lignin suffer a thermal breakdown by flash pyrolysis, leading to the formation of a huge number of compounds, ranging from low molecular simple gaseous products to condensable monomeric and oligomeric chemicals with different functional groups, molecular sizes and weight. Pyrolysis products composition depends on pyrolysis conditions, on the type of feedstock and pretreatment procedures. Pyrolysis yield is around 75 % w/w bio-oil from dry ash free biomass and contains around 35% of oxygen (Bridgwater, 2001). Because of this pyrolysis products are unsuited for normal oil refineries, they may be used for chemicals production or as fuels, but due to their different characteristics, their utilization in devices designed for oil derived fuels has not yet been fully proved.

Pyrolysis products may be upgraded and converted into synthetic biofuels by hydroprocessing (Maggi et al. 1997), or by zeolitic cracking (Huber et al. 2006). However, further research is still needed to prove the viability of these processes. Without upgrading, pyrolysis oils are physically and chemically unstable, because they contain many reactive components that may react between them or between the oil components and oxygen, leading to the formation of deposits and gums, a decrease of volatility, an increase in viscosity, phase separation and other undesirable changes during storage or utilization in boilers, engines or for other applications.

The challenge for future research is a full understanding of the bio-oil chemistry to prevent or minimize the processes that degrade oil quality and limit its applications.

6.4. Biofuels Contribution for Energy Production

Fuels resulting from plant materials are usually named as biofuels. They have the advantage of being carbon neutral, as the carbon released to the atmosphere during their utilization is compensated by the carbon absorbed by the plant during its growing. Biofuels are also renewable, as new supplies can be grown as needed, in many different environments. Biofuels may include a wide range of products with many different characteristics (they may be solid, liquid or gaseous) and different potentials for decreasing greenhouse gas emissions. Different feedstocks may be used for biofuels production. Some are also used in human and farm animal food, like starch, wheat and sugar, which may have numerous economic and environmental impacts, others are energy crops, as for instance: switchgrass and miscanthus or short rotation forests that can supply lignocellulose feedstocks for bioenergy and biofuel production. The use of feedstocks that are also used in human and farm animal food should be avoided or done with most care, as they may introduce great rises of cereals market price and consequently famine, especially in poorer countries and communities.

Biofuels may be also produced from co-products of cereal production, like: corn stover, straws from wheat, barley, oats, rice, cane bagasse and sorghum or from agricultural and forestry wastes, such as: cereal straw several parts of food crops that are not included in the food chain. Plant oils are also a feedstock for biofuels production, like: soybean, rapeseed, sunflower seed and jatropha, used for biofuels production. A lowest cost biodiesel may be also produced from recycled cooking oil and waste animal grease. Apart from all these feedstocks, biofuels may be also obtained from marine macroalgae, which allows the production of biogas, through anaerobic digestion of Municipal Solid Waste (MSW).

From this wide range of feedstocks a large variety of biofuels: solid, liquid or gaseous, may be produced with different environmental, social and economic impacts. Feedstock conversion into biofuels can be done through different processes: biological, chemical and thermal. Thermal conversion processes include gasisification, methanol synthesis or Fischer-Tropsch synthesis (described in 6.1.2 and 6.1.4) and pyrolysis, referred in 6.3.

6.4.1. Bioethanol Production

As shown in Figure 9, bioethanol can be produced by three different processes: biological fermentation, gasification followed by ethanol synthesis and gasification followed by biological fermentation. Chemical and thermal processes are very much dependent on a clean source of syngas from biomass gasification. Syngas produced by gasification may be used to produce ethanol by a chemical process, similar to that described previously in 6.1.4. Another option could be the use of syngas in a biological fermentation process to produce ethanol in presence of micro-organisms. However, some issues still need to be solved, like the effective dissolution of syngas in the aqueous solution of biological fermentation process without affecting the physiology and viability of the fermentation micro-organisms (The Royal Society, 2008).

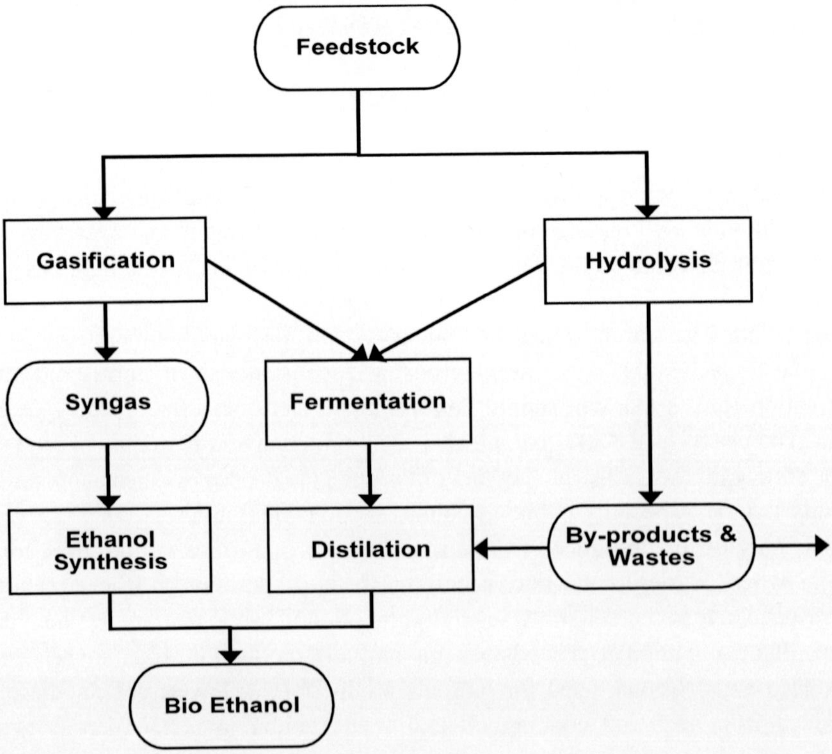

Figure 9. Bioethanol production by thermal and biological processes (The Royal Society, 2008).

Sugars and starches from food materials are more suitable to be used in biological processes. While sugars like sugar cane and sugar beet can go directly into the fermentation process, starches like corn or grain, need to be hydrolysed to convert the starch into sugars prior to fermentation in presence of an enzyme mixture named as amylases. During fermentation, sucrose and glucose are converted into bioethanol, in presence of yeast or the bacterium Zymomonas mobilis, which allows obtaining higher specific productivity, ethanol yield and high alcohol tolerance. Neither of these organisms can ferment the pentoses obtained from hemicellulose, which would increase ethanol yield. The achievement of high efficiencies in pentoses fermentation is a challenge for research work. Ethanol is usually in a dilute solution, due to the low alcohol concentrations that micro-organisms normally tolerate. Therefore, ethanol separation involves a primary distillation, followed by either further distillation or by using molecular sieves, which increases process costs. This problem may be overtaken by thermophilic fermentation at high temperatures with nitrogen stripping to remove ethanol from the fermenter, which is being concentrated, as the overhead vapor condense, leading to a ethanol rich liquid and cheaper to purify (The Royal Society, 2008). However, some issues still need to be improved, such as: increasing ethanol tolerance of micro-organisms, improving the fermentation process to enlarge ethanol yields and improving the capacity of organism to process lignocellulose derived polymers.

6.4.2. Biodiesel

Biodiesel is a mixture of methyl esters, whose chemical composition properties of the final product depend on the plant used and is different from conventional diesel. Biodiesel is produced through the trans-esterification of different plant oils like: rapeseed, soy bean or palm oil, using methanol, usually obtained from fossil fuels, producing a liquid with improved consistency and miscibility with diesel and lower viscosity. Bioethanol may be also used for the transesterification process to obtain ethyl esters that are less viscous. Another possibility, under active investigation, is the direct production of fatty acid ethyl esters from lignocellulose by organisms such as Actinomycetes (The Royal Society, 2008). According to EU legislation (Directive 2003/30/EC) biodiesel content in conventional diesel should reach values around 5%, and this limit should increase to 10% until 2020 (COM(2006) 848). Prior to incorporation of biodiesel in conventional diesel fuels they should be hydrogenated to avoid the presence of triglycerides and to hydrogenate the fat acids.

6.4.3. Synthetic biofuels

Synthetic biofuels are produced from biomass through gasification or fast-pyrolysis processes, as shown in Figure 10, from which it is possible to produce syngas or liquid fuels, such as: diesel, petrol and in principle aviation fuel. These processes were already described in 6.1.2, 6.1.4 and 6.3.2. Synthetic biofuels are usually cleaner than traditional fuels, due to the need of removing the contaminants that will poison the catalysts used in biofuels synthesis processes. A range of chemical products may be also produced from syngas or pyrolysis liquids, like ethers, dimethyl ether, ethyl tetrabutyl ether or methyl tetrabutyl ether.

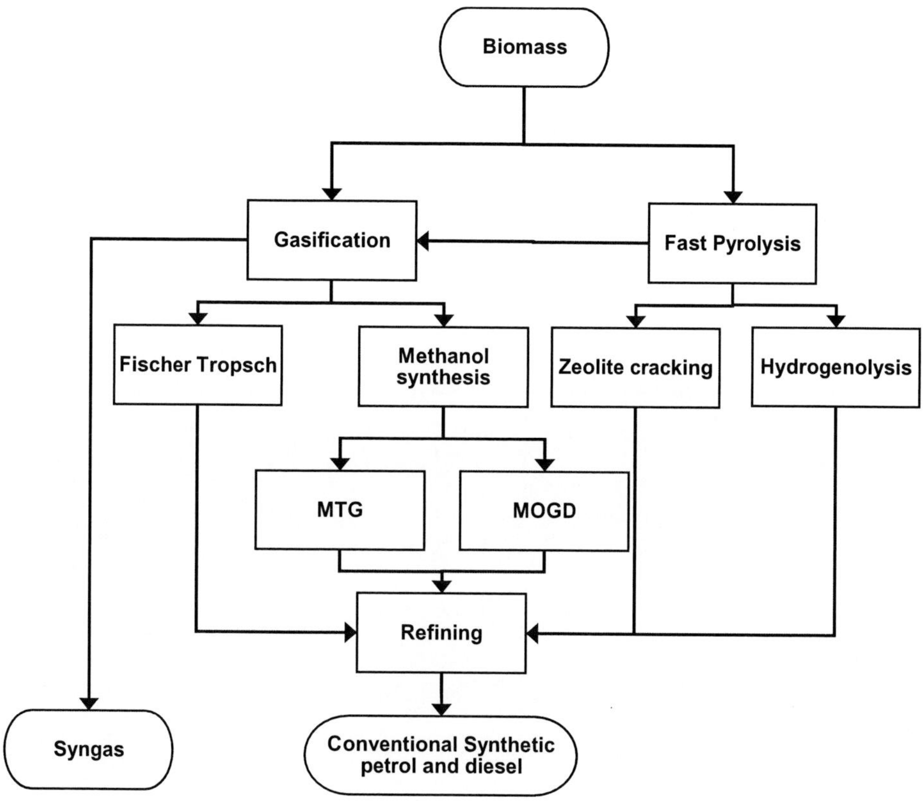

Figure 10. Thermal biomass processes for synthetic biofuels production (The Royal Society, 2008).

Some of the processes presented in Figure 10 still need some technical advances and require to be demonstrated at small and large scales to prove their technical, economical and environmental viability to produce biofuels.

6.4.4. Biorefineries

The concept of biorefinery integrates biological and thermal processes to produce biofuels and raw chemical materials, with the aim of optimizing the use of resources and minimizing the production of wastes. In Figure 11 is presented one possible concept for a biorefinery diagram. Apart from syngas, biorefineries should produce chemicals, as it will increase the economic viability of a biorefinery. Chemicals production depends on the feedstock and on the processing strategies. Biorefineries should be designed to be flexible and modular, in order to process a wider range of feedstocks, which are available at different times of the year and to be adapted to changes in specific chemicals demand, without huge capital costs.

Research Needs

Biofuels are expected to make a significant contribution to global transport fuel needs and ethanol is expected to account for most of the raise in biofuels use worldwide. But before achieving a great utilization of biofuels, further research is still needed in the following areas:

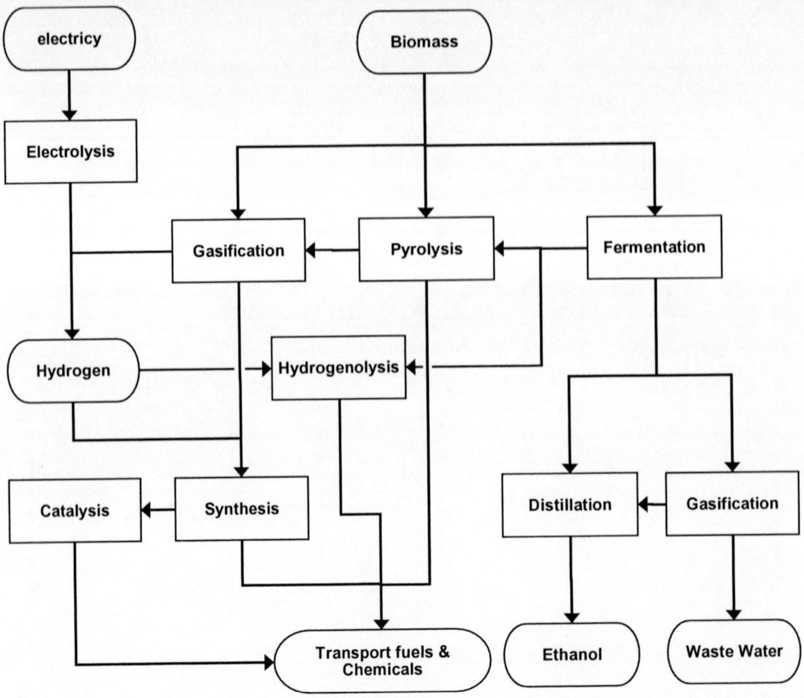

Figure 11. Biorefinery process with the integration of biological and thermal technologies (The Royal Society, 2008).

a) increasing biomass production, by enhancing yield per hectare of feedstock, while decreasing negative environmental impacts, by developing new species that can grow in more hostile environments, b) biotechnology application to biofuels (biodiesel and bio-ethanol), c) biomass thermochemical conversion technologies (cleaning of gasification gas, Fischer-Tropsch process), d) pyrolysis of waste materials, namely plastic wastes to recuperate the hydrocarbons, e) valorization of by-products from these processes. Significant technological challenges still need to be achieved bifore biofuels production technologies may become commercially viable; for instance: development of new microbial strains to process directly hemicellulose and cellulose, micro-organisms tolerance to alcohols needs to be improved, catalyst development is required for several processes, development and demonstration of integrated biorefineries and integration of biofuel development with engine development.

Although biofuels have limited capacities to replace fossil fuels in the near future their contribution to energy production is being promoted by several policy measures. Trade and financial supporting policies will determine which feedstocks and technologies will be used for biofuels production in the future and will help to promote energy diversity and reduction of carbon-dioxide emissions.

6.5. Renewable Energy

According to IEA, 2006[b] definition renewable energy is an energy derived from natural processes that are restocked continuously. Renewable energy is applied to a large range of energy natural sources including: solar, hydro, wind, wave, and biomass, geothermal, tidal and ocean currents. Renewable energy technologies presently provide 13.3% of the world's primary energy needs, through well known technologies like hydropower and biomass (IEA, 2006[c]). In the future renewable energy may play an important part in achieving low carbon emissions and should increase its contribution to worldwide energy production.

6.5.1. Bioenergy

Bioenergy includes ancient utilizations of biomass in rural economies, but also the modern technologies referred in 6.4.1 to 6.4.4. Bioenergy is the renewable energy with the largest contributions to total primary energy supply in IEA member countries, 2.9% in 2003, representing 53% of the total renewable energy supply (IEA, 2006[c]). The main research needs for further developing these technologies were already commented in 6.4. For achieving further utilization of bioenergy it is necessary to increase biomass production to guarantee its availability at large scale and at a relatively low cost, develop standards and norms to ensure biomass quality and further raising the conversion of new processes with lower costs.

6.5.2. Hydropower

Hydropower contribution to renewable energy supply was 37% in 2003, representing 2.0% of the total primary energy supply in IEA member countries (IEA, 2006[c]). Hydropower is a well developed technology, therefore, there is no need of so great technological advances, large expenses in RD&D or major changes in the development of resources as it happens with other renewable energies. However, permanent upgrading of technology, maximization of existing resources, additional RD&D and governments support to promote and ensure sustainability of new projects are still needed. RD&D activities should concentrate in the following: improve efficiency of large hydropower and small hydropower facilities, reduce equipment, operating and maintenance costs, development of innovative and environment friendly technologies, integration with other renewables and development of hybrid systems, with for instance hydrogen, as an additional product.

6.5.3. Geothermal

Geothermal power is also an established technology and the number of countries that uses geothermal energy to generate electricity has been increasing in the last decades. Geothermal contribution to total primary energy supply in IEA member countries was 0.4% in 2003, being its share in total renewable energy supply of 7.1% (IEA, 2006[c]). Growth of this share is mainly limited through local geologic characteristics that support this renewable energy source.

Although geothermal costs are among the lowest cost of renewables, costs can be further reduced to accelerate its growth. Other RD&D areas could also include: sustainable use, improve conversion efficiency cycles, development of better exploration ways and deeper geothermal resources, extension of the technology to other applications, both for power generation and direct heat use (geothermal co-generation) and covering larger regions, even farther away from tectonic plate boundaries.

6.5.4. Wind energy

In 2003, the share of wind energy in renewable energy supply was 1.8%, accounting for 0.1% of total primary energy supply in IEA countries (IEA, 2006[c]). Due to successful RD&D, wind energy market is in fast development. In the last years, industry RD&D efforts concentrated in developing larger and more effective wind energy systems and allowed decreasing production costs of wind turbines by a factor of four, thus, making wind energy cost competitive with other forms of electricity (IEA, 2006[c]). However, RD&D activities should continue to reduce cost and uncertainty of extensive exploitation, increase the knowledge about how aerodynamics, electrical generation and extreme wind situations affect wind turbine operation, about forecasting techniques, minimization of environmental impacts and increasing public acceptance.

6.5.5. Photovoltaics

Photovoltaics are now commercial available for several utilizations, namely: for illuminating signs, for water pumping, and for lighting isolated places. In 2003, photovoltaics accounted for 0.007% of total electricity production and for 0.05% of total renewable electricity produced in the world (IEA, 2006[c]). Before photovoltaics may be used for electricity production at large scale, it is fundamental a to achieve substantially reduction of photovoltaics costs. In fact, photovoltaics costs are higher than other forms of electricity generation, but in last years photovoltaics market has been increasing with growth rates higher than 30% per year (IEA, 2006[c]) and great reductions in photovoltaics cost are expected, due to RD&D efforts to improve materials, processes, conversion efficiency, design and improving synergies with other technological sectors. Considerable cost decreases can also be achieved by increasing manufacturing volume, the size of components and plants and economies of scale.

Photovoltaics produce electricity without emissions and with no noise, however, depending on the technology used, photovoltaics production installations use toxic, flammable and explosive gases, like: silane, phosphine or germane and toxic metals like cadmium. R&D activities should also focus on decreasing wastes and emissions during photovoltaics production, to keep emissions lower than those of coal based generation (IEA, 2006[c]). Another important aspect is improving recycling technologies for cell materials.

6.5.6. Solar Heating and Cooling

Solar heating and cooling technologies may substitute oil and gas presently used for heating, which is responsible for more than one-third of global energy use (IEA, 2006[c]). In solar heating and cooling may be included a wide range of technologies: solar water heating, active solar space heating, and passive solar heating and cooling, commercially available for more than 30 years and the emerging combi-systems that join water and space heating (IEA, 2007[e]). During the last ten years, the market implementation of these systems, including the combi-systems doubled and a reduction of 20% in installation costs was achieved (IEA, 2006[c]).

RD&D attempts are needed to develop competitiveness by achieving technical advances in materials and components, storage, scaling up and increasing efficiency. Architectural aspects also should be considered to achieve a full integration of solar thermal collectors in the buildings to enlarge its penetration in the market and achieve environmental benefits.

6.5.7. Concentrating Solar Power

Concentrating solar power may have an important role in the future to provide clean renewable energy. Concentrating solar power technologies have the following basic parts: concentrator, receiver, transport, storage, and power conversion. The concentrator captures and concentrates solar radiation and sends it to the receiver that absorbs the concentrated radiation and transfers its heat to a working fluid. The transport-storage system transmits the fluid from the receiver to the power conversion system, part of the thermal energy may be stored for later use. Some concentrating solar power systems have been successfully demonstrated like: Rankine, Brayton, combined or Stirling cycles, while others are emerging (IEA, 2006[c]).

Concentrating solar power may be used in remote villages and for grid connected applications, however, before this, considerable cost reductions are needed, which may be achieved through RD&D efforts by: a) increasing scale, which also increases performance and decreases investment, operation and maintenance costs, b) improved technology for concentrator performance and cost, c) improved storage, and d) new deployment efforts.

6.5.8. Ocean energy

Oceans contain an enormous amount of power that may be exploited to produce useful energy. Ocean energy includes: tidal energy, thermal energy, marine currents and ocean waves. Technologies available are at an early stage, when compared to other renewable energies. Wave energy systems may be of two types: fixed devices in the coastline or in off-shore platform fixed on the seafloor and floating devices connected to the shore through a high voltage power cable. For each type there is a wide range of technologies under development, which need basic knowledge about wave behavior and hydrodynamics of wave absorption, structure and hull design methods, mooring techniques to ensure that the system is safely kept in position no matter the weather conditions, power take off systems that converts incoming energy into electricity and deployment methods. RD&D efforts should fulfill these requirements and ensure that new technologies may accomplish low cost, efficient energy conversions, reliability, simplicity, safety and low environmental impact. RD&D should also contribute to diminish the large technical and financial risks faced by device developers, due to the ruthless environment of strong currents and waves.

7. Conclusions

It is predictable that energy demand will have a great increase in the years to come due to increasing needs in life quality, development of emerging economies of Asian countries like India and China and the growth of world population. Nowadays predictions show that fossil fuels will continue to have a great contribution to energy production, particularly to power generation, which means that CO_2 and other greenhouse gases emissions will continue to rise, threatening the world as we know it, due to global warming and climate change.

Therefore, it is urgent to promote the utilization of alternative fuels and new technologies, especially those already available, even if at reasonable costs. To attain this goal, the appropriate policy approach is needed, together with vigorous and concerted government actions supported by increased funding from both public and private sources.

The main areas that still need further scientific and technological developments to ensure sustainable security of energy supply in the future were identified and the main R&D needs of each process or technology analyzed were highlighted. But, the sustainable security of energy supply can not be achieved without EU regulations, governmental political measures and R&D strategies supported by research financing programs.

Several studies and analysis, some of them with the support of IEA, have demonstrated that more sustainable energy future is within our reach, and that technology is the key issue. The role of science is very important and crucial for the development of new robust technologies to ensure sustainable security of energy, by increasing energy efficiency, developing alternative fuels and new technologies, increasing CO_2 capture and storage and renewables energy utilization. Besides R&D, a key issue is technology transfer. RD&D activities have played a major role in the successful development and commercialization of a range of new renewable energy technologies in recent years and should continue in the future to ensure larger utilization of these technologies.

RD&D efforts have been made in Europe and in more industrialized countries to develop reliable, commercially viable, clean, efficient and low carbon dioxide emissions technologies. More industrialized countries should share the attained knowledge and should help under developing countries and emerging economies to use the new technologies as long as they are going to be developed, otherwise the target of reducing the dependency on fossil fuels, ensuring energy supply security and environment protection will never be achieved.

In spite of the RD&D efforts, significant technical and non-technical barriers to deployment of new technologies and alternative fuels still remain, being the latter often more difficult to deal with. These are often related to institutional frameworks, perceptions, socio-economic issues or commitments to related technology sectors. Unlike research questions, non-technical barriers are usually not soluble only by developing technical knowledge or expertise, behavior change and governmental political support are most needed.

8. References

Bounaceur R., Nancy L., Denis R., Cecile V., Eric F., (2006), Membrane processes for post-combustion carbon dioxide capture: A parametric study, Energy, 31, 2556–2570.

Bridgwater A. V., (1999), Fast Pyrolysis of Biomass - A Handbook, Volume 1, CPL Press Science Publishers.

Bridgwater A. V., (2001), Fast Pyrolysis of Biomass - A Handbook, Volume 2, CPL Press Science Publishers.

Bridgwater, A. V., Toft A. J., Brammer J. G., (2002), A techno-economic comparison of power production by biomass fast pyrolysis with gasification and combustion, Renewable and Sustainable Energy Reviews, 6, 181-248.

Buchner Barbara, (2007), Policy Uncertainty, Investment and Commitment Periods, IEA.

Buekens A., (2006), Introduction to Feedstock Recycling of Plastics, Feedstock Recycling and Pyrolysis of Waste Plastics Edited by J. Scheirs and W. Kaminsky, John Wiley and Sons, Ltd.

Buhre B.J.P., Elliott L.K., Sheng C.D., Gupta R.P., Wall T.F., (2005), Oxy-fuel combustion technology for coal-fired power generation, Progress in Energy and Combustion Science 31, 283–307.

COM(97) 514; A Community Strategy to Promote Combined Heat and Power (CHP) and to Dismantle Barriers to Its Development; Commission of the European Communities, Brussels, 15 October 1997.

COM(2000) 769; Towards a European Strategy for Energy Supply Security - Green Paper; Commission of the European Communities, Brussels; 29 November 2000.

COM(2001) 547; Communication on the promotion of alternative fuels for road transportation and on a set of measures to promote the use of biofuels for transport alternative fuels for road transportation; Commission of the European Communities, Brussels; 7 November 2001.

COM(2005) 265; Doing More With Less – Green Paper on Energy Efficiency; European Commission; Luxembourg; 22 June 2005.

COM(2005) 628; EU Biomass Action Plan (BAP); Commission of the European Communities, Brussels; 7 December 2005.

COM(2006) 34; EU Strategy for Biofuels; Commission of the European Communities, Brussels; 8 February 2006.

COM(2006) 105; Green Paper on a European Strategy for Sustainable, Competitive and Secure Energy; Commission of the European Communities, Brussels; 8 February 2006.

COM(2006) 583; Global Energy Efficiency and Renewable Energy Fund (GEEREF); Commission of the European Communities, Brussels; 6 October 2006.

COM(2006) 545; Action Plan for Energy Efficiency "Saving 20% by 2020"; Commission of the European Communities, Brussels; 19 October 2006.

COM(2006) 848; Renewable Energy Road Map – Renewable energies in the 21st century: building a more sustainable future; Commission of the European Communities, Brussels; 10 January 2007.

COM(2007) 1; An Energy Policy for Europe; Commission of the European Communities, Brussels; 10 January 2007.

COM(2007) 723; A European Strategic Energy Technology Plan (Set-Plan)–"Towards a low carbon future", Brussels; 22 November 2007.

COM(2008) 19; Promotion of the use of energy from renewable sources; Commission of the European Communities, Brussels; 23 January 2008.

Competitiveness and Innovation Programme (CIP); 13855/06 (Presse 284); Council of the European Union; Luxembourg; 12 October 2006.

Corella José, Toledo José M., Padilla Rita, (2004), Catalytic Hot Gas Cleaning with Monoliths in Biomass Gasification in Fluidized Beds. 1. Their Effectiveness for Tar Elimination, *Ind. Eng. Chem. Res.*, 43, 10, 2433 -2445.

Corella José, Toledo José M., Padilla Rita, (2005), Catalytic Hot Gas Cleaning with Monoliths in Biomass Gasification in Fluidized Beds. 3. 3. Their Effectiveness for Ammonia Elimination, *Ind. Eng. Chem. Res.*, 44, 7, 2036 -2045.

Directive 2001/77/EC; Promotion of electricity produced from renewable energy sources in the internal electricity market; Directive of the European Parliament and of the Council of 27 September 2001; *Official Journal of the European Union*; L283; 27 October 2001.

Directive 2003/30/EC; Promotion of the use of biofuels or other renewable fuels for transport; Directive of the European Parliament and of the Council of 8 May 2003; *Official Journal of the European Union;* L123; 17 May 2003.

Directive 2004/8/EC; Promotion of cogeneration based on a useful heat demand in the internal energy market; Directive of the European Parliament and of the Council of 11 February 2004; *Official Journal of the European Union*; L52; 21 February 2004.

Directive 2006/32/EC; Directive on energy end-use efficiency and energy services; Directive of the European Parliament and of the Council of 5 April 2006; Official Journal of the European Union; L114; 27 April 2006.

Dolan M. D., Dave N. C., Ilyushechkin A. Y., Morpeth L. D., McLennan K. G., 2006, Composition and operation of hydrogen-selective amorphous alloy membranes, Journal of Membrane Science, 285, 30–55.

European Communities, (2006), Biofuels in the European Union. A vision for 2030 and beyond, Final report of the Biofuels Research Advisory Council, Directorate-General for Research, Directorate J – Energy, Unit 3 – New and Renewable Energy Sources.

Gray D., Tomlinson G., (2002), Hydrogen from coal, Mitretek Technical Paper, July.

Gray M. L., Soong Y., Champagne K. W., Stephens R. W., Toochinda P., Chuang S.S.C., (2002), Prepr. Am. Chem. Soc. Div. of Fuel Chem., 47(1), 63.

Huber G.W., Iborra S., Corma A., (2006), Synthesis of transportation fuels from biomass: chemistry, catalysts and engineering. Chemical Reviews 106, 4044–4098.

IEA, 2004, Renewable energy – market and policy trends in IEA countries, International Energy Agency,OECD/IEA, Paris.

IEA, 2005, Projected Costs of Generating Electricity - 2005 Update, International Energy Agency,OECD/IEA, Paris.

IEA, 2006[a], World Energy Outlook, International Energy Agency, OECD/IEA, Paris.

IEA, 2006[b], Global renewable energy policy and measures database. International Energy Agency, OECD/IEA, Paris.

IEA, 2006[c], Renewable Energy: R&D priorities. Insights from IEA Technology programmes, International Energy Agency, OECD/IEA, Paris.

IEA, 2007[a], World Energy Outlook, International Energy Agency, OECD/IEA, Paris.

IEA, 2007[b], Clean Coal Technologies, Accelerating Commercial and Policy Drivers for Deployment, International Energy Agency, OECD/IEA, Paris.

IEA, 2007[c], CO_2 Emissions from Fuel Combustion, International Energy Agency, OECD/IEA, Paris.

IEA, 2007[d], CO_2 Capture Ready Plants, Technical Study, Report Number: 2007/4, May, International Energy Agency, OECD/IEA, Paris.

IEA, 2007[e], Nuclear Power - Energy Technology Essentials, IEA, March, International Energy Agency, OECD/IEA, Paris.

IEA, 2007[f], Renewables for Heating and Cooling, International Energy Agency, OECD/IEA, Paris.

Jackson S., Hlozek B., (2003), Louisiana gas plant hikes production with quick solvent changeout, Oil and Gas Journal, 101, 23, June 09.

Loiacono O., (2001), Consuming hydrogen, Hydrocarbon Engineering, November.

Kaldis S.P., Skodras G., Sakellaropoulos G.P., (2004), Energy and capital cost analysis of CO_2 capture in coal IGCC processes via gas separation membranes, Fuel Processing Technology, 85, 337– 346.

Kaminsky W., Schmidt H., Simon C.M., (2000), Recycling of mixed plastics by pyrolysis in a fluidised bed, *Macromol.* Symp. 152, 191-199.

Kaminsky W., Eger C., (2001), Pyrolysis of filled PMMA for monomer recovery, *J. of Anal. and Appl. Pyrolysis* 58-59, 781-78.

Kaminsky W., Mennerich C., (2001), Pyrolysis of synthetic tire rubber in a fluidised-bed reactor to yield 1,3-butadiene, styrene and carbon black, *J. of Anal. and Appl. Pyrolysis* 58-59, 803-811.

Kaminsky W., Predel M., Sadiki A., (2004), Feedstock recycling of polymers by pyrolysis in a fluidised bed, Polymer Degradation and Stability 85, 1045-1050.

Kanniche M., Bouallou C., (2007), CO_2 capture study in advanced integrated gasification combined cycle, *Applied Thermal Engineering* 27, 2693–2702.

Kubek D., Sharp C., (2002), Recent Selexol, PolySep and PolyBed operating experience with gasification for power and hydrogen. UOP LLC, Gasification Technologies Conference, October.

Maggi R.E., Elliott D.C., (1997), Upgrading overview. In: Developments in thermochemical biomass conversion. Bridgwater AV and Boocock DGB (Eds). Blackie Aand P.

McDonald D., De Vault D., Varagani R., (2007), Oxy-Combustion in Pulverized Coal Power Plants for Carbon Dioxide Concentration, Electric Power Conference, May 1-3, 2007, Chicago, Illinois, U.S.A.

McKee Barbara, (2002), Solutions for the 21st Century, Zero Emissions Technologies for Fossil Fuels, Technology Status Report, Technologies WPFF-International Energy Agency, May.

Nuclear Power - IEA Energy Technology Essentials, IEA, March 2007.

Philibert Cédric, Ellis Jane, Podkanski Jacek, (2007), Carbon Capture and Storage. In The CDM, IEA.

Pinto F., Gulyurtlu I., Costa P., Cabrita I., (1999), Pyrolysis of Plastic Waste. 1. Effect of Plastic Waste Composition on Product Yield, *Journal of Analytical and Applied Pyrolysis,* 51, 1-2, 39-55.

Pinto F., Lopes H., André R. N., Gulyurtlu I., Cabrita I., (2007), Effect of Catalysts in the Quality of Syngas and By-Products Obtained by Co- Gasification of Coal and Wastes. 1. Tars and Nitrogen Compounds Abatement, Fuel, 86, 14, 2052-2063.

Pinto F., Lopes H., André R. N., Gulyurtlu I., Cabrita I., (2008), Effect of Catalysts in the Quality of Syngas and By-Products Obtained by Co- Gasification of Coal and Wastes. 2. Heavy Metals, Sulphur and Halogen Compounds Abatement, Fuel, 87, 7, 1050-1062.

Radtke K. R., Heinritz-Adrian M., Marsico C., (2006), New Wave of Coal-to-liquids, VGB PowerTech 5/2006.

Rostrup-Nielsen J. R., (2002), Syngas in perspective, Catalysis Today, Vol. 71, 243-247.

Schäfer B., Mather A.E., Marsh K.N., (2002), Enthalpies of solution of carbon dioxide in mixed solvents, Fluid Phase Equilibria 194–197, 929–935.

Scheirs J., Kaminsky W., (2006), Feedstock Recycling and Pyrolysis of Waste Plastics: Converting Waste Plastics into Diesel and Other Fuels, Chem. Tec. Publishing, ISBN 978-0-470-02152-1.

Sircar S., Golden T. C., Rao M. B., (1996), Activated carbon for gas separation and storage. Carbon, 34, 1, 1-12.

Siriwardane R. V., Shen M., Fisher E. P., Poston J., (2001), Adsorption of CO2 on molecular sieves and activated carbon. Energy and Fuels, 15, 279-284.

Stiegel G. J., Ramezan M., (2006), Hydrogen from coal gasification: An economical pathway to a sustainable energy future, *International Journal of Coal Geology*, 65 173-190.

Stiegel G. J., Maxwell Russell C., (2001), Gasification technologies: the path to clean, affordable energy in the 21st century, Fuel Processing Technology, 71, 79–97.

The Royal Society (2008), Sustainable biofuels: prospects and challenges, Policy document 01/08, January.

Tukker, A., de Groot, H., Simons, L., S. Wiegersma, (1999), Chemical Recycling of Plastics Waste (PVC and other resins), TNO report for DGIII (STB-99-55 Final), December.

Vasileiadis S., Ziaka-Vasileiadou Z., (2004), Efficient catalytic reactors-processors for fuel cells and synthesis applications, Separation and Purification Technology, 34, 213–225.

Vosloo A.C., (2001), Fischer-Tropsch: a futuristic view, Fuel Processing Technology, Vol. 71, 149-155.

Xu X., Andresen J.M., Song C., Miller B.G., Scaroni A.W., (2002), *Prepr. Am. Chem. Soc.* Div. of Fuel Chem., 47(1), 67.

In: Energy in Europe: Economics, Policy and Strategy
Editors: F. L. Magnusson and O. W. Bengtsson
ISBN: 978-1-60456-829-5

Chapter 8

EUROPEAN GAS REGULATION – A THEORETICAL CRITIQUE

***Aldo Spanjer*[1]**
Economics Department, Leiden University
P.O. Box 9520 2300 RA Leiden, The Netherlands

ABSTRACT

Due to a number of structural market shifts, a new context for European gas regulation has arisen. In this new context, regulation must ensure adequate investments in a gas market that is dominated by producers, where political considerations are becoming more important, and where supply security and sustainability are top policy priorities. This requires a regulatory policy which emphasizes criteria like irreversible investments, risks, uncertainties, and governance. This contrasts starkly with the old situation where a main emphasis for regulation was to sweat existing assets.

This chapter's main question is whether existing gas regulation, which is based on neoclassical guiding principles, is adequate in this new context. This chapter argues it is not, because the neoclassical perspective does not explicitly acknowledge the above criteria. These criteria are explicitly treated by transaction costs economics. This chapter therefore argues from a theoretical perspective that inclusion of the transaction cost perspective will likely result in better outcomes in terms of market behavior. Hence, designing a proper regulatory regime in the new regulatory context requires that existing regulation be complemented with the transaction cost perspective. This implies a liberalized European gas market which is different from the one prescribed by the neoclassical perspective as well as the one envisaged in existing gas regulation.

Furthermore, due to its perfect competition guiding principles, existing regulation is tautological. That is, by assuming the eventual arrival of a perfectly competitive internal gas market, the option that the current shortcomings in the development of the internal gas market are attributable to a discrepancy between the existing regulation and the structural characteristics of the European gas market is ruled out by definition. The transaction cost perspective implies that the latter may well be the case, however.

[1] Phone ++31 71 527 7754, a.r.spanjer@law.leidenuniv.nl.

1. Introduction

This introduction provides the background to the issues addressed in this chapter. Section 1.1 sets out the traditional, pre-liberalization structure of the European gas market. Section 1.2 provides some reasons why this structure came under increasing pressure and why liberalization was thought to be a proper prescription. Section 1.3 sets out the questions addressed in this chapter.

1.1. The Traditional Market Structure

Six main types of players can be identified on the European gas market: 1) producers; 2) national transmission companies; 3) large industrial consumers; 4) local – often municipal – distribution companies (LDCs); 5) small consumers and businesses; and 6) the government.

The European gas market developed at both the national and the European level (Estrada et al., 1988; Percebois, 1999; Correljé et al., 2003; and Finon, 2004). The national level was characterized by national monopolies. These monopolies, which on the Continent were dominated by the government, developed 'their' own national market, taking into account market specifics such as resource endowment, energy mix, etcetera. This chapter focuses on the European context. The producers, mostly government controlled, sometimes in cooperation with international oil companies such as Shell and ExxonMobil, were responsible for exploration and production activities. They sold their gas to the national transmission companies via the upstream pipeline network under long-term contracts which usually ranged from fifteen to twenty years. Two types of transmission companies can be distinguished: those of the exporting and the importing countries.

The exporting transmission companies – predominantly Russia's Gazprom, Algeria's Sonatrach, Norway's GFU, Holland's Gasunie,[2] and UK's British Gas – held a national export monopoly and handled the volumes of domestically produced gas. On a European scale, they formed an oligopoly, which Finon (2004, pp. 185) refers to as the oligopoly of sellers or producers. Finon also identifies an oligopoly of purchasers which consists of the importing transmission companies of Continental Europe. They imported their gas and were in a monopoly (or quasi-monopoly) position for wholesale supply in their country. Examples are Germany's Ruhrgas,[3] Belgium's Distrigas, France's Gaz de France, Italy's SNAM, Austria's OMV, Spain's Gas Natural, and Holland's Gasunie.[4] The European upstream gas market could thus be characterized as a two-sided oligopoly.

The national transmission companies played a pivotal role in the traditional structure (Stern, 1998). The national transmission companies were monopsony buyers of gas from the producers. They were monopoly sellers of their gas via the high-pressure transmission system to LDCs and large industrial consumers. This was arranged through medium-term contracts which ranged one to five years. All Member States, with the exception of Germany, had one

---

[2] From September 2006 on, Gasunie is the name of the infrastructure company which has been unbundled from the trading company Gasterra.

[3] EON/Ruhrgas after 2003.

[4] Gasunie – now Gasterra – appears in both the oligopoly of purchasers and of sellers, since it held statutory monopolies as exporter, importer, and wholesale supplier on the Dutch natural gas market.

transmission company.[5] The LDCs used their low-pressure distribution system to deliver the gas to the small consumers and businesses on the basis of short-term supply agreements. The LDCs were often captive to the national transmission companies, urging them to cooperate in order to develop an appropriate status quo.

In addition, government involvement on the European gas market was substantial. This is due to at least two factors (Radetzki, 1999, pp. 18). First, the energy market situation of the 1970s and 1980s contributed to government involvement. The two oil price shocks and the perception of gas scarcity in this period resulted in fears concerning security of supply. The fear of being dependent on producers that were perceived as unstable or unreliable, led to a situation where government or political involvement was considered essential to guarantee sufficient supplies and to establish stable consumer-producer relationships.[6] A second reason for government involvement in the energy sector was satisfying other domestic policy objectives such as employment or redistributive purposes (see below). The government owned or was involved in almost all transmission companies, giving it the possibility the use gas rents for social or wider economic ends.

The European gas market was also heavily monopolized. The main reason for this was the gas network's natural monopoly status. When costs are subadditive over the relevant production range, the first operator obtains a natural monopoly position, since he will be able to undercut prices of potential entrants (cf. Baumol, 1977). Hence, the lowest cost solution is to have one dominant company supplying the entire market with gas. To this end, in all Member States, the transmission and distribution companies had been granted a local monopoly for pipeline operation and distribution. The powerful position of the European transmission companies was also regarded essential because their control of the national markets granted them a relatively strong bargaining position against the producers. It enabled them to conclude long-term contracts with purchase obligations with the producers. These long-term contracts contained three speeicfic clauses which facilitated the investments necessary to develop the European gas market.

The first contractual clause is the take-or-pay clause, compelling the transmission company to pay for the gas also if less than the contractually agreed quantity is taken from the system. Accordingly, volume risk was borne by the transmission companies. Due to their monopolistic position on their home markets, competition came from substitutes the consumers could switch to rather than new entrants. This resulted in the early 1960s, after the 1959 discovery of the Groningen field, in a fixing of gas prices at the level of substitute fuels in order to obtain and retain a market share for gas. Since oil was the main substitute available, gas prices were linked to those of oil – hence the term oil price linkage. Residential gas prices were linked to those of heating oil and industrial prices to heavy fuel oil. Producer revenues were derived from the end-user prices through net-backing, i.e., subtracting the costs of transportation, storage, and ancillary services from the end-user price of the competing fuels in order to obtain the commodity price of gas. This explains why the oil price linkage is often referred to as net-back market value approach. Maisonnier (2006, pp. 26) provides the following example of a typical European gas pricing formula:

---

[5] Germany had about 15 regional transmission companies (some of which had direct access to import and production) and over 700 local distributors, or 'stadtwerke'.

[6] The latter was to be achieved by, amongst others, the Energy Charter process, which was designed to export the benefits of Europe's liberalization policy beyond European borders.

$$P = P_0 + A^*(G-G_0) + B^*(F-F_0),$$

where P is the monthly price paid for gas from a producer; G denotes the average heating oil price over 3, 6, or 9 months; and F indicates the average heavy fuel oil price over the same intervals. The index 0 indicates the values at the initial date of contract implementation.

Note that the spot market price does not enter the equation, so that the commodity price remains unaltered if spot gas prices fall. The lower revenues are passed on to the producers, indicating that price risk is borne by them.[7] The third and final clause is the final destination clause. A final destination clause prohibits transmission companies from reselling purchased gas, which guarantees that this gas could not be sold elsewhere (in competition with a producer's other gas supplies) at a higher profit. This reinforced the producers' market power and effectively partitioned national markets, obstructing inter-Member State trade and de facto creating a patchwork of national energy markets implementing to a large extent their own energy policies.

These contracts successfully facilitated the large scale investments necessary to develop a European gas market (Stern, 1998, pp. 16, 18). They provided the producers with security of demand and revenues, facilitating producer investments the production and exploration of gas. Furthermore, the diversified portfolio of long-term contracts and sole access to pipelines ensured investments from the transmission companies too. Moving downward along the value chain, looking at the LDCs, essentially the same situation applied in that their monopoly position facilitated their investments in the distribution networks.[8]

In addition to these vertical contractual clauses, vertical integration was a widespread phenomenon in the industry too. The international oil companies responsible for gas production on the Continent were also present further down the gas chain in for example transportation and resale companies, in order to capture part of the downstream profits. Examples are Shell and ExxonMobil: they were responsible for the production of gas in the Netherlands and the North Sea and also held important downstream ownership positions in for instance Gasunie and BEB, a German pipeline and storage operator. National transmission companies were partly owned by the producers. This vertical integration in combination with the fact that these owners were relatively indifferent between up- or downstream profits, provides one explanation for the observation that the national transmission companies did not, despite their strong bargaining position and contrary to expectations, bargain down the price for the gas they imported (Radetzki, 1999, pp. 19).[9] The monopoly franchises themselves were also typically vertically integrated to reap economies of scale and scope.

All in all, the above structure allowed gas to enter the market at competitive prices and has facilitated investments up-, mid-, and downstream. This created a mature and secure European gas supply system in which the risks associated with the high capital costs were reduced and spread along the gas value chain. There were at least three additional reasons for this structure. First, energy was and is considered vital from an economic (by contributing to

---

[7] Dailami and Hauswald (2000) provide empirical evidence on risk sharing. They have studied the Ras Gas LNG Project in Qatar, which delivers LNG on a long-term take-or-pay basis to South Korean Kogas (who resells it to the Korean Electric Power Company). It is shown that risks are shared in that volume risk is borne by Kogas, while price risk remains with Ras Gas.

[8] In a few European countries, such as France and the UK, distribution had been integrated with transmission (IEA, 2000, pp. 24).

[9] A second reason put forward by Radetzki is that several national transmission companies were publicly owned utilities, required to provide a 'normal' return on capital rather than maximizing profits.

GDP and employing thousands of people), political (as a redistributive vehicle) and strategic (controlling strategic infrastructures in case of a war or crisis) perspective, providing incentives to organize everything within the national government's realm (Geradin, 2006, pp. 6, 7). Second, it was relatively easy to secure the public service obligations (PSOs). It was customary to grant monopoly franchises to a utility delivering gas. In exchange for this monopoly franchise, the utility was either a regulated private player, as in the United States, or under state ownership, as for the greater part of Europe, with a mandate to operate in the public interest (Newbery, 1999, pp. 86). These franchise monopolies were able to recoup investment costs by passing them on to consumers.

1.2. Liberalization

An important impetus for this change in economic focus was the early experiences of liberalization in the United States in the late 1970s, and the United Kingdom in the early 1980s. With some delay, Europe followed suit. In the mid-1980s, the policy emphasis shifted from governments and national transmission companies controlling the gas markets towards reducing government involvement and unleashing competitive forces. The energy sectors were thought to operate more efficiently when subjected to a market reform that focused on liberalization, privatization and competition.

The market environment significantly contributed to this competition emphasis coming into fashion. The excess supply situation in Europe (the gas bubble) made gas widely available and cheap, shifting the balance of power towards the buyers. Consequently, as opposed to the situation after the oil crises, no government involvement to guarantee supply security was considered necessary. The investments that were facilitated by the traditional structure had created a matuire gas market. The public monopolies had created a highly reliable gas infrastructure in which initial investments had long been amotized. Stimulating investments in order to develop the market no longer necessary; new investments were primarily required in expansion and interconnection of pipelines, rather than constructing enw ones. Gas supply companies face virtually no volume risk. Accordingly, an alternative policy was required (Ellis et al., 2000; IEA, 2000). Because the economics of the gas markets had changed, policy had to follow suit. Consequently, the emphasis shifted from stimulating investments in new assets towards more efficiently deploying (i.e., sweating) existing assets.

Due to the combination of these factors, the traditional market organization lost its glitter and made introducing competitive forces through liberalization the new panacea. The European Commission's drive to liberalize European gas markets in order to create an internal European market, reflects this view (CEC, 1985, 1988). Liberalization is supposed to transform the rigid and managed gas market indicated above into a truly European internal gas market on which unfettered competitions reigns.

A few anticipated benefits of liberalization were that asset sweating would increase consumer surplus by driving down network costs as close as possible to marginal operating costs. In the buyer's market that had developed, such a view was quite understandable and liberalizing in order to reach as close as possible the perfect competition outcome of prices at marginal costs would indeed be welfare enhancing. Furthermore, in a competitive environment public monopolies would increasingly lose their grip on the national markets. Besides network operations, government-initiated investments would be unnecessary.

Competitive forces would also induce operators to only undertake efficient investments. Finally, after seeing the benefits that liberalization entails to Europe, the producers were expected to welcome European foreign direct investment and technological knowledge into their energy sectors in order to develop their reserve potential. This would guarantee access to reserves and secure future supplies. In order to liberalize the European gas markets, two EU Gas Directives have been introduced; a third legislative package one had been proposed in September 2007 (CEC, 1998, 2003, 2007c).

1.3. Outline

However, liberalization is currently not working as anticipated (CEC, 2004, 2005a). The European Commission interprets these problems as a sign of insufficient competition. In other words, it assumes that all existing problems will be overcome once the internal gas market has developed. Hence, the prescription is to more vehemently pursue competition. However, the unsatisfactory state of liberalization can also be interpreted as an indication that the current competition-oriented approach does not square with the characteristics of European gas markets. The latter option is ruled out by definition in the current liberalization process. This chapter approaches this issue from a theoretical perspective by examining whether the current regulatory approach is the right one on which to base gas liberalization. To answer this question, four subquestions are specified.

1) How has the international gas market changed recently and what are the consequences for regulation?
2) What are the theoretical underpinnings of current regulation and what does it try to achieve?
3) What kind of doubts are there concerning this regulatory scheme and why do these require an institutional perspective?
4) What are the implications for gas liberalization of an institutional perspective?

This chapter is structured according to these subquestions. The first question emanates from the observation that any regulatory scheme is dependent on the market it is applied to as well as the environment in which the market operates. Therefore, section two discusses a number of structural market shifts at the international gas market. It is argued that these shifts reshape the gas market. In fact, the shifts create new context for gas regulation of which the criteria are provided. The issue comes down to whether existing regulation is able to adapt to this new context while upholding its current underpinnings. Existing gas regulation builds on the neoclassical view towards economics. Therefore, section three sets out the neoclassical perspective, its implications for regulatory intervention as well as the structure of a liberalized European gas market that emanates from a neoclassical regulatory scheme. Section four discusses existing gas regulation and shows the strong link with neoclassical economics. This answers the second subquestion. Section five answers the third subquestion by arguing that the criteria for a regulatory scheme that emanate from the market shifts imply that the neoclassical perspective must be complemented with institutional economics. The fourth subquestion is answered by setting out this institutional perspective, and its consequences for regulation as well as for a liberalized European gas market. It illustrates that the institutional perspective implies substantial revisions to the neoclassical prescriptions, and by extension to the internal European gas market. Finally, section six provides the main conclusions. The conclusion provides the foundations for the empirical research needed to complement this chapter's theoretical critique. In particular, it sets out the research agenda that answers the question whether recent amendments to gas regulation (allowing long-term import contracts and the enlarged scope for exemptions from the third party access regime) which indicate a movement from the neoclassical perspective towards the institutional one, are sufficient in light of the structural changes that have taken place.

2. Market Shifts

2.1. Introduction

As indicated in the introduction, the purpose of this chapter is to assess existing gas regulation from a theoretical perspective. The efficiency and efficacy of any regulatory scheme hinges on the environment a market is operating in. An important argument of this chapter is that the international gas market is experiencing a number of shifts which together fundamentally reshape the European gas market. Sections 2.2 to 2.7 discuss these shifts, after which section 2.8 argues that these shifts create a new context for gas regulation.

2.2. Import Dependence

The excess supply referred to in the introduction made the issue of importing gas supplies for producers outside Europe relatively unimportant. However, anxieties have flared up because European import dependence is growing. Table 1 illustrates the issues.

Table 1. Europe and its gas suppliers, 2006

	Russia	Iran	Qatar	Nigeria	Algeria	Norway	EU-25	Rest world
Proven reserves								
- percent world[a]	26.3	15.5	14.0	2.9	2.5	1.6	1.3	35.9
- bcm	47650	28130	25360	5210	4500	2890	2430	
Production								
- percent world[a]	21.3	3.7	1.7	1.0	2.9	3.0	6.6	59.2
- bcm	612.1	105.0	49.5	28.2	84.5	87.6	190.0	
R/P	77.8	>100	>100	>100	53.3	33.0	12.8	
Consumption								
- percent world[a]	15.1	3.7	0.7	Na	0.8	0.2	16.3	63.2
- bcm	432.1	105.1	19.5	Na	23.7	4.4	467.4	

Source: Own calculations from BP (2007).

a) percent world: percentage of world total.

The table shows that EU-25 consumption more than doubles production and that proven reserves[10] are low. Hence, Europe is quite import dependent (with some Eastern Member States completely dependent on Russian supplies). Long-term projections aggravate matters.

In fact, CEC (2007d, pp. 13) shows that without policy changes EU gas import dependence will rise from 51 percent in 2000 to 81 percent in 2030. Projections of the International Energy Agency (IEA, 2007, pp. 85-88) confirm the picture of rising import dependence. According to its reference scenario, between 2000 and 2030 OECD Europe's[11] gas demand will rise from 477 to 771 bcm. Maturing European gas fields cause gas production to level off and then decline. IEA projects OECD Europe's gas production to fall from 304 bcm in 2000 to 251 bcm in 2030. The gap between consumption and production, which rises from 173 bcm in 2000 to 520 bcm in 2030, will need to be filled by imports (consequently, imports will make up 67 percent of gas demand in 2030).

The reserves to production (R/P) ratio shows how many years current proven reserves will last with current production levels: if current production remains unchanged, all currently proven EU reserves will be depleted in 12.8 years. Some producers have abundant reserves, as indicated by the R/P ratios in the table. Globally, the reserve picture is quite positive (BP, 2007, pp. 26). The global amount of proven reserves suffices to satisfy global gas demand for approximately another 65 years. Hence, the global amount of reserves does not yet pose a real threat. This is even more so because the undiscovered, potential reserves are considered to be very large.[12]

The location of reserves may introduce some concerns, however. By far the most of the proven – and also undiscovered – reserves are located outside Europe (BP, 2007, pp. 26). For instance, the Middle East possess by far the largest R/P ratio of over 200 years, quadrupling that of Europe and Eurasia combined which is just over 50 years. In consequence, Europe will increasingly have to rely on imported gas supplies, the geographical spread of which may create concerns. These are exacerbated by political risk and ownership concerns: Röller et al. (2007, pp. 7) argue that around 70 percent of global reserves are located in medium and high risk areas. They also argue that Russian gas is of a (much) higher political risk than indigenous volumes. Hence, Europe's increasing import dependence will increase anxieties concerning (political) security of supply risks. It also draws attention to the behavior of the producers.

2.3. Producer Behavior

A second shift pertains to producer behavior. Two main developments in this regard are increasing state ownership and the prospect of producer cooperation.

---

10 Proven reserves are reserves which are shown by geological and engineering data to be recoverable with reasonable certainty in future years from known reservoirs under existing economic and operating (BP, 2007, pp. 22)

11 OECD Europe comprises all European Member States of the OECD, i.e., Austria, Albania, Belarus, Belgium, Bosnia and Herzegovina, Bulgaria, Croatia, Czech Republic, Cyprus, Denmark, Estonia, Finland, France, Germany, Greece, Hungary, Iceland, Ireland, Italy, Latvia, Lithuania, Luxembourg, The Former Yugoslav Republic of Macedonia, Moldova, Netherlands, Norway, Poland, Portugal, Romania, Serbia and Montenegro, Slovak Republic, Slovenia, Spain, Sweden, Switzerland, Turkey, Ukraine and United Kingdom.

12 This is confirmed by Adelman and Lynch (2002), who argue that in the very long-term supply fears due to geological constraints are overstated.

State Ownership

Obtaining access to gas reserves from outside of Europe was not considered to be a significant problem until recently, because the producer countries were expected to welcome European foreign direct investment and technological knowledge into their energy sectors in order to develop their reserve potential. Furthermore, producers were thought to embrace Europe's liberalization movement and privatize their energy incumbents. However, table 2 paints a very different picture.

The table shows that in 2004 European indigenous production accounted for a mere 37 percent of demand, the rest being provided by government-dominated exporters. Current developments indicate an increasing state ownership in producing countries. Producers' national oil companies (NOCs) largely control reserves; the international oil majors (IOCs) have direct access to about one third of oil and gas reserves (CIEP, 2006, pp. 7). High levels of state ownership create apprehensions of energy being deployed as a strategic asset, or even weapon, to further political, non-energy goals to the detriment of European consumers.

Table 2. State ownership of gas exporters to Europe, 2004

	% of demand	Main exporting company	Status	State ownership %
Norway	17	Statoil	Monopoly	70
Russia	29	Gazprom	Monopoly	51
Algeria	13	Sonatrach	Monopoly	100
Nigeria	1	BBOC	Monopoly	100
Qatar	1	Qatargas	Monopoly	65
EU production	37			

Source: Röller et al. (2007, pp.7).

Al-Moneef (2006) downsizes the concerns by arguing that energy security is a shared concern of both producer and consumer nations and that there is ample room for fruitful cooperation between them. Cooperation between IOCs and governments/NOCs is not a zero sum game. Rather, both sides can benefit because the producers have the subsoil resources the IOCs are looking for, while the IOCs have the technical and management skills producers generally lack to optimally exploit the resources. Hence, NOCs and IOCs entering into partnerships and undertaking adequate and timely investments potentially mitigates concerns at both sides.

This positive view is not universally shared, however. Marcel (2006) argues that the interests of a typical NOC differ from those of an IOC.[13] For instance, an NOC wants to optimize resource development. This is a long-term planning decision which may in the short-run require below-capacity production. Furthermore, a government's responsibility for its citizens may urge it to subsidize domestic gas prices (as for instance in Russia). This facilitates an inefficiently high domestic gas demand, to the detriment of export volumes. These are unlikely the goals of an IOC, because IOCs typically have incentives to develop an existing resource base as fast as possible – for economic reasons, but also to expose themselves to political risks as shortly as possible. Hence, an NOC will typically implement a more conservative depletion policy than an IOC or producer government. Such differences create tensions that may hamper cooperation.[14]

All in all, there is significant uncertainty on the consequences of increasing state ownership at the producer's side. Current developments in Russia, Europe's largest gas supplier, illustrate this. Russia's renationalization may be based on economic considerations: according to Reynolds and Kolodziej (2007), it may be prompted by the soaring oil and gas prices. However, it may also be partly politically or nationalistically motivated. According to Helm (2007b), Russia's increasing energy assertiveness is a rational attempt to maximize both economic and political leverage, which increases uncertainty at the European side.

Producer Cooperation

The fears regarding Europe's increasing dependence are aggravated by the gas exporting countries forum (GECF). GECF is an informally structured group comprising the world's leading gas producers which is aimed at representing and promoting their mutual interests.[15] collectively, GECF currently controls around 74 percent of global proven reserves, contributes to 45 percent of production, and takes care of 91 percent of LNG trade.[16] Importers fear it may develop into a gas-OPEC or Gaspec. A recent, April 2007, GECF meeting, where a number of members – predominantly Venezuela and Iran – explicitly expressed interest in creating a Gaspec, has fuelled anxieties. Producers' views on a Gaspec are quite diverse: Russia, Egypt and Qatar, as opposed to Venezuela and Iran, expressly denied intentions towards creating a Gaspec, stressing that the GECF's main impetus is and will be to cooperate with consumers in a transparent manner.

Views on the likelihood of a Gaspec differ too. Hallouche (2006) argues that GECF might become more pro-active in regulating oversupply, but that it is currently improbable (but not impossible) that it will eventually evolve into a Gaspec. Others are more worried. According to Odell (2003), producers become increasingly likely to collude in response to EU liberalization. Wagbara (2007) argues that a gas cartel is quite likely to emerge.

In conclusion, despite a lack of consensus on the likelihood and severity of the concerns, the increasing external dependence constitutes a structural break with the past in several respects – a break gas regulation will have to reckon with.

2.4. More Long Distance Supply

While so far Europe has relied largely on domestic production to satisfy demand, future gas supplies will increasingly come from countries outside Europe. Gas will need to travel long distances, which increases transport and supply costs. Also, gas transport is very capital intensive which means that fixed costs have a large share in total costs. The larger the

[13] Marcel considers the case of oil. Nevertheless, the basics of the IOC-NOC relationship are applicable to gas too.

[14] This is not the whole story, however. According to Zanoyan (2004, pp. 8), objectives may differ between a producer government and its own NOC. For example, where NOCs are conservative and try to maximize the resource base and retain control over resources, national governments may have stronger preferences towards short-term revenue generation.

[15] Membership has fluctuated, but members include Algeria, Bolivia, Brunei, Egypt, Equatorial Guinea, Indonesia, Iran, Libya, Malaysia, Nigeria, Oman, Qatar, Russia, Trinidad & Tobago, the United Arab Emirates, and Venezuela. Turkmenistan has been involved in the past, and Norway participates as an observer.

[16] Calculated from BP (2007).

transport distance, the longer and bigger pipelines must be, increasing the share of fixed costs. This urges producers to reduce spare capacity and utilize a pipeline to its maximum capacity in order to spread the fixed costs over as large a volume of gas as possible. The result is a low ability to offer flexible off-takes. Flexibility is vital to both the national transmission companies as well as the local distribution companies (LDCs) whose small consumers have a low ability to adapt gas. Without flexibility it will be very hard, if not impossible, to satisfy the capricious gas demand. In addition to unsatisfied customers, this situation could also result in high imbalance fines. Since the producers can not be expected to provide this flexibility, the transmission companies and/or LDCs need to obtain flexibility by themselves which will increase the demand for flexibility in Europe. Seasonal gas storages may be built to this end. A projected 45 to 60 bcm of seasonal storage working volume will be needed up to 2025, requiring substantial additional investments (CIEP, 2005).

Another consequence of gas traveling larger distances is that it stimulates the development of liquefied natural gas (LNG).[17] As pipelined gas becomes more expensive, LNG becomes more attractive as an alternative. Jensen (2004, pp. 7) provides estimates of the costs of LNG versus pipelines. He shows that LNG becomes more competitive with distance, until after about 3000 miles building an LNG train is the cheapest option available. LNG's attractiveness furthermore increases because of 1) its lower transit risks compared to pipeline gas trade (which usually transits several countries before reaching its destination); 2) the emphasis on diversifying gas supplies; and 3) the large reduction in LNG costs. LNG may partly offset the above import dependence concerns by increasing flexibility and enabling consumers to diversify suppliers. LNG is much more flexible than pipelined gas, which is mostly dedicated, because in theory LNG can be delivered by ship to any receiving terminal. In future, price developments could determine where an LNG ship delivers its cargo.[18] For this reason LNG is regarded as an important vehicle to increase competition and integration between consumer centers such as Europe, the United States, and Asia. There is, however, still a long way to go for full integration to develop (if at all). Note that LNG's flexibility also benefits producers as it allows them to diversify between consumers. Finally, as noted by Jensen (ibid., pp. 1), long-term contracts will remain the mainstay of the LNG business, limiting flexibility.

2.5. More Transit

LNG trade will grow, but is not expected to overtake pipeline trade. Pipelines will remain the mainstay of the European gas industry. All importing pipelines must pass through third countries before reaching the European border. Hence, increasing imports will increase transit. Russia is the most prominent example: more Russian pipeline gas will, given Russia's existing pipeline system, result in more transit through Ukraine and Byelorussia. Gas transit leads to disputes between Russia and the transit countries, which threatens Europe's gas

[17] See Jensen (2003, 2004); and CIEP (2003).

[18] Obviously, there can and will be bottlenecks at sea too, such as the Suez Canal or the Strait of Malacca. Therefore, besides transport costs, arbitrage opportunities will also be influenced by the ability to reach a certain port with favorable conditions.

deliveries.[19] Transit problems are not only harmful to Europe; Russia also wants to avoid transit problems as much as possible. An interruption in Russian gas supplies could have detrimental consequences for Russia's reputation as a secure gas supplier and would result in lower Russian revenues since Europe is a premium[20] market for Russian gas. In the long-run, transit problems could urge Europe to increasingly undertake measures to move away from Russian gas (through for instance supply diversification or energy efficiency measures). The significance of transit for Russia is illustrated by the Nordstream and Southstream pipelines, both of which are to a large extent built to avoid transit.

2.6. Delinking and Relinking

Structural shifts have occurred at the demand side too. The first relates to the link between GDP growth and energy demand.

Prior to the oil crises of the 1970s, energy demand growth was linked to GDP growth. As a consequence, projections for investments in new infrastructure were based on GDP growth figures. The crises prompted European economies to rethink their energy policies. One spearhead became to lower energy intensity. As an illustration, current German energy intensity has declined 30 percent from the 1970s, while the French managed a 40 percent reduction (CEC, 2005b, pp. 12). For Europe as a whole, energy intensity has decreased on average by 1.6 percent per year until 2002 (ibid., pp. 47).

In addition, the recession which followed the crises facilitated economic restructuring. As a result, economic activity shifted away from the relatively unattractive energy intensive industries towards less energy intensive service sectors like finance. Furthermore, rising energy prices lowered energy demand which enhanced energy efficiency. These factors have lowered energy demand growth to around 1 percent per year which is much lower than GDP growth of approximately 2.4 percent per year (ibid., pp. 47). Note that these shifts have lowered but not obviated the increase in European energy demand. This is confirmed by IEA demand projections which over the last few years show a decline in the still rising trend (Honoré and Stern, 2007, pp. 226).

Global gas demand growth, on the other hand, has not been delinked from GDP growth. Rather, several countries see their economic growth, and their energy use on the back of that, rise to unprecedented heights. Prime examples are found in Asia – especially China, India, and Indonesia – and Africa, but Latin America and the Middle-East grow too. These countries are reaching an industrialization phase in which mobility is taking-off, resulting in a sharply rising energy demand. Hence, global energy demand growth has become relinked to GDP growth (Shell, 2005, pp. 28).

The global picture is one of growing energy demand – predominantly in Asia and Africa, but also, though to a much lesser extent, in Europe. For Europe, the upshot is a shift towards increased competition for imports, since the Middle-Eastern, Caspian, and Russian supplies are within economic reach of all importing regions. The expected rise of LNG makes gas

---

[19] There have been a few transit incidents, for example between Russia and Byelorussia (Bruce, 2005) and Russia and the Ukraine (Stern, 2006).

[20] Russian gas is currently sold to Europe at much higher prices than to Russian or FSU (former Soviet Union) customers, cf. Stern (2005); and Spanjer (2007).

trade more flexible and expands the range of gas deliveries also to the US, further increasing competition among consumers.

2.7. Sustainability

Economic growth exhibits an increasing carbon intensity. Whereas up to 1997 carbon intensity in GDP was clearly declining as a consequence of for example the declining importance of the coal industry and improvement in energy efficiency, this trend has reversed since (Helm, 2007a, pp. 14). The countries mentioned in the previous section are expected to achieve the highest growth rates. Their industrialization will not only create an explosive growth in energy demand but also in carbon emissions, which reinforces the awareness of emissions that has developed due to the greenhouse effect. To illustrate the anxieties, atmospheric CO_2 concentrations have risen from pre-industrial levels of 270 parts per million to currently 350 ppm, while extrapolating the current trend puts CO_2 concentrations in the realm of 750 ppm at the end of this century compared to the safety threshold which is estimated at around 450 ppm (Helm, 2005).

This will in turn result in energy policy being concerned not only with the volume of supplies but to an increasing extent also with their content. Consequently, climate change is pushing up the stairs of energy policy goals again.

2.8. Power Generation Gas Demand

Power generation gas demand has occupied Europe for over thirty years (Söderholm, 1998). After the 1973 oil shock, gas was considered a premium fuel, available to power generation only if it could not be used for other higher-value uses such as residential or commercial consumption. Increasing dependency on Russian gas and protection of the indigenous coal industry were additional reasons for curbing gas use (Newbery, 1999, pp. 150). Changing market fundamentals – especially the gas bubble during the 1980s – obviated this policy. This has enlarged the scope for power generation to generate electricity with gas. Since then power generation gas demand has increased substantially, and this trend is expected to continue in the coming years. Power generation takes up an increasing chunk of gas demand, changing the composition of gas demand. Honoré and Stern (2007, pp. 227) argue that approximately 70 percent of the total demand increase until 2030 is expected to be attributable to power generation. They identify three reasons for this dash for gas: 1) the economics and efficiency – lower capital costs, lower construction lead times, and lower economies of scale – of new combined cycle gas turbine power plants; 2) the low emissions of gas; and 3) the adaptability, flexibility, and availability of gas in an open power sector. However, we must note that these positive projections for power generation gas demand are still quite uncertain. In fact, as argued in Honoré (2006) and Honoré and Stern (2007), current projections only materialize at substantially lower gas prices than currently.

2.9. A New Context for Regulation

This section infers the implications for regulation from the above market shifts. These shifts can broadly be categorized in four main categories:

- a shift in international relations between producers and consumers – the emergence of a seller's market rather than a buyer's market;
- a shift in energy policy objectives with security of supply having become the top priority, while climate change has climbed up the ladder too;
- an increasing need for investments along the entire value chain. Total EU-27 investment requirements amount to roughly 1800 billion Euro of which around 12 percent – 216 billion Euro – is required for gas (CEC, 2007a, pp. 17); and
- an increasing influence of politics on energy relations.

Hence, the issue for regulation is how to ensure adequate investments in a gas market which is dominated by sellers, where political considerations are becoming more important, and where supply security and sustainability have become the top priorities for energy policy. This is substantial change compared to the traditional structure where investment risks were transferred to consumers through public monopoly franchises. The next two sections consider the consequences of the market shifts for regulation.

2.9.1 Investment Characteristics

Investments in gas markets are required along the entire value chain which includes exploration and production, transmission, distribution, trade, and ancillary services. Investments in exploration, production, and transmission are predominantly irreversible (sunk). Examples are costs for project identification, investment planning, and construction (Hubert, 2007, pp. 63). These costs imply that the economic viability of a project is not guaranteed beforehand. In addition, as indicated in section 1.1, transmission and distribution are considered a natural monopoly due to economies of scale and scope which create cost subadditivity. Other segments exhibit scale and scope economies too. In descending order we have gas storage and blending (i.e., quality conversion); exploration and production; and gas trading.

Many new investments will concern production and transport of gas. Their irreversible nature creates investment risks. For instance, a gas pipeline is very specific to a particular trading relationship, because once laid has very limited, if any, alternative use. This creates a quasi-rent which is the difference between an investment's pay-off in its current use and its highest alternative use. Furthermore, the decision to build a pipeline is usually based on negotiations between a specific consumer and a specific producer. This locks both parties into a bilateral dependency which changes through time and which affects the appropriation of the quasi-rent. Prior to the investment, the producer/investor has a relatively strong bargaining position, as the consumer or its regulator depends on him for undertaking the investment.[21] Ex post, however, the limited alternative use of its sunk investment ties the investor to the market for the foreseeable future, which effectively shifts the bargaining power to the regulator. This provides him with the incentives to adapt his policy in order to increase his

[21] We refer to investor and regulator throughout this study.

own or society's rents at the expense of the investor's through arrogating the quasi-rent.[22] This can be achieved by renegotiation or expropriation. The former entails a regulator using information it has obtained during the regulatory process to increase welfare in the subsequent period(s) at the expense of the investor. Expropriation means that a regulator creams-off investor profits via for instance determining low prices or by cheaply or freely permitting entry. Both may obstruct or hold-up investments. The key element in this regard is regulatory risk.

The bilateral dependency and lock-in discussed above are primarily due to the specific nature of the concerned assets (or asset specificity, see section 5 for a more elaborate discussion). Gas markets exhibit several forms of asset specificity. The first is locational specificity, which is due to the spread of gas reserves. Because potential gas producers are limited, importers often largely depend on a specific producer. Second, the network-bound character of gas creates physical asset specificity, which most clearly manifests itself in the distribution segment where consumers are commonly connected to appliances that support a specific range of gas specifications. Another example pertains to gas storage. Third, dedicated assets are created by investments which are particular to a specific consumer-producer relationship. As indicated, pipelines that connect a producer with a consumer are good examples.

In the traditional market structure these considerations were not very important. As indicated, stimulating investments was not necessary due to the system's maturity, while an oversupply of gas rendered import dependency issues relatively unimportant. The market shifts change the picture. Many new investments are required. These will to a large extent be in new pipelines, LNG terminals and storage facilities, all of which are irreversible. Hence, the market shifts increase regulatory risks, which makes regulatory credibility a vital parameter for future energy regulation. LNG may reduce asset specificity, especially locational specificity, because LNG creates a possibility to diversify between suppliers. Also, an LNG terminal does not need to be constructed for a specific producer/consumer which makes the investment less dedicated. However, pipelines will remain the mainstay of the gas business. Therefore, the increasing import dependence requires new pipelines from the producers to Europe, which increases asset specificity. Furthermore, as indicated, the longer transport distances will allow a level of flexibility which necessitates additional investments in European storage which will increase the natural monopoly problem.

2.9.2. Uncertainty

Uncertainty is another vital parameter when regulating in order to facilitate investments. Investment under uncertainty is what Dixit and Pindyck (1994) are concerned with in their real options theory. With uncertainty, delaying an investment in order to wait for more certainty to arrive, may be beneficial. Uncertainty impacts the investment through the option value: the higher is uncertainty, the larger the option value, and the higher is the hurdle rate (i.e., the price against an investor is willing to immediately invest). Most empirical studies of the impact of uncertainty on investments take a short-run perspective and find a negative impact (Altug et al., 2000, pp. 2, 3). Most empirical studies examining investment under

[22] Society's rents increase because most regulatory models specify a social welfare function that attaches a higher weight to consumer benefits than to producer benefits. See Baron and Myerson (1982); and Laffont and Tirole (1986).

uncertainty conclude that uncertainty generally lowers investment, also in the long-run (Serven, 1997; Carruth et al., 2000).

Dixit and Pindyck (1994, pp. 18) distinguish two types of uncertainty: aggregate uncertainty that affects all firms in an industry and idiosyncratic or firm-specific uncertainty that affects a particular firm. If industry-wide demand increases, investing becomes attractive for every firm in the industry. Depending on the competitive status of the industry, the associated output expansion – through entry of new firms or capacity expansion of existing firms – lowers prices somewhat and accordingly lowers the profit a particular firm can reap. When, on the other hand, industry demand falls, all firms again are affected. However, the lack of exit possibilities due to the irreversibility of investment precludes a cushioning of this effect. Hence, the response is asymmetric: positive shocks to industry profitability are scaled down, whereas negative shocks are not. Profits are truncated, which makes a firm cautious to invest, the argument goes.

Idiosyncratic uncertainty, on the other hand, changes the effect of a positive demand shock. It now affects just one firm and increased profitability is not shared with others. There is therefore no risk of profits being eroded by entry or capacity expansion, mitigating the above asymmetry and improving investment incentives. However, the investment timing decision is still influenced because higher risk increases the option value which may induce the investor to postpone his investment.

One particularly important form of uncertainty is regulatory or political uncertainty (Altug et al., 2000; and Buckland and Fraser, 2001). An example is the urge of some governments, prompted by record high gas prices on the back of those of oil, to impose a tax on energy company profits in order to redistribute some of these profits back to consumers. This increases regulatory risk, which energy companies will incorporate into their calculations. Ultimately, these higher risk levels will feed through into higher consumer prices, at least partly contradicting the goal of changing the rent distribution.

The option value of waiting also increases if investors doubt the sustainability of a policy which is beneficial to them (Sérven, 1997, pp. 14). The value of waiting increases and investments are postponed until the return on capital rises sufficiently to compensate the losses to the investor should a policy reversal take place (of course, when the necessary hurdle rate is never reached, this investment pause is perpetual and no investment is undertaken at all). An example pertains to renewable energy. Currently, most energy policy makers are concerned with climate change, which makes a reform towards a policy more conducive renewables likely. If an investor doubts this reform's sustainability if the emphasis on climate change were to weaken again, the investment response to the policy reform may be relatively muted.

Hence, uncertainty hampers investments. The source of uncertainty is important, because the effect on investments of aggregate shocks is larger than that of an idiosyncratic shock. Regulatory credibility is once again an important parameter. Rodrik (1991) shows that the option value is quite high even with high credibility, implying high hurdle rates to investment anyway.

The market shifts create additional uncertainties of which we provide three examples. First, increasing external dependence creates price risks through the uncertainty concerning producer collusion. This is possible due to the leading role of many producer governments, but also because Europe's supply portfolio is not very well diversified. In fact, very few Member States are self-sufficient or net-exporters while most are heavily import dependent.

Only Northwestern Europe has a diversified supply portfolio. Hence, increasing import dependence creates fears of opportunistic producer behavior. Transit problems are a second source of uncertainty. Examples are 1) the postponement of the start of Algerian supply because of internal political and safety problems as well as difficult relationships with Tunisia and Morocco (Stern, 2002, pp. 14); 2) the cut-off of Russian gas via Byelorussia (to Poland and Germany) after a conflict between the Byelorussian government and Russian Gazprom in early 2004 (Bruce, 2005, pp. 19); and 3) the recent gas dispute between Russia and Ukraine in early 2006 (Stern, 2006). These two examples illustrate the uncertainty which is created by the increasing influence of geopolitics on the gas market. Third, climate change increases uncertainty because its future developments are unclear. Unresolved issues are 1) what the second commitment period of the Kyoto Protocol is going to look like; 2) if there will still be the political will and commitment to keep reducing emission levels when the low-cost, easy to achieve options are exhausted; and 3) the prospects on nuclear and sustainable energy.

In sum, the market shifts have far-reaching consequences. They increase risk and uncertainty on the European gas markets. The paramount issue for regulation shifts to facilitating investments under these circumstances. This requires a regulatory policy that focuses on creating a governance structure in which risks are allocated to the party best able to undertake them in an uncertain world. That is, the regulator will have to choose which governance form, for example, spot markets, vertical integration, or an intermediary solution like long-term contracts, is the most efficient solution. Regulatory credibility is vital in this setting. We highlight three additional considerations (CIEP, 2006, pp. 18, 19). First, the roles of the relevant parties may change along the way. A prominent example pertains to the producers. If they integrate downwards into the European distribution sector, their role changes compared to their current role of supplier. Unbundling integrated incumbents is another example, because the role of trading companies without a network is very different from that of an integrated incumbent, for example in terms of negotiating power vis-à-vis the producers. Second, the demand side deserves more attention. Consumers can lower import dependence and lessen price spikes by actively participating in gas markets. Third, regulation must consider that the value chain is fragmented over different jurisdictions, each with its own policies and approaches.

This section has indicated that the shifts create a different perspective from which to design regulation. Hence, European gas regulation needs to adapt to the changed circumstances. The remainder of this chapter analyzes from a theoretical perspective whether existing regulation will be able to do so. Answering this question requires us to contrast two economic perspectives towards regulation: the neoclassical and the transaction cost perspective. The next section discusses the neoclassical perspective, its main assumptions and the implications for regulation.

3. The Neoclassical Perspective

3.1. Introduction

This section starts by setting out the main assumptions underlying the neoclassical perspective in section 3.2. After that, section 3.3 discusses the implications of the neoclassical perspective for regulatory intervention, while section 3.4 infers the neoclassical prescriptions for liberalizing the European gas market.

3.2. The Neoclassical Methodology

The hardcore of the neoclassical perspective towards economics consists of three main axioms: methodological individualism; methodological instrumentalism; and methodological equilibration (Arnsperger and Varoufakis, 2006).

3.2.1. Methodological Individualism

Methodological individualism states that all actions are performed by individuals. In consequence, all social phenomena can be explained in individual terms, which implies that all analyses must ultimately be couched in terms of the behavior of individuals. Hodgson (1988, pp. 55-62; 2007, pp. 211) points out that this requires purposeful individual action which is a sufficient cause for all social action. Arnsperger and Varoufakis (2006) argue that the analytical method employed to this end consists of three steps: 1) the socio-economic phenomenon is to be analyzed by focusing on the goals, plans, and actions of individuals who have brought about this phenomenon; 2) understanding fully their working at the individual level; and 3) synthesizing the individual-level knowledge to explain the phenomenon at hand.

By assuming individual action as the sole determinant of social phenomena, other, possibly complementary macro explanations for human behavior like norms, habits, or social institutions, are ruled out by definition. While recent advances in neoclassical economics allow a link between individual behavior and these complementary explanations, according to Arnsperger and Varoufakis, the latter are explained rather than being explanatory variables.

3.2.2. Methodological Instrumentalism

Methodological instrumentalism assumes that all behavior is preference-driven, that is, individual behavior maximizes some objective function. The individual is specified as a bundle of preferences who, given these preferences, undertakes those actions that maximize its expected utility. This requires a number of assumptions, in particular perfect and costless information, complete and transitive preferences, and the ability of the individual to act as a lightning calculator able to make all required calculations. The preferences themselves are not explained. They are exogenous to the analysis, i.e., separate from the structure of interaction in which the individual is involved, either because they are psychological phenomena which should be explained by psychologists rather than economists, or because individual purposefulness is entirely indeterminate (Hodgson, 1988, pp. 60, 66). It is rather assumed that people behave as if they possessed consistent preferences which guide their behavior. That is, the ends are given and the individual chooses the means to reach its objectives. According to Arnsperger and Varoufakis (2006), the fundamental point of individual instrumentalism is that the preferences fully determine an individual's actions. Therefore, even the current emphasis of some neoclassicists towards endogenizing preferences does not refute methodological instrumentalism as the preferences remain fully determining.

3.2.3. Methodological Equilibration

Methodological equilibration posits that human behavior always tends towards an equilibrium, which in economic theory refers to a situation where price adjustments in the process of competition tend to equalize supply and demand of a good or service. Arnsperger and Varoufakis (2006) define the concept by specifying the three steps: 1) discover an equilibrium; 2) assume that agents will find themselves at that equilibrium; and 3) show that in equilibrium small changes will not move self-interested behavior away from the equilibrium. In effect, the presence of an equilibrium is assumed rather than explained.

3.3. Regulatory Intervention in the Neoclassical World

In this neoclassical world, utility and profit maximizing actions of individuals and firms create the optimal, perfect competition equilibrium, rendering regulatory intervention unnecessary. This provides straightforward guidance for regulatory intervention: a market situation that deviates from the characteristics of perfect competition is a candidate for regulatory intervention. Hence, in the neoclassical world, regulators intervene in order to correct a market failure and restore the perfect competition equilibrium. Regulation is employed from a public, rather than private, interest perspective. The three most prominent examples of market failure on the European gas market are 1) failure of competition; 2) public goods; and 3) externalities.

Under the neoclassical assumptions, competition is perfect and individual suppliers have no scope to influence the market price. If competition fails, however, they do have that influence. Competition on European gas markets may fail due to 1) producer domination which impedes an importer's choice of supplier; 2) the natural monopoly status of the gas network; and 3) the creation of national champions by governments – think of the mergers between E.ON and Ruhrgas or Suez and Gaz de France. Governments usually intervene through price regulation of the network or use competition policy to improve competition.

Public goods are a second source of market failure. They are non-rivalrous, which means that consumption of the good by one person does not obstruct another person from enjoying the benefits of using the good. Their second property is non-exclusivity, which means that nobody can be excluded from enjoying the good. If the market is to function perfectly, those persons who are not willing or able to pay for the good must be excluded from its use and benefits. Non-exclusivity, however, by definition negates the possibility of exclusion and by extension the possibility to ask a price. As a result, a private producer will not have adequate incentives to produce the good because others can free ride on his expenses. This leads to the conclusion that private markets will not produce public goods, which is rationale for the government to step in and for example produce the good herself.

Security of supply is a public good. If an extra EU-citizen profits from an enhanced security of supply, the benefit of other citizens will not deteriorate. Security of supply, moreover, can not be split up in parts and sold on a market, so a private entrepreneur will not be able ask a price for this good. Security of supply is thus both non-rivalrous and non-excludable. In addition, the environment is a textbook example of a pure public good. No one can be excluded from clean air, and an extra person enjoying it will not make a difference for the rest. Therefore, both supply security and environmental issues exhibit market failures which may provoke regulatory intervention.

The third main market failure is the presence of externalities. Externalities arise when the actions of one person or firm affect other persons or firms, negatively or positively, without adequate compensation. In case of negative externalities (e.g., polluting emissions) the negative consequences of producing a good are not incorporated in its price. Social costs outweigh private costs, and as a result, price is too low and output too high from a social point of view. On the other hand, positive externalities (e.g., security of supply) arise when production generates benefits that are not incorporated in the price. Social benefits outweigh private benefits, which renders output too low and price too high from a social perspective. All in all, externalities, positive and negative, distort the market allocation away from the perfect competition optimum at which point in the neoclassical world the regulator should step in. The most vivid example of regulatory intervention in case of externalities are the Kyoto Protocol and the resulting European Emissions Trading (ETS) scheme. The rationale for both initiatives is essentially to internalize the external effect of energy use by putting a price on carbon.

3.4. Liberalizing the European Gas Market According to the Neoclassical Perspective

The sections above have set out the implications of the neoclassical perspective for regulatory intervention. This section completes the picture by tracing the implications for the structure of a liberalized European gas market. It draws in part from Joskow (1996) who applies the recommendations of the standard neoclassical prescription to electricity.

The gas value chain comprises exploration and production, transmission, distribution, and (wholesale and retail) trade activities. Exploration and production and trade are potentially competitive, while the infrastructure links, transmission and distribution, are natural monopolies for which monopoly provision is the lowest-cost solution. This market failure should be corrected by permanent regulation of the infrastructure operator. In the traditional market structure, vertical integration has extended into the potentially competitive links as well. Accordingly, regulation has been introduced in segments which should be governed by market forces. Horizontal integration within a link has resulted in a typically low number of competitors. Hence, in order for real competition to develop, significant horizontal and vertical restructuring is necessary.

3.4.1. Restructuring

Restructuring can occur between different links of the value chain or within a specific link. These are referred to as vertical and horizontal restructuring, respectively. We start with vertical restructuring; horizontal restructuring predominantly applies to the trading and production segments and is discussed afterwards. Competitors need access to the infrastructure on fair and non-discriminatory terms to be able to effectively compete with the incumbent operator. Vertical integration creates competitive hazards in this regard. A fear is that an incumbent infrastructure operator which also possesses competitive activities, may distort competition in the latter through cross-subsidization and discriminatory access. The neoclassical assumption regarding vertical integration is that firms integrate vertically in response to a market failure of some kind. Because in the neoclassical world the costs associated with vertical integration are not recognized, any market failure is a candidate for vertical integration (Joskow, 2005, pp. 325). Furthermore, in the neoclassical perspective, the boundaries of a firm are assumed to be determined by technological considerations. Accordingly, there can be no technical or physical reason for expanding these boundaries, which implies that vertical integration has no efficiency rationale. This explains neoclassical economics' inhospitable view towards vertical integration: vertical integration is seen as a response to existing market power or as an effort create market power (Williamson, 1985, pp. 26).

Because the indicated competitive disadvantages can by definition not be offset by efficiency advantages, the policy prescription is to prevent vertical integration as much as possible. If vertical integration is already present, public policy should try to remove it. Mitigating the competitive hazards if vertical integration is already present, requires some sort of separation of the competitive from the non-competitive parts. There are five main options to this end (OECD, 2001; CPB, 2005).[23]

Structural Separation

The transmission and distribution assets are vertically divested from the competitive activities and organized into different companies. The network operator is not allowed to have any financial ties to production or trading. This is supposed to severe any link with production and trading activities. Accordingly, no incentive or possibility to distort competition in these segments is possible.

Operational Separation

Operational separation separates network ownership from control. Control is transferred to an independent entity (an Independent System Operator, or ISO) while the ownership remains with the integrated company. The prime advantage is that the integrated company will no longer have an incentive to abuse the non-competitive segment to obstruct competition in the competitive segment. The outcome depends on the composition of the independent entity. For example, if the entity consists of representatives of the competitive firms, the situation corresponds to club ownership; if the regulator takes on the task of the independent entity, the situation resembles regulated access which is discussed below. Another important consideration is whether the entity receives a part of the profits of the non-competitive activity. If not, it might have few incentives to stimulate efficient operation of the non-competitive segment. The following three approaches have less impact (Hardt, 1995).

Accounting Separation

Accounting separation obliges a firm to keep separate accounts of regulated and non-regulated activities. The integrated firm has to charge itself the same internal prices for use of the non-regulated activities as it does other parties. The activities are conducted within the

[23] The OECD study also discusses club ownership and separation of the non-competitive part into reciprocal parts as potential solutions. These options are unlikely for a sector like gas. Club ownership allows firms to jointly own the network assets. This is unlikely for gas, because governments are in practice very reluctant to give up their control over energy networks. The overall efficiency of separation of the non-competitive component into reciprocal parts hinges on network efficiencies offsetting the incentive to deny interconnection. Because gas possess some positive but very small network effects, this option is likely inefficient.

same company. This is a reason why accounting separation usually fails to remove fears of anti-competitive behavior of the integrated firm.

Functional Separation

Functional separation goes a step further by requiring that different activities are grouped into different divisions of the firm. In addition to charging identical prices, the integrated firm should rely on the same information as other firms do. Also, the employees of both activities must be physically separated from each other. Once more, all activities are conducted within the same integrated company.

Corporate Separation

Corporate separation, finally, goes another step further and requires that different activities are grouped into different corporate entities, as opposed to different divisions. As with operational separation, ownership and control are separated. The difference is that with corporate separation, everything remains within the integrated company.

In a neoclassical world the choice between these options is clear. Recall the assumed absence of efficiencies related to vertical integration. In that case the most efficient option is structural separation because it provides the best assurance for the removal of competitive distortions.

The neoclassical perspective towards horizontal restructuring is quite similar. Perfect competition hinges on the presence of a large number of competitors that vehemently compete with each other. This is not only contingent on the presence of competition between links as above, but also on competition within a link. Introducing competition within the network link is unwise due to its natural monopoly status (see below). On the other hand, production and trading activities may become more efficient if more competitors arise. This may require horizontal restructuring. For example, divesting production or trading assets increases the number of competitors and by extension should promote competition. As in the vertical case, the neoclassical assumption is that firms integrate horizontally in response to or anticipation of market power. Without efficiency rationale, the straightforward prescription is to combat horizontal integration as much as possible. If removal appears unfeasible, regulation is required.

3.4.2. Access

In addition to vertical restructuring which prepares the sector for competition, the second main requirement is the actual introduction of competition into the sector. Network access is no issue if competition between infrastructures, where each entrant possesses its own gas network and is able to compete against the incumbent's, materializes. However, the natural monopoly character of gas infrastructure renders this option inefficient and likely unviable. Accordingly, rather than between infrastructures, competition should take place on the infrastructure. This requires entrants to be granted fair and non-discriminatory access to the single network. There are two basic ways to grant access: negotiated and regulated third party access (NTPA and RTPA, respectively). NTPA relies on negotiations between the parties involved, the entrant seeking access and the network operator, concerning the access conditions. If no agreement is reached, the regulator should intervene and settle the dispute. Without structural separation, criticism on NTPA usually concentrates on the entrant's dependence on the incumbent, as there is a threat of the incumbent abusing its advantageous position. Also, it is difficult for an entrant to adequately determine whether the incumbent's tariffs are fair, i.e., based only on cost considerations. Moreover, a regulator's information deficit (see section 4.2.2) hampers dispute resolution, for instance because the regulator will have to depend on information provided by the incumbent. The advantage of NTPA is that if all goes well, no government interference is required and the access conditions may be relatively easily adapted if conditions change.

TPA can also be provided on a regulated basis. RTPA requires a regulator to initially set the tariffs or the way the tariffs should be calculated in order to prevent the incumbent from abusing its advantageous position. The entrant's position is important when choosing between both. If entrants are in a relatively favorable negotiating position, the market power objection to NTPA becomes less significant. Then NTPA, with its lower government interference, may be preferred. In the opposite case, RTPA will be preferred. Under structural separation, the market power threat dissipates because the network operator has no incentives to obstruct entrants from accessing its network. Furthermore, the full information assumption removes any informational problems regarding dispute resolution. Hence, the neoclassical prescription to liberalize the European gas market is to implement negotiated TPA on a structurally separated gas network, while letting perfect competition flourish in the other links.

4. Existing Regulation

This section sets out existing gas regulation. We start with recognizing that gas regulation depends on the interaction between European institutions and European and national stakeholders. Another constraint on regulatory intervention is that regulation will inherently fail, as a consequence of which she will generally be unable to implement the optimal scheme in case of a market failure. This is discussed in section 4.2. Sections 4.3 and 4.4 discuss the latest regulatory initiatives for the European gas market. The final section illustrates the connection between the anticipated internal gas market and the neoclassical perspective.

4.1. European Institutions and Stakeholders

Understanding how the European Union is moving towards an internal gas market, requires a recognition that there is not a single European policymaker with a single objective, but rather that policy is influenced and shaped by a host of different stakeholders each acting in accordance to their own objectives. This section discusses some of these stakeholders. It sets out the European institutions which together are responsible for the regulatory measures as well as other stakeholders that exert influence on European gas regulation and its implementation into national law. We start with the European institutions. For our purpose, the main decision-making institutions of the European Union are: The European Commission; The Council of Ministers; The European Parliament; and The European Court of Justice.

The European Commission is the party which is most visible when new European legislation is put forward. She has different legislative instruments at her disposal. She may issue a Regulation which is directly binding in all Member States of the European

Communities. She may also issue a Directive which fixes a binding objective but leaves the method of achieving this up to the Member States. Hence, a Directive allows more scope for national sovereignty, which has been an important reason for issuing Directives in energy as will be pointed out below. The consequence is that a Directive has to be transposed into Member State law, providing scope for national stakeholders to influence regulation. Furthermore, a Directive is generally the result of a lengthy consultation process involving interest groups, politicians, and other stakeholders. Thirdly, the Commission may also put out Decisions which deal with specific issues and are binding on those to whom they are addressed. This may include Member States, but also companies or individuals. Finally, she may issue Recommendations and Opinions which do not have binding force.

The Council of Ministers is the main decision-making body of the EU. It comprises the ministers of the Member States (plus one representative of the Commission) and is the only institution that directly represents the Member States. It has the power to accept or reject Commission proposals. Decisions are taken by a vote of Ministers from the Member States, which can take three forms: majority (for procedural decisions); qualified majority (for many decisions concerning the internal market, economic affairs and trade); and unanimity (for foreign policy, defence, judicial and police cooperation, and taxation).

The European Parliament is primarily a consultative body. It shares legislative power equally with the Council, implying that it is empowered to agree, reject, or amend European legislation. In addition, it has the power of political initiative, i.e., it may demand that the Commission issues legislative proposals for laws to the Parliament. Another relevant right is that the Parliament may dismiss a single Commissioner or even the whole Commission.

Finally, the Court of Justice interprets Community law and settles disputes that arise from the interpretation of Treaties or related legislation. EU law takes precedence over national law, hence the Court of Justice can overrule national courts and governments. It has the power to impose fines on countries that do not fulfil the obligations imposed on them by Community law.

These European-level institutions help us understand how the Gas Directives and accompanying legislation come about. European regulation is influenced by other stakeholders as well, because Commission proposals are usually the result of lengthy consultations with different stakeholders like industry representatives and interest groups. This allows these stakeholders to influence the contents of regulation before adoption. Players from all links of the value chain have created associations to influence the European regulatory process: regulators (CEER, ERGEG), producers (OGP), the gas industry (Eurogas, Marcogaz), infrastructure (i.e., transmission, storage, and LNG) operators (GIE), independent distribution companies (GEODE), industrial energy users (IFIEC), and energy traders (EFET).

Finally, a Directive must be transposed into Member State law. A Directive's provisions must be reflected in national law, inconsistencies of existing national laws need to be removed, and structures for application and enforcement need to be created (Mastenbroek, 2007, pp. 19, 20). This process will have winners and losers. There will consequently be national stakeholders and institutions that want to influence European regulation ex post through impacting the implementation of the European Gas Directives into national legislation. Among national institutions are the relevant ministry, national regulatory authorities, and national competition authorities. The ministry is responsible for translating the European Directives into sector-specific and competition legislation; the regulatory and competition authorities are responsible for implementation. These national institutions may be inclined to adapt some provisions to a Member State's market specifics, which is positive. Other stakeholders are the national counterparts of the above European associations. They may try to influence the national institutions into moulding the provisions into a fashion more favourable to them.

In sum, therefore, the impact of the Gas Directives will likely differ between Member States according to their interests and endowments. Nevertheless, the objectives specified in the Directives are supposed to be implemented by everyone. We can therefore conclude that the Gas Directives will substantially influence but not determine the outcome in a particular Member State. Without going into detail, we point out that liberalization of European gas markets entails substantial adaptations to the traditional structure and therefore will have many winners as well as losers. The opposition has created a torturous liberalization process which has resulted in a plethora of legislation, regulations and policy initiatives. Table 3 below provides some landmarks in European energy liberalization policy.

Table 3. An overview of European liberalization policy

1987	Single European Act enters into force and largely excludes energy
May 1988	Publication Commission's paper 'The Internal Energy Market'
July 1990	Price Transparency Directive enters into force
December 1990	Electricity Transit Directive enters into force
May 1991	Gas Transit Directive enters into force
May 1994	Hydrocarbons Directive enters into force
February 1997	First Electricity Directive enters into force
August 1998	First Gas Directive enters into force
September 1999	First meeting Madrid Forum
August 2000	Gas market opening begins
July 2003	Second (Acceleration) Gas and Electricity Directives.
November 2003	European Regulators Group for Electricity and Gas established.
July 2004	Market opening following the second Directives begins
June 2005	DG COMP Energy Sector Inquiry launched
November 2005	Gas Regulation comes in to force
January 2007	Publication of Energy Package and DG COMP report on the energy Sector Inquiry
July 2007	Full gas and electricity market opening
September 2007	Proposal of a third legislative package

Source: adapted from Honoré and Stern (2007, pp. 244).

Nonetheless, despite the many problems and after severe opposition, the adoption of the Electricity and Gas Directives does signal the dismantling of the old market structure into one that ultimately should develop into a genuine internal energy market. Before discussing the existing regulatory regime for gas, we must appreciate that reaching consensus on regulation is only one step. The next is to implement a proper regulatory scheme. The next section highlights some general problems in this regard. It illustrates that regulatory intervention is a very difficult task, inherently prone to failure.

4.2. Regulatory Failure

The neoclassical perspective towards regulatory intervention set out in the previous section prescribes regulation in response to market failure. The regulatory task is to remove this market failure and restore the perfect competition equilibrium. This only works to the extent that regulation does not fail too. Agency theory, however, illustrates that regulation is likely to fail. It poses regulation as an implicit contract between a regulator/government and a firm. It focuses on ex ante incentive alignment. Regulation is stated as a game in which a principal – a regulator – tries to design a regulatory contract in order to provide the proper incentives to an agent – a regulated firm.[24] This section discusses the insights for regulatory intervention that emanate from agency theory.

4.2.1. Objectives

A regulator's objectives determine which kind of regulation will be imposed eventually. Typically, a regulator is assumed to act in society's interest and maximize social welfare, which consists of both consumer's rents as well as those of the firms. In its simplest form, this amounts to the following formula:

$$SW = f(S; \alpha R)$$

Where SW stand for expected social welfare; S is an indication of consumer rents (or surplus); R indicates producer rents (or surplus); and α is a parameter denoting the relative weight attached to producer surplus by the regulator or society. If producer and consumer rents are weighed equally, α is 1. Most regulatory models assume a regulator to favor consumer rents over producer rents. There are two general approaches to this end. First, one might simply assume an α smaller than 1 (Baron and Myerson, 1982). This may be due to regulators being more concerned with the utility of local consumers, instead of that of company owners, who are not all local residents (Baron, 1989, pp. 1362). A second approach emphasizes the social cost of transfer payments to a firm (Laffont and Tirole, 1986). In this case, a regulator is able to directly compensate a firm, in order to reimburse his costs and/or to provide a net transfer, through payments that draw from the government budget. These transfer payments carry with them a social cost due to distortions caused by the taxes that are used to generate these funds. Therefore, one dollar of transfer payment requires more than one dollar of taxes to raise it. To illustrate this, Laffont and Tirole (1986) introduce the parameter $\Lambda \geq 0$. That is, one dollar of transfer payment results in one + Λ dollar of social costs. Laffont (2005, pp. 1, 2) estimates Λ to be around 0.3 for developing countries.

Both approaches are similar in that welfare is increased by transferring producer's rents to consumers. The optimal outcomes differ, however. In the former case, the optimal, perfect information outcome is marginal cost pricing with transfer payments compensating the firm's fixed costs. However, in the latter case, transfer payments to the firm are costly.[25] This

---

[24] Posing regulation as a principal-agent problem follows the "new regulatory economics" approach that was developed in the early 1990s. Cf. Laffont and Tirole (1993); and Laffont (1994).

[25] This may be due to distortions created by subsidies or by an insufficient amount of public funds.

renders Ramsey-Boiteux pricing, rather than direct compensation, optimal (Ramsey, 1927; and Boiteux, 1956).[26]

4.2.2. Information

The information at a regulator's disposal to a large extent determines its ability to regulate. Regulation is straightforward under the neoclassical assumption of perfect information, for if a regulator can simply dictate the optimal regulatory scheme to the monopolist, no complications arise and all monopolists will act efficiently according to a regulator's (and society's) objectives. This would create the perfect competition equilibrium. However, a regulator is generally unable to perfectly observe neither a firm's costs nor its managerial efforts (Agrell and Bogetoft, 2004, pp. 6; and Joskow, 2006, pp. 4, 5). A firm will intrinsically have better information on these issues. This asymmetric information creates a classical principal-agent problem in regulation (given a divergence of the interests of the regulator and the regulatee). This induces strategic behavior which severely complicates regulation.

Unobservability of costs provides a firm with an incentive to portray itself as a higher cost firm than it actually is, because this will generally induce a regulator to set or allow higher prices. This in itself decreases welfare. It also shifts rents from consumers to the firm, which further decreases welfare (as indicated above). A regulator thus faces an adverse selection problem in trying to distinguish high cost firms from low cost firms (Baron and Myerson, 1982). Observing a firm's realized costs for a certain period and then adjusting the prices accordingly, may mitigate the adverse selection problem.

However, this creates a second information problem. Auditing a firm's costs lowers the prospect of a firm to earn a rent as prices will be adjusted to costs (with a delay). This might give the managers an incentive decrease their efforts because the imperfect observability of managerial effort implies that high managerial effort will not be rewarded adequately and low effort not punished sufficiently. Accordingly, by solving the adverse selection problem, the regulator has exposed itself to a moral hazard problem (Laffont and Tirole, 1993, pp. 35, 36). Moral hazard results in managers exerting as low an amount of effort as possible. This managerial slack and accompanying x-inefficiencies (Leibenstein, 1966) pushes up costs higher than necessary, which increases consumer prices to above-optimal levels. Hence, the regulator faces a trade-off between cost reduction (moral hazard) and rent extraction (adverse selection), which constitutes the basis for the wide variety of approaches and instruments for regulating network industries that can be observed in practice.

Asymmetric information creates problems in other respects too. First, it creates scope for regulatees or interest groups to influence a regulator to undertake actions preferential to them but not necessarily to society as a whole (Stigler, 1971). Such regulatory capture significantly afflicts regulatory policies. In consequence, regulatory design focuses to a large extent on how to overcome capture (Helm, 2006, pp. 174).[27] Second, asymmetric information rules out the optimal equilibrium. Crew and Kleindorfer (2006) submit that theories which conclude that regulatory policy might result in optimal prices, always assume some sort of common

---

[26] Ramsey-Boiteux pricing proposes a pricing rule based on the inverse elasticity rule. The price of the relatively inelastic good must be higher than that of its more elastic counterpart. Intuitively, it is optimal to increase prices to consumers which are not likely to switch because this increases profits, while consumers prone to switch must be charged lower prices to prevent them from moving to another supplier.

[27] See Dal Bó (2006) for a review of empirical and theoretical literature on regulatory capture.

knowledge. Hence, asymmetric information renders it unlikely that a regulatory policy will result in optimal prices. Third, a firm's informational advantage implies that a firm can not be motivated to disclose its privileged information unless it receives a (substantial) rent from doing so. In particular, Armstrong and Sappington (2006, pp. 330, 331) argue that the regulated firm needs to receive the full surplus of its activities in order to behave efficiently, that is, to operate at minimum costs and to satisfy consumer demand and needs. However, the huge producer rents inherent in such a scheme usually justify sacrificing some efficiency for distributional purposes.

4.2.3. Dynamic Interaction

A regulator and a regulated firm will often repeatedly interact with each other. The above considerations apply to a static, one period setting. Two main issues arise when dynamic interactions are taken into account.

First, a regulator might learn about the regulated firm(s) when interacting more than once. As indicated, this might reduce its informational deficit as it could use newly obtained information in the next regulatory period. It creates a risk for the regulatee if we consider a regulator's objectives. Because a regulator typically attaches a higher value to consumer rents than to firm profits, it has incentives to increase welfare by transferring rents from firms to consumers. In regulatory terminology: a regulator has limited commitment powers not to renegotiate the regulatory contract. If the firms realize this, they will change their behavior accordingly. The regulated firm faces a trade-off between better performance in a particular period and resulting higher performance standards in future periods. This could induce the firm to underproduce, at least at the end of the regulatory period, in order to avoid more demanding future regulatory schemes – the so-called ratchet effect (Weitzman, 1980; Freixas et al., 1985; and Dalen, 1997). A related dynamic consideration is expropriation. Expropriation arises in the presence of sunk (or irreversible) investments which are particularly relevant to network industries (Williamson, 1975). Section 2.9.1 has indicated that such investments tie the investor to the market it has invested in, creating scope for ex post renegotiation. If a regulator has limited commitment powers, it may choose to do so. It might choose not to compensate the firm for its investment, but rather to deliver all benefits to the consumers via for instance imposing low prices or encouraging entry via providing free or cheap access to the sunk investment. This changes an investor's ex ante behavior. It might in the extreme case choose not to make any up front investments at all. The negative impact on investments of limited regulatory credibility is referred to as the hold-up problem. Hold-up of investments is a key parameter in TCE and will therefore feature prominently in section 5.

A second issue arising in the context of the dynamic interaction between a regulator and a regulatee is that the composition of investments may change. Investors will choose a less capital intensive production technique to mitigate the risk of being expropriated (Baumol and Klevorick, 1970, pp. 179, 180; and Gilbert and Newbery, 1994, pp. 551). If cost-reducing technologies require sunk investments, this effect moves the market equilibrium away from the optimum. Dynamic interactions not necessarily reduce investment incentives, however. In a model where investments last forever, thus with a very large risk of expropriation, the desired level of investments is reached when a firm gradually builds up its asset base (Salant and Woroch, 1992). In this setting, if a regulator displays ex post opportunism, the firm can credibly threaten to abandon its investment plan.

In sum, regulation will fail predominantly due to a regulator's information deficit which results in moral hazard and adverse selection problems as well as regulatory capture. Ex post regulatory opportunism is another source of problems. Two important implications arise. First, limiting regulatory discretion has advantages because it lowers both a regulator's opportunity to behave opportunistically as well as mitigates the adverse consequences if a regulator is captured. On the other hand, limited regulatory flexibility also lowers the scope for welfare-enhancing adaptations to regulation which may arise due to changing market circumstances, new forecasts, or advancing knowledge of market parties. Hence, unless opportunistic regulatory behavior can be separated from efficient adaptations to regulation, there is a fundamental trade-off between discretion and commitment. Second, a regulator will generally be unable to implement the optimal regulatory regime. Even if an optimal regulatory scheme would be possible, its information deficit obliges a regulator to allow the firm some rents to induce it to reveal its superior information. To limit these rents and increase consumer benefits, some inefficiency must consequently be allowed. Hence, any regulatory policy will be inefficient to a certain degree.

These intrinsic inefficiencies inherently provide scope to argue that competition improves matters by lowering the need for regulation. Accordingly, regulatory failure must be weighed against the market failure it is supposed to correct in order to determine whether regulatory intervention is the right course of action.[28] This has led Martimort (2006, pp. 7) to argue that an important lesson from optimal regulation literature is that efficiency should not be the sole criterion on which to assess regulation. The relevant concept is rather interim inefficiency, i.e., reaching an optimal balance between efficiency and limiting a monopolist's rents from its privileged information.

4.2.4. Regulatory Instruments

A regulator's objectives, its informational deficit, and the interaction with the regulatees determine which regulatory instruments to deploy. There are basically five options to regulate network industries:[29]

1) A regulator can change the external organization of a network sector by integrating or breaking up companies either vertically or horizontally;
2) The ownership structure can be altered;
3) The mode of competition can be determined or influenced;
4) Tariffs and prices can be regulated; and
5) Additional regulation can be implemented.

Options 1 and 2 are structural reform measures which change the structure of the market. Option 3 is a liberalization measure which tries to free a market from obstructions and

---

[28] This is one of the main arguments of the institutional perspective towards regulation which will be discussed in section 5.

[29] Laffont and Tirole (1993) provide a standard reference work. Armstrong et al. (1994) analyze what economic theory has to say about reforming utilities (network industries). Aalbers et al. (2002) provide an overview of what "one can say based on the economic literature about reform and privatization in network industries". A recent and extensive overview of normative regulation theories on utilities (electricity, gas, telecommunications, and water) is provided by Jamison et al. (2005).

limitations through introducing competition. Options 4 and 5 are conduct regulation measures that focus on the behavior of market players.

The next sections discuss existing gas regulation, with an emphasis on the second Gas Directive and the recent proposal for the third legislative package. These prescribe structural reform and liberalization measures; conduct regulation is left primarily to the member States or is addressed by other Directives. We will accordingly focus our attention on these categories. Furthermore, as privatization measures are not included in the Gas Directives either, we will not explicitly consider option 2. Finally, the Gas Directives do not emphasize price regulation. Hence, the next sections predominantly focus on options 1 and 3.

4.3. The Second Gas Directive

The second Gas Directive sets out to fill the gaps left by the first Directive and to increase the pace of liberalization. It is supposed to make it practically impossible for Member States to allow their own, shielded market to exist. It tries to create the necessary regulatory framework to tackle the remaining barriers to the completion of the internal market. We highlight the changes compared to the first Directive regarding market opening; access; unbundling; public service obligations; regulatory authorities; and long-term contracts.

Market Opening

First, according to Article 23 of the second Directive, the gas market must be opened up to competition at an accelerated speed: for non-residential customers 1 July 2004 at the latest and full market opening for all customers at 1 July 2007. This is a substantial improvement considering the absence of a definitive date for full market opening in the first Directive. This is supposed to solve the problem of uneven market opening among Member States. Another measure to create a more level European playing field, is the reciprocity clause in Article 23(2). This is a temporary clause which allows a Member State that has opened its market more than required by the Directives to refuse access to the network of an eligible consumer to a supplier from a Member State in which the targeted consumer is not yet eligible. This clause allows Member States to liberalize more quickly than required without exposing them to unfair competition from less liberalized Member States. The Commission is also allowed to impose measures to ensure external reciprocity to guarantee an equal opening of downstream and upstream markets.

Access

Another important change in the second Directive concerns the access regime. The NTPA regime was considered unsatisfactory. Therefore, the choice between NTPA and RTPA in the first Directive is in Article 18 of the second replaced by RTPA. That is, TPA to both transmission and distribution has to be granted based on published and regulated tariffs. Gas storage is recognized as a special case in that on the one hand storage is considered an ancillary service providing flexibility to new entrants, which implies that access should be provided for when requested. However, on the other hand, the importance of storage for matching supply and demand, optimizing the system, and creating a level playing field is recognized, which calls for a more restrictive access regime. Therefore, Article 19(1) allows access to storage facilities, including line pack, to be granted based on RTPA, NTPA, or both. Of course, the access criteria must in any case be objective, transparent, and non-discriminatory. One exception to the storage rule, provided by Article 19(2), is that ancillary services and temporary storage related to LNG facilities and which are necessary for re-gasification and subsequent delivery of gas to the transmission system fall under the RTPA regime. Another exception follows from the Commission's interpretation as laid down in a non legally binding Interpretation Note (DG TREN, 2004a): storage facilities that are exclusively reserved to transmission system operators (TSOs) for carrying out their functions and storage facilities that are used for production operations may also be exempted from the TPA provisions.

Regarding access to upstream pipeline networks,[30] Article 20 stipulates that the access provisions apply to the extent that these pipelines are not used for local production. This grants Member States considerable discretion. According to Article 21(1), access may be refused if this results in serious economic and financial difficulties with TOP contracts (see also Article 27 regarding derogations based on TOP contracts). Article 21(2) allows access to be refused due to a lack of capacity. The same Article states that Member States may take measures to ensure necessary improvements to the pipeline network insofar as it is economic to do so or when a consumer is willing to pay for them. Finally, Article 22 permits granting exemptions from TPA to major new infrastructure investments like interconnectors, LNG, and storage facilities.

Unbundling

The third main requirement for developing the internal market is unbundling. The second Directive goes substantially further than the first: rather than unbundling of accounts, the emphasis in Articles 9 and 13 of the second Directive is on legal unbundling. This means that TSOs and DSOs (distribution system operators) that are part of a vertically integrated undertaking must be independent at least in terms of their legal form, organization, and decision-making from other activities not related to transmission and distribution.[31] No change of assets is implied, however. Legal unbundling must therefore be carefully distinguished from ownership or structural unbundling. Legal unbundling is limited to the natural monopoly (the network business); all other activities, like supply and production, may still be operated within a single company. According to Article 13 and Recital 11, Member States may be exempted from legally unbundling DSOs when a disproportionate financial and administrative burden on small distribution companies would result. Articles 13 and 33(2) state that for small DSOs serving less than 100,000 customers the exemption is not time limited, as opposed to larger DSOs for whom unbundling may be postponed only until 1 July 2007. In addition to the above three conditions for creating an internal market, the public service obligations, the status of national regulatory agencies, and the presence of long-term contracts matter. We start with the former.

---

[30] Defined by Article 2(2) as "pipelines or a network of pipelines operated and/or constructed as part of an oil/gas project, or used to convey natural gas from one or more projects to a processing plant or terminal or final coastal landing terminal".

[31] See DG TREN (2004b) for the Commission's interpretation.

Public Service Obligations

A glance at Annex A and Article 3 of the Directive reveals that the provisions concerning the public service obligations (PSOs) and consumer protection have been strengthened. The measures to this end can be classified in three groups (Cameron, 2005, pp. 24). First, there are obligations imposed upon the Member States: 1) ensuring that gas companies respect the Directives' requirements and do not discriminate; 2) protect consumers, especially the vulnerable consumers; 3) notify measures taken to achieve the PSOs; 4) publish the PSOs; and 5) ensure easy switching between suppliers for consumers. Second, there are objectives to be pursued by the Member States, like the PSOs. The Directive allows Member States to define these objectives and impose them on natural gas undertakings, provided these are clearly defined, transparent, non-discriminatory and verifiable. Article 3(2) defines the PSOs as "security, including security of supply, regularity, quality and price of supplies and environmental protection, including energy efficiency and climate protection". Following Articles 5 and 26, other objectives to be pursued are protection of final consumers and social and economic cohesion. Member States are allowed monitor and intervene based on security of supply considerations.[32] The third group pertains to Member States being allowed to establish a supplier of last resort and to protect remote consumers.

Regulatory Authorities

Regarding regulatory authorities, two improvements have been implemented. First, Article 25(1) obliges Member States to set up one or more competent regulatory authorities which are wholly independent from the industry (but not from the government). This improves clarity and transparency compared to the first Directive which only prescribed that appropriate and efficient mechanisms for regulation, control, and transparency must be created in order to avoid any abuse of a dominant position. To ensure as much of a level playing field as possible, a second innovation, to be found in Recital 13, is that a number of regulatory competences are formulated in terms of harmonization, subsidiarity notwithstanding. This applies in particular to network access and setting and approving network tariffs or the methodology underlying their calculation. Important in this respect is Recital 14's emphasis on regulatory cooperation and coordination in Europe, which is considered vital to prevent the patchwork of regulatory strategies (Eberlein, 2005, pp. 65). To facilitate cooperation and coordination, a number of informal institutions have been set up. The idea is to bring together regulators and stakeholders and allow them to informally discuss regulatory issues. For gas, the Madrid Forum has been set up to this end. It was established in 1999 and provides an informal EU-level framework for discussing issues concerning, and exchange experience of, the establishment of a competitive internal gas market. Madrid is also the platform to discuss (mostly technical) issues relating to the creation of an internal gas market which are not addressed in the Directive(s).[33] Furthermore, the Forum concept has also been developed as a way to overcome Member State reluctance to give up control over their energy sectors. Instead of vertically delegating regulatory powers to the Community, the Madrid Forum allows a horizontal sharing of powers via national regulatory agencies that cooperate with each other and the Commission (Eberlein, 2005, pp. 63, 81).

[32] See DG TREN (2004c).

The Madrid Forum has reached an agreement on a few elementary principles concerning for example cross-border transactions, cost-reflective tariffs and non-discrimination procedures. However, as a consequence of diverging interests of Member States, incumbents, and new entrants as well as its informal status, the Forum proved unable to convert these agreements into formal economic and technical rules. As a consequence, and in order to harmonize the work of different national regulators, the European Regulators Group for Electricity and Gas (ERGEG) has been set up in November 2003 to ensure a consistent application across all Member States of the adopted Directives.[34] While ERGEG diminishes the influence of the Madrid Forum, Madrid will nevertheless remain an important discussion platform between regulators and stakeholders. The Gas regulation below highlights the importance of the Madrid Forum for regulation, its informal status notwithstanding.

Long-Term Contracts

Finally, the second Gas Directive considers long-term gas contracts. Given the potential usefulness of long-term contracts for facilitating sunk investments, Article 18(3) explicitly mentions that these contracts should not be precluded, insofar as they are compatible with the Directive and the competition rules of the Community.

4.4. The Proposal for a Third Legislative Package

The proposed third package (which is currently debated in and possibly amended by both the Council and Parliament) is supposed to remove all structural failings. It proposes amendments in six main areas: unbundling; cooperation among national regulators; powers and independence of national regulators; cooperation between TSOs; improving the functioning of markets; and security of supply.

Unbundling

Regarding unbundling, the view is that the existing provisions are insufficient to create a well-functioning market. Legal and functional separation of network operations from supply and production allow 1) a system operator to treat its own affiliates more favorable than others; 2) does not guarantee non-discriminatory access to information; and 3) retains an integrated company's distorted investment incentives. The Commission's preferred solution is ownership unbundling (i.e., structural separation in section 3.4.1). Note that the unbundling requirements only apply to the transmission grids; for distribution system operators, legal and functional unbundling is considered sufficient. Member States must ensure that a person who exercises control over a supply undertaking can not at the same time hold any interest in or exercise any control over a transmission system. While this is the preferred option, an alternative is provided for those Member States that do not wish to go down the ownership unbundling path. The requirement for this alternative is that network independence and investment incentives are similar to ownership unbundling. The Independent System Operator option (i.e., operational separation in section 3.4.1) allows a vertically integrated undertaking

[33] The Madrid Forum comprises officials from the Commission, national regulatory authorities and EU Energy Ministries, international organizations, and associations representing the gas industry and the gas and electricity consumers.

to retain the ownership of its network assets, but requires that it is managed by an independent system operator which is completely separate from the vertically integrated undertaking.

The package also considers undertakings from non-European countries (like Gazprom from Russia, which explains why this clause is sometimes called the "Gazprom clause"). These companies must demonstrably and unequivocally comply with the unbundling requirements. Furthermore, the Commission proposes that third country undertakings can acquire control over a Community transmission system or TSO only if this is permitted by an agreement between the EU and the third country.

Cooperation Among National Regulators

A second legislative change follows from cross-border issues. Cross-border issues have been one reason for setting up the Madrid Forum and ERGEG. While both have been useful, a regulatory gap has always existed. In order to create true supra-national gas markets, the technical rules need to be harmonized. This is out of the Commission's reach as well as ERGEG's or Madrid's. Accordingly, a new entity, independent of the Commission, must be devised that bundles the expertise of all 27 national regulators. An Agency for the Cooperation of Energy Regulators (ACER) is consequently proposed. ACER complements the actions of national regulators. Its main tasks are: allowing national regulators to cooperate; providing regulatory oversight of the cooperation between transmission system operators (see below); handling specific cross-border issue like exemption requests for interconnectors; and advising the Commission.

Powers And Independence of National Regulators

As set out in the second Directive, Member States must establish national regulators that oversee and ensure the proper functioning of the market. Some national regulators are well-established, have strong powers, and are independent. In some Member States, however, such regulators have been established just recently, are not independent, and/or lack the required powers. It is therefore proposed to strengthen the powers of national regulators by giving them a clear mandate to cooperate at the European level, in close cooperation with ACER and the Commission. Second, their regulatory powers are strengthened by for instance allowing regulators to review investment plans; monitor compliance with TPA rules; and impose sanctions. Furthermore, regulators must have access to information on the operational decisions of the firms. The requirements for regulatory independence are strengthened. Rather than being independent from only industry interests, a regulator must now be independent from any other public or private entity.

Cooperation between TSOs

Market integration requires that national TSOs cooperate effectively. For instance, network access and operational rules need to be compatible for supra-national markets to develop. The Commission considers voluntary cooperation initiatives in this regard, like Gas Transmission Europe, to be insufficient. Cooperation must therefore be strengthened. Highlighted areas of cooperation are the development of market and technical codes; research

---

[34] Lavrijssen (2004) provides a discussion on ERGEG.

and innovation initiatives of common interest; coordination of grid operation; and investment planning.

Improving Market Functioning

Market functioning needs to be improved according to the Commission. This relates to TPA; transparency; access rules for storage and LNG; and retail markets.

Regarding TPA, the third package sets out to streamline the procedure for applying and granting exemptions from the TPA regime, as well as to clarify some of the conditions. An example is to provide general minimum conditions for capacity allocation and congestion management of an exempted infrastructure, rather than determining these on a case-by-case basis. The transparency requirements imposed on incumbents must be increased in order to create a level playing field for new entrants. In particular, in addition to publishing network capacities, it is proposed that transparency requirements also cover gas stocks; demand and supply projections; and balancing and trading costs. The strengthened national regulators must ensure and oversee this process. The Commission still considers access to storage and LNG facilities to be insufficient. For example, the second Directive leaves too much discretion to Member States to define their storage regime. The Commission proposes a better implementation of the voluntary Guidelines for Good Practice for Storage System Operators, agreed upon by stakeholders in the Madrid Forum, as a solution. These Guidelines should become legally binding; storage system operators must be legally and functionally unbundled; national regulators must have more powers to oversee access to storages; and there must be clarity over which regulatory regime applies to storage facilities (as opposed to the choice of access regime in the second Directive). LNG access also needs to become clearer. The second Directive's requirement that access must be regulated is considered too vague and creates room for diverging Member State interpretations. Therefore, legally binding guidelines must be designed which spell out more clearly defined LNG access rules. A final initiative to improve market functioning is aimed at retail markets. The Commission considers setting up a retail forum (by analogy with Madrid and Florence) which would allow all stakeholders concerned to discuss specific retail problems.

Security of Supply

Ensuring security of supply requires sufficient storage and import capacity to be in place in order to satisfy a peak demand. It is proposed that the above network of European TSOs monitors supply security and makes system adequacy forecasts for summers, winters, as well as the long-term. A European outlook needs to be construed which takes account of the possibilities to export and import gas in case of peak demand. Finally, cooperation among Member States is seen as an important element of supply security. The current coordination platform is supplemented with additional transparency requirements on the level of commercial stocks (e.g., a storage operator must disclose on a daily basis the volume of working gas in its storages) and a solidarity mechanism between Member States is proposed.

All in all, the third legislative package tries to improve matters by a strict separation of supply and production from transmission. Furthermore, the abilities of regulators need to be strengthened, both nationally as well as at the European level. Cooperation and coordination needs to be improved – between national regulators, TSOs, and Member States. Operations all along the value chain need to become more transparent. Finally, Member States will benefit from greater solidarity through a more secure gas supply.

The regulatory framework specified above has been developed in order to create an internal gas market. At each stage, problems regarding the competitiveness or the development of the internal market have occurred. New proposals were suggested that were supposed to remove the impediments. In consequence, the changes to regulation have been substantial. Note that amendments have been made in the second Directive that detract from the neoclassical perspective. For instance, long-term import contracts are allowed, while certain investments are allowed to be exempted from the third party access regime. We get back to these amendments in section six.

As set out in section 4.2, there is a trade-off between regulatory flexibility and credibility. To be credible, regulation must be as stable and predictable as possible, but there must be room to implement changes if necessary. Therefore, the third package is communicated as the final one capable of creating the anticipated internal gas market.

4.5. The Internal Gas Market and Neoclassical Economics

The ultimate goal of the regulatory provisions set out above is to create an internal European gas market. The Madrid Forum has laid out a road map towards the internal market (Madrid Forum, 2002). It presents a long-term vision of the European gas market. It also issues a step-by-step plan to achieve the long-term vision. It contains the following high-level objectives:[35]

- real supply-side competition;
- contractual flows becoming decoupled from physical flows in many locations;
- entry-exit tariff arrangements for access to transmission networks;
- liquid hub-based trading;
- a robust and comprehensive regulatory framework; and
- effective mechanisms for investments in cross-border infrastructure.

A second road map builds on the first and acknowledges the following elements (ERGEG, 2005, pp. 6): "effective competition delivering real benefits for gas consumers throughout the EU; a stable regulatory framework facilitating efficient levels of investment; secure supplies; choice; and gas suppliers able to market their services to all consumers across the EU. This vision also now includes the availability of new sources of gas supply to Europe, notably LNG."

The ultimate objective of a fully integrated single (internal) gas market is also stressed by the Lisbon Council in 2000. The first step is the creation of effective regional gas markets based on liquid hub-to-hub trading.[36] These regional markets should eventually be integrated to create a single European natural gas market.

---

[35] This vision sets out the regulators' views. Some issues, such as the political framework necessary to facilitate commercial agreements for supplying gas between producers outside Europe and the EU, are touched upon only briefly.

[36] See http://www.energy-regulators.eu/portal/page/portal/EER_HOME/EER_INITIATIVES/GRI for an overview of the regional market initiatives.

The link with the neoclassical perspective is obvious. The internal market logic is that eventually an atomized gas market with a large number of suppliers will be created. The suppliers vehemently compete with each other on an equal footing and a short-term basis, creating unfettered supply competition. This supply competition will lower consumer prices, increase service levels, and facilitate efficient investments. Efficient demand and efficient investments help alleviate supply security anxieties. Furthermore, in this perfect setting, gas prices internalize the costs of emissions, which mitigates sustainability issues. This is nothing short of a direct translation of perfect competition. Therefore, the neoclassical perspective towards regulation is the guiding principle of existing gas regulation. This perspective implies that any impediment to liberalization or the development of the internal gas market is a sign of insufficient competition. This explains why the Commission's solution to the current unsatisfactory state of affairs is to more strongly pursue competition in Member States' gas markets (CEC, 2005c, 2007c).

This builds on another important assumption, namely, that by changing the structure of the market it is possible to change market participant's conduct in the desired direction. This assumption follows from the neoclassical structure-conduct-performance paradigm (Bain, 1956), which states that market structure (the number and relative sizes of firms in an industry) determines firm conduct (output decisions and pricing behavior), which yields an industry's overall performance (for example, its efficiency and profitability). For example, the lack of competition at the transmission level can according to the third legislative package be mitigated if we change the market structure through ownership unbundling. After ownership unbundling, firm conduct will be much more competitive, yielding a more efficient outcome and improving welfare. The next section provides a critique on these rather simplistic assumptions.

Furthermore, a consequence of the perfect competition assumption is that once the internal market has developed, the invisible hand will dissipate, or have dissipated, all problems. This argument is tautological: by assuming that the market will eventually work perfect, all problems that may occur are considered temporary and are therefore effectively assumed away. This, in turn, by definition rules out the option that the current state of affairs may be due to a structural discrepancy between full competition and gas market specifics. The next section provides the contrasting view that such a discrepancy may well be created if existing regulation is upholded in its new context.

5. The Institutional Perspective

5.1. Introduction

This section discusses the institutional perspective towards regulating gas markets. It starts with a few critiques on the neoclassical perspective which have spurred the development of institutional alternatives. Section 5.3 sets out the New Institutional Economics and its relation with neoclassical economics. Thereafter section 5.4 discusses the Transaction Cost Economics perspective. After having set out its analytical framework, section 5.5 infers its implications for regulatory intervention compared to those of neoclassical economics. This section finishes by providing the prescriptions for liberalizing the European gas market from a transaction costs perspective and contrasts these with the neoclassical prescriptions.

5.2. Alternatives to the Neoclassical Perspective

As indicated in sections 3.2 and 3.3, the neoclassical axioms result in a perfectly competitive market in which competition itself is sufficient to steer firm and consumer behavior in the optimal direction. Hence, no government guidance is required. The firm is considered a black box or production function that transforms resources into products, which are subsequently sold on a (spot) market. In consequence, following to Joskow (2006, pp. 4), firms and markets are seen as complementary, while in fact they are substitute governance structures (Coase, 1937). Neoclassical economics fails to provide insights into the boundaries of the firm. That is, it does not fully explain the plethora of governance structures that lie between markets and firms, like long-term contracts, that can be observed in practice.

In addition, its black box assumption precludes a consideration of a firm's internal organization. In fact, organization of activities within a firm is assumed to be costless. Neoclassical approaches are also criticized for their neglect of the influence of institutions on economic behavior. By analyzing matters through a perfect competition lens, neoclassical economics only implicitly allow for the presence of institutions. However, even if assuming perfect competition, one simultaneously assumes the presence of a set of institutional constraints guiding market exchange. Hence, institutions matter and must be considered explicitly.

These considerations have spurred a substantial body of research amending the neoclassical approach. It is broadly coherent in arguing that institutions have to be incorporated into the analysis in order to adequately explain economic performance, and that institutions shape and define the actions of different players (Nelson and Sampat, 2001, pp. 39). However, the proposed amendments to neoclassical theory differ between three main approaches. Therefore, in order to delineate our own approach, we set out the main premises of these approaches below.

Closest to the neoclassical fundament is the research programme sometimes labeled Neoinstitutional economics (cf. Alchian and Demsetz, 1972; de Alessi, 1983; and Eggertson, 1990).[37] It upholds the neoclassical core of methodological individualism, methodological instrumentalism, and methodological equilibration, and generalizes neoclassical economics by adding a few extra ingredients – information, the structure of property rights, and transaction costs – to the mix. Given these additional constraints and with given preferences, an individual optimizes his welfare. This eventually creates an equilibrium outcome, possibly characterized by an alternative organizational form.

This approach contrasts with the New Institutional Economics (NIE) approach which rejects parts of the neoclassical core, most importantly, the rational-choice model which follows from the methodological instrumentalism axiom. Simon's (1957) work which challenges the rationality assumption is important in this regard. Simon argues that the human

[37] Several classifications have been made. We use the one found in Eggertson (1990), who contrasts Neoinstitutional economics with New Institutional Economics, or NIE. An alternative classification is to identify old institutional economics – along the lines of Veblen and Hodgson below – with NIE, as in Hodgson (1989); Langlois (1989); and Rutherford (1989).

mind has a limited ability to formulate and solve all complex problems required for rationality; rationality is bounded in consequence. Individuals will display satisficing rather than maximizing behavior: they try to attain acceptable minima rather than optima. Building on this and by pulling together all threads into a coherent framework, Williamson (1975) is widely regarded as a landmark in NIE, which has grown considerably in size and substance. NIE is primarily based on Transaction Cost Economics (TCE). The discussion below emphasizes the TCE line of reasoning. By rejecting the rational choice postulate and replacing it with bounded rationality, NIE and TCE detract somewhat from neoclassicism. They nevertheless retain large parts of it – for instance upholding optimizing behavior by way of cost minimization and following strict methodological individualism.

Bounded rationality has also given rise to a more forceful rejection of neoclassical economics. For instance, Hodgson (1988), also building on Simon (1957) but incorporating contributions by earlier institutionalist scholars like Veblen (1909) who questions the conception of economic man central to neoclassical economics, goes further than Williamson. In fact, chapter 9 of his work provides a number of critiques on the Williamson approach, of which we highlight three. Hodgson (ibid., pp. 203) argues that Williamson incorrectly treats information in a positivist sense, i.e., he ignores that information on (transaction) costs has to be interpreted by the economic actors whose interpretations will differ due to differing cognitive frameworks. Furthermore, Hodgson's treatment of the market differs from Williamson's: rather than a state of nature for Williamson, for Hodgson the market consists of mechanisms to structure, organize, and legitimate exchange (pp. 174); the market is a social institution itself (pp. 206). In consequence, rather than explaining the presence of the firm, the issue becomes just as much to explain the presence of markets. A third departure from the Williamson approach is that according to Hodgson, the existence of the firm must be explained not merely from the perspective of cost-minimization and efficiency, but also from the view of the firm as an institution of power (pp. 214).

5.3. New Institutional Economics

As indicated, a fundamental criticism on neoclassical economics was that it ignores the important institutional determinants that explain alternative forms of economic organization. As a solution, current regulation should be enriched both to enable it to account for the costs of internal organization, as well as to recognize that in real life there is a wide array of governance structures through which transactions can be mediated (see Estache and Martimort, 1999; Berg, 2001; CIEP, 2006; and Dassler, 2006 for recent contributions). To this end, academic research has focused on theories of organization, also referred to as the economics of institutions, consisting of property rights theory, agency theory, and Transaction Cost Economics (TCE) (Menard, 2005, pp. 281). New Institutional Economics (NIE) has developed primarily out of Transaction Cost Economics (TCE) and is based on three pillars: transaction costs, contracts, and property rights (see Klein, 1999, for a survey of the very broad NIE literature). These pillars have created two distinct lines alongside which NIE has developed, both of which originate from Coase's research referred to above. Williamson (1998, pp. 24) puts it like this:

"The New Institutional Economics comes in two parts. Part one deals with the institutional environment – the rules of the game – and traces its origins to Ronald Coase's 1960 paper on 'The Problem of Social Cost'. Part two deals with the institutions of governance – the play of the game – and originates from Coase's 1937 paper on 'The Nature of the Firm'".

The first part was further developed by North (1981, 1990), the second part by Williamson (1971, 1975, 1985, 1996). Figure 1 below illustrates the relationship between the NIE, TCE, and the traditional neoclassical and agency theories of economics. It also shows the impact of these policies on the European gas market.

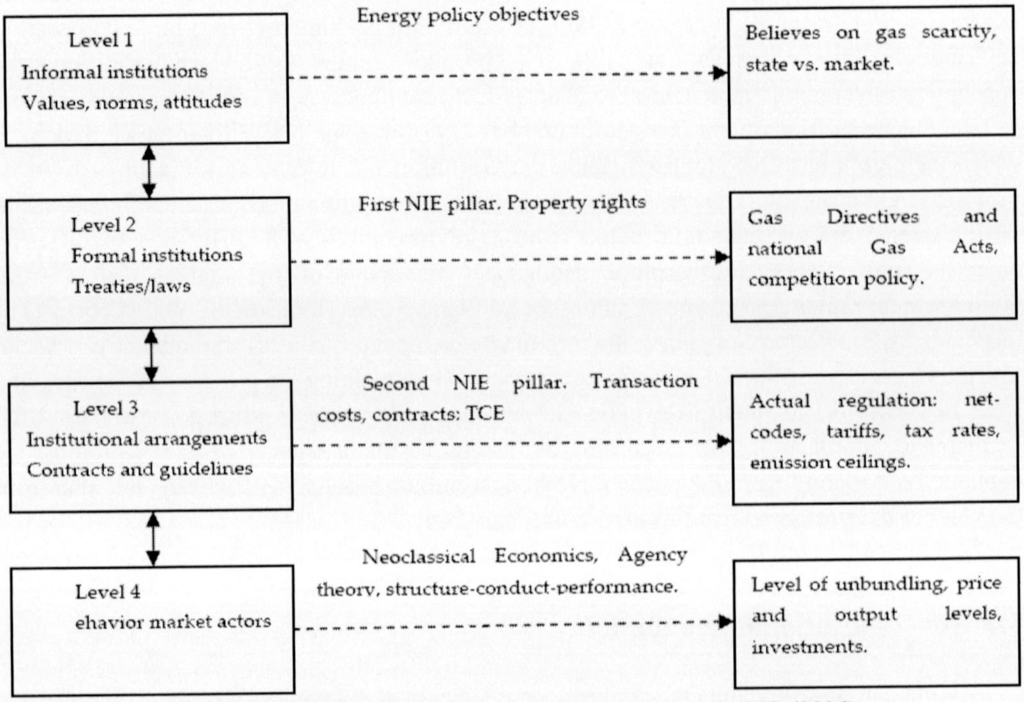

Source: adapted from Williamson (1998); Groenewegen (2005); and CIEP (2006).

Figure 1. New Institutional Economics, Transaction Cost Economics and gas policy .

Level one comprises the informal institutions where the basic societal values and beliefs are located. One could think of values, norms, and traditions that create the level of social embeddedness for individuals. These informal institutions are the underlying determinants of a society's basic view on market reform and energy policy. For gas, issues such as believes on resource scarcity and opinions about state versus markets are relevant. Changes in these institutions occur very slowly and may take centuries to millennia.

The informal institutions determine the formal institutions at level two where the polity, judiciary, and bureaucracy are located. This is the property rights part of NIE which determines the rules of the game. One could think of (inter)national treaties and laws laying down the ground rules for the behavior of market actors. Hence, the Gas Directives and the national Gas Acts that emanate from the transposition of the Directives into national law as well as competition policy provisions, belong to this level. According to Williamson (1998, pp. 27), first-order economizing – getting the institutional environment right – takes place at this level. Formal institutions change very sluggishly too: changes in the institutional environmental can take up to decades or centuries to materialize. Levels one and two comprise what Davis and North (1971) refer to as the institutional environment.

Getting the institutional environment right is not an easy task, because the traditional structure of the European gas market was not conducive to market competition. The vertical and horizontal integration as well as the shielded and managed national gas markets, prevented competitive forces from spreading across Europe. Therefore, liberalization requires significant restructuring of the European gas market as a consequence of which there will be winners (presumably the consumers) and losers (the incumbents). Section 4.1 has pointed out the important role of interest groups and stakeholders in shaping and moulding gas regulation. Regulation in order to implement liberalization can only be properly devised and implemented if there is sufficient support among stakeholders. Glachant and Perez (2007) provide interesting insights into the consequences for network industry regulation. They argue that if the pro reform coalition is not sufficiently strong, the practical effect may be much less than anticipated. For instance, vested interests may be exempted from some provisions in order to launch the reforms. This implies that it is difficult in practice to promptly restructure the European gas market into one governed by competitive forces. The result is that liberalization initiatives are scaled down initially in order to limit opposition. Accordingly, the liberalization process is spread out over many years and implemented in different successive stages. European gas regulation reflects this approach by proposing to open the market in progressive stages and allowing implementation to the Member States. Another example is the enlarged scope in the second Directive to implement exemptions from the TPA regime. Glachant and Perez raise the point that this splitting up of the process may impact the course of liberalization: if a certain stage in liberalization has been reached, it may be hard to progress from there on because of the opposition of (new) stakeholders.

In addition, restructuring alone will be insufficient. For instance, vertical separation may improve matters by removing anti-competitive incentives. However, its effect will be severely limited if it is copied onto the traditional market structure without additional measures. Removing vertical integration may be necessary but not sufficient: changing the institutional environment by rearranging existing property rights in order to create an institutional structure that aligns with the separated entities is necessary.

Changing institutions is not without problems, however. This can be seen by considering North's (1990) analysis of the process of institutional change. According to North, institutional changes are necessary because no one really knows what will happen in the longer term.[38] He argues that existing institutions do not change overnight, and will accordingly impact the present structure as well as the possibilities for structural change. Therefore, institutional changes will not be radical adaptations to the existing structure, but rather incremental ones at the margin (ibid., pp. 101). In other words, changes are marginal and build on the existing institutional structure. Hence, when the formal institutions are changed, for instance through issuing a new Gas Directive, a disequilibrium is created because the more tenacious informal institutions do not change accordingly. In North's (1994,

[38] This contrasts with the Williamson view which emphasizes that institutional changes are necessary because individuals are unable to conclude complete contracts that specify all possible contingencies (see the next section). Accordingly, some unforeseen future development may require institutional changes.

pp. 367) own words: "It is adaptive rather than allocative efficiency which is the key to long-run growth".

From a slightly different perspective, Aoki (2001) points out that new or changed institutions are implemented in a world where numerous other institutions already exist. These existing institutions (for instance, long-term take-or-pay contracts) partly lock-in the current situation.[39] This will impact, either positively or negatively, the development and prospects of the new institutions. A consequence is that a discrepancy between the sluggish existing institutional environment and the institutional changes implemented in order to materialize the competitive reforms, may arise. This creates problems and tensions which, for instance, increase opposition to and lower the feasibility of the reforms. In consequence, the new, adjusted institutional structure may well be less far-reaching than originally intended by the new Directive.

The rules of the game determine the play of the game which is located at the third level. These institutional arrangements are the focus of TCE. The laws, treaties, and property rights developed at the second level are at this level transposed by Member States into institutional arrangements such as contracts or guidelines. This is where the governance structure – where issues like spot market trading, the level of unbundling, or state ownership are being addressed – is determined (Williamson, 1985, 1996).[40] Getting the governance structure right is referred to as second-order economizing. Changes in governance can take up to a decade to materialize. TCE analyses take the institutional environment as given. That is, TCE determines the most efficient mode of governance (in terms of transaction costs, as indicated below) given the prevailing property rights, rule of law, and regulations. The next section elaborates on the TCE framework which is central to this chapter's analysis.

At the fourth and final level, the institutional arrangements drive the behavior of market actors. This is what neoclassical economics is concerned with. The issue is to get the marginal conditions right which is referred to as third-order economizing. Changes at this level concern prices and quantities, which more or less occur continuously in response to changing market conditions. As indicated, the neoclassical emphasis towards regulation is illustrated by the structure-conduct-performance paradigm (Bain, 1956). From this perspective a firm's only function is to transform inputs into outputs. In contrast, NIE and TCE regard the firm as an organization which itself must be subjected to economic analysis (Klein, 1999, pp. 463, 464).

Hence, the value of the neoclassical approach notwithstanding, Figure 1 clearly indicates that this viewpoint towards regulation is not apt anymore.[41] Accordingly, a main message of this chapter is that the neoclassical approach needs be complemented with an institutional

---

[39] Different reasons have been put forward. Couwenberg and Woerdman (2006) analyze lock-in on the Dutch natural gas market and argue that the old transactions create a lock-in that makes implementation of full competition impossible. Without explicitly analyzing path dependence, Polo and Scarpa (2007) provide a similar argument by submitting that the presence of take-or-pay commitments might negate the pro-competitive effects of stimulating entry. They claim that take-or-pay commitments might create a segmented market, i.e., a market in which each supplier targets a different segment of the market. The resulting entry without competition would harm liberalization's efficacy. From a different perspective, Mañé Estrada (2006, pp. 3784) "can think only of two possible answers" to the observation that the EU appears to be locked-in to its current competition approach: 1) the current policies are the only viable ones in today's EU; or 2) statesman of the EU have ceased to observe the actual world around them. The first answer corresponds with lock-in being caused by the ongoing integration movement, while the second one implies that the market shifts from section two may not be adequately recognized by European policy makers.

[40] See Groenewegen (2005, pp. 10) for a number of options.

framework in order to explicitly recognize the important role of transaction costs, contracts, and property rights in the new context for gas regulation. Note that while the above description of the four levels implies a one-way causality, the arrows indicate otherwise. The causality also runs from bottom to top, for instance because actions of market players and stakeholders influence the regulation that is being imposed on a market.

5.4. Transaction Cost Economics

As indicated, TCE is concerned with institutional arrangements such as contracts and guidelines. In short, TCE submits that most transactions create contractual hazards which must be mitigated through an appropriate governance structure. It studies comparative contractual arrangements and examines which of the feasible alternatives is best suited given the transactions at hand. This makes TCE ideally suited to analyze the European gas market in the new regulatory context set out in section 2.9.

The ease of contracting is an important difference between neoclassical economics (NCE) and TCE. NCE with its perfect competition postulate, assumes that contracts are complete, i.e., that every possible contingency can be foreseen, specified, and agreed upon in a contract. This does not imply the absence of uncertainty, however. Rather, the future is not known but the probability distributions of all possible future events are (Klein, 1999, pp. 466). In Knight's (1921) terminology, NCE assumes the presence of risks rather than uncertainties, which allows contracts to foresee and anticipate all possible contingencies. Under these assumptions, there is no efficiency argument for coordinating transactions outside the market. Hence, recourse to the spot market is the best option. TCE takes a different perspective. Figure 2 provides the analytical framework from which TCE approaches the matter.

TCE acknowledges the presence of uncertainties. Consequently, the available contracting options are limited because nobody knows exactly what will happen in the future. This translates into contractual incompleteness, because a contract can not take account of ex post adaptations that can not be foreseen ex ante. Even if we assume the possibility of contractual completeness, the costs of writing, monitoring, verifying, and enforcing a complete contract are likely prohibitive. If we consider a very simple transaction like buying an off-the-shelf component, the uncertainty is not very important and spot transactions are a proper solution. However, more complicated, specialized transactions which are common on gas markets, require more sophisticated modes of governance.

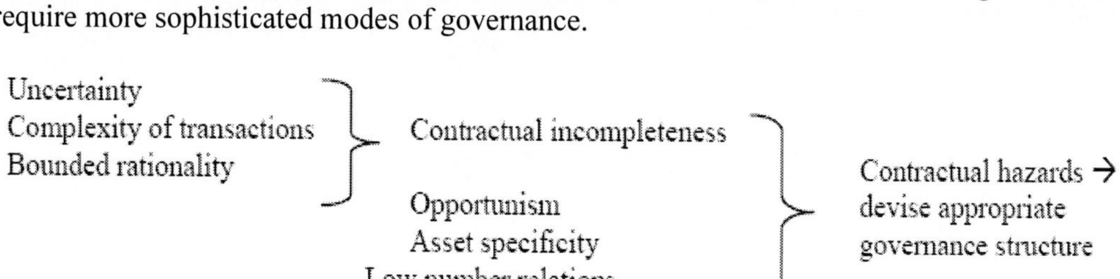

Own elaborations

Figure 2. The analytical framework of TCE

[41] Crew and Kleindorfer (2006), for example, reach a similar conclusion by emphasizing information problems and distribution issues.

According to Williamson (1975, pp. 4), in addition to uncertainty and complexity, a third main cause for contractual incompleteness is bounded rationality. That is, an agent intends to behave rational but will do so to a limited degree, for instance because human beings are limited in knowledge, foresight, skill, and time (creating the moral hazard problem in section 4.2). This explains the left-hand column of Figure 2: the inherent presence of uncertainty combined with the complexity of transactions and bounded rationality of individuals, renders most contracts unavoidably incomplete.

Contractual incompleteness invites opportunism which is defined by Williamson (1996, pp. 6) as "self-interest seeking with guile" (creating the adverse selection problem in section 4.2). The gaps present in any contract can be abused by an agent to achieve its own personal goals which not necessarily coincide with those of the society as a whole – think of sub-goal pursuit, shirking, etcetera.[42]

In addition to contractual incompleteness and opportunism, asset specificity is another important cornerstone of TCE. It takes one of several forms.[43] For European gas markets, common forms are locational and physical asset specificity and especially dedicated assets, as indicated in section 2.9.1. Dedicated assets are arguably the most important in gas. Dedicated assets sink investments into a market and create a bilateral relationship between the parties involved. These then have an incentive to bargain about the rents that emanate from the sunk investment. This can have negative effects both on ex ante investment and ex post efficiency, as well as increase transaction costs. In fact, the costs of a transaction are determined to a large extent by the degree of asset specificity (Williamson, 1985, chapter 4). Furthermore, asset specificity is closely related to low numbers bargaining. This is because asset specificity creates a condition of bilateral dependency, transforming what may have been a large number supply situation at the outset, into small number exchange relation thereafter.

The third column of the figure indicates that the interplay between bounded rationality, opportunism, and asset specificity determines how a transaction should be governed in practice (Newbery, 1999, pp. 54). For example, with bounded rationality and opportunism but no asset specificity, the market is contestable and competition is the solution. In case of asset specificity and opportunism without bounded rationality, devising appropriate – complete – contracts solves the problem. Without opportunism, promises are always kept and simple contracts suffice. Hence, only the simultaneous presence of all three characteristics creates contractual hazards. Such hazards require costly safeguards to be devised. In addition, these hazards render market exchange costly too, which opens the door for a comparative analysis of alternative governance forms in order to assess their respective advantages. Hence, the TCE perspective employs an ex-post governance approach (Williamson, 2005, pp. 44).

Empirical Evidence

We emphasize three areas of the institutional approach that have been extensively tested, namely, vertical integration, transactions cost economics (TCE), and long-term contracts (Klein, 1999, pp. 470-478 provides an overview).

---

[42] Hodgson (1998, pp. 205) objects to the central role that Williamson apportions to opportunism. According to Hodgson, the fundamental issue regarding contract compliance is not opportunism or self-interest, but rather uncertainty (which may or may not be due to opportunism).

[43] These include 1) site or location specificity; 2) physical asset specificity; 3) human-capital specificity; 4) dedicated assets; 5) brand name capital; and 6) temporal specificity. See Williamson (1996, pp. 59, 60); and Creti and Villeneuve (2003, pp. 4).

The first two are closely related, because TCE is usually tested by analyzing to what extent asset specificity increases vertical integration. The common denominator is that there is indeed a positive relationship between asset specificity and vertical integration (see the survey articles of Joskow, 1988a, 2005; Shelanski and Klein, 1995; Crocker and Masten, 1996; and Lyons, 1996). This corroborates the general view that TCE performs well empirically. In fact, according to Williamson (1996, pp. 55), "TCE is an empirical success story".

Nevertheless, the role of asset specificity has created and still creates considerable controversy within TCE. Klein et al. (1978) have argued that asset specificity creates hold-up problems which in turn are a rationale for vertical integration. This has been challenged by Coase. Coase agrees that high levels of asset specificity may indeed create hold-up problems, but objects to the assertion that this is likely to result in vertical integration rather than the intermediate solution of concluding long-term contracts. Menard (2005, pp. 310) argues that this controversy – which, as illustrated by Coase (2006) and Klein (2007), is far from resolved – indicates a need for additional empirical research. Specifically, uncertainty and the role of human assets needs to be more thoroughly studied and tests must be conducted at the firm or inter-firm, rather than sector, level.

Regarding the solution of long-term contracts to which Coase refers, Joskow (1985, 1987, 1988b, 1990) argues that for coal, contract duration tends to increase when relation-specific investments are at stake. Crocker and Masten (1988) corroborate this result for the gas market and add the observation that adaptation to changing economic conditions is vital when concluding long-term contracts. They furthermore argue that agents are sensitive to relatively small adaptations in contract terms. Crocker and Masten (1991) furthermore argue that the longer a contract becomes, the more likely agents are to conclude relational contracts, as opposed to the mechanistic contracts often assumed in economic literature.[44] In their view, due to its inherent incompleteness, a contract should be considered as a means to structure the inherent negotiations rather than as an attempt to stipulate all possible contingencies. Recent empirical studies show that liberalization initially decreases contract length but that asset specificity may increase contract length again (Neumann and Von Hirschhausen, 2004, 2006; and Neuhoff and Von Hirschhausen, 2006).

Hence, the majority of empirical research is consistent with the insights from TCE. However, in addition to Menard's remarks above, Klein (1999, pp. 477, 478) identifies a number of important difficulties of the above empirical tests. Two of these are related to measurement and confusion about the definitions of key variables, especially uncertainty. In addition, Hubbard and Weiner (1991) and Joskow (1991) argue that alternative hypotheses that could also fit the data – in their papers, (neoclassical) market power explanations – are rarely stated and compared. A final problem is that most empirical studies do not establish causal relations, but rather observe correlations.

In sum, TCE provides a richer framework of analysis than the neoclassical perspective. At this point we refer back to Figure 1, which indicated that current regulation's perfect competition focus – located at level 4 – fails to adequately take account of the institutional organization of the gas sector.

Recall that section 2.9 has pointed out that the market shifts currently taking place at the international gas market create a new context for regulation in which the predominant issue

[44] Often referred to as contingent claims contracts.

has become to stimulate investments in a gas market which is dominated by producers, where political considerations are gaining importance, and where supply security and sustainability have move to the top of the policy agenda. Section 2.9 has furthermore indicated that in such an environment, regulatory policy must focus on investments and their interplay with irreversibility, risks, uncertainty, and governance. The neoclassical perspective is unsuited for this task, because it explicitly recognizes none of the indicated criteria. This section has pointed out that the TCE perspective, on the other hand, takes account of the relevant criteria in the new context. Hence, order to adequately analyze the European gas market and its regulation in the new context, we need to complement the traditional neoclassical perspective towards regulation with the view that emanates from the transaction cost theory.

5.5. Regulatory Intervention in the Transaction Cost World

TCE provides a different perspective towards regulation than the neoclassical approach.[45] Without the intention of being exhaustive, we provide four important implications when viewing regulation through a TCE rather than neoclassical lens.

Firstly, compared to the neoclassical perspective, TCE jettisons rational choice but retains optimizing behavior. That is, the governance structure that entails the lowest transaction costs will eventually be chosen. The fundamental difference with neoclassical economics is that that due to, amongst others, bounded rationality, not necessarily the first-best optimum is chosen. Rather, TCE advocates a comparative institutional analysis that chooses the best out of a set of feasible (not necessarily optimal) alternatives. This observation has consequences for the scope for regulatory intervention. Recall that the neoclassical perspective advocates regulatory intervention in response to a market failure, i.e., a deviation from perfect competition. The costs of market failure obviously must be weighed against the costs of regulatory failure (for instance due to capture).

In contrast, regulatory intervention from the TCE perspective is justified only if its transaction costs – comprising the direct contracting costs as well as the costs related to ex post contracting hazards – are lower than for private ordering. If not, the best solution is to refrain from regulatory intervention, the market failure notwithstanding. This implies that regulatory intervention is warranted only in case of complex externalities, i.e., if the transaction costs of private internalization become prohibitively expensive because for instance many actors are involved, the sources of externalities are diffuse, or when information is asymmetric. In neoclassical settings, where all information on costs and benefits is reflected in prices, the costs of regulatory intervention are quite low. In the TCE perspective, however, the transaction costs of intervention are much higher. For example, the direct contracting costs are very high, internalizing these externalities is very expensive,

[45] TCE also detracts from the capture and agency theories which appeared in section four. According to capture theories, regulation is supplied in response to private interests, which assumes a passive regulator. TCE sees regulators as active, for instance, by behaving opportunistically. Regarding the latter theories, Spiller and Tomassi (2005, pp. 516) argue that the principal-agent contracting schemes that are required to provide the proper ex ante incentives depend on the institutional environment. That is, rather than an implicit contract between a government/regulator and a firm, they see regulation as an implicit contract between policy makers which directly impacts firm's incentives and performance (pp. 517, 538). Furthermore, as indicated, TCE emphasizes ex-post governance whereas the principal-agent theories analyze matters from an ex ante incentives alignment perspective (Williamson, 2005, pp. 44).

information is asymmetric, and regulators may behave opportunistically. Hence, in this setting, the scope regulatory intervention is likely lower.

Second, TCE treats a government of regulator differently than neoclassical economics does. In contrast to the neoclassical perspective, TCE assumes a regulator to be non-benevolent and generally unable to implement an optimal regulatory scheme.

Third, TCE rejects the neoclassical notion that state ownership and state regulation are substitutes (Menard and Shirley, 2005, pp. 13). According to TCE, they are different modes of governance with their own characteristics, efficiency properties, and transaction costs. Their respective feasibility depends on the characteristics of the transaction at hand.

Fourth, from a TCE perspective, no governance structure should be ruled out beforehand. This implication is very well illustrated by the current discussions on vertical integration and unbundling (see below).

5.6. Liberalizing the European Gas Market from a Transaction Cost Perspective

This section revisits the neoclassical prescriptions set out in section 3.4 in order to illustrate the differences between the neoclassical and the transaction cost perspective regarding the structure of a liberalized European gas market.

Restructuring/Unbundling

Regarding vertical restructuring, the neoclassical recommendation was clear: structurally separate integrated firms, because vertical integration is without efficiency rationale while likely stifling competition. The TCE framework is much less inhospitable to vertical integration. We follow Joskow (2005) who applies the TCE framework to vertical integration. TCE criticizes the neoclassical perspective for 1) ignoring both the costs of internal organization and more complicated hybrid governance forms; and 2) assuming that vertical integration has no efficiency rationale. Generally, the comparative assessment between the transaction costs of external (market) exchange – discovering prices, and costs of negotiating and enforcing contracts – and those of internal exchange – costs of providing incentives to managers, and bureaucratic costs of monitoring and evaluating performance – is what determines the governance structure. Hence, if a market is unable to mitigate hold-up, contractual arrangements will be devised as a solution. Vertical integration is the most far-reaching alternative available. With low asset specificity, vertical integration entails relatively modest benefits, since transactions without integration are not very problematic. Market provision or an intermediate contractual solution will suffice. High asset specificity, on the other hand, implies that the relationship between parties is very specific, which results in high transaction costs of market exchange. In addition, contractual solutions will not dissipate all ex post hazards.

Mitigating these hazards through vertical integration then results in large benefits to the parties involved. Gas transmission and distribution assets are predominantly of a highly specific nature. This implies that for the gas market, vertical integration definitely entails benefits because it lowers the inefficiencies associated with lock-in. Of course, as set out above, this efficiency rationale must be weighed against the cost of internal organization in order to decide if vertical integration will improve efficiency. If so, then vertical integration with accompanying behavioral regulation is preferable to vertical separation. If not, then it might still be the case that one of the alternative separation options set out in section 3.4.1 is the best option. The upshot is that, as indicated in the previous section, no governance structure can be ruled out beforehand. In particular, vertical integration should not be removed (through ownership unbundling) solely based on its anti-competitive effects. Instead, the efficiencies of retaining integration must be weighed against the costs in terms of obstructions to competition.

We can be concise regarding horizontal restructuring, since the issues are once again similar to those of its vertical counterpart. TCE departs from the neoclassical perspective by 1) recognizing that horizontal integration may have an efficiency rationale through economizing on transaction costs; and 2) explicitly acknowledging the costs of internal organization and hybrid arrangements. Accordingly, a decision to horizontally separate production or trading firms must be based on a comparative assessment between the benefits of enhanced competition and the costs of, for instance, impeded coordination.

In sum, the specific characteristics of the European gas market and its transactions critically determine its preferred organizational structure. This illustrates the departure of TCE from the neoclassical emphasis: rather than as a response to market power or an effort to create market power, vertical and horizontal integration can also occur because they mitigate ex post hazards which might lower transaction costs compared to market exchange or an intermediate contracting solution.

Access

This observation has repercussions for the access regime. With a structurally separated, ownership unbundled network, competitive anxieties are minor and negotiated access is preferred. However, if a less far-reaching separation proves to be the option with the lowest inefficiencies (i.e., transaction costs), the increased fears of competitive abuse may well be best mitigated through a regulated access regime.

This section has argued that the criteria for a proper regulatory scheme that emanate from the market shifts are explicitly incorporated in the TCE perspective in contrast to the neoclassical perspective. Hence, we need to complement the current neoclassically-inspired gas regulation with the TCE insights in order to devise an adequate regulatory regime in the new context. This section has furthermore illustrated that incorporating TCE into the analysis implies substantial revisions to the neoclassical prescriptions, and by extension to the internal European gas market.

6. CONCLUSION

Due to a number of structural market shifts, a new context for regulation has arisen. In this new context, regulation must ensure adequate investments in a gas market that is dominated by producers, where political considerations are becoming more important, and where supply security and sustainability are top policy priorities. In consequence, we require a regulatory policy of which investments, and relatedly, risks, uncertainties, and governance are the pillars. This contrasts starkly with the old situation where a main emphasis for regulation was to sweat existing assets.

Our main question is whether existing gas regulation, which is based on neoclassical guiding principles, will be able to adapt in order to be applicable also in this new context. This chapter argues it does not, because the neoclassical perspective does not explicitly acknowledge the criteria which arise in this new context. Because these criteria are explicitly incorporated in the Transaction Costs Economics (TCE) perspective, this chapter argues that designing a regulatory regime with adequate theoretical underpinnings requires inclusion of the TCE framework into the analysis. Liberalizing from this perspective implies a radically different European gas market than the one implied by the neoclassical perspective as well as the one that is envisaged in existing gas regulation.

Furthermore, due to its perfect competition guiding principles, existing regulation is tautological. That is, by assuming the eventual arrival of a perfectly competitive internal gas market, the risk is that all shortcomings in the development of the internal gas market are attributed to a lack of competition instead of a discrepancy between the internal market and the structural characteristics of the European gas market. The TCE perspective implies that the latter may well be the case, however.

This chapter argues from a theoretical perspective that inclusion of TCE will likely result in better outcomes in terms of market behavior given the new context for regulation. We need to recognize that this theoretical critique is a first step, however. This theoretical critique must be substantiated by empirical observations, because the superiority of the TCE perspective over the current neoclassical perspective can only be proven by illustrating that it better explains and predicts actual regulatory behavior. The above analysis allows us to determine which empirical observations are required. Section 4.3 illustrates that the second Gas Directive has implemented some amendments that may imply a shift towards the TCE perspective: the possibility to exclude certain investments from the third party access provisions and the allowance of a certain extent of long-term contracting. Case studies on both amendments are required that examine the actual regulatory process in order to trace the arguments voiced by the different stakeholders, allowing us to infer which specific changes to regulation have been implemented and why these changes were made. This indicates whether the amendments imply a step back from the neoclassical perspective in actual practice.

If both amendments can nevertheless be explained by the neoclassical perspective, then current regulation is appropriate in the new context, and possible adaptations to gas regulation would be marginal (i.e., build on the neoclassical guiding principles). If not, however, gas regulation's guiding principles should be changed, which requires more fundamental adaptations. The issue then becomes whether the TCE perspective better explains the amendments. If so, this study's theoretical critique is corroborated. The final step would then be to infer whether the current amendments sufficiently incorporate the TCE perspective into European gas regulation.[46]

---

[46] These empirical analyses are outside the scope of this paper. Interested readers are referred to Spanjer's forthcoming Ph.D. thesis which provides empirical case studies on both regulatory amendments that show that both amendments can indeed be considered adaptations to the neoclassical perspective but insufficiently incorporate the TCE perspective due to their unaltered neoclassical guiding principles.

REFERENCES

Aalbers, R.F.T., Dijkgraaf, E., Varkevisser, M. and Vollebergh, H.R.J. (2002). Welvaart en de regulering van netwerksectoren. The Hague: Ministry of Economic Affairs.

Adelman, M.A., and Lynch, M.C. (2002). Natural Gas Supply to 2100. International Gas Union and DRI-WEFA.

Agrell, P., and Bogetoft, P. (2004). Evolutionary Regulation: From CPI-X towards contestability. SUMICSID AB.

Alchian, A.A., and Demsetz, H. (1972). Production, Information Costs, and Economic Organization. *American Economic Review, 62(5)*, 777-795.

Alessi, L. de (1983). Property Rights, Transaction Costs, and X-Efficiency: An Essay in Economic Theory. *American Economic Review, 73(1)*, 64-81.

Al-Moneef, M.A. (2006). A Producer's Perspective of Oil and Gas Supply Security. *IAEE Newsletter*, 15, International Association for Energy Economics.

Altug, S., Demers, F.S., and Demers, M. (2000). Political Risk And Irreversible Investment: Theory And An Application To Quebec. Centre for Economic Policy Research Discussion Paper, No. 2405, London.

Aoki, M. (2001). *Towards a Comparative Institutional Analysis*. Cambridge, Massachusetts: MIT Press.

Armstrong, M., Cowan, S.W. and Vickers, J. (1994). *Regulatory Reform: Economic Analysis and British Experience*. Cambridge, Massachusetts: MIT Press.

Armstrong, M., and Sappington, D.E.M. (2006). Regulation, Competition, and Liberalization. *Journal of Economic Literature, 44(2)*, 325-366.

Arnsperger, C., and Varoufakis, Y. (2006). What is Neoclassical Economics?. *Post-Autistic Economics Review, 38(1)*, 2-13.

Bain, J. (1956). *Barriers to New Competition: their character and consequences in manufacturing industries*. Cambridge, Massachusetts: Harvard University Press.

Baron, D.P. (1989). Design of Regulatory Mechanisms and Institutions. In R. Schmalensee, and R.D. Willig (eds.), *Handbook of Industrial Organization* (volume 2, pp. 1347-1447). Amsterdam, North-Holland: Elsevier

Baron, D.P., and Myerson, R.B. (1982). Regulating a Monopolist with Unknown Costs. *Econometrica, 50(4)*, 911-930.

Baumol, W.J. (1977). On the Proper Cost Tests for Natural Monopoly in a Multiproduct Industry. *American Economic Review, 67(5)*, 809-822.

Baumol, W.J. (1982). Contestable Markets: An uprising in the theory of industrial structure. *American Economic Review, 72(1)*, 1-15.

Baumol, W.J., and Klevorick, A.K. (1970). Input Choices and Rate-of-Return Regulation: an Overview of the Discussion. *Bell Journal of Economics and Management Science, 1(2)*, 162-190.

Berg, S.V. (2001). Infrastructure Regulation: Risk, Return, and Performance. *Global Utilities, 1*, 3-10.

Boiteux, M. (1956). Sur la Gestion des Monopoles Publics Astreints à l'Equilibre Budgetaire. *Econometrica, 24(1)*, 22-40. Translated as Boiteux, M. (1971). On the Management of Public Monopolies Subject to Budgetary Constraints. *Journal of Economic Theory, 3(3)*, 219-240.

BP (2007). BP Statistical Review of World Energy. London: British Petroleum p.l.c. At: www.bp.com/productlanding.do?categoryId=6848andcontentId=7033471.

Bruce, C. (2005). Fraternal Friction or Fraternal Fiction? The Gas Factor in Russian-Belarussian Relations. OIES NG 8. Oxford: Oxford Institute for Energy Studies.

Buckland, R., and Fraser, P. (2001). Political and Regulatory Risk: Beta Sensitivity in U.K. Electricity Distribution. *Journal of Regulatory Economics, 19(1)*, 5-25.

Cameron, P.D. (2005). Completing the internal market in energy: An introduction to the new legislation. In P.D. Cameron (ed.), *Legal aspects of EU energy regulation: implementing the new directives on electricity and gas across Europe* (pp. 7-39), Oxford: Oxford University Press.

Carruth, A., Dickerson, A., and Henley, A. (2000). What do we know about investment under uncertainty?. *Journal of Economic Surveys, 14(2)*, 119-153.

CIEP (2003). The Role of Liquefied Natural Gas (LNG) in the European Gas Market. CIEP 03/2003, The Hague: Clingendael International Energy Programme.

CIEP (2005). The European Market for Seasonal Gas Storage. The Hague: Clingendael International Energy Programme.

CIEP (2006). The paradigm change in international natural gas markets and the impact on regulation. The Hague, The Netherlands: International Gas Union/ The Clingendael Institute

Coase, R. (1937). The Nature of the Firm. *Economica, 4(16)*, 386-405.

Coase, R. (1960). The Problem of Social Cost. *Journal of Law and Economics, 3(1)*, 1-44.

Coase, R. (2006). The Conduct of Economics: The Example of Fisher Body and General Motors. *Journal of Economics and Management Strategy, 15(2)*, 255-278.

CEC (Commission of the European Communities) (1985). Completing the Internal Energy Market. COM(85) 310 final, Brussels: CEC.

CEC (1988). The Internal Energy Market (Commission Working Document). COM(88) 238 final, Brussels: CEC.

CEC (1998). Directive 98/30/EC of the European Parliament and of the Council of 22 June 1998 concerning common rules for the internal market in natural gas. OJ L 204, 1-12, Brussels: EC

CEC (2003). Directive 2003/55/EC of the European Parliament and of the Council of 26 June 2003 concerning common rules for the internal market in natural gas and repealing Directive 98/30/EC. OJ L 176, 57-78, Brussels: EC.

CEC (2004). Third Benchmarking report on the implementation of the internal electricity and gas market. Commission Draft Staff Working Paper, Brussels: CEC

CEC (2005a). Fourth Benchmarking report on the implementation of the internal electricity and gas market. Annual Report on the implementation of the Gas and Electricity Internal Market. COM (2004)863 Final, Brussels: CEC.

CEC (2005b). Green Paper on Energy efficiency or doing more with less. COM(2005) 265 final, Brussels: CEC.

CEC (2005c). Regulation (EC) No 1775/2005 of the European Parliament and of the Council of 28 September 2005 on conditions for access to the natural gas transmission networks. OJ L 289, 1-13, Brussels: CEC.

CEC (2007a). Prospects for the internal gas and electricity market. SEC(2007) 12, Brussels: CEC.

CEC (2007b). DG Competition Report on Energy Sector Inquiry. SEC(2006) 1724, Brussels: EC.

CEC (2007c). Proposal for a Directive of the european parliament and of the council amending Directive 2003/55/EC concerning common rules for the internal market in natural gas. COM(2007) 529 final, Brussels: CEC.

CEC (2007d). EU energy policy data. SEC(2007) 12, Brussels: CEC.

Correljé, A.F., van der Linde, C., and Westerwoudt, T. (2003). *Natural Gas in the Netherlands. From Cooperation to Competition?*. Amsterdam/The Hague: Oranje-Nassau/Clingendael International Energy Programme.

Couwenberg, O., and Woerdman, E. (2006). Shifts in Gas Market Governance: Path-Dependent Institutional Innovation in the Netherlands. In A. Prinz, E. Steenge, and J. Schmidt (eds.), *Innovation: Technical, Economic and Institutional Aspects* (pp. 25-64), Berlin, Germany: Lit Verlag.

CPB Netherlands Bureau for Economic Policy Analysis (2005). Vertical separation of the energy-distribution industry. An assessment of several options for unbundling. CPB Document No. 84, The Hague, The Netherlands.

Creti, A., and Villeneuve, B. (2003). Long-term contracts and take-or-pay clauses in natural gas markets. *Energy Studies Review, 13(1)*, 75-94.

Crew, M.A., and Kleindorfer, P.R. (2006). Regulation, pricing and social welfare. In M.A. Crew, and D. Parker (eds.), *International Handbook on Economic Regulation* (pp. 63-81). Cheltenham, United Kingdom: Edward Elgar.

Crocker, K.J., and Masten, S.E. (1988). Mitigating Contractual Hazards: Unilateral Options and Contract Length. *Rand Journal of Economics, 19(3)*, 327-343.

Crocker, K.J., and Masten, S.E. (1991). Pretia Ex Machina? Prices and Process in Long-Term Contracts. *Journal of Law and Economics, 34(1)*, 69-99.

Crocker, K.J., and Masten, S.E. (1996). Regulation and Administered Contracts Revisited: Lessons from Transaction-Cost Economics for Public Utility Regulation. *Journal of Regulatory Economics, 9(1)*, 5-39.

Dailami, M., and Hauswald. R. (2000). Risk Shifting and Long-Term Contracts. Evidence from the Ras Gas Project. *World Bank Policy Research Working Paper, No. 2469*, Washington, D.C.

Dal Bó, E. (2006). Regulatory Capture: A Review. *Oxford Review of Economic Policy, 22(2)*, 203-225.

Dalen, D.M. (1997). Regulation of Quality and the Ratchet Effect: Does Unverifiability Hurt the Regulator?. *Journal of Regulatory Economics, 11(2)*, 139-155.

Dassler, T. (2006). Combining theories of regulation – Proposing a framework for analyzing regulatory systems worldwide. *Utilities Policy, 14(1)*, 31-43.

Davis, L.E, and North, D.C. (1971). *Institutional Change and American Economic Growth*. Cambridge, New York: Cambridge University Press.

Directorate General for Energy and Transport (DG TREN) (2004a). Third Party Access to Storage Facilities. Interpretation Note, Brussels: DG TREN

DG TREN (2004b). The Unbundling Regime. Interpretation Note, Brussels: DG TREN.

DG TREN (2004c). Security of Supply Provisions for Gas. Interpretation Note, Brussels: DG TREN.

Dixit, A.K., and Pindyck, R.S. (1994). *Investment under Uncertainty*. Princeton, New Jersey: Princeton University Press.

Eberlein, B. (2005). Regulation by cooperation: the 'third way' in making rules for the internal energy market. In P.D. Cameron (ed.). *Legal aspects of EU energy regulation: implementing the new Directives on electricity and gas across Europe* (pp. 59-88), Oxford: Oxford University Press.

Eggertson, T. (1990). *Economic behavior and institutions*. Cambridge, United Kingdom: Cambridge University Press.

Ellis, A., Bowitz, E., and Roland, K. (2000). Structural Change in Europe's Gas Markets: three scenarios for the development of the European gas market to 2020. *Energy Policy, 28(5)*, 297-309.

Estache, A., and Martimort, D. (1999). Politics, Transaction Costs, and the Design of Regulatory Institutions. World Bank Policy Research Working Paper, No. 2073, Washington.

Estrada, J., Bergesen, H.O., Moe, A. and Sydnes, A.K. (1988). *Natural Gas in Europe. Markets, Organisation and Politics*, London: Pinter Publishers.

European Regulators Group for Electricity and Gas (ERGEG) (2005). Roadmap for a competitive single gas market in Europe. E05-SEM-13-03, Brussels: ERGEG.

Finon, D. (2004). European Gas Markets: Nascent Competition and Integration in a Diversity of Models. In D. Finon, and A. Midtun (eds.). *Reshaping European Gas and Electricity Markets. Regulation, Markets and Business Strategies* (183-235), Amsterdam, North-Holland: Elsevier.

Freixas, X., Guesnerie, R., and Tirole, J. (1985). Planning under Incomplete Information and the Ratchet Effect. *Review of Economic Studies, 52(2)*, 173-191.

Geradin, D. (2006). Twenty Years of Liberalization of Network Industries in the European Union: Where Do We Go Now. Inaugural Lecture, Tilburg Law and Economics Center.

Gilbert, R.J., and Newbery, D.M. (1994). The Dynamic Efficiency of Regulatory Constitutions. *Rand Journal of Economics, 25(4)*, 538-554.

Glachant, J.-M., and Perez, Y. (2007). Institutional Economics and Network Industry Deregulation Policy. *Groupe Résaux Jean-Monnet Working Paper, 2007-7*, Paris.

Groenewegen, J.P.M. (2005). Designing Markets in Infrastructures: From Blueprint to Learning. Inaugural Lecture, 27 May, Section Economy of Infrastructures, Faculty TPM, Delft University of Technology

Hallouche, H. (2006). The Gas Exporting Countries Forum: Is it really a Gas OPEC in the Making?. OIES NG 13, Oxford: Oxford Institute for Energy Studies.

Hardt, M. (1995). The non-equivalence of accounting separation and structural separation as regulatory devices. *Telecommunications Policy, 19(1)*, 69-72.

Helm, D (2005). Credible Carbon Policy. In: D. Helm (ed.), *Climate-Change Policy* (pp. 322-339). Oxford: Oxford University Press.

Helm, D. (2006). Regulatory Reform, Capture, and the Regulatory Burden. *Oxford Review of Economic Policy, 22(2)*, 169-185.

Helm, D. (2007a). The New Energy Paradigm. In D. Helm (ed.), *The New Energy Paradigm* (pp. 9-36), Oxford: Oxford University Press.

Helm, D. (2007b). The Russian dimension and Europe's external energy policy. www.dieterhelm.co.uk.

Hodgson, G.M. (1988). *Economics and Institutions: A Manifesto for a Modern Institutional Economics*. Cambridge, United Kingdom: Polity Press and Basil Blackwell.

Hodgson, G.M. (1989). Institutional economic theory: the old versus the new. *Review of Political Economy, 1(3)*, 249-269.

Hodgson, G.M. (2007). Meanings of Methodological Individualism. *Journal of Economic Methodology*, 14(2), 211-226.

Honoré, A. (2006). Natural Gas Demand in Europe – The role of the power sector. Oxford: Oxford Institute for Energy Studies.

Honoré, A., and Stern, J.P. (2007). A Constrained Future for Gas in Europe?. In D. Helm (ed.), *The New Energy Paradigm* (pp. 223-255). Oxford, United Kingdom: Oxford University Press.

Hubbard, R.G., and Weiner, R.J. (1991). Efficient Contracting and Market Power: Evidence From the U.S. Natural gas Market. *Journal of Law and Economics, 34(1)*, 25-76.

Hubert, F. (2007). Strategic investment in international gas transport systems. *EIB Papers, 12(2)*, 62-81. Brussels, Belgium: European Investment Bank Research Series.

International Energy Agency (IEA) (2000). Regulatory Reform: European Gas, Paris, France: IEA.

International Energy Agency (IEA) (2007). World Energy Outlook 2007. Paris, France: IEA.

Jamison, M.A., Berg, S.V., Gasmi, F., and Távara, J.I. (2005). The Regulation of Utility Infrastructure and Services: An Annotated Reading List. Developed for: The World Bank and The Public Private Infrastructure Advisory Board (PPIAF). New York: The World Bank.

Jensen, J.T. (2003). The LNG Revolution. *Energy Journal, 24(2)*, 1-46.

Jensen, J.T. (2004). *The Development of a Global LNG Market. Is it Likely? If so, When?*. Oxford: Oxford Institute for Energy Studies.

Joskow, P.L. (1985). Vertical Integration and Long-Term Contracts: The Case of Coal-Burning Electric Generating Plants. *Journal of Law, Economics and Organization, 1(1)*, 33-80.

Joskow, P.L. (1987). Contract Duration and Relation-specific Investments: Empirical Evidence from Coal Markets. *American Economic Review, 77(1)*, 168-185.

Joskow, P.L. (1988a). Asset Specificity and the Structure of Vertical Relationships: Empirical Evidence. *Journal of Law, Economics and Organization, 4(1)*, 95-117.

Joskow, P.L. (1988b). Price Adjustment in Long-Term Contracts: The case of Coal. *Journal of Law and Economics, 31(1)*, 47-83.

Joskow, P.L. (1990). The Performance of Long-Term Contracts: Further Evidence form the Coal Markets. *Rand Journal of Economics, 21(2)*, 251-274.

Joskow, P.L. (1991). The Role of Transaction Cost Economics in Antitrust and Public Utility Regulatory Policies. *Journal of Law, Economics and Organization, 7 (special issue)*, 53-83.

Joskow, P.L. (1996). Introducing Competition into Regulated Network Industries: from Hierarchies to Markets in Electricity. *Industrial and Corporate Change, 5(2)*, 341-382.

Joskow, P.L. (2005). Vertical Integration. In C. Menard, and M.M. Shirley (eds.), *Handbook of New Institutional Economics* (pp. 319-349). Amsterdam, North-Holland: Elsevier.

Joskow, P.L. (2006). Incentive Regulation in Theory and Practice: Electric Distribution and Transmission. Paper prepared for the National Bureau of Economic Research regulation project. At http://econ.mit.edu/faculty/ download_pdf.php?id=1220.

Klein, B. (2007). The Economic Lessons of Fisher Body-General Motors. *International Journal of the Economics of Business, 14(1)*, 1-36.

Klein, B., Crawford, C., and Alchian, A. (1978). Vertical integration, appropriable rents and the competitive contracting process. *Journal of Law and Economics, 21(2)*, 297-326.

Klein, P.G. (1999). New Institutional Economics. In B. Bouckaert, and G. de Geest (eds.), *Encyclopedia of Law and Economics. The History and Methodology of Law and Economics* (volume 1, pp. 456-489). Cheltenham, United Kingdom: Edward Elgar.

Knight, F.H. (1921). *Risk, Uncertainty, and Profit.* New York: Hart, Schaffner and Marx.

Laffont, J.J. (1994). The New Economics of Regulation Ten Years Later. *Econometrica, 62(3)*, 507-537.

Laffont, J.J. (2005). *Regulation and Development.* Cambridge, United Kingdom: Cambridge University Press.

Laffont, J.J., and Tirole, J. (1986). Using Cost Observation to Regulate Firms. *Journal of Political Economy, 94(3) Part 1*, 614-641.

Laffont, J.J., and Tirole, J. (1993). *A Theory of Incentives in Procurement and Regulation.* Cambridge, Massachusetts and London, England: MIT Press.

Langlois, R.M. (1989). What was wrong with the old institutional economics (and what is still wrong with the new)?. *Review of Political Economy, 1(3)*, 270-298.

Lavrijssen, S.A.C.M. (2004). De European Energy Regulators Group; een panacee voor good governance?. *Het Nederlands Tijdschrift voor Energierecht, 3(5)*, 176-184.

Leibenstein, H. (1966). Allocative efficiency vs. "X- efficiency"". *The American Economic Review, 56(3)*, 392-415.

Lyons, B.R. (1996). Empirical Relevance of Efficient Contract Theory: Inter-firm Contracts. *Oxford Review of Economic Policy, 12(4)*, 27-52.

Madrid Forum (2002). A Long-term vision of a fully operational single market for gas in Europe-a Strategy Paper. Joint Working Group of the European Gas Regulatory Forum, Madrid: EGRF.

Maisonnier, G. (2006). The Ties Between Natural Gas and Oil Prices. *IAEE Newsletter, 15*, 26-29, International Association for Energy Economics.

Mañé Estrada, A. (2006). European energy security: Towards the creation of the *geo-energy* space. *Energy Policy, 34(18)*, 3773-3786.

Marcel, V. (2006). Investment in Middle East oil: Who needs whom?. London, United Kingdom: Chatham House.

Martimort, D. (2006). An agency perspective on the costs and benefits of privatization. *Journal of Regulatory Economics , 30(1)*, 5-44.

Mastenbroek, E. (2007). *The politics of compliance.* Ph.D. thesis, Leiden: Leiden University Press.

Menard, C. (2005). A New Institutional Approach to Organization. In C. Menard, and M.M. Shirley (eds.), *Handbook of New Institutional Economics* (pp. 281-319), Dordrecht, The Netherlands: Springer.

Menard, C., and Shirley, M.M. (2005). Introduction. In C. Menard, and M.M. Shirley (eds.), *Handbook of New Institutional Economics* (pp. 1-18), Dordrecht, The Netherlands: Springer.

Nelson, R.R., and Sampat, B.N. (2001). Making sense of institutions as a factor shaping economic performance. *Journal of Economic Behavior and Organization, 44(1)*, pp. 31-54.

Neuhoff, K., and Von Hirschhausen, C. (2006). Long-Term vs. Short-Term Contracts: A European Perspective on Natural Gas. Cambridge Working Papers in Economics 0539/Electricity Policy Research Group 05, Cambridge.

Neumann, A., and Von Hirschhausen, C. (2004). Less Long-Term Gas to Europe? A Quantitative Analysis of European Long-Term Gas Supply Contracts. *Zeitschrift für Energiewirtschaft, 28(3)*, 175-182.

Neumann, A., and Von Hirschhausen, C. (2006). Long-term Contracts and Asset Specificity Revisited – An Empirical Analysis of Producer-Importer Relations in the Natural Gas Industry. Electricity Policy Research Group 06/10 and MIT Centre for Energy and Environmental Policy Research 06-010, Cambridge.

Newbery, D.M. (1999). *Privatization, Restructuring, and Regulation of Network Utilities*. Cambridge, Massachusetts: MIT Press.

North, D.C. (1981). *Structure and Change in Economic History*. New York, London: Norton.

North, D.C. (1990). *Institutions, Institutional Change and Economic Performance*. Cambridge, Massachusetts: Cambridge University Press.

North, D.C. (1994). Economic Performance Through Time. *American Economic Review, 84(3)*, 359-368.

Odell, P. (2003). European Gas: Prospects. In: A. Correljé, C. van der Linde, and T. Westerwoudt, *Natural Gas in the Netherlands. From Cooperation to Competition?* (pp. 141-144). Amsterdam/The Hague: Oranje-Nassau Groep/Clingendael International Energy Programme.

OECD (2001). Structural Separation in Regulated Industries. DAFFE/CLP(2001)1. Paris, France: Organization for Economic Co-operation and Development.

Percebois, J. (1999). The gas deregulation process in Europe: economic and political approach. *Energy Policy, 27(1)*, 9-15.

Polo, M., and C. Scarpa, C. (2007). Take or Pay Contracts and Market Segmentation. Paper presented at the Conference on The Economics of Energy Markets, Toulouse, France.

Radetzki, M. (1999). European natural gas: market forces will bring about competition in any case. *Energy Policy, 27(1)*, 17-24.

Ramsey, F.P. (1927). A contribution to the Theory of taxation. *The Economic Journal, 37(1)*, 47-61.

Reynolds, D.B., and Kolodziej, M. (2007). Institutions and the supply of oil: A case study of Russia. *Energy Policy, 35(2)*, 939-949.

Rodrik, D. (1991). Policy uncertainty and private investment in developing countries. *Journal of Development Economics, 36(2)*, 229-242.

Röller, L.-H., Delgado, J., and Friederiszick, H.W. (2007). *Energy: Choices for Europe*. Bruegel Blueprint Series, Brussels: Breughel.

Rutherford, M. (1989). What is wrong with the new institutional economics (and what is still wrong with the old)?. *Review of Political Economy, 1(3)*, 299-318.

Salant, D.J., and Woroch, G.A. (1992). Trigger Price Regulation. *Rand Journal of Economics, 23(1)*, 29-51.

Servén, L. (1997). Uncertainty, Instability, and Irreversible Investment. Theory, Evidence, and Lessons for Africa. World Bank Policy Research Working Paper, 1722. Washington, D.C.: World Bank.

Shelanski, H.A., and Klein, P.G. (1995). Empirical Research in Transaction Cost Economics: A Review and Assessment. *Journal of Law, Economics, and Organization, 11(2)*, 335-361.

Shell (2005). Shell global Scenarios to 2025. The future business environment: trends, trade-offs and choices. London: Shell International Limited.

Simon, H. (1957), *Models of Man*, New York: John Wiley and Sons.

Söderholm, P. (1998). Fuel choice in West European power generation since the 1960s. *OPEC Review, 22(2)*, 201-231.

Spanjer, A.R. (2007). Russian gas price reform and the EU-Russia gas relationship: Incentives, consequences and European security of supply. *Energy Policy, 35(5)*, 2889-2898.

Spanjer, A.R. (forthcoming). *Structural and regulatory reform on the European gas market – Does the current approach secure the public service obligations?*. Ph.D. thesis, Leiden: Leiden University Press.

Spiller, P.T., and M. Tomassi, M. (2005). The Institutions of Regulation: An Application to Public Utilities. In C. Menard, and M.M. Shirley (eds.), *Handbook of New Institutional Economics* (pp. 515-543). Amsterdam, North-Holland: Elsevier.

Stern, J.P. (1998). Competition and Liberalization in European Gas Markets. A Diversity of Models. London: Royal Institute of International Affairs.

Stern, J.P. (2002). Security of European Natural Gas Supplies: The impact of import dependence and liberalization. London: Royal Institute of International Affairs.

Stern, J.P. (2005). *The Future of Russian Gas and Gazprom*. Oxford: Oxford University Press.

Stern, J.P. (2006). The Russian-Ukrainian gas crisis of January 2006. OIES Comment 01/06. Oxford: Oxford Institute for Energy Studies.

Stigler, G.J. (1971). The Theory of Economic Regulation. *Bell Journal of Economics and Management Science, 2(1)*, 3-21.

Veblen, T.B. (1909). The Limitations of Marginal Utility. *Journal of Political Economy, 17(9)*, 620-636.

Wagbara, O.N. (2007). How would the gas exporting countries forum influence gas trade?. *Energy Policy, 35(2)*, 1224-1237.

Weitzman, M.L. (1980). The "Ratchet Principle" and Performance Incentives. *Bell Journal of Economics, 11(1)*, 302-308.

Williamson, O.E. (1971). The Vertical Integration of Production: Market Failure Considerations. *American Economic Review, 61(2)*, 112-123.

Williamson, O.E. (1975). *Markets and Hierarchies: Analysis and Antitrust Implications*. New York: Free Press.

Williamson, O.E. (1985). *The Economic Institutions of Capitalism*. New York: Free Press.

Williamson, O.E. (1996). *The Mechanisms of Governance*. New York, Oxford: Oxford University Press.

Williamson, O.E. (1998). Transaction Cost Economics: How it works; Where it is headed. *De Economist, 146(1)*, 23-58.

Williamson, O.E. (2005). Transaction Cost Economics. In C. Menard, and M.M. Shirley (eds.), *Handbook of New Institutional Economics* (pp. 41-65). Amsterdam, North-Holland: Elsevier.

Zanoyan, V. (2004). Institutional Cooperation in oil and Gas: Governments, Companies and the Investments Climate. Paper presented at the 9th International Energy Forum, 22-24 May, Amsterdam, The Netherlands.

In: Energy in Europe: Economics, Policy and Strategy ISBN: 978-1-60456-829-5
Editors: F. L. Magnusson and O. W. Bengtsson.

Chapter 9

EFFICIENT HEAT RESOURCE UTILISATION IN ENERGY SYSTEMS

Maria Danestig[1] and Dag Henning[2]
1. Department of Management and Engineering (IEI), Energy Systems Linköping University, SE-581 83 Linköping, Sweden
2. Optensys Energianalys, Örngatan 8C, SE-582 37 Linköping, Sweden

ABSTRACT

In cold-climate countries, energy for heating of buildings is essential and heating options that interplay with the power system through electricity use or generation have potential for efficiency improvements. In Sweden, district heating is used extensively, especially in large buildings but to a growing extent also for small houses. Since rising electricity prices in recent years cause problems for the extensive use of electric heating, further switching to district heating should be a possible option. To be economically favourable, district-heating systems require a certain heat load density, which depends on heat production cost. New low-energy houses and energy-efficiency measures in existing buildings decrease the heat demand in buildings and, thus, in district heating systems. In addition, higher outdoor temperatures due to the enhanced greenhouse effect may reduce heat demand and district-heating sales substantially. Decreased heat demand should decrease the use of natural resources, such as fossil fuels, and enable redirections of bio-fuels into other fields of applications, such as production of automotive fuel. But the consequences of changed heat demand are complex when the heat is supplied from resources that can be seen as surplus, for instance heat from waste incineration or industrial waste heat. Increased electricity prices and policy instruments that promote renewable electricity generation have caused much interest in investments in new combined heat and power plants in the already well-developed Swedish district-heating systems. The opportunities for combined heat and power production rearrange the view of how the energy system should be evaluated. If district heating is produced in a combined heat and power plant, the heat demand enables electricity generation. Such effects can be studied with an energy system approach, which tries to go beyond the details in order to get a grip of the most important factors. Optimisation models have been used in several studies of large and complex energy systems. With the models, scenarios with changing policy instruments and changed consumer behaviour can be analysed. New applications for waste heat, such as using heat for absorption cooling, and switching from electricity and oil to district heating in industrial processes can also be studied. Energy efficiency measures as well as good conditions for efficient electricity generation, which can replace old and inefficient plants, are needed to reduce carbon dioxide emissions from the energy sector.

INTRODUCTION

This chapter aims at a systems approach to energy supply and use, focussing district heating (DH). Combined heat and power (CHP) couples DH and electricity and makes them dependent of each other at the same time as energy resources are very well utilised. The electricity sales market in Sweden is deregulated and the aim is to create a competitive deregulated common electricity market in Europe (Directive 2005/89/EC, Regulation (EC) No 1228/2003, Directive 2003/54/EC).

The efficiency in CHP production means that about 70-95% of the supplied fuels will be used in the forms of heat and electricity, where the electricity share varies between 20 and 50% depending on the fuels and technologies chosen. In conventional electricity generation in condensing plants, the efficiency is about 35-55% with electricity as the single useful output.

District heating can also be derived from other resources that would have been wasted otherwise, for instance, industrial surplus heat and heat from waste combustion. Different *waste* or surplus heat resources may be available and may *compete* with each other to cover the demand in a district heating system and that is one issue where an energy systems approach can be useful by an analysis.

Several energy studies have been done that apply a systems approach to broaden the view of, for instance, economic conditions and physical boarders within a building, municipality or nation. Energy system models can be useful tools when analysing how changes due to, for example, new policy instruments affect an energy system. Applying a systems approach with models can help analysing possible changes in global carbon dioxide (CO_2) emissions due to local energy-system changes and provides knowledge of, for instance, problems when means for decreased CO_2 emissions in one sector lead to increased emissions in another sector. Hughes et al. (2005) described how the shift from oil to electricity in the residential sector decreases emissions from residential energy use by 4.1% but at the same time increases emissions in the electricity and heat supply sector by 8.2%. This is due to an increase in fossil-fuelled electricity generation (Hughes et. al. 2005). Electricity generation in Sweden is basically free from CO_2 emissions but when assuming a common European electricity market, the entire system's electricity generation is dominated by fossil fuels and, thus, has major CO_2 emissions. Because the energy sector is one of the largest anthropogenic causes to greenhouse gas (GHG) emissions, through CO_2, means for decreasing wasted energy resources should be important when considering the problems with global warming.

The Intergovernmental Panel on Climate Change reported in November 2007 (IPCC, 2007) that it is now evident that warming of the climate system is unmistakable. There is very high confidence that human activities, since 1750, have caused warming. Most of the observed increase in globally-averaged temperatures since the mid-20th century is very likely due to the observed increase in anthropogenic GHG concentrations. It is likely that there has been significant anthropogenic warming over the past 50 years averaged over each continent (except Antarctica). Continued GHG emissions at or above current rates would cause further warming and induce many changes in the global climate system during the 21st century that would very likely be larger than those observed during the 20th century. Effects of global warming is contraction of snow cover area, decrease in sea ice extent, increases in thaw depth over most permafrost regions, increase in frequency of hot extremes, heat waves, and heavy precipitation, as well as increase in tropical cyclone intensity. Precipitation increases in high latitudes and likely decreases in most subtropical land regions etc. (IPCC, 2007).

The leaders in the European Council have agreed on a comprehensive package of measures, concerning climate change and energy policy, which points out that the global warming problem is taken seriously. The targets is to: Reducing greenhouse gas emissions in the EU by 20% by 2020, and by 30% if an international agreement is reached, improving energy efficiency by 20% by 2020, raising the share of renewable energy to 20% by 2020 and increasing the level of bio-fuels in transport fuel to 10% by 2020.

Heating and Interplay with Electricity Supply and Use

In Euroheat and Power's survey (Euroheat and Power, 2007) of DH and cooling, the total heat demands in the surveyed European Union[1] (EU) and EFTA[2] countries represent 5425 TWh/year. Germany has the largest national share with about one fourth of the total EU and EFTA heat demand (Euroheat and Power, 2007). District heating covers 10% of the total net heat demand in the industrial, residential and service sectors in EU and EFTA and the average share of the DH that is produced in CHP plants is 63% (Euroheat and Power, 2007).

In EU member countries the heat demands for the residential and industrial sectors have the same range of magnitude and account for 43% and 39% respectively with a slightly higher share for the residential than the industrial sector in the old member states. The service sector accounts only for 13% of the total heat demand while the agricultural sector represents only 5% (Euroheat and Power, 2007).

District Heating

District heating means that one or several larger plants (e.g. woodchips-fired boilers) produce hot water that is distributed through a pipe network to consumers who buy the heat from the supplier. District heating can be used for space heating, hot-water preparation and low-temperature heat demand in industrial processes, but also for cooling generation through absorption cooling.

There have been DH systems in the United States for 130 years and DH has recently expanded significantly in Korea and China. Global DH supply is 2 800 TWh/year and much of it takes place in Russia (Werner, 2006b).

---

[1] EU; Austria, Belgium, Bulgaria, Cyprus, Czech Republic, Denmark, Estonia, Finland, France, Germany, Greece, Hungary, Ireland, Italy, Latvia, Lithuania, Luxembourg, Malta, Netherlands, Poland, Portugal, Romania, Slovakia, Slovenia, Spain, Sweden and United Kingdom.

[2] EFTA; Iceland, Liechtenstein, Norway and Switzerland

Werner (2006b) has investigated the DH in the EU, Croatia, Iceland, Norway and Switzerland in 2003. District heat supplies 530 TWh/year in more than 5000 DH systems in the surveyed region. Sixty percent of the DH is delivered to residential buildings, and service premises and industries each consume about 20% of the heat (Werner, 2006a).

DH is more common in Eastern Europe than in western. Poland (95 TWh/year) and Germany have the largest DH sales (Euroheat and Power, 2007). Much DH is produced per capita in Denmark, Finland, Sweden, the Baltic countries and the Czech Republic. In Iceland, DH covers almost 90% of the total heat demand (Euroheat and Power, 2007). Figure 1 shows the fraction of the *total* heat demand in residences, service and industry that is covered by DH in some countries.

Natural gas is the most common energy source for DH in most European countries except Poland, Slovenia and the Czech Republic where coal dominates, Switzerland, Norway and Italy utilising waste incineration, Iceland using almost exclusively geothermal heat and Sweden having the most diversified supply with biofuel as the number one source. Biofuels also make significant contributions in the Baltic States. In the rest of Eastern Europe, little renewable sources are used. In Germany, coal and natural gas each produce almost one-half of the DH. Denmark has a mixture of natural gas, coal, waste and biofuel (Euroheat and Power, 2007).

Oil supplements the main DH sources in most countries. Twenty percent of the district heat comes from heat-only production with fossil fuels, especially in Eastern Europe and France.

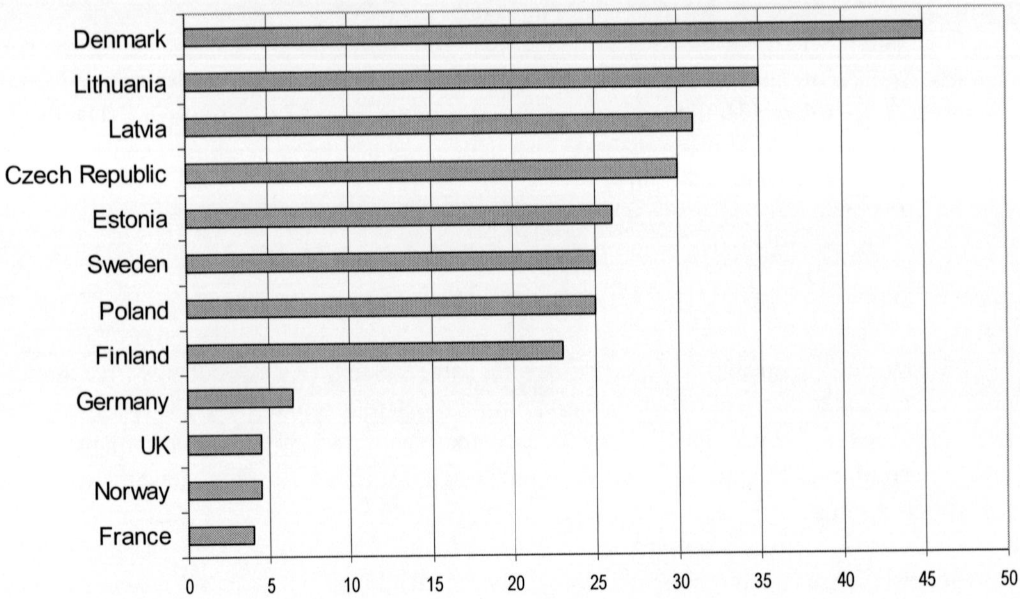

Figure 1. Share (%) for DH of the total heat demand (Euroheat and Power, 2007).

Electricity is used for DH production in large heat pumps and even boilers in countries that have had low electricity prices due to extensive hydropower production (Norway,

Sweden, Iceland). Industrial surplus heat is almost only used for DH in Sweden and comes primarily from metal, pulp and paper and chemical industries (Werner, 2006b).

DH expanded somewhat in many countries in Western Europe during the 1990s but not in Germany and France (Werner, 2006b). Now, DH grows significantly in, for example, Austria, Italy and Norway. In eastern-Europe, DH deliveries decrease due to reduced industrial demand and more energy-efficient buildings (Euroheat and Power, 2007).

The CO_2 emissions per kWh of delivered DH are low where CO_2 lean energy sources are used, such as waste in Switzerland and Norway, geothermal heat in Iceland and biofuel in Austria and Sweden. CO_2 emissions are high in eastern European countries with much coal combustion and high distribution losses (e.g. Poland, Czech Republic, Werner, 2006b).

In Sweden, district heating is used extensively, especially in large buildings but to a growing extent also for small houses. Figure 2 shows how DH was produced in Sweden in 2004. There are several sources, which mainly are renewable or sources that are difficult to utilise without a DH system, such as waste heat and heat from waste incineration. Sweden has had high taxes on fossil fuels for heat-only production for many years. Therefore, less than 20% of the DH derives from fossil fuels. In total 60 TWh of fuels etc. were used in 2004 to supply 48 TWh of heat to consumers, which was somewhat lower than normal due to higher outdoor temperatures. Twenty-five TWh of the heat were supplied to multi-family buildings, 4 TWh to detached houses, 5 TWh each to industries and public premises and the rest to other customers (e.g. shops, SDHA, 2006). DH covers one-half of the heat demand within households and service (SEA, 2007).

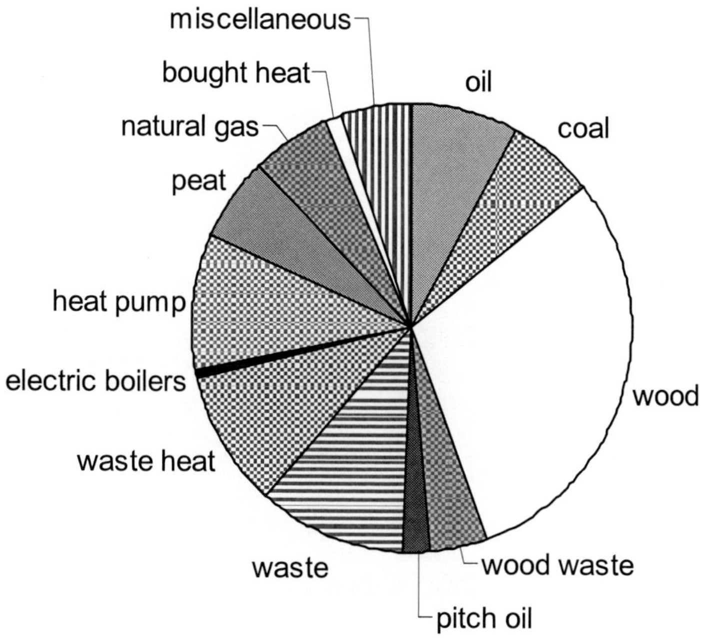

Figure 2. Swedish DH production in 2004 (SDHA, 2006).

The characteristics of a typical DH demand curve in Sweden is dominated by a heat-load component depending on outdoor temperature and a socially dependent heat-load component.

Heat demand increases while outdoor temperature decreases and the heat load demand curve has a typical characteristic following the seasons during a year. The social component depends on the habits that different heat users have, primarily hot-water use. The design of DH systems in Sweden has been developed according to the types of customers and the fact that most of the heat is used for heating of buildings. This means that the system in principle has base heat load production plants which are designed to be run as many hours as possible, with low running costs and often high investment costs. In Figure 3 it is waste incineration. The high investment cost limits the capacities for base load plants and the DH system has to be complemented with other plants, which normally have a merit order according to cost. A wood and coal fuelled CHP plant and a wood-fired heat-only boiler are shown in Figure 3. Peak heat load plants are run only a few days a year, in the coldest winter days (oil-fired boilers in Figure 3). These plants must have low investment costs because the running time is short and the running costs are high due to use of expensive refined fuels.

Normally, a Swedish district-heating company was founded by the municipality where it operates but now the ownerships of many companies have been transferred to large private or state-owned energy companies, such as Eon and Vattenfall, respectively. Irrespective of owner, the individual DH company operates in a local area, where price setting is based on local conditions. Customers can choose among several heating techniques but mostly there is just one DH supplier. It gives the DH market (but not the heat market) the characteristics of a monopoly, often called a natural monopoly because the large investments in the infrastructure of pipelines and energy conversion units make it impossible to build parallel systems.

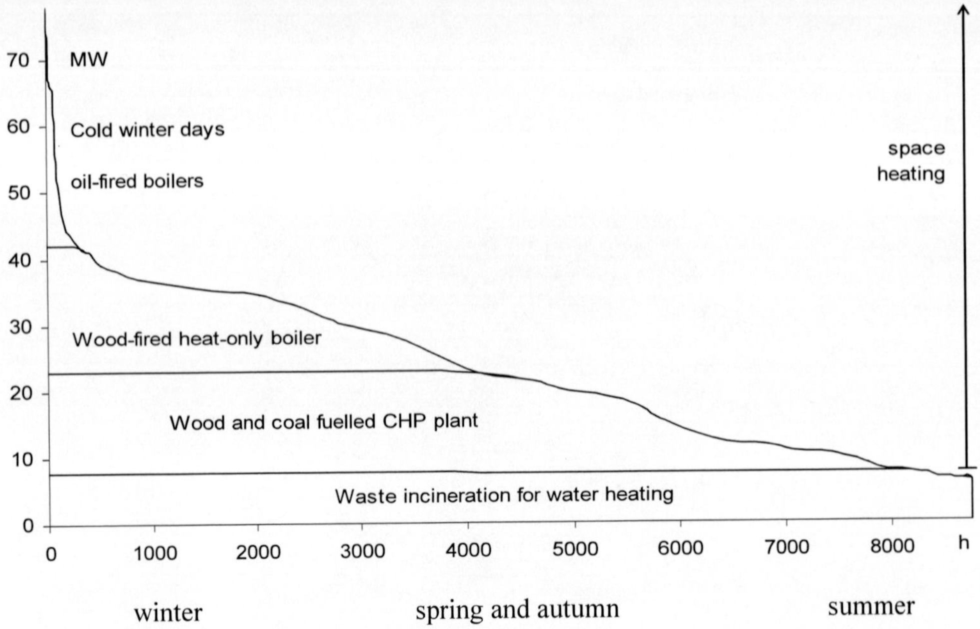

Figure 3. Demand duration curve for a Swedish district-heating system and how heat can be produced during a year.

The economic conditions for district heating are dependent on the density of the heat demand in a geographical area

A recent project was carried out in co-operation between a local district heating utility, spatial-planning researchers, energy researchers, as well as municipal employees and politicians, and aimed at integrating energy issues in the spatial planning process in a municipality (Henning and Danestig, 2007, 2008). District heating dependent parameters and typical spatial planning factors were compared. The geographical area of a municipality was divided in 65 districts (Figure 4) according to the characteristics of the sub areas (districts). The division was, for instance, based on type, age and style of buildings. Multi- and single family houses etc. were separated. In the studied municipality, most houses are small. If multi-family houses can be built in new areas, it may be possible to expand the present district heating network. Each district's present heating solutions and heat load density (i.e. heat demand divided with geographical area, kWh/m^2,year) were analysed.

The heat load density is crucial for a local district heating utility because it indicates if it is economically possible to introduce district heating in an area. The density that is required for making DH profitable depends on whether the heat can be produced for a low cost through, for example, waste incineration or if expensive heat-only production with fossil fuels dominates. The heat load density is coupled to the floor space index (total floor area divided by land area for a district). For each district, heat load density and floor space index were compared (Figure 4). The varying relations between the two parameters can be due to building year and building and energy standards but also the user's preferred hot tap-water use and indoor temperature. Low-energy buildings and extensive energy-efficiency measures in certain districts should increase deviations (Henning and Danestig, 2007, 2008).

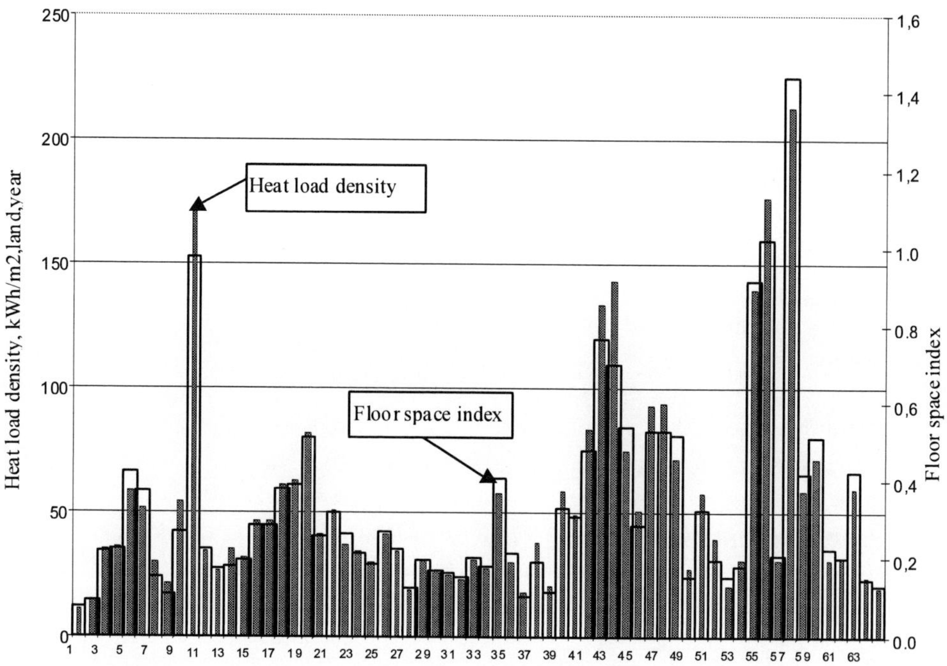

Figure 4. Heat load density and floor space index for 65 districts in the Swedish municipality Ulricehamn (Henning and Danestig, 2008).

ELECTRICITY GENERATION, MARKET AND USE

An ambition in the EU is to establish a common market for electricity, which emphasises the continuous interplay among electricity supply and electricity use in Sweden and the surrounding world. Decreased electricity use and enhanced power generation in Sweden should in this system lead to reduction of operation and CO_2 emissions from coal-fired condensing power plants in continental Europe. This is because such plants are mostly the units with highest operation costs that are committed in northern Europe, which makes them the marginal power generation method. When old coal-fired condensing power plants are being phased-out, the marginal generation will be natural-gas-fuelled condensing plants. Coal now covers about 30 % of the total electricity generation in Europe, Figure 5.

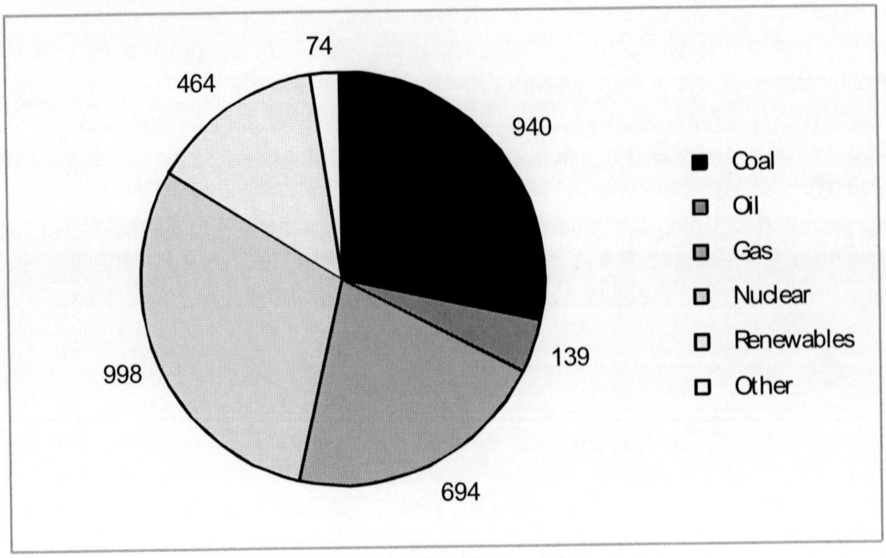

Figure 5. Electricity generation in EU-27, TWh in 2005 (EC, 2008).

Electricity generation in Sweden consists of about 50% nuclear and 50% hydro generation. From the mid 1970s, the electricity supplies have been good, which has been the basis for a large per-capita use of electricity due to low electricity prices. For example in the pulp and paper industry, electricity-based mechanical manufacturing processes have been developed and used instead of chemical processes, which are most common in the rest of the world. It has also been shown that electricity use in Swedish industry is more than twice the use in similar industries in other European countries (Nord-Ågren, 2002). Industry uses about 40% of the total electricity in Sweden, Figure 6. Electricity is also common for heating of buildings in Sweden, even in the cold climate areas with high heating demand. About 14 % of the total Swedish electricity consumption is used for heating of buildings, mainly through direct electric heating and heat pumps. Since Sweden became a member of the European Union, the energy sector has also been affected by the work in EU. The opening up of the electricity market to increase the competitiveness according to EU ambitions has already affected and will probably affect the Swedish electricity market drastically. The Swedish electricity market is only about 5 % of the total European electricity market.

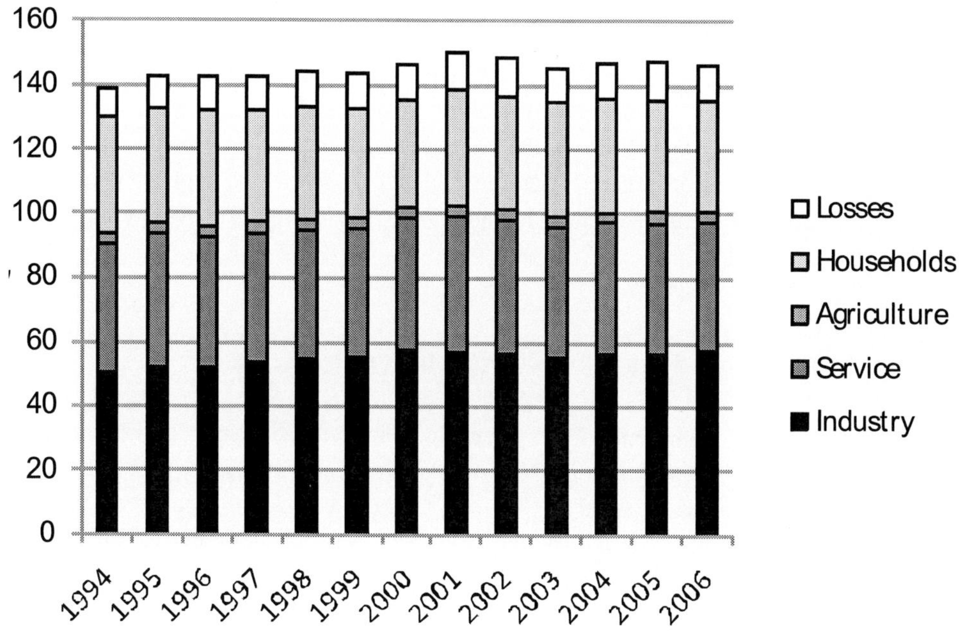

Figure 6. Electricity use in Sweden including distribution losses (TWh/year, SCB, 2008).

District heating has during the 1980s and 1990s partly been supplied from electricity but this is now less common since the electricity prices have increased in recent years. Because electricity prices in Sweden historically have been low, electricity generation in combined heat and power (CHP) plants has not been developed in the same way as in other countries with a high degree of district heating.

Combined Heat and Power (CHP)

The opportunities for combined heat and electricity generation in CHP plants can increase the efficiency of power generation from 35-55 % up to 70-95 % since both heat and electricity output can be utilised. CHP expansion has also the advantage that the energy supply now used for heating can be avoided or used for other purposes. The current annual heat losses from electricity generation in the European Union, Croatia, Iceland, Norway and Switzerland are 4 400 TWh to be compared with the total heat demand in the EU and EFTA of 5,425 TWh (Euroheat and Power, 2007). In Sweden, the heat demand and electricity use are substantially higher in winter than in summer, which gives a mutual benefit for the CHP production. New useful heat applications in the warm seasons would increase the benefits from CHP.

Only about 30% of the district heating comes from CHP in Sweden, despite the beneficial conditions for heat demand. The same pattern can be seen in, for instance, France, which in common with Sweden has a high fraction of nuclear electricity. In the UK, Portugal and Benelux a high fraction of the district heating is produced in CHP plants. In total, 200 TWh of electricity are annually generated in CHP plants in the European Union, Croatia, Iceland, Norway and Switzerland (Werner, 2006b).

Among EU-15 countries, Netherlands and Germany also have high CHP shares in district heat generation (95% and 83%, respectively), mainly due to existing support schemes in these countries. In Central and Eastern Europe, the countries with the lowest shares of CHP heat in the district-heating production still are Estonia and Latvia while the highest percentage of CHP in district heating is registered in Hungary and the Czech Republic (Euroheat and Power, 2007).

AN ENERGY SYSTEMS APPROACH

Churchman (1968) recommended a "system approach" to be able to solve a number of problems that in principle can be solved by modern technology but so far has remained unsolved. They are unsolved because we are not organised to solve them even though the problems often are interconnected and overlapping each other. System analyses are a way of thinking about a total system and its components, an approach to the whole system, not a number of methods. To make use of a system approach, the studied system has to be defined. Churchman described a system as consisting of sets of components that work together for the overall objective of the whole, and he outlined five basic considerations that must be considered when defining a system:

- The total system's objective, specifically the performance measures of the system
- The systems environment, the constrains that cannot be affected by the system
- The resources within the system
- The components of the system, their activities, goals and measures of performance
- The management of the system

In energy studies, it is important to encompass such a large energy system as possible. A solution that seems suitable for one single building may not be sustainable for the whole system. A system boundary is sometimes defined as the border where what is inside the system depends on what an actor controls and the parts that he does not control are outside the system boundary.

When a specific issue (e.g. sustainability) is in focus, the first task must be to define what this expression means. Sustainability can be defined by several assumptions, such as, that it is supported by renewable energy sources, decreased CO_2 emissions, acceptable economic conditions, security of supply etc. When these assumptions are set, the work with how to define the system being studied can be done. In addition to the technical components in the system, questions on what can be affected in the society as well as in private will now also affect where the system boundary is being set. The research area in socio-technical systems addresses this type of questions and also encompasses the field of human nature.

Energy system studies according to a systems approach referring to Churchman has, for instance, defined the system boundary as a European energy system, mainly based on the facts that the European countries have a common commitment under the Kyoto protocol and several common rules concerning the electricity market are prescribed by the EU (Trygg, 2006). In some studies, the analyzed energy systems are Swedish industries and in some studies the system consists of both industries and energy suppliers. A generally formulated objective of the energy system can be to find the most cost-effective energy supply with low environmental pollution (Sjödin, 2003). Different perspectives on a system can change who is the manager of the system. It can be delicate to identify the management of the system if the objective is analysing what is most suitable for society as a whole. In this case, a system study can offer guidance to decision makers on a national level and increase general knowledge (Holmgren, 2006b). The system's environment consists, for instance, of institutional regulations, such as taxes, fees, fuel prices, emission trading and electricity certificates that cannot be affected by the system manager.

ENERGY SYSTEMS FOR HEATING

System boundaries can be different for actors with different interests in the system, for instance, when considering economy. The technical energy system for buildings can be described as in Figure 7. Economy for a private person is at the system border 1 where information about the economy for investing in surplus insulation or in new energy efficient windows can be experienced and evaluated. However, when considering economy and environmental effects for the society as whole other system boundaries can be necessary (Persson et al., 2005). It is recommended to use two different system boundaries: number 1 and 5 in Figure 7. These two boundaries can complement each other. Number 1 is used to compare energy efficiency of the building constructions for different buildings and number 5 gives information about the type of primary energy supply used.

Required primary energy for various energy supply alternatives should be compared. Primary energy could be approximated by the energy content in fuels, if energy required for refining and transportation is neglected, but energy losses by electricity generation are substantial. Hydropower losses can be neglected but in condensing power plants only about one-third of the fuel energy is utilised. Therefore, the primary energy use is three times the electricity use for European marginal electricity, which normally comes from coal-fired condensing plants. For the same reason, the primary energy is three times the electricity use for the half of Swedish electricity that comes from nuclear plants. Hydroelectricity equals the primary energy required. Therefore required primary energy on average is twice the electricity use for Swedish electricity.

The perspective with system boundary 5 illustrates the potential for energy efficiency in buildings as well as the possibilities for efficiency of energy supply.

The geographical boundaries can, when district heating is present, be at the same borders as the possible development of district heating. Outside the system boundary, there are factors that cannot be affected by the management of the system but will have effects on the system. This can, for example, be energy prices, EU legislation, national policies etc. Energy systems for heating have different preconditions, such as possibilities for utilisation of industrial surplus heat and heat from waste combustion, as well as a more or less well developed district heating infrastructure.

Many problems with local energy systems are related to the issue that it is difficult to say who the manager of the system is. If we limit the system only to comprise a building, it is mostly the owner of the building who decides which heating alternative that will be chosen.

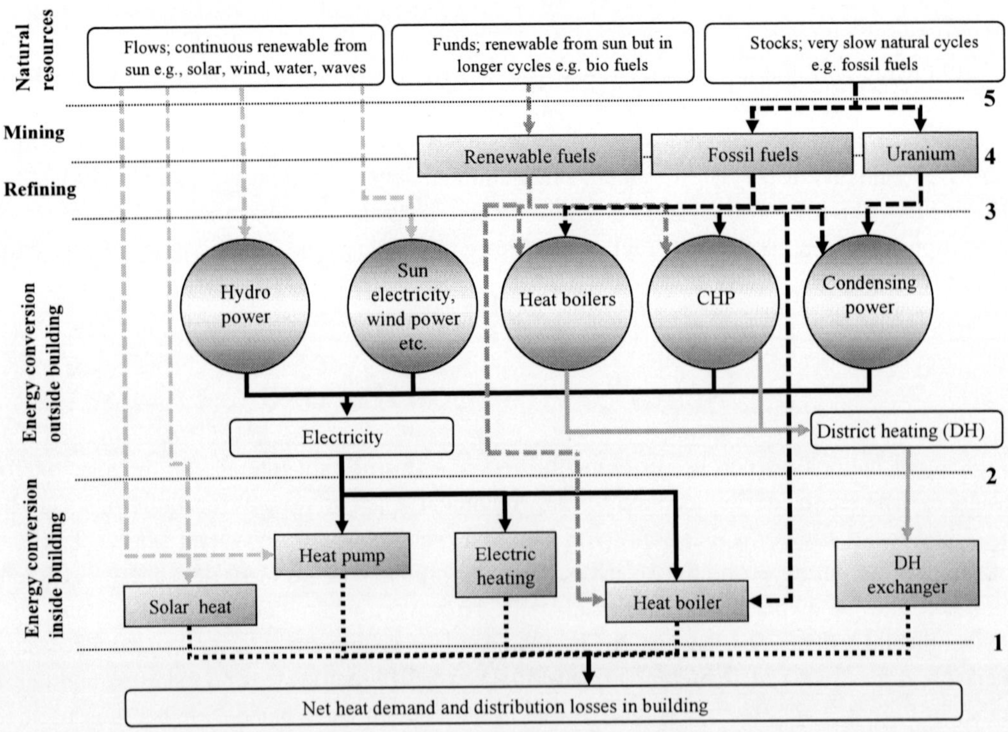

Figure 7. Alternative heating and system borders (cf. Persson et al., 2005).

If the studied system is restricted only to individual buildings, it is not possible to see the consequences concerning, for example, CO_2 emissions. What arises in , for example, a district heating system as consequences of the choices that different real-estate owners make is neither visible. A larger energy system, for example within a whole municipality, has several different owners that influence the system in various ways. A local government may contemplate how their officers best influence various actors in the municipality.

Analysing with Models

Through systems analysis, it may be determined how the available resources within a system boundary should be utilised to satisfy the aims in the best possible way. A systems analysis should include goal definition, information gathering, identification and comparison of alternatives and normally model calculations and result interpretations. Models are advantageous tools for investigating various possibilities under the influence of different conditions. Computer models can treat complex correlations between phenomena and helps in finding beneficial system designs (cf. Henning, 1999). By model construction, the important parameters and their relations must be known and it must be assumed that they remain valid or how they change. With respect to the issues to elucidate, model input must be complete, correct and consistent. It must be scrutinised that the output is reasonable, especially for prevailing or extreme conditions.

An energy model may reflect the planned economy of a single company or a market of many consumers that may chose among energy supply options and who may find energy conservation, energy-carrier switching and load management profitable compared to the energy supply cost. Environmental impact can influence model results through emission limitations and estimated external costs (e.g. Carlson, 2003). Through scenarios and sensitivity analyses, the influence of varying conditions should be analysed and robust strategies may be found. A model interface should facilitate testing of unlikely conditions that may show favourable but not obvious solutions, such as co-operation beneficial to many actors. Besides results, modelling gives insights to the involved persons and improves the understanding of system behaviour (cf. Wene and Rydén, 1988), which may even be more valuable than the, after all, uncertain result numbers with limited lifetimes. Long-term studies are naturally problematic because the future is unknown. For example, the foundations for costs may include new elements (e.g. environmental externalities). The normally used discounting with an interest rate reduces the cost and, hence, the importance of future activities (Groscurth and Schweiker, 1995). To be useful, a model must prove to be an adequate tool for solving the issues under study but many factors outside models must certainly be considered by decision making (cf. Henning, 1999).

Optimisation models have been used in several studies of large and complex energy systems. Energy system optimisation models can have great technical detail of energy sources and supply and demand-side technologies (Henning, 1999). Such models can (consider many component combinations) make a simultaneous consideration of the multitude of component configurations that may be constructed with the available menu of equipment. Energy system optimisation models mostly use the optimisation method linear programming to select technologies and operation according to least total cost, under consideration of constraints, such as limitations on capacities and fuel supplies. The models present the system design and operation that best serve the objective, which normally is minimum cost for satisfying useful energy demand (Henning, 1999). Some energy system optimisation models are MARKAL (e.g. Unger and Ekvall, 2003), EFOM (e.g. Holttinen and Tuhkanen, 2004), TIMES (e.g. Remme et al., 2001) and MODEST (Henning, 1999; Gebremedhin, 2003; Henning et al., 2006a).

MODEST minimises the cost of satisfying demand. The model can consider diurnal, weekly, seasonal and long-term time fluctuations of, for example, demand. MODEST has been used for analyses of electricity and DH production in many towns (e.g. Gebremedhin and Moshfegh, 2004; Sjödin and Henning, 2004). Studies often consider policy instruments (Henning et al., 2006b), such as taxes (Holmgren and Gebremedhin, 2004), electricity certificates (Henning et al., 2006a) and emission allowances (Holmgren, 2006a).

SCENARIOS

Through work with scenarios, pictures that differ from today's system can be created.

Because conditions vary among different places, such as municipalities, countries, regions, industries etc, general recommendations can be difficult to use. Adapted scenarios can serve as better starting points for discussions and further investigations. Börjeson et al. (2006) describe three questions that scenario users can pose: What will happen? , responded to by predictive scenarios. What can happen? , responded to by explorative scenarios. How can a specific target be reached? , responded to by normative scenarios. In the Swedish Energy Agency's project Sustainable municipalities (Henning and Danestig, 2007, 2008) the aim has been increased sustainability, which addresses the third question. But in practice it was found to be important to address all of the three questions. Probable alternatives were investigated, which included predictions of what can be realised in a municipality. This means that scenarios that are partly predictive and partly explorative can be created. They can increase knowledge, initiate interaction among researchers and practitioners and establish an understanding of sustainability. Scenarios could show sustainable solutions that are acceptable for local stakeholders, but the necessary steps toward the desired development must also be described and discussed.

FUTURE HEATING SYSTEMS

In the former Sections, the possibilities for increased efficient use of resources for heating in large have been glanced at. In the following, experiences and learning from different studies are presented in order to support how to work with efficient energy solutions and present further possibilities.

DISTRICT HEATING AND CHP

District heating (DH) can expand significantly in Europe. Werner (2006b) investigated the possible prospects for DH in the EU, Croatia, Iceland, Norway and Switzerland. He estimated the potential additional annual district heat sales in these countries to 1 900 TWh. This expansion should primarily take place in the UK, France and Germany with low current market shares for DH.

The five major heat source options (cf. Werner, 2006b), which can be utilised, are:

- heat from electricity generation through CHP production (cogeneration)
- heat form waste incineration
- surplus heat from industrial processes
- geothermal heat
- combustible renewable energy sources, primarily biomass (e.g. wood), which are difficult to handle in small boilers

The current heat losses from electricity generation are 4 400 TWh/year in the analysed countries and can be substantially reduced through expansion of CHP production (Werner, 2006b). CHP can most easily expand in countries with much DH but little CHP, such as Sweden and the Baltic countries. DH from waste could primarily expand where a lot of heat from existing waste incineration is not utilised, for example in France, Belgium and the Netherlands, and in countries where waste incineration is little used at all, such as Poland and the Baltic countries. In total, at least another 60 TWh of DH should annually be possible to derive from waste in the studied area . If industrial surplus heat would be utilised in Europe to the same extent as in Sweden, there are annually 300 TWh of available heat, of which at least 60 TWh should be possible to use as DH supply (Werner, 2006b). A potential future source is the excess heat from plants for automotive biofuel production.

In the EU, Croatia, Iceland, Norway and Switzerland, there are annually at least 80 000 TWh of geothermal heat with sufficient temperature for direct use as DH that are not used today, for example in Hungary, Germany and France. Annually, 7 TWh or more thereof should be possible to utilise. The annual wood growth in forests in the area under study corresponds to 1 200 TWh but is needed also for pulp and timber. There should be forest residues available as DH fuel in, for example, Ireland and the Czech Republic. At least another 90 TWh of biofuel could annually produce DH in the EU, Croatia, Iceland, Norway and Switzerland (Werner, 2006b). Waste and wood fuel should preferably be used in CHP plants. Another potential DH source is solar heating.

The main benefits with DH are (cf. Werner, 2006b):

- reduced primary energy supply due to higher energy efficiency through enhanced utilisation of surplus heat from electricity generation and industrial processes
- improved security of supply due to reduced import dependency and a higher degree of self-sufficiency through utilisation of domestic renewables
- lower CO_2 emissions due to the higher efficiency, less fossil-fuel use and increased used of renewables.

District heating expansion in Europe could reduce CO_2 emissions substantially. Werner (2006b) estimated that doubled DH sales would decrease CO_2 emissions with 400 million tons a year due to increased use of renewables and reduced energy losses through utilisation of heat from industries and power generation. Knowledge on how to utilise the various heat resources for DH to achieve the benefits needs to be transferred from countries where the different possibilities already are used to places with favourable but less exploited conditions for sustainable DH solutions.

Increased electricity prices and policy instruments that promote renewable electricity generation have caused much interest in investments in new CHP plants in the already well-developed Swedish district-heating systems. Several new investments in bio-fuelled CHP have been planned recently in Sweden, mostly due to the possible benefits from green electricity certificates. The investments are, however, largely affected by changes in other markets, such as the steel market, and it is uncertain how much of the plans that will be realised. In a study of the CHP possibilities in the Stockholm area with the MODEST model, it was shown that natural-gas-based combined cycle CHP could reduce the CO_2 emissions more than bio-fuelled CHP if electricity prices are high and there were no green electricity certificates (Danestig et al., 2007). This is under the additional condition that coal condensing plants supply the marginal electricity generation and that no other barriers for reducing CO_2 emissions are present. If the fixed cap for European emission allowances will not be reduced, it would hinder this reduction but reallocate the emissions to the most valuable purpose.

Electricity generation from CHP in Sweden can be increased from 8% in 2004 to 16% only through expansion in the Stockholm region (Danestig et al., 2007). When electricity prices are low, bio-fuelled CHP is most profitable due to the support from green certificates.

Because the heat demand in the district heating network sets the limit for possible electricity generation, a CHP alternative with a high electricity-to-heat ratio will be more profitable when electricity prices are high. The potential for CO_2 emission reductions in Stockholm exceeds Sweden's undertakings under the Kyoto protocol and national reduction goals (Danestig et al. 2007).

To be economically favourable, district-heating systems require a certain heat load density. To utilise the high investments in DH systems further, additional heat loads outside the cold seasons and not following the characteristics of the most common social dependent heat loads should be beneficial. When calculating for new base heat load plant investments, possible expansions of district heating systems to new customers and geographical areas should be considered.

SWITCHING ENERGY SUPPLY IN BUILDINGS

By energy-carrier switching, one energy carrier (e.g. electricity) is replaced by another energy carrier, such as a fuel. Such measures are most easily used for heat production. But increasing demand for comfort, more indoor waste heat from electric appliances and climate change increase the demand for cooling. Normally, electricity is used to generate cooling but heat can be used as energy supply to absorption cooling installations, which is further illustrated in section 4.4.

In a study of the Swedish municipality Solna, several scenarios were presented as bases for discussions on sustainability. Figure 8 shows the impact on CO_2 emissions due to changing energy supply in the district heating system from heat pumps to bio-fuels in different scenarios for a new development area. The lowest CO_2 emissions are when new buildings are assumed to be like passive houses for Swedish conditions, Figure 8. The low energy use in this case may normally be too small to motivate district heating, but since district heating is already present in the area and there are many houses in it, the economic preconditions for this solution may be sufficient.

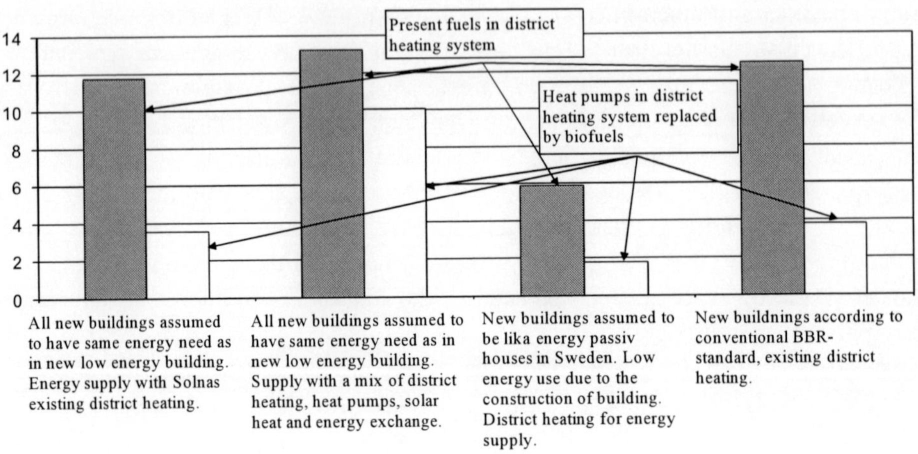

Figure 8. Increased CO2 emissions due to space heating and hot tap water for a new settlement in Solna, now and in four scenarios, thousands of tonnes of CO2 a year (Henning and Danestig, 2008).

A detailed survey of heat demand, heat sources and energy-carrier use was made for a coastal village in northern Sweden with primarily small houses built during the 1920-1980 period (Henning and Danestig, 2007, 2008). One-half of the heat demand is now covered by electricity, mainly through electric boilers but also some resistance heating and a few heat pumps. Oil and wood (pellets and firewood) each supply one-fourth of the heat. Possibilities for energy-carrier switching were investigated. In an extreme scenario, all boilers and heat pumps are replaced by DH but some electric resistance heating was considered too difficult to exchange. The DH is produced with wood but during a few cold days oil is also required. The switching to DH would reduce primary energy use due to the reduction of electricity consumption and the associated energy losses in nuclear and coal-fired condensing power plants (Henning and Danestig, 2007, 2008).

In practice, all houses in the village cannot be connected to a DH network and all house-owners do not want DH. District heating should firstly be built in areas with large buildings and high heat load density. The heat density may be too low for DH, at least in some districts in the village. In some rather heat dense districts, there are recently installed heat pumps, which may hamper DH (Henning and Danestig, 2007, 2008).

If the establishment or expansion of a district-heating network is delayed in an area, the DH supplier should offer temporary solutions, such as wood-pellets boilers, to make more customers chose DH and secure a basis for a district-heating system. In areas with low heat density it is especially important to attract as many DH customers as possible because if too many heat consumers choose other heating options, the demand for DH may become too low to make it economically possible to build a DH system. Initially, a small boiler and heating network may be established, which may expand or become connected to a larger network (Henning and Danestig, 2007, 2008).

Impact of Reduced Heat Demand in Buildings

Energy conservation possibilities were also analysed for the coastal village in the previous section (Henning and Danestig, 2007, 2008). Extensive additional insulation of walls and attic, new windows with low heat transfer and installation of ventilation with heat recovery could bring heat demand to a calculated minimum of 90 kWh (per m^2 of floor space and year) for space heating and hot tap water in detached houses (Figure 9). This extensive energy conservation is estimated to reduce the heat demand by 50%. In addition, all roofs have solar collectors, which reduce external energy supply, and oil is switched to electric heat pumps. In total, electricity, wood and solar heating would each cover about one-third of the heat demand in the village (Figure 9). Only one-third of present primary energy use remains due to the extensive energy conservation and because solar energy covers a substantial share of heat demand. The relative primary energy reduction is independent of the kind of electricity generation that is avoided but in absolute numbers the reduced heat losses in condensing plants dominate. Condensing plant losses in Figure 9 arise in nuclear plants if electricity is Swedish but if European power is considered the additional coal plant losses also arise. In practice, the outlined extensive restructuring of buildings takes a long time and the, partly radical, measures should be taken when renovation is required. The possibilities for energy conservation and energy-carrier switching vary among districts with buildings of different ages (Henning and Danestig, 2007, 2008).

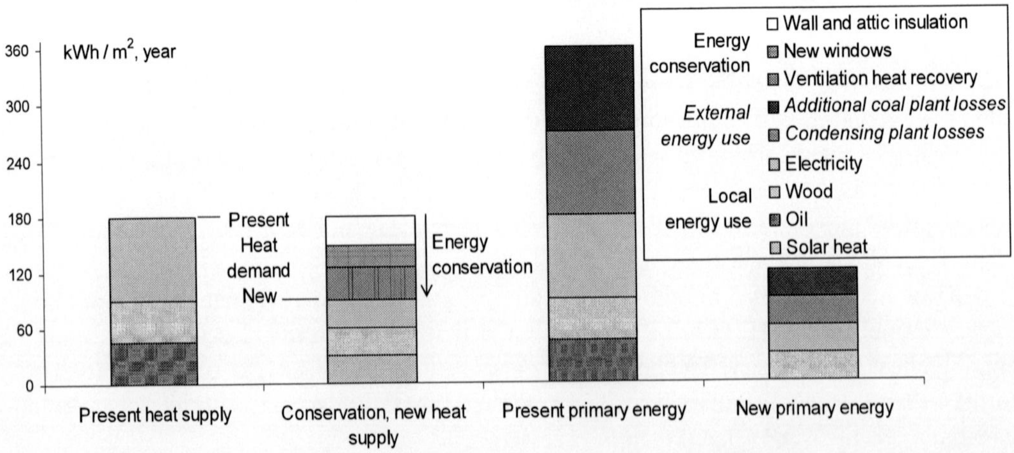

Figure 9. Present and outlined future heat supply and primary energy use, and possible impact of energy conservation, for an average detached house in a Swedish village (kWh per m2 of floor space and year)

CO_2 emission due to local oil burning and from coal condensing plants that are assumed to run to supply electricity were calculated for this energy-conservation scenario and the district-heating scenario above. CO_2 emissions are lower than today for both scenarios irrespective of electricity origin. Low energy use due to extensive energy conservation causes the highest CO_2 emissions if heating partly is electric and the interplay in the European electricity market is considered. Switching from electricity to biofuel would reduce the emissions. The previously described biofuelled DH supplied to houses of current standard cause lower CO_2 emissions despite a higher energy demand because very little oil and electricity is used and wood fuel is considered not to enhance the greenhouse effect (Henning and Danestig, 2007, 2008).

The consequences of changed heat demand are complex when the heat is supplied from resources that can be seen as surplus, for instance heat from waste incineration or industrial surplus heat. When conditions in a specific municipality or region are favourable for utilising industrial waste heat or heat from CHP plants in district heating systems, this must also be considered in scenarios and studies of energy conservation measures that reduce heat demand.

In addition to the emerging of more energy-efficient buildings, higher outdoor temperatures due to the enhanced greenhouse effect may reduce heat demand and district-heating sales substantially. The space-heating demand in southern Sweden is estimated to be reduced by 20% until 2040 compared to the 1961-1990 period (SMHI, 2008). Decreased heat demand should decrease the use of natural resources, such as fossil fuels, and enable redirections of biofuels into other fields of applications, such as production of automotive fuel. Biomass is a renewable but not an infinite resource. The limited biofuel supplies must not be wasted but used efficiently, primarily for direct or indirect replacement of fossil fuel, that is for example, oil-fired boilers and electricity from coal condensing plants, respectively.

A modest utilisation of local renewable energy resources was estimated for some Swedish municipalities. In a metropolitan area, the only possible local renewable energy supply may be to cover a small fraction of the energy demand through solar energy installations on roofs. In a less populated region with favourable wind conditions, the potential wind power production may exceed total energy demand. In a municipality with much heavy pulp and paper industry, the total energy demand may correspond to using all forest and arable land for wood fuel production. Along rivers, local electricity generation in hydropower stations may take place (Henning and Danestig, 2008).

District Heating Supply Also for Cooling and Industrial Processes

The total energy end use in Swedish industry is 160 TWh/year where electricity supplies about 35% and district heating 4%. It has been shown in several energy audits that energy-carrier switching from oil and electricity to district heating is possible for manufacturing processes if the DH temperature is sufficient.

In a study of energy-carrier switching in 34 industries where a district heating connection was present or available, it was possible to decrease the space heating part of the total district heating demand from 88% to 47% due to increased DH use in industrial processes (Danestig et al., 2008). The process that increased district heating most was switching from conventional compressor cooling to absorption cooling. Absorption cooling uses heat to run the cooling process and can decrease electricity use for cooling substantially (Trygg and Amiri, 2007). Heating and drying of goods also contributed notably to the increased district heating demand (Danestig et al., 2008). Electricity demand and replacement possibilities in non-heavy lines of business (e.g. food, vehicles) that use one-third of the electricity in Swedish industry have also been investigated. One terawatt-hour of electricity annually can be replaced by district heating in the studied lines of business without decreasing the benefit of energy use (Henning, 2005; Henning and Trygg, 2008).

The seasonal demand fluctuations in district heating systems are changed when heat supply for industrial processes is switched to district heating. This increases the possibilities for enhanced electricity generation in CHP plants during warm seasons when district heating demand for heating in buildings is low. An increased district-heating demand due to larger DH supply to buildings and industries should, hence, be utilised for electricity generation in efficient combined heat and power production plants, which can displace condensing power plants and their heat losses. Figure 10 outlines possible future supply and use for a DH system (cf. Figure 3). The base load consists of hot-water preparation and DH deliveries to industrial manufacturing processes and is covered by surplus heat from industries. The DH demand that depends on outdoor temperature is covered by a wood-fired CHP plant and, during the coldest winter days, oil-fired boilers. The latter are preferably fired with some renewable oil but petroleum products could be acceptable because the fuel consumption and CO_2 emissions are small. Investing in the large CHP plant capacity shown in Figure 10 would not be profitable if the plant only would supply the DH demand but is made possible through the additional steam deliveries to an adjacent industry, which substantially prolongs the annual CHP operation. A plant according to this concept is now being built in the Swedish town Örnsköldsvik (cf. Danestig and Henning, 2004). Steam supply does not take place when the industrial plant is closed for holiday and maintenance during July to the very right in Figure

10. This concept presupposes that there is supplementing steam production in winter and the total annual efficiency of the joint DH and steam supply must be considered. By integration of DH supply with automotive bio-fuel production, steam deliveries and excess heat utilisation may be combined.

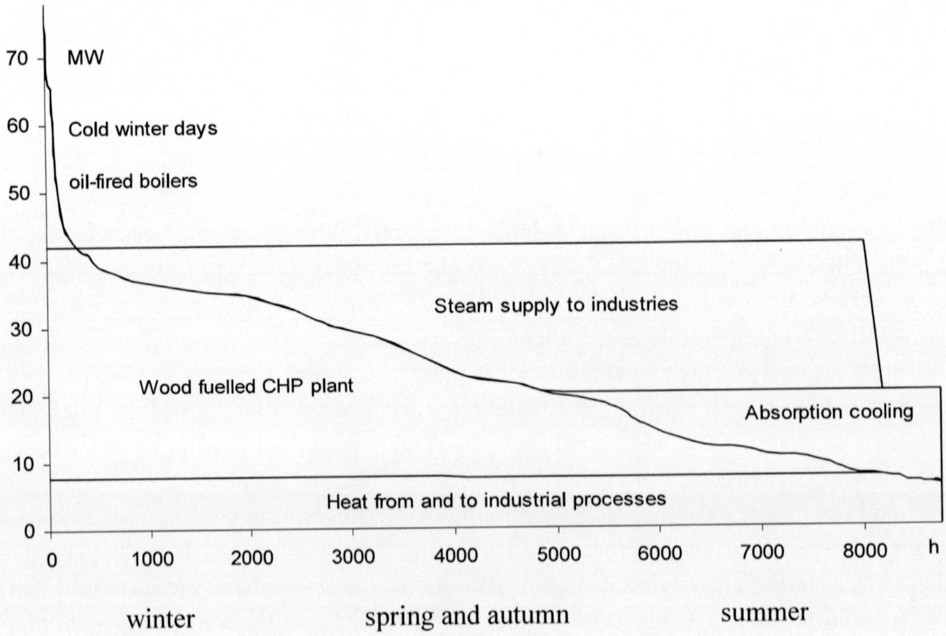

Figure 10. Possible future district-heating supply and use during a year.

In Figure 10, DH is also used for cooling generation through absorption cooling. The cooling demand primarily occurs in summer and supplements nicely the climate-dependent heat demand. Besides the absorption cooling indicated in Figure 10, part of base load may consist of cooling of offices, shops etc. that is demanded throughout the year. Switching from electric to absorption cooling or introducing the latter instead of the former can influence electricity supply and demand. Electricity that is replaced by DH from a CHP (CHP) plant has a dual impact on the electricity system through reduced demand and increased electricity generation.

The 1 TWh/year of electricity that can be replaced by DH in non-heavy lines of business that use one-third of Sweden's industrial electricity use (See above) are partly used for manufacturing processes that are operating rather evenly throughout the year. That is therefore a favourable heat sink for CHP production. A systems analysis of Swedish power production, trade, demand and conservation with the MODEST model showed that the DH should primarily be produced in CHP plants that also generate 300 GWh of electricity a year. The switching from electricity to DH, thus, makes 1.3 TWh of electricity from other sources available for alternative purposes (Henning, 2005; Henning and Trygg, 2008).

Increased power production in Swedish CHP plants makes larger electricity exports possible, which can replace expensive power from coal-fired condensing plants in the European electricity market. When foreign electricity trade influences operation of coal-fired condensing plants, switching from electricity to DH from CHP plants can reduce net CO_2

emissions substantially. This can be a cost-efficient contribution to the reduction of CO_2 emissions within the EU that the emission allowances aim at. Energy system optimisations, thus, show that it is better to use efficient CHP plants than condensing plants with low efficiency and large emissions (e.g. Henning et al., 2006a, Danestig et al. 2007, Henning and Trygg, 2008).

DISCUSSION

When assigning a system approach in studies of energy systems, the results are mostly different from detailed studies and this emphasises that the purpose with a certain study must be clear. When a study has goals, such as decreasing CO_2 emissions or increasing *sustainability*, a study solely based on the economics of today may not lead to the wished goals in the future. System studies may instead imply, for policymakers, what the necessary economic changes to reach the desired goals are. If, for example, district heating in combination with combined heat and power production have possibilities to reduce CO_2 emissions, policies that promote this should be introduced. In Sweden, green electricity certificates are supporting bio-fuelled CHP. This is, however, disadvantageous for natural-gas-fuelled combined heat and power, which can generate more electricity based on the same heat demand. When using the reasoning that new electricity generation can replace coal-condensing generation in Europe, natural gas would have the largest potential for CO_2 reduction. There are obviously more considerations taken than the single goal of CO_2 reduction. Electricity certificates aim at increasing renewable electricity generation but conflict in this case with desired CO_2 emission reductions. This also brings us to the issue security of supply. There are large domestic supplies of biofuels in Sweden but dependency on import for natural gas. Uncertainty about how political decisions on the acceptance for natural gas will affect future elections may also influence Swedish energy policy. Since system studies can not cover everything, uncertainties will always remain, but results from system studies should make it easier to avoid several mistakes and increase the awareness of the surroundings of systems.

Reduced heat demand should decrease the use of natural resources, such as fuels, and enable redirections of fuels into other fields of applications. But the consequences of changed heat demand are complex when the heat is supplied from resources that can be seen as surplus. In order to be economically favourable, district heating needs a certain heat load and this may lower the incentive for energy efficiency measures. The discrepancy of what is surplus and what is unnecessary use is vague for heat. The large unused surplus (or waste) heat from electricity generation in Europe should lead to decreased use of fuels in the heating market if it could be utilised. It is important that large demands for heat are supplied by more sustainable solutions such as DH and CHP. High heating demand is also necessary to make investments in DH systems beneficial. It is obvious that the total energy use must be reduced, but as long as condensing power plants dominate European power supply, resources may better be used for other purposes than wall insulation etc. that hamper utilisation of waste heat from electricity generation through DH systems. On the other hand, buildings that are erected now may be used in their present shapes beyond the decommissioning of many CO_2-rich power plants. This implies that new buildings should be as energy efficient as possible while old buildings with district heating should try to use heat in as many applications as possible, instead of other energy sources. In studied municipalities, the reality is not always that simple because when a municipality (or region) has DH and CHP it is still beneficial to connect new building areas to existing DH networks.

CONCLUSION

Electricity generation from CHP in Sweden can be doubled just through expansion in Stockholm, primarily with bio-fuelled CHP plants. It could reduce Swedish CO_2 emissions substantially. This is when applying an energy system that works as intended and where new electricity generation can replace old. Since CHP is dependent on a useful heat demand, additional heat loads outside the cold seasons should be beneficial. When considering new base load plants, such as CHP plants, possible expansion of district heating systems to new customers and areas should be evaluated. A potential for this is when switching from electricity use to district heating, which reduces primary energy use due to the reduction of heat losses in nuclear and coal-fired condensing power plants but also increases the possibilities for effective CHP generation due to increased heat demand in district heating systems. Energy-carrier switching from oil and electricity to district heating is also possible for manufacturing processes, such as heating and drying of goods, which often are operating rather evenly throughout the year and, therefore, are a favourable heat sink for CHP production. Especially, switching from electric heating and cooling to district heating and absorption cooling, driven by district heating that comes from a combined heat and power (CHP) plant, has a dual impact on the electricity system through reduced electricity demand and increased electricity generation. Cooling demand is larger in summer and can supplement the space heating demand. Absorption cooling increases the possibilities for enhanced electricity generation in CHP plants during warm seasons when district heating demand for heating in buildings is low. The introduction of district-heating use for other new proposes can increase the annual running hours of efficient combined heat and power production plants, which can displace condensing power plants and their heat losses.

A problem in areas with low heat load density is that it is especially important to attract as many district-heating customers as possible because if too many heat consumers choose other heating options, the demand for district heating may become too low to make it economically possible to build a district-heating system. If the heat load density for district heating is too low to motivate DH in an area, the possibilities for increased energy efficiency in buildings are large. Extensive additional insulation of walls and attic, new windows and ventilation with heat recovery could reduce the heat demand by 50% in an average Swedish detached house.

Therefore, both energy efficiency measures as well as good conditions for efficient electricity generation are needed to reduce carbon dioxide emissions from the energy sector.

Energy efficiency in buildings and district heating systems with CHP should, though, not be combined at the same place. Because district heating is unfavourable in districts with low heat load density, energy efficiency measures should be done there in the first place. If other heat resources that are difficult to utilise in other ways, for example, industrial surplus heat and heat from waste incineration, can be used in areas without CHP this could be effective.

But systems with CHP and electricity generation may enhance the efficiency and reduce the environmental impact of the total energy system more effectively.

ACKNOWLEDGEMENT

The work has been carried out under the auspices of The Energy Systems Programme, which is primarily financed by the Swedish Energy Agency.

REFERENCES

Börjeson, L., Höjer, M., Dreborg, K.-H., Ekvall, T. and Finnveden, G. (2006). Scenario types and techniques: Towards a user's guide, Futures, Vol. 38, No. 7, pp 723-739.

Carlson, A. (2003) Energy systems and the climate dilemma—reflecting the impact on CO_2 emissions by reconstructing regional energy systems. Energy Policy 31, 951–959.

Churchman, C.W. (1968). The Systems Approach. New York: Dell Publishing.

Danestig M. and Henning D. (2004). Increased system benefit from cogeneration due to cooperation between district heating utility and industry, Proceedings of the 9th international symposium on district heating and cooling, Espoo, Finland, 30-31 August.

Danestig, M, Gebremedhin, A. and Karlsson, B. (2007). Stockholm CHP potential—An opportunity for CO2 reductions? Energy Policy, Volume 35, Issue 9, September 2007, Pages 4650-4660.

Danestig, M, Difs, K and Trygg, L (2008). Industrial district heating applications – The importance of heat load characteristics. Manuscript in preparation.

Directive 2005/89/EC of the European Parliament and of the Council of 18 January 2006 concerning measures to safeguard security of electricity supply and infrastructure investment.

Directive 2003/54/EC of the European Parliament and of the Council of 26 June 2003 concerning common rules for the internal market in electricity and repealing Directive 96/92/EC.

EC (2008). European Commission, Energy and Transport : Figures and Main Facts. Directorate-General for Energy and Transport. Homepage; http://ec.europa.eu/dgs/energy_ transport/figures/pocketbook/2007_en.htm. Visited 2008-03-09.

Euroheat and Power (2007). District heating and Cooling, country by country/2007 survey. Brussels, Belgium: Euroheat and Power, 2007.

Gebremedhin A. (2003). Regional and Industrial Co-operation in District Heating Systems, Dissertation No. 849, Linköping institute of technology, Linköping, Sweden.

Gebremedhin A., Moshfegh B., (2004). Modelling and optimisation of district heating and industrial energy system. An approach to a locally deregulated heat market. The International Journal of Energy Research 28, 411–422.

Groscurth H.-M. and Schweiker A. (1995). Contribution of Computer Models to Solving the Energy Problem, Energy Sources, Vol. 17, No. 2, pp. 161-177.

Henning, D. (1999). Optimisation of local and national energy systems: development and use of the MODEST model. Dissertation 559, Linköping Institute of Technology, Linköping, Sweden.

Henning, D. (2005). El till vad och hur mycket i svensk industri (Electricity for what and how much in Swedish industry). IKP Energy Systems, Linköping University and Optensys Energianalys, Linköping, Sweden. Available from www.energimyndigheten.se (in Swedish).

Henning D., Amiri S., Holmgren K. (2006a). Modelling and optimisation of electricity, steam and district heating production for a local Swedish utility, European Journal of Operational Research, Vol. 175, No. 2, pp. 1224-1247.

Henning D. och Trygg L. (2008). Reduction of Electricity Use in Swedish Industry and its Impact on National Power Supply and European CO_2 Emissions, Energy Policy, in press.

Henning D, Danestig M (2008). Energifrågor i fysisk planering: Förutsättningar och scenarier för energitillförsel och energihushållning, Energimyndigheten, Eskilstuna, to be published.

Henning, D., Danestig, M., Holmgren, K., Gebremedhin, A. (2006b). Modelling the impact of policy instruments on district heating operations – experiences from Sweden, in: Lectures, 10th International Symposium on District Heating and Cooling, Hanover, Germany, 3-5 September. AGFW-VDEW, Frankfurt a M, Germany.

Henning, D. and Danestig, M. (2007). Local development possibilities for sustainable energy supply and use in Sweden, VHU conference Science for Sustainable development, Linköping, 6-7 September.

Holmgren K. and Gebremedhin A. (2004). Modelling a district heating system: introduction of waste incineration, policy instruments and co-operation with an industry, Energy Policy, Vol. 32, No. 16, pp. 1807-1817.

Holmgren, K., (2006a). The role of a district heating network as a user of waste heat supply from various sources—the case of Göteborg. Applied Energy 83, 1351–1367.

Holmgren, K. (2006b). A system perspective on district heating and waste incineration. Linköpings universitet, Sweden.

Holttinen, H., Tuhkanen, S., 2004. The effect of wind power on CO_2 abatement in the Nordic countries. Energy Policy 32, 1639–1652.

Hughes, L, Bohan, K, Good, J and Jafapur K, 2005. Calculating residential carbon dioxide emissions-a new approach. Energy Policy 33, 1865-1871.

IPCC, 2007. Intergovernmental Panel on Climate Change Fourth Assessment Report Climate Change 2007: Synthesis Report. November 2007, Valencia, Spain.

Nord-Ågren, E., 2002. The Electricity Usage Pattern for Cooker Manufacturing in Three European Countries: A Benchmarking study of Electrolux Factories. Report LIU-TEK-LIC-2002:60, Linköpings universitet, Linköping, Sweden.

Persson, A., Rydstrand, C., Hedenskog, P. (2005). Allt eller inget – Systemgränser för byggnaders uppvärmning, ÅF Energi and Miljö AB, Stockholm.

Regulation (EC) No 1228/2003 of the European Parliament and of the Council of 26 June 2003 on conditions for access to the network for cross-border exchanges in electricity

Remme, U., Goldstein, G.A., Schellmann, U., Schlenzig, C., 2001. MESAP/TIMES—advanced decision support for energy and environmental planning. In: Chamoni, P., Leisten, R., Martin, A., Minnemann, J., Stadtler, H. (Eds.), Operations Research Proceedings 2001—Selected Papers of the International Conference on Operations

Research (OR 2001), 3–5 September. Springer, Duisburg, Germany, pp. 59–66. (www.uni-duisburg.de/or2001).

Sjödin J. and Henning D. (2004) Calculating the marginal costs of a district heating utility, Applied Energy, Vol. 78, No. 1, pp. 1-18.

Sjödin J. (2003. Swedish district heating systems and a harmonised European energy market. Means to reduce carbon dioxide emissions. Linköpings universitet, Sweden.

SCB, 2008. Statistics Sweden. Swedish official statistics homepage (www.scb.se) Visited 2008-03-09.

SDHA (2006). Statistik 2004, Svensk fjärrvärme, Stockholm (www.svenskfjarrvarme.se).

SEA (2007). Energy in Sweden, ET 2007:49, Swedish Energy Agency, Eskilstuna.

SMHI (2008). Sveriges klimat i framtiden, Swedish meteorological and hydrological institute, Norrköping (www.smhi.se)

Trygg, L. (2006). Swedish industrial and energy supply measures in a European system perspective. Linköpings universitet, Sweden.

Trygg, L., and Amiri, S. (2007). Absorption cooling in a European perspective – a case study from Norrköping. Applied Energy, Volume 84, Issue 12, December 2007, Pages 1319-1337.

Unger, T., Ekvall, T. (2003). Benefits from increased cooperation and energy trade under CO_2 commitments—the Nordic case. Climate Policy 3, 279–294.

Wene C.-O. and Rydén B. (1988). A comprehensive energy model in the municipal energy planning process, *European Journal of Operational Research*, Vol. 33, No. 2, pp. 212-222.

Werner S (2006a). The European heat market, Ecoheatcool work package 1, Euroheat and power, Brussels, Belgium (www.euroheat.org/ecoheatcool).

Werner S (2006b). Possibilities with more district heating in Europe, Ecoheatcool work package 4, Euroheat and power, Brussels, Belgium (www.euroheat.org/ecoheatcool).

In: Energy in Europe: Economics, Policy and Strategy
Editors: F. L. Magnusson and O. W. Bengtsson.
ISBN: 978-1-60456-829-5

Chapter 10

DEVELOPMENT OF INTEGRATED BIOENERGY DEVICES FOR IMPROVEMENT OF QUALITY OF LIFE OF POOR PEOPLE IN DEVELOPING COUNTRIES

Abdeen Mustafa Omer[1]

ABSTRACT

There is strong scientific evidence that the average temperature of the earth's surface is rising. This was a result of the increased concentration of carbon dioxide (CO_2), and other greenhouse gases (GHGs) in the atmosphere as released by burning fossil fuels. This global warming will eventually lead to substantial changes in the world's climate, which will, in turn, have a major impact on human life and the environment. Energy use reductions can be achieved by minimising the energy demand, by rational energy use, by recovering heat and the use of more green energies. This study was a step towards achieving this goal. The adoption of green or sustainable approaches to the way in which society is run is seen as an important strategy in finding a solution to the energy problem. The key factors to reducing and controlling CO_2, which is the major contributor to global warming, are the use of alternative approaches to energy generation and the exploration of how these alternatives are used today and may be used in the future as green energy sources. Even with modest assumptions about the availability of land, comprehensive fuel-wood farming programmes offer significant energy, economic and environmental benefits. These benefits would be dispersed in rural areas where they are greatly needed and can serve as linkages for further rural economic development. The nations as a whole would benefit from savings in foreign exchange, improved energy security, and socio-economic improvements. With a nine-fold increase in forest – plantation cover, the nation's resource base would be greatly improved. The international community would benefit from pollution reduction, climate mitigation, and the increased trading opportunities that arise from new income sources. The non-technical issues, which have recently gained attention, include: (1) Environmental and ecological factors e.g., carbon sequestration, reforestation and revegetation. (2) Renewables as a CO_2 neutral replacement for fossil fuels. (3) Greater recognition of the importance of renewable energy, particularly modern biomass energy carriers, at the policy and planning levels.

[1] 17 Juniper Court, Forest Road West, Nottingham NG7 4EU, United Kingdom.

(4) Greater recognition of the difficulties of gathering good and reliable biomass energy data, and efforts to improve it. (5) Studies on the detrimental health efforts of biomass energy particularly from traditional energy users. This study discusses a comprehensive review of biomass energy sources, environment and sustainable development. This includes all the biomass energy technologies, energy efficiency systems, energy conservation scenarios, energy savings and other mitigation measures necessary to reduce climate change.

Keywords: *Biomass energy, biofuels, biogas, bioheat, utilisation and development.*

1. INTRODUCTION

Energy is an essential factor in development since it stimulates, and supports economic growth and development. Fossil fuels, especially oil and natural gas, are finite in extent, and should be regarded as depleting assets, and efforts are oriented to search for new sources of energy. The clamour all over the world for the need to conserve energy and the environment has intensified as traditional energy resources continue to dwindle whilst the environment becomes increasingly degraded. Alternatively energy sources can potentially help fulfill the acute energy demand and sustain economic growth in many regions of the world. Bioenergy is beginning to gain importance in the global fight to prevent climate change. The scope for exploiting organic waste as a source of energy is not limited to direct incineration or burning refuse-derived fuels. Biogas, biofuels and woody biomass are other forms of energy sources that can be derived from organic waste materials. These biomass energy sources have significant potential in the fight against climate change.

The sources to alleviate the energy situation in the world are sufficient to supply all foreseeable needs. Conservation of energy and rationing in some form will however have to be practised by most countries, to reduce oil imports and redress balance of payments positions. Meanwhile development and application of nuclear power and some of the traditional solar, wind, biomass and water energy alternatives must be set in hand to supplement what remains of the fossil fuels.

The encouragement of greater energy use is an essential component of development. In the short term it requires mechanisms to enable the rapid increase in energy/capita, and in the long term we should be working towards a way of life, which makes use of energy efficiency and without the impairment of the environment or of causing safety problems. Such a programme should as far as possible be based on renewable energy resources.

Large scale, conventional, power plant such as hydropower, has an important part to play in development. It does not, however, provide a complete solution. There is an important complementary role for the greater use of small scale, rural based-power plants. Such plant can be used to assist development since it can be made locally using local resources, enabling a rapid built-up in total equipment to be made without a corresponding and unacceptably large demand on central funds. Renewable resources are particularly suitable for providing the energy for such equipment and its use is also compatible with the long-term aims.

In compiling energy consumption data one can categorise usage according to a number of different schemes:

- Traditional sector- industrial, transportation, etc.
- End-use- space heating, process steam, etc.
- Final demand- total energy consumption related to automobiles, to food, etc.
- Energy source- oil, coal, etc.
- Energy form at point of use- electric drive, low temperature heat, etc.

2. Bioenergy Development

Bioenergy is energy from the sun stored in materials of biological origin. This includes plant matter and animal waste, known as biomass. Plants store solar energy through photosynthesis in cellulose and lignin, whereas animals store energy as fats. When burned, these sugars break down and release energy exothermically, releasing carbon dioxide, heat and steam. The by-products of this reaction can be captured and manipulated to create power, commonly called bioenergy. Biomass is considered renewable because the carbon is taken out of the atmosphere and replenished more quickly than the millions of years required for fossil fuels to form. The use of biofuels to replace fossil fuels contributes to a reduction in the overall release of carbon dioxide into the atmosphere and hence helps to tackle global warming. The range of waste treatment technologies that are tailored to produce bioenergy is growing. There are a number of key areas of bioenergy from wastes including (but not limited to) biogas, biofuels and bioheat. When considering using bioenergy, it is important to take into account the overall emission of carbon in the process of electricity production.

The biomass energy resources are particularly suited for the provision of rural power supplies and a major advantage is that equipment such as flat plate solar driers, wind machines, etc., can be constructed using local resources and without the high capital cost of more conventional equipment. Further advantage results from the feasibility of local maintenance and the general encouragement such local manufacture gives to the build up of small scale rural based industry. Table 1 lists the energy sources available.

Table 1. Sources of energy

Energy source	Energy carrier	Energy end-use
Vegetation	Fuel-wood	Cooking Water heating Building materials Animal fodder preparation
Oil	Kerosene	Lighting Ignition fires
Dry cells	Dry cell batteries	Lighting Small appliances
Muscle power	Animal power	Transport Land preparation for farming Food preparation (threshing)
Muscle power	Human power	Transport Land preparation for farming Food preparation (threshing)

Currently the 'non-commercial' fuels wood, crop residues and animal dung are used in large amounts in the rural areas of developing countries, principally for heating and cooking; the method of use is highly inefficient. Table 2 presented some renewable applications.

Table 3 lists the most important of energy needs.

Considerations when selecting power plant include the following:

- Power level- whether continuous or discontinuous.
- Cost- initial cost, total running cost including fuel, maintenance and capital amortised over life.
- Complexity of operation.
- Maintenance and availability of spares.
- Life.
- Suitability for local manufacture.

Table 2. Renewable applications

Systems	Applications
Water supply	Rain collection, purification, storage and recycling
Wastes disposal	Anaerobic digestion (CH_4)
Cooking	Methane
Food	Cultivate the 1 hectare plot and greenhouse for four people
Electrical demands	Wind generator
Space heating	Solar collectors
Water heating	Solar collectors and excess wind energy
Control system	Ultimately hardware
Building fabric	Integration of subsystems to cut costs

Table 3. Energy needs in rural areas

Transport e.g., small vehicles and boats
Agricultural machinery e.g., two-wheeled tractors
Crop processing e.g., milling
Water pumping
Small industries e.g., workshop equipment
Electricity generation e.g., hospitals and schools
Domestic e.g., cooking, heating, lighting

Table 4 listed methods of energy conversion.

The internal combustion engine is a major contributor to rising CO_2 emissions worldwide and some pretty dramatic new thinking is needed if our planet is to counter the effects. With its use increasing in developing world economies, there is something to be said for the argument that the vehicles we use to help keep our inner-city environments free from waste, litter and grime should be at the forefront of developments in low-emissions technology. Materials handled by waste management companies are becoming increasingly valuable. Those responsible for the security of facilities that treat waste or manage scrap will testify to the precautions needed to fight an ongoing battle against unauthorised access by criminals and crucially, to prevent the damage they can cause through theft, vandalism or even arson. Of particular concern is the escalating level of metal theft, driven by various factors including the demand for metal in rapidly developing economies such as India and China.

Table 4. Methods of energy conversion

Muscle power	Man, animals
Internal combustion engines	
Reciprocating	Petrol- spark ignition
	Diesel- compression ignition
	Humphrey water piston
Rotating	Gas turbines
Heat engines	
Vapour (Rankine)	
Reciprocating	Steam engine
Rotating	Steam turbine
Gas Stirling (Reciprocating)	Steam engine
Gas Brayton (Rotating)	Steam turbine
Electron gas	Thermionic, thermoelectric
Electromagnetic radiation	Photo devices
Hydraulic engines	Wheels, screws, buckets, turbines
Wind engines (wind machines)	Vertical axis, horizontal axis
Electrical/mechanical	Dynamo/alternator, motor

There is a need for greater attention to be devoted to this field in the development of new designs, the dissemination of information and the encouragement of its use. International and government bodies and independent organisations all have a role to play in biomass energy technologies. Environment has no prcise limits because it is in fact a part of everything. Indeed, environment is, as anyone probably already knows, not only flowers blossoming or birds singing in the spring, or a lake surrounded by beautiful mountains. It is also human settlements, the places where people live, work, rest, the quality of the food they eat, the noise or silence of the street they live in. environment is not only the fact that our cars consume a good deal of energy and pollute the air, but also, that we often need them to go to work and for hoildays. Obviously man uses energy just as plants, bactria, mishrooms, bees, fish and rats do (Figure 1). Man largely uses solar energy- food, hydropower, wood- and thus participates harmoniously in the natural flow of energy through the environment. But man also uses oil, gas, coal and nuclear power. By using such sources of energy, man is thus modifying his environment.

Economic importance of environmental issue is increasing, and new technologies are expected to reduce pollution derived both from productive processes and products, with costs that are still unknown. This is due to market uncertainty, weak appropriability regime, lack of a dominant design, and difficulties in reconfiguring organisational routines. The degradation of the global environment is one of the most serious energy issues. Various options are proposed and investigated to mitigate climate change, acid rain or other environmental problems.

3. ENERGY USE AND THE ENVIRONMENT

Energy use is one of several essential components for developing countries:

- The overall situation and the implications of increased energy use in the future.
- The problem of the provision of power in rural areas, including the consideration of energy resources and energy conversion.

In addition to the drain on resources, such an increase in consumption consequences, together with the increased hazards of pollution and the safety problems associated with a large nuclear fission programmes. This is a disturbing prospect. It would be equally unacceptable to suggest that the difference in energy between the developed and developing countries and prudent for the developed countries to move towards a way of life which, whilst maintaining or even increasing quality of life, reduce significantly the energy consumption per capita. Such savings can be achieved in a number of ways:

- Improved efficiency of energy use, for example better thermal insulation, energy recovery, and total energy.
- Conservation of energy resources by design for long life and recycling rather than the short life throwaway product and systematic replanning of our way of life, for example in the field of transport.

Energy ratio is defined as the ratio of:
Energy content of the food product/Energy input to produce the food (1)

Table 5. Annual GHG emissions from different sources of power plants

Primary source of energy	Emissions		Waste (x 10^3 metric tons)	Area (km^2)
	Atmosphere	Water		
Coal	380	7-41	60-3000	120
Oil	70-160	3-6	negligible	70-84
Gas	24	1	-	84
Nuclear	6	21	2600	77

Table 6. Energy consumption per population

Region	Population (millions)	Energy (Watt/m^2)
Africa	820	0.54
Asia	3780	2.74
Central America	180	1.44
North America	335	0.34
South America	475	0.52
Western Europe	445	2.24
Eastern Europe	130	2.57
Oceania	35	0.08
Russia	330	0.29

Table 7. Summary of material recycling practices in the construction sector [1]

Construction and demolition material	Recycling technology options	Recycling product
Asphalt	Cold recycling: heat generation; Minnesota process; parallel drum process; elongated drum; microwave asphalt recycling system; finfalt; surface regeneration	Recycling asphalt; asphalt aggregate
Brick	Burn to ash, crush into aggregate	Slime burn ash; filling material; hardcore
Concrete	Crush into aggregate	Recycling aggregate; cement replacement; protection of levee; backfilling; filter
Ferrous metal	Melt; reuse directly	Recycled steel scrap
Glass	Reuse directly; grind to powder; polishing; crush into aggregate; burn to ash	Recycled window unit; glass fibre; filling material; tile; paving block; asphalt; recycled aggregate; cement replacement; manmade soil
Masonry	Crush into aggregate; heat to 900°C to ash	Thermal insulating concrete; traditional clay
Non-ferrous metal	Melt	Recycled metal
Paper and cardboard	Purification	Recycled paper
Plastic	Convert to powder by cryogenic milling; clopping; crush into aggregate; burn to ash	Panel; recycled plastic; plastic lumber; recycled aggregate; landfill drainage; asphalt; manmade soil
Timber	Reuse directly; cut into aggregate; blast furnace deoxidisation; gasification or pyrolysis; chipping; moulding by pressurising timber chip under steam and water	Whole timber; furniture and kitchen utensils; lightweight recycled aggregate; source of energy; chemical production; wood-based panel; plastic lumber; geofibre; insulation board

4. Combined Heat and Power (CHP)

The atmospheric emissions of fossil fuelled installations are mostly aldehydes, carbon monoxide, nitrogen oxides, sulpher oxides and particles (i.e., ash) as well as carbon dioxide. Table 5 shows estimates include not only the releases occurring at the power plant itself but also cover fuel extraction and treatment, as well as the storage of wastes and the area of land required for operation (Table 6). A review of the potential range of recyclables is presented in Table 7.

Combined heat and power (CHP) installations are quite common in greenhouses, which grow high-energy, input crops (e.g., salad vegetables, pot plants, etc.). Scientific assumptions for a short-term energy strategy suggest that the most economically efficient way to replace the thermal plants is to modernise existing power plants to increase their energy efficiency and to improve their environmental performance.

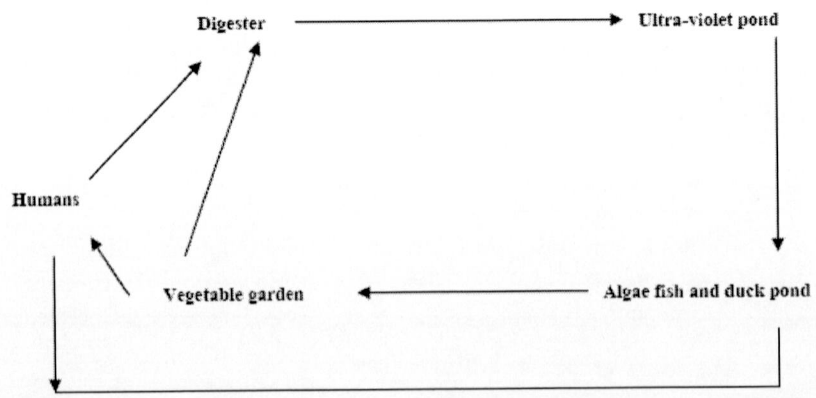

Figure 1. Biomass utilisation concept.

However, utilisation of wind power and the conversion of gas-fired CHP plants to biomass would significantly reduce the dependence on imported fossil fuels. Although a lack of generating capacity is forecast in the long-term, utilisation of the existing renewable energy potential and the huge possibilities for increasing energy efficiency are sufficient to meet future energy demands in the short-term.

A total shift towards a sustainable energy system is a complex and long process, but is one that can be achieved within a period of about 20 years. Implementation will require initial investment, long-term national strategies and action plans. However, the changes will have a number of benefits including: a more stable energy supply than at present and major improvement in the environmental performance of the energy sector, and certain social benefits (Figure 2). A vision used a methodology and calculations based on computer modelling that utilised:

- Data from existing governmental programmes.
- Potential renewable energy sources and energy efficiency improvements.
- Assumptions for future economy growth.
- Information from studies and surveys on the recent situation in the energy sector.

The main advantages are related to energy, agriculture and environment problems, are foreseeable both at national level and at worldwide level and can be summarised as follows:

- Reduction of dependence on import of energy and related products.
- Reduction of environmental impact of energy production (greenhouse effect, air pollution, waste degradation).
- Substitution of food crops and reduction of food surpluses and of related economic burdens.

- Utilisation of marginal lands and of set aside lands and reduction of related socio-economic and environmental problems (soil erosion, urbanisation, landscape deterioration, etc.).
- Development of new know-how and production of technological innovation.

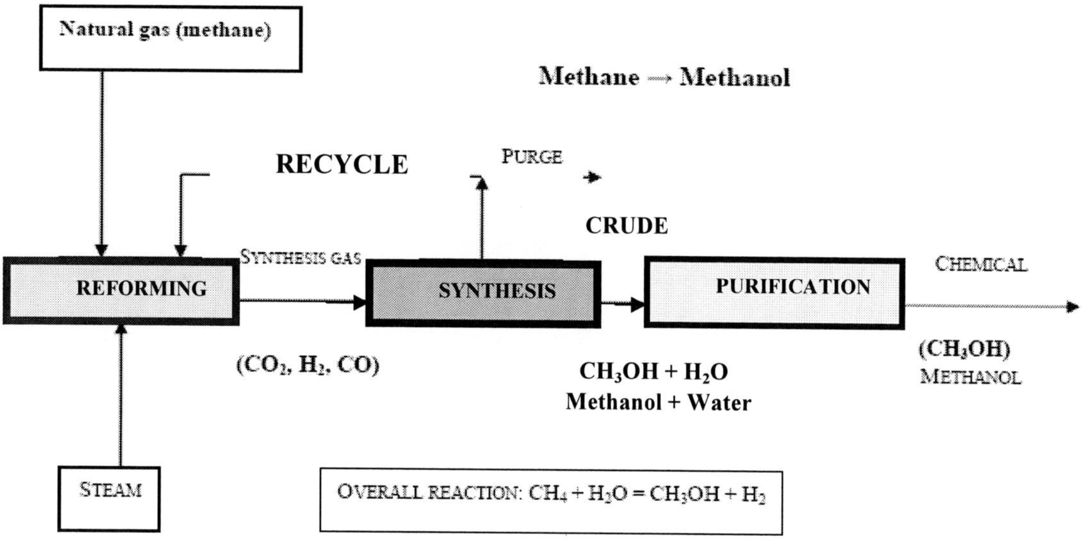

Figure 2. Schematic process flowsheet.

In some countries, a wide range of economic incentives and other measures are already helping to protect the environment. These include:

- Taxes and user charges that reflect the costs of using the environment e.g., pollution taxes and waste disposal charges.
- Subsidies, credits and grants that encourage environmental protection.
- Deposit-refund systems that prevent pollution on resource misuse and promote product reuse or recycling.
- Financial enforcement incentives, e.g., fines for non-compliance with environmental regulations.
- Tradable permits for activities that harm the environment.

District Heating (DH), also known as community heating can be a key factor to achieve energy savings, reduce CO_2 emissions and at the same time provide consumers with a high quality heat supply at a competitive price. DH should generally only be considered for areas where the heat density is sufficiently high to make DH economical. In countries like Denmark DH may today be economical even to new developments with lower density areas due to the high level of taxation on oil and gas fuels combined with the efficient production of DH. To improve the opportunity for DH local councils can adapt the following plan:

- Analyse the options for heat supply during local planning stage.

- In areas where DH is the least cost solution it should be made part of the infrastructure just like for instance water and sewage connecting all existing and new buildings.
- Where possible all public buildings should be connected to DH.
- The government provides low interest loans or funding to minimise conversion costs for its citizens.
- Use other powers, for instance national legislation to ensure the most economical development of the heat supply and enable an obligation to connect buildings to a DH scheme.

Denmark has broadly seen three scales of CHP which where largely implemented in the following chronological order [3]:

- Large scale CHP in cities (>50 MWe).
- Small (5 kWe – 5 MWe) and medium scale (5-50 MWe).
- Industrial and small scale CHP.

Most of the heat is produced by large CHP plants (gas-fired combined cycle plants using natural gas, biomass, waste or biogas). DH is energy efficient because of the way the heat is produced and the required temperature level is an important factor. Buildings can be heated to temperature of 21°C and domestic hot water (DHW) can be supplied with a temperature of 55°C using energy sources that are most efficient when producing low temperature levels (<95°C) for the DH water. Most of these heat sources are CO_2 neutral or emit low levels. Only a few of these sources are available to small individual systems at a reasonably cost, whereas DH schemes because of the plant's size and location can have access to most of the heat sources and at a low cost. Low temperature DH, with return temperatures of around 30-40°C can utilise the following heat sources:

- Efficient use of CHP by extracting heat at low calorific value (CV).
- Efficient use of biomass or gas boilers by condensing heat in economisers (Table 8).
- Efficient utilisation of geothermal energy.
- Direct utilisation of excess low temperature heat from industrial processes.
- Efficient use of large-scale solar heating plants.

Heat tariffs may include a number of components such as: a connection charge, a fixed charge and a variable energy charge. Also, consumers may be incentivised to lower the return temperature. Hence, it is difficult to generalise but the heat practice for any DH company no matter what the ownership structure can be highlighted as follows:

- To develop and maintain a development plan for the connection of new consumers.
- To evaluate the options for least cost production of heat.
- To implement the most competitive solutions by signing agreements with other companies or by implementing own investment projects.
- To monitor all internal costs and with the help of benchmarking, improve the efficiency of the company.

- To maintain a good relationship with the consumer and deliver heat supply services at a sufficient quality.

Table 8. Final energy projections including biomass (Mtoe) [4]

Region	Biomass	Conventional Energy	Total	Share of Biomass (%)
1995				
Africa	205	136	341	60
China	206	649	855	24
East Asia	106	316	422	25
Latin America	73	342	415	18
South Asia	235	188	423	56
Total developing countries	825	1632	2457	34
Other non-OECD countries	24	1037	1061	2
Total non-OECD countries	849	2669	3518	24
OECD countries	81	3044	3125	3
World	930	5713	6643	14
2020				
Africa	371	266	637	59
China	224	1524	1748	13
East Asia	118	813	931	13
Latin America	81	706	787	10
South Asia	276	523	799	35
Total developing countries	1071	3825	4896	22
Other non-OECD countries	26	1669	1695	2
Total non-OECD countries	1097	5494	6591	17
OECD countries	96	3872	3968	2
World	1193	9365	10558	11

Installing DH should be pursued to meet the objectives for improving the environment through the improvement of energy efficiency in the heating sector. At the same time DH can serve the consumer with a reasonable quality of heat at the lowest possible cost. The variety of possible solutions combined with the collaboration between individual companies, the district heating association, the suppliers and consultants can, as it has been in Denmark, be the way forward for developing DH in the United Kingdom.

5. Biomass Utilisation and Development of Conversion Technologies

Sustainable energy is energy that, in its production or consumption, has minimal negative impacts on human health and the healthy functioning of vital ecological systems, including the global environment. It is an accepted fact that renewable energy is a sustainable form of energy, which has attracted more attention during recent years. A great amount of renewable energy potential, environmental interest, as well as economic consideration of fossil fuel consumption and high emphasis of sustainable development for the future will be needed. Explanations for the use of inefficient agricultural-environmental polices include: the high cost of information required to measure benefits on a site-specific basis, information asymmetries between government agencies and farm decision makers that result in high implementation costs, distribution effects and political considerations [2]. Achieving the aim of agric-environment schemes through:

- Sustain the beauty and diversity of the landscape.
- Improve and extend wildlife habitats.
- Conserve archaeological sites and historic features.
- Improve opportunities for countryside enjoyment.
- Restore neglected land or features, and
- Create new habitats and landscapes.

5.1. Efficient Bio-Energy Use and Improvement

The data required to perform the trade-off analysis simulation can be classified according to the divisions given in Table 9: the overall system or individual plants, and the existing situation or future development.

The effective economic utilisations of these resources are shown in Table 10, but their use is hindered by many problems such as those related to harvesting, collection, and transportation, besides the photo-sanitary control regulations. Biomass energy is experiencing a surge in interest stemming from a combination of factors, e.g., greater recognition of its current role and future potential contribution as a modern fuel, global environmental benefits, its development and entrepreneurial opportunities, etc. Possible routes of biomass energy development are shown in Table 11.

The key to successful future appears to lie with successful marketing of the treatment by products. There is also potential for using solid residue in the construction industry as a filling agent for concrete. Research suggests that the composition of the residue locks metals within the material, thus preventing their escape and any subsequent negative effect on the environment.

The use of biomass through direct combustion has long been, and still is, the most common mode of biomass utilisation as shown in Tables (9-11). Examples for dry (thermo-chemical) conversion processes are charcoal making from wood (slow pyrolysis), gasification of forest and agricultural residues (fast pyrolysis – this is still in demonstration phase), and of course, direct combustion in stoves, furnaces, etc. Wet processes require substantial amount of water to be mixed with the biomass. Biomass technologies include:

- Briquetting.
- Improved stoves.
- Biogas.
- Improved charcoal.
- Carbonisation.
- Gasification.

Table 9. Classifications of data requirements

	Plant data	System data
Existing data	Size Life Cost (fixed and var. O and M) Forced outage Maintenance Efficiency Fuel Emissions	Peak load Load shape Capital costs Fuel costs Depreciation Rate of return Taxes
Future data	All of above, plus Capital costs Construction trajectory Date in service	System lead growth Fuel price growth Fuel import limits Inflation

Table 10. Effective biomass resource utilisation

Subject	Tools	Constraints
Utilisation and land clearance for agriculture expansion	Stumpage fees Control Extension Conversion Technology	Policy Fuel-wood planning Lack of extension Institutional
Utilisation of agricultural residues	Briquetting Carbonisation Carbonisation and briquetting Fermentation Gasification	Capital Pricing Policy and legislation Social acceptability

The increased demand for gas and petroleum, food crops, fish and large sources of vegetative matter mean that the global harvesting of carbon has in turn intensified. It could be said that mankind is mining nearly everything except its waste piles. It is simply a matter of time until the significant carbon stream present in municipal solid waste is fully captured. In the meantime, the waste industry needs to continue on the pathway to increased awareness and better optimised biowaste resources. Optimisation of waste carbon may require widespread regulatory drivers (including strict limits on the landfilling of organic materials), public acceptance of the benefits of waste carbon products for soil improvements/crop enhancements and more investment in capital facilities. In short, a significant effort will be required in order to capture a greater portion of the carbon stream and put it to beneficial use. From the standpoint of waste practitioners, further research and pilot programmes are necessary before the available carbon in the waste stream can be extracted in sufficient quality and quantities to create the desired end products. Other details need to be ironed out too, including measurement methods, diversion calculations, sequestration values and determination of acceptance contamination thresholds.

Table 11. Agricultural residues routes for development

Source	Process	Product	End use
Agricultural residues	Direct	Combustion	Rural poor Urban household Industrial use
	Processing	Briquettes	Industrial use Limited household use
	Processing	Carbonisation (small scale)	Rural household (self sufficiency)
	Carbonisation	Briquettes Carbonised	Urban fuel
	Fermentation	Biogas	Energy services Household Industry
Agricultural, and animal residues	Direct	Combustion	(Save or less efficiency as wood)
	Briquettes	Direct combustion	(Similar end use devices or improved)
	Carbonisation	Carbonised	Use
	Carbonisation	Briquettes	Briquettes use
	Fermentation	Biogas	Use

5.1.1. Briquette

Charcoal stoves are very familiar to African society. As for the stove technology, the present charcoal stove can be used, and can be improved upon for better efficiency. This energy term will be of particular interest to both urban and rural households and all the income groups due to the simplicity, convenience, and lower air polluting characteristics. However, the market price of the fuel together with that of its end-use technology may not enhance its early high market penetration especially in the urban low income and rural households.

Briquetting is the formation of a char (an energy-dense solid fuel source) from otherwise wasted agricultural and forestry residues. One of the disadvantages of wood fuel is that it is bulky with a low energy density and is therefore enquire to transport. Briquette formation allows for a more energy-dense fuel to be delivered, thus reducing the transportation cost and making the resource more competitive. It also adds some uniformity, which makes the fuel more compatible with systems that are sensitive to the specific fuel input.

5.1.2. Improved Cook Stoves

Traditional wood stoves can be classified into four types: three stone, metal cylindrical shaped, metal tripod and clay type. Another area in which rural energy availability could be secured where woody fuels have become scarce, are the improvements of traditional cookers and ovens to raise the efficiency of fuel saving. Also, by planting fast growing trees to provide a constant fuel supply. The rural development is essential and economically important since it will eventually lead to better standards of living, people's settlement, and self sufficient in the following:

- Food and water supplies.
- Better services in education and health care.
- Good communication modes.

5.1.3. Biogas

Biogas technology can not only provide fuel, but is also important for comprehensive utilisation of biomass forestry, animal husbandry, fishery, agricultural economy, protecting the environment, realising agricultural recycling, as well as improving the sanitary conditions, in rural areas. The introduction of biogas technology on wide scale has implications for macro planning such as the allocation of government investment and effects on the balance of payments. Factors that determine the rate of acceptance of biogas plants, such as credit facilities and technical backup services, are likely to have to be planned as part of general macro-policy, as do the allocation of research and development funds [5].

Biogas is a generic term for gases generated from the decomposition of organic material. As the material breaks down, methane (CH_4) is produced as shown in Figure 3. Sources that generate biogas are numerous and varied. These include landfill sites, wastewater treatment plants and anaerobic digesters. Landfills and wastewater treatment plants emit biogas from decaying waste. To date, the waste industry has focused on controlling these emissions to our environment and in some cases, tapping this potential source of fuel to power gas turbines, thus generating electricity. The primary components of landfill gas are methane (CH_4), carbon dioxide (CO_2), and nitrogen (N_2). The average concentration of methane is ~45%, CO_2 is ~36% and nitrogen is ~18%. Other components in the gas are oxygen (O_2), water vapour and trace amounts of a wide range of non-methane organic compounds (NMOCs).

For hot water and heating, renewables contributions come from biomass power and heat, geothermal direct heat, ground source heat pumps, and rooftop solar hot water and space heating systems. Solar assisted cooling makes a very small but growing contribution. When it comes to the installation of large amounts of PV, the cities have several important factors in common. These factors include:

- A strong local political commitment to the environment and sustainability.
- The presence of municipal departments or offices dedicated to the environment, sustainability or renewable energy.
- Information provision about the possibilities of renewables.
- Obligations that some or all buildings include renewable energy.

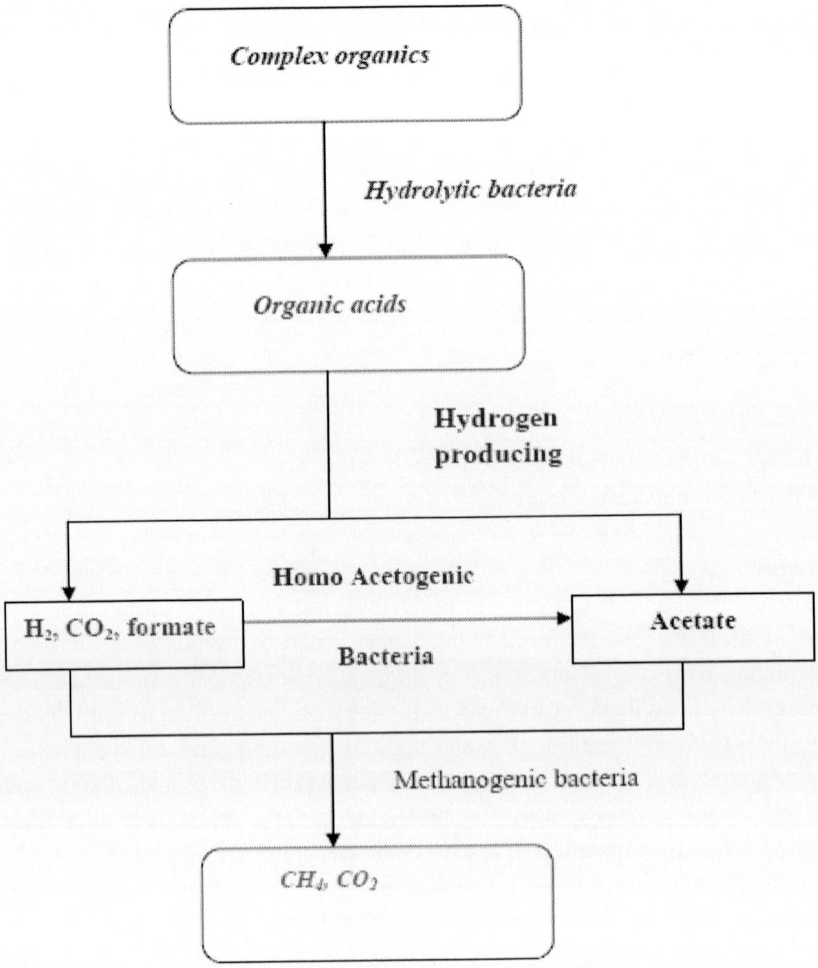

Figure 3. Biogas production process.

5.1.4. Improved Forest and Tree Management

Dry cell batteries are a practical but expensive form of mobile fuel that is used by rural people when moving around at night and for powering radios and other small appliances. The high cost of dry cell batteries is financially constraining for rural households, but their popularity gives a good indication of how valuable a versatile fuel like electricity is in rural area. Dry cell batteries can constitute an environmental hazard unless they are recycled in a proper fashion. Direct burning of fuel-wood and crop residues constitute the main usage of biomass, as is the case with many developing countries. However, the direct burning of biomass in an inefficient manner causes economic loss and adversely affects human health. In order to address the problem of inefficiency, research centres around the world have investigated the viability of converting the resource to a more useful form, namely solid briquettes and fuel gas (Figure 4). Biomass resources play a significant role in energy supply in all developing countries. Biomass resources should be divided into residues or dedicated resources, the latter including firewood and charcoal can also be produced from forest residues (Table 12).

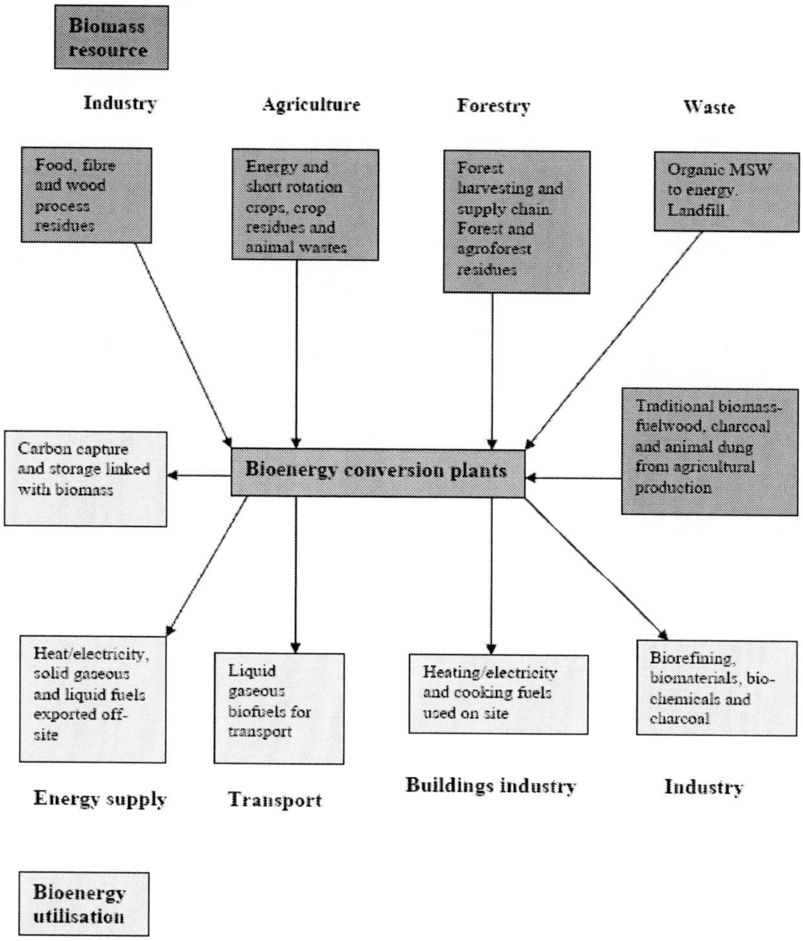

Figure 4. Biomass resources from several sources are converted into a range of products for use by transport, industry and building sectors [6].

5.1.5. Gasification

Gasification is based on the formation of a fuel gas (mostly CO and H_2) by partially oxidising raw solid fuel at high temperatures in the presence of steam or air. The technology can use wood chips, groundnut shells, sugar cane bagasse, and other similar fuels to generate capacities from 3 kW to 100 kW. Three types of gasifier designs have been developed to make use of the diversity of fuel inputs and to meet the requirements of the product gas output (degree of cleanliness, composition, heating value, etc.). The requirements of gas for various purposes, and a comparison between biogas and various commercial fuels in terms of calorific value, and thermal efficiency are presented in Table 13.

Sewage sludge is rich in nutrients such as nitrogen and phosphorous. It also contains valuable organic matter, useful for remediation of depleted or eroded soils. This is why untreated sludge has been used for many years as a soil fertiliser and for enhancing the organic matter of soil. A key concern is that treatment of sludge tends to concentrate heavy metals, poorly biodegradable trace organic compounds and potentially pathogenic organisms (viruses, bacteria and the like) present in wastewaters. These materials can pose a serious threat to the environment. When deposited in soils, heavy metals are passed through the food chain, first entering crops, and then animals that feed on the crops and eventually human beings, to whom they appear to be highly toxic. In addition they also leach from soils, getting into groundwater and further spreading contamination in an uncontrolled manner. European and American markets aiming to transform various organic wastes (animal farm wastes, industrial and municipal wastes) into two main by-products:

- A solution of humic substances (a liquid oxidate).
- A solid residue.

Table 12. Biomass residues and current use

Type of residue	Current use
Wood industry waste	Residues available
Vegetable crop residues	Animal feed
Food processing residue	Energy needs
Sorghum, millet, wheat residues	Fodder, and building materials
Groundnut shells	Fodder, brick making, direct fining oil mills
Cotton stalks	Domestic fuel considerable amounts available for short period
Sugar, bagasse, molasses	Fodder, energy need, ethanol production (surplus available)
Manure	Fertiliser, brick making, plastering

Table 13. Comparison of various fuels [7]

Fuel	Calorific value (kcal)	Burning mode	Thermal efficiency (%)
Electricity, kWh	880	Hot plate	70
Coal gas, kg	4004	Standard burner	60
Biogas, m^3	5373	‘’ ‘’	60
Kerosene, l	9122	Pressure stove	50
Charcoal, kg	6930	Open stove	28
Soft coke, kg	6292	‘’ ‘’	28
Firewood, kg	3821	‘’ ‘’	17
Cow dung, kg	2092	‘’ ‘’	11

Biomass is a raw material that has been utilised for a wide variety of tasks since the dawn of civilisation. Important as a supply of fuel in the third world, biomass was also the first raw material in the production of textiles. The gasification of the carbon char with steam can make a large difference to the surface area of the carbon. The corresponding stream gasification reactions are endothermic and demonstrate how the steam reacts with the carbon char [8].

$$H_2O\ (g) + C_x\ (s) \rightarrow H_2\ (g) + CO\ (g) + C_{x-1}\ (s) \quad (2)$$

$$CO\ (g) + H_2O\ (g) \rightarrow CO_2\ (g) + H_2\ (g) \quad (3)$$

$$CO_2\ (g) + C_x\ (s) \rightarrow 2\,CO\ (g) + C_{x-1}\ (s) \quad (4)$$

Agricultural wastes are abundantly available globally and can be converted to energy and useful chemicals by a number of microorganisms. The organic matter was biodegradable to produce biogas and the variation show a normal methanogene bacteria activity and good working biological process as shown in Figures 5-7. The success of promoting any technology depends on careful planning, management, implementation, training and monitoring. Main features of gasification project are:

- Networking and institutional development/strengthening.
- Promotion and extension.
- Construction of demonstration projects.
- Research and development, and training and monitoring.

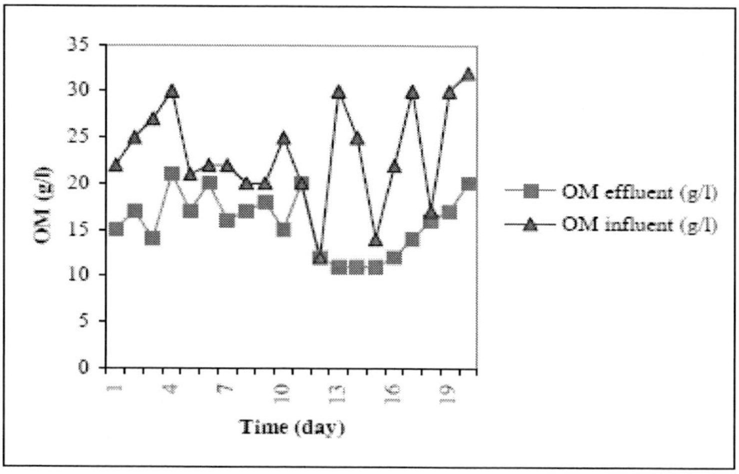

Figure 5. Organic matters before and after treatment in digester.

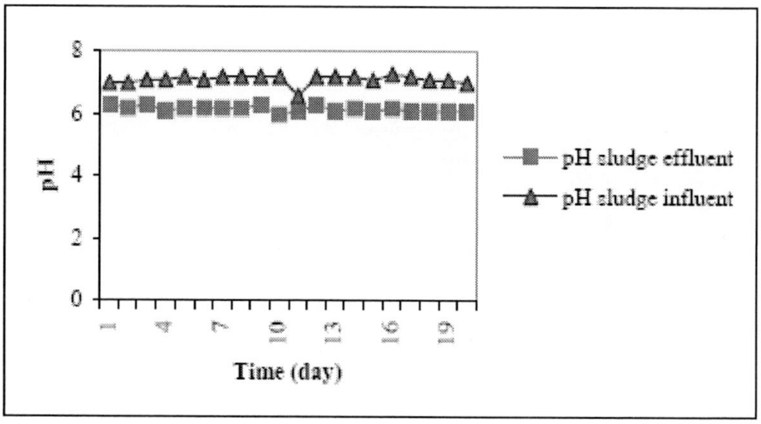

Figure 6. pH sludge before and after treatment in the digester.

6. BIOHEAT

Bioenergy is a growing source of power that is playing an ever-increasing role in the provision of electricity. The potential contribution of the waste industry to bioenergy is huge and has the ability to account for a source of large amount of total bioenergy production. Woody biomass is usually converted into power through combustion or gasification. Biomass can be specially grown in the case of energy crops. Waste wood makes up a significant proportion of a variety of municipal, commercial and industrial waste streams. It is common practice to dispose of this waste wood in landfill where it slowly degraded and takes up valuable void space. This wood is a good source of energy and is an alternative to energy crops.

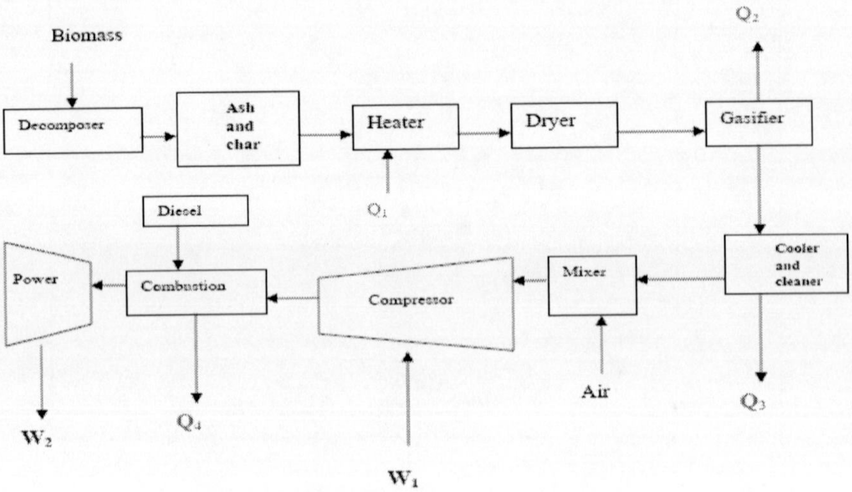

Figure 7 Advanced biomass power with diesel engine

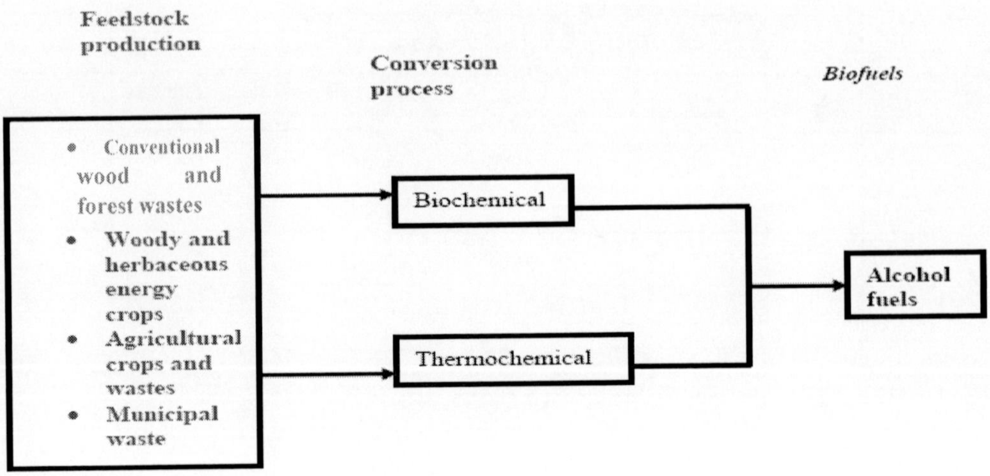

Figure 8. Biofuel pathways for renewable alcohol fuels.

The biomass directly produced by cultivation can be transformed by different processes into gaseous, liquid or solid fuels (Table 14). The whole process of production of methyl or ethyl esters (biodiesel) is summarised in Figures 8-9.

Table 14. Biomass conversions to energy

Feedstock	Crops	Conversion process	End product
Wood-cellulosic biomass	Short rotation forest (poplar, willow), plant species (sorghum, mischantus, etc.), fibre-crops (cynara, kenaf, etc.)	Direct combustion Gasification Pyrolysis	Heat Methane Hydrogen Oil
Vegetable oils	Oleaginous crops (rapeseed, soybean, sunflower, etc.)	Direct combustion Esterification	Heat Biodiesel
Sugar/starch	Cereals, root and tuber crops, grape, topinambour, etc.	Fermentation	Ethanol

6.1. Waste Policy in Context

In terms of solid waste management policy, many NGOs have changed drastically in the past ten years from a mass production and mass consumption society to 'material-cycle society'. In addition to national legislation, municipalities are legally obliged to develop a plan for handling the municipal solid waste (MSW) generated in administrative areas. Such plans contain:

- Estimates of future waste volume.
- Measures to reduce waste.
- Measures to encourage source separation.
- A framework for solid waste disposal and the construction and management of solid waste management facilities.

Landfilling is in the least referred tier of the hierarchy of waste management options: Waste minimisation, reuse and recycling, incineration with energy recovery, and optimised final disposal. The key elements are as follows: construction impacts, atmospheric emissions, noise, water quality, landscape, visual impacts, socio economics, ecological impacts, traffic, solid waste disposal and cultural heritage.

6.2. Energy from Agricultural Biomass

The main advantages are related to energy, agriculture and environment problems, are foreseeable both at regional level and at worldwide level and can be summarised as follows:

- Reduction of dependence on import of energy and related products.

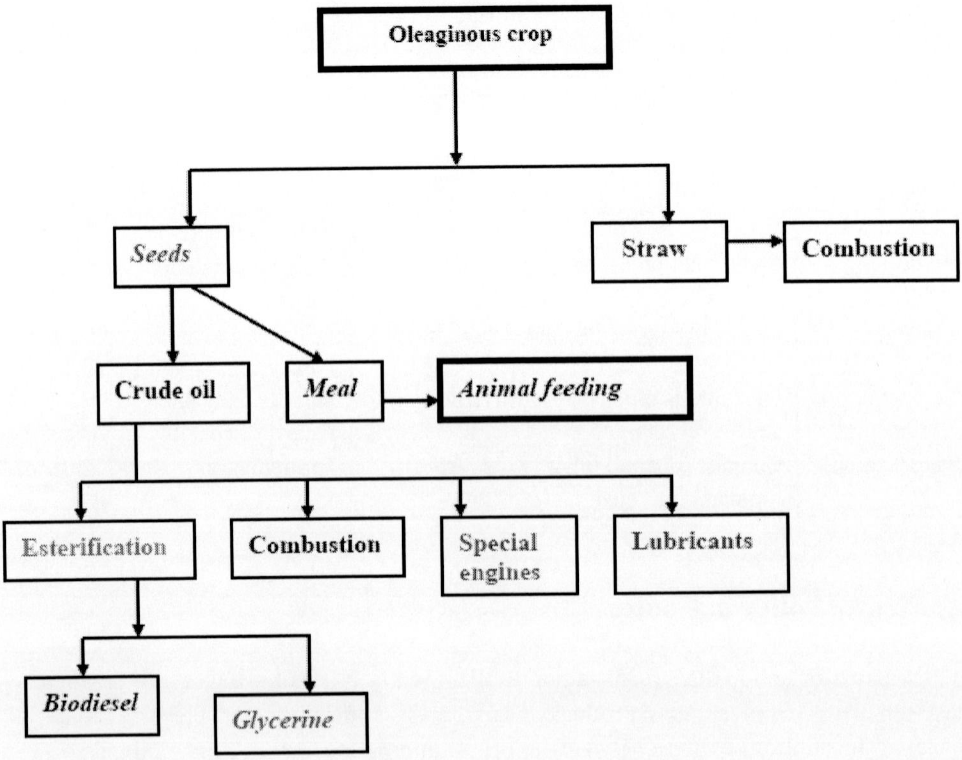

Figure 9. Flow chart of biodiesel production.

- Reduction of environmental impact of energy production (greenhouse effect, air pollution, waste degradation).
- Substitution of food crops and reduction of food surpluses and of related economic burdens.
- Utilisation of marginal lands and of set aside lands and reduction of related socio-economic and environmental problems (soil erosion, urbanisation, landscape deterioration, etc.).
- Development of new know-how and production of technological innovation.

A study [8] individuated on the basis of botanical, genetical, physiological, biochemical, agronomical and technological knowledge reported in literature some 150 species potentially exploitable divided as reported in Table 15.

7. Role of Chemical Engineering

Turning to chemical engineering and the experience of the chemical process industry represents a wakening up but does not lead to an immediate solution to the problems. The traditional techniques are not very kind to biological products, which are controlled by difficulty and unique physico-chemical properties such as low mechanical, thermal and chemical stabilities. There is the question of selectivity. The fermentation broths resulting from microbial growth contain a bewildering mixture of many compounds closely related to the product of interests. By the standards of the process streams in chemical industry, fermenter is highly impure and extremely dilutes aqueous systems (Table 16).

The disadvantages of the fermentation media are as the following: mechanically fragile, temperature sensitive, rapidly deteriorating quality, harmful if escaping into the environment, corrosive (acids, chlorides, etc.), and troublesome (solids, theological, etc.), and expensive.

Table 15. Plant species potentially exploitable for production of agricultural biomass for energy or industrial utilisations [9]

Groups of plants	Number of species
Plants cultivated for food purposes that can be reconverted to new uses	9
Plants cultivated in the past, but not in culture any more	46
Plants cultivated in other world areas	46
Wild species, both indigenous and exotic	47
Total	148
Plant product	Number of species
Biomass	8
Sugars and polysaccharides	38
Cellulose	17
Hydrocarbons	3
Polymeric hydrocarbons	5
Gums and resins	12
Tannins and phenolic compounds	3
Waxes	7
Vegetable oils	38
Total	131

Thus, pilot plants for scale-up work must be flexible. In general, they should contain suitably interconnected equipment for: fermentation, primary separation, cell disruption fractionalises and clarifications, purification by means of high-resolution techniques and concentration and dry. The effects of the chlorofluorocarbons (CFCs) molecule can last for over a century.

Table 16. Typical product concentrations exiting fermenters [9]

Product	Concentration (kg/m^3)
Ethanol	70-120
Organic acids (e.g., citric)	40-100
Vitamin B12	0.02
Interferon	50-70
Single-cell protein	30-50
Antibiotics (e.g., Penicillin G)	10-30
Enzyme protein (e.g., protease)	2-5

7.1. Fluidised Bed Drying

An important consideration for operators of wastewater treatment plants is how to handle the disposal of the residual sludge in a reliable, sustainable, legal and economical way. The benefits of drying sludge can be seen in two main treatment options:

- Use of the dewatered sludge as a fertiliser or in fertiliser blends.
- Incineration with energy recovery.

Use as a fertiliser takes advantage of the high organic content 40%-70% of the dewatered sludge and its high levels of phosphorous and other nutrients. However, there are a number of concerns about this route including:

- The chemical composition of the sludge (e.g., heavy metals, hormones and other pharmaceutical residues).
- Pathogen risk (e.g., salmonella, escherichia coli, prionic proteins, etc.).
- Potential accumulation of heavy metals and other chemicals in the soil.

Sludge can be applied as a fertiliser in three forms: liquid sludge, wet cake blended into compost, and dried granules.

The advantages of energy recovery sludge include:

- The use of dewatered sludge is a 'sink' for pollutants such as heavy metals, toxic organic compounds and pharmaceutical residues. Thus, offering a potential disposal route for these substances provided the combustion plant has adequate flue gas cleaning.
- The potential, under certain circumstances, to utilise the inorganic residue from sludge incineration (incinerator ash), such as in cement or gravel.
- The high calorific value (similar to lignite) of dewatered sludge.
- The use of dewatered sludge as a carbon dioxide neutral substitute for primary fuels such as oil, gas and coal.

7.2. Energy Efficiency

Energy efficiency is the most cost-effective way of cutting carbon dioxide emissions and improvements to households and businesses. It can also have many other additional social, economic and health benefits, such as warmer and healthier homes, lower fuel bills and company running costs and, indirectly, jobs. Britain wastes 20 per cent of its fossil fuel and electricity use. This implies that it would be cost-effective to cut £10 billion a year off the collective fuel bill and reduce CO_2 emissions by some 120 million tones. Yet, due to lack of good information and advice on energy saving, along with the capital to finance energy efficiency improvements, this huge potential for reducing energy demand is not being realised. Traditionally, energy utilities have been essentially fuel providers and the industry has pursued profits from increased volume of sales. Institutional and market arrangements have favoured energy consumption rather than conservation. However, energy is at the centre of the sustainable development paradigm as few activities affect the environment as much as the continually increasing use of energy. Most of the used energy depends on finite resources, such as coal, oil, gas and uranium. In addition, more than three quarters of the world's consumption of these fuels is used, often inefficiently, by only one quarter of the world's population. Without even addressing these inequities or the precious, finite nature of these resources, the scale of environmental damage will force the reduction of the usage of these fuels long before they run out.

Throughout the energy generation process there are impacts on the environment on local, national and international levels, from opencast mining and oil exploration to emissions of the potent greenhouse gas carbon dioxide in ever increasing concentration. Recently, the world's leading climate scientists reached an agreement that human activities, such as burning fossil fuels for energy and transport, are causing the world's temperature to rise. The Intergovernmental Panel on Climate Change has concluded that ''the balance of evidence suggests a discernible human influence on global climate''. It predicts a rate of warming greater than any one seen in the last 10,000 years, in other words, throughout human history. The exact impact of climate change is difficult to predict and will vary regionally. It could, however, include sea level rise, disrupted agriculture and food supplies and the possibility of more freak weather events such as hurricanes and droughts. Indeed, people already are waking up to the financial and social, as well as the environmental, risks of unsustainable energy generation methods that represent the costs of the impacts of climate change, acid rain and oil spills. The insurance industry, for example, concerned about the billion dollar costs of hurricanes and floods, has joined sides with environmentalists to lobby for greenhouse gas emissions reduction. Friends of the earth are campaigning for a more sustainable energy policy, guided by the principal of environmental protection and with the objectives of sound natural resource management and long-term energy security. The key priorities of such an energy policy must be to reduce fossil fuel use, move away from nuclear power, improve the efficiency with which energy is used and increase the amount of energy obtainable from sustainable, renewable sources. Efficient energy use has never been more crucial than it is today, particularly with the prospect of the imminent introduction of the climate change levy (CCL). Establishing an energy use action plan is the essential foundation to the elimination of energy waste. A logical starting point is to carry out an energy audit that enables the assessment of the energy use and determine what actions to take. The actions are best categorised by splitting measures into the following three general groups:

(1) High priority/low cost:

These are normally measures, which require minimal investment and can be implemented quickly. The followings are some examples of such measures:

- Good housekeeping, monitoring energy use and targeting waste-fuel practices.
- Adjusting controls to match requirements.
- Improved greenhouse space utilisation.
- Small capital item time switches, thermostats, etc.
- Carrying out minor maintenance and repairs.

- Staff education and training.
- Ensuring that energy is being purchased through the most suitable tariff or contract arrangements.

(2) Medium priority/medium cost:

Measures, which, although involve little or no design, involve greater expenditure and can take longer to implement. Examples of such measures are listed below:

- New or replacement controls.
- Greenhouse component alteration e.g., insulation, sealing glass joints, etc.
- Alternative equipment components e.g., energy efficient lamps in light fittings, etc.

(3) Long term/high cost:

These measures require detailed study and design. They can be best represented by the followings:

- Replacing or upgrading of plant and equipment.
- Fundamental redesign of systems e.g., CHP installations.

This process can often be a complex experience and therefore the most cost-effective approach is to employ an energy specialist to help.

7.3. Policy Recommendations for a Sustainable Energy Future

Sustainability is regarded as a major consideration for both urban and rural development. People have been exploiting the natural resources with no consideration to the effects, both short-term (environmental) and long-term (resources crunch). It is also felt that knowledge and technology have not been used effectively in utilising energy resources. Energy is the vital input for economic and social development of any country. Its sustainability is an important factor to be considered. The urban areas depend, to a large extent, on commercial energy sources. The rural areas use non-commercial sources like firewood and agricultural wastes. With the present day trends for improving the quality of life and sustenance of mankind, environmental issues are considered highly important. In this context, the term energy loss has no significant technical meaning. Instead, the exergy loss has to be considered, as destruction of exergy is possible. Hence, exergy loss minimisation will help in sustainability. In the process of developing, there are two options to manage energy resources: (1) End use matching/demand side management, which focuses on the utilities. The mode of obtaining this is decided based on economic terms. It is, therefore, a quantitative approach. (2) Supply side management, which focuses on the renewable energy resource and methods of utilising it. This is decided based on thermodynamic consideration having the resource-user temperature or exergy destruction as the objective criteria. It is, therefore, a qualitative approach. The two options are explained schematically in Figure 10. The exergy-based energy, developed with supply side perspective is shown in Figure 11.

The following policy measures had been identified:

- Clear environmental and social objectives for energy market liberalisation, including a commitment to energy efficiency and renewables.
- Economic, institutional and regulatory frameworks, which encourage the transition to total energy services.
- Economic measures to encourage utility investment in energy efficiency (e.g., levies on fuel bills).
- Incentives for demand side management, including grants for low-income households, expert advice and training, standards for appliances and buildings and tax incentives.
- Research and development funding for renewable energy technologies not yet commercially viable.
- Continued institutional support for new renewables (such as standard cost-reflective payments and obligation on utilities to buy).
- Ecological tax reform to internalise external environmental and social costs within energy prices.
- Planning for sensitive development and public acceptability for renewable energy.

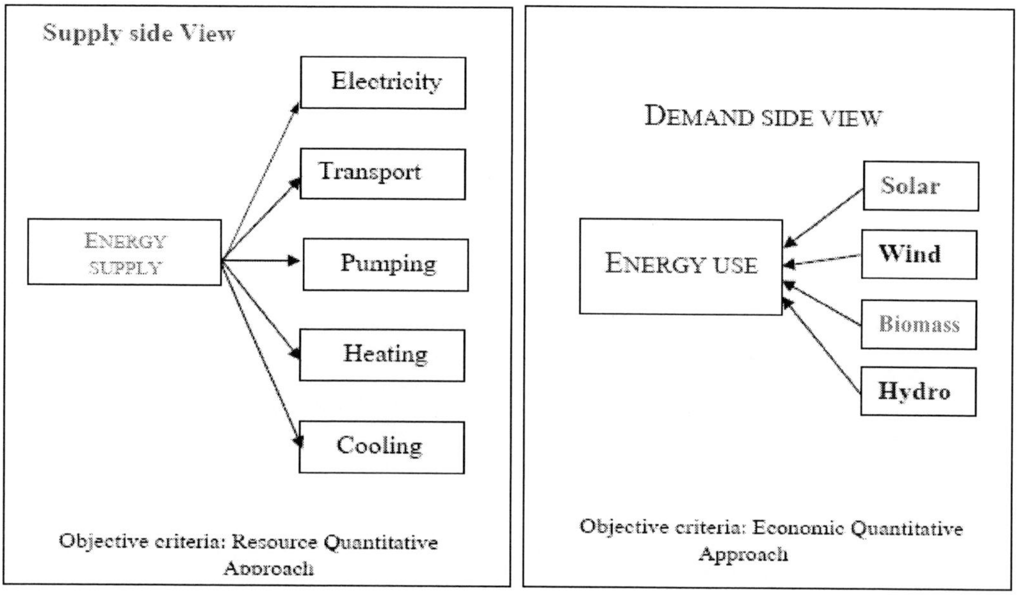

Figure 10. Supply side and demand side management approach for energy.

Energy resources are needed for societal development. Their sustainable development requires a supply of energy resources that are sustainably available at a reasonable cost and can cause no negative societal impacts. Energy resources such as fossil fuels are finite and lack sustainability, while renewable energy sources are sustainable over a relatively longer term. Environmental concerns are also a major factor in sustainable development, as activities, which degrade the environment, are not sustainable. Hence, as much as environmental impact is associated with energy, sustainable development requires the use of energy resources, which cause as little environmental impact as possible. One way to reduce the resource depletion associated with cycling is to reduce the losses that accompany the transfer of exergy to consume resources by increasing the efficiency of exergy transfer between resources i.e., increasing the fraction of exergy removed from one resource that is transferred to another [10].

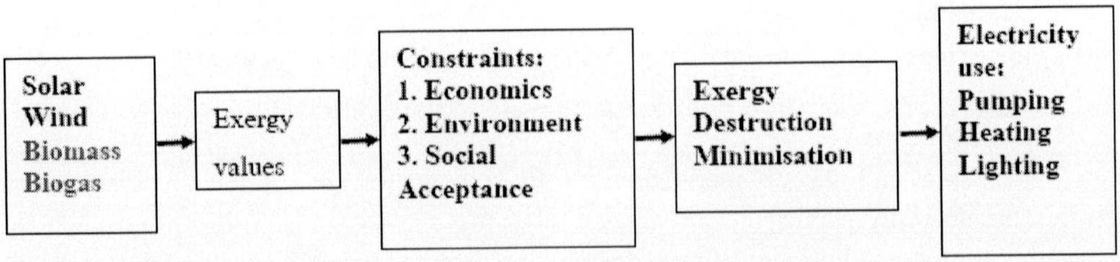

Figure 11. Exergy based optimal energy model.

As explained above, exergy efficiency may be thought of as a more accurate measure of energy efficiency that accounts for quantity and quality aspects of energy flows. Improved exergy efficiency leads to reduced exergy losses. Most efficiency improvements produce direct environmental benefits in two ways. First, operating energy input requirements are reduced per unit output, and pollutants generated are correspondingly reduced. Second, consideration of the entire life cycle for energy resources and technologies suggests that improved efficiency reduces environmental impact during most stages of the life cycle. Quite often, the main concept of sustainability, which often inspires local and national authorities to incorporate environmental consideration into setting up energy programmes have different meanings in different contexts though it usually embodies a long-term perspective. Future energy systems will largely be shaped by broad and powerful trends that have their roots in basic human needs. Combined with increasing world population, the need will become more apparent for successful implementation of sustainable development.

Heat has a lower exergy, or quality of energy, compared with work. Therefore, heat cannot be converted into work by 100% efficiency. Some examples of the difference between energy and exergy are shown in Table 17.

Table 17. Qualities of various energy sources

Source	Energy (J)	Exergy (J)	CQF
Water at 80°C	100	16	0.16
Steam at 120°C	100	24	0.24
Natural gas	100	99	0.99
Electricity/work	100	100	1.00

The terms used in Table 17 have the following meanings:

Carnot Quality Factor (CQF) = $(1-T_o/T_s)$ (5)

Exergy = Energy (transferred) x CQF (6)

where T_o is the environment temperature (K) and T_s is the temperature of the stream (K).

Various parameters are essential to achieving sustainable development in a society. Some of them are as follows:

- Public awareness.
- Information.
- Environmental education and training.
- Innovative energy strategies.
- Renewable energy sources and cleaner technologies.
- Financing.
- Monitoring and evaluation tools.

The development of a renewable energy in a country depends on many factors. Those important to success are listed below:

(1) Motivation of the population

The population should be motivated towards awareness of high environmental issues, rational use of energy in order to reduce cost. Subsidy programme should be implemented as incentives to install biomass energy plants. In addition, image campaigns to raise awareness of renewable technology.

(2) Technical product development

To achieve technical development of biomass energy technologies the following should be addressed:

- Increasing the longevity and reliability of renewable technology.
- Adapting renewable technology to household technology (hot water supply).
- Integration of renewable technology in heating technology.
- Integration of renewable technology in architecture, e.g., in the roof or façade.
- Development of new applications, e.g., solar cooling.
- Cost reduction.

(3) Distribution and sales

Commercialisation of biomass energy technology requires:

- Inclusion of renewable technology in the product range of heating trades at all levels of the distribution process (wholesale, retail).
- Building distribution nets for renewable technology.
- Training of personnel in distribution and sales, training of field sales force.

(4) Consumer consultation and installation

To encourage all sectors of the population to participate in adoption of biomass energy technologies, the following has to be realised:

- Acceptance by craftspeople, marketing by them.
- Technical training of craftspeople, initial and follow-up training programmes.
- Sales training for craftspeople.
- Information material to be made available to craftspeople for consumer consultation.

(5) Projecting and planning
Successful application of biomass technologies also require:

- Acceptance by decision makers in the building sector (architects, house technology planners, etc.).
- Integration of renewable technology in training.
- Demonstration projects/architecture competitions.
- Biomass energy project developers should prepare to participate in the carbon market by:

- Ensuring that renewable energy projects comply with Kyoto Protocol requirements.
- Quantifying the expected avoided emissions.
- Registering the project with the required offices.
- Contractually allocating the right to this revenue stream.

- Other ecological measures employed on the development include:

- Simplified building details.
- Reduced number of materials.
- Materials that can be recycled or reused.
- Materials easily maintained and repaired.
- Materials that don't have a bad influence on the indoor climate (i.e., non-toxic).
- Local cleaning of grey water.
- Collecting and use of rainwater for outdoor purposes and park elements.
- Building volumes designed to give maximum access to neighbouring park areas.
- All apartments have visual access to both backyard and park.

(6) Energy saving measures
The following energy saving measures should also be considered:

- Building integrated solar PV system.
- Day-lighting.
- Ecological insulation materials.
- Natural/hybrid ventilation.
- Passive cooling.
- Passive solar heating.
- Solar heating of domestic hot water.
- Utilisation of rainwater for flushing.

Improving access for rural and urban low-income areas in developing countries must be through energy efficiency and renewable energies. Sustainable energy is a prerequisite for development. Energy-based living standards in developing countries, however, are clearly below standards in developed countries. Low levels of access to affordable and environmentally sound energy in both rural and urban low-income areas are therefore a predominant issue in developing countries. In recent years many programmes for development aid or technical assistance have been focusing on improving access to sustainable energy, many of them with impressive results [11].

Apart from success stories, however, experience also shows that positive appraisals of many projects evaporate after completion and vanishing of the implementation expert team. Altogether, the diffusion of sustainable technologies such as energy efficiency and renewable energies for cooking, heating, lighting, electrical appliances and building insulation in developing countries has been slow.

Energy efficiency and renewable energy programmes could be more sustainable and pilot studies more effective and pulse releasing if the entire policy and implementation process was considered and redesigned from the outset. New financing and implementation processes are needed which allow reallocating financial resources and thus enabling countries themselves to achieve a sustainable energy infrastructure. The links between the energy policy framework, financing and implementation of renewable energy and energy efficiency projects have to be strengthened and capacity building efforts are required.

8. Conclusions

Even with modest assumptions about the availability of land, comprehensive fuel-wood farming programmes offer significant energy, economic and environmental benefits. These benefits would be dispersed in rural areas where they are greatly needed and can serve as linkages for further rural economic development. The nations, as a whole would benefit from savings in foreign exchange, improved energy security, and socio-economic improvements. With a nine-fold increase in forest – plantation cover, the nation's resource base would be greatly improved. The international community would benefit from pollution reduction, climate mitigation, and the increased trading opportunities that arise from new income sources. The aim of any modern biomass energy systems must be:

- To maximise yields with minimum inputs.
- Utilisation and selection of adequate plant materials and processes.
- Optimum use of land, water, and fertiliser.
- Create an adequate infrastructure and strong R and D base.

Furthermore, investigating the potential to make use of more and more of its waste. Household waste, vegetable market waste, and waste from the cotton stalks, leather, and pulp; and paper industries can be used to produce useful energy either by direct incineration, gasification, digestion (biogas production), fermentation, or cogeneration. Therefore, effort has to be made to reduce fossil energy use and to promote green energies, particularly in the building sector. Energy use reductions can be achieved by minimising the energy demand, by rational energy use, by recovering heat and the use of more green energies. This study was a step towards achieving that goal. The adoption of green or sustainable approaches to the way in which society is run is seen as an important strategy in finding a solution to the energy problem. The key factors to reducing and controlling CO_2, which is the major contributor to global warming, are the use of alternative approaches to energy generation and the exploration of how these alternatives are used today and may be used in the future as green energy sources. Even with modest assumptions about the availability of land, comprehensive fuel-wood farming programmes offer significant energy, economic and environmental benefits. These benefits would be dispersed in rural areas where they are greatly needed and can serve as linkages for further rural economic development. The nations as a whole would benefit from savings in foreign exchange, improved energy security, and socio-economic improvements. With a nine-fold increase in forest – plantation cover, a nation's resource base would be greatly improved. The international community would benefit from pollution reduction, climate mitigation, and the increased trading opportunities that arise from new income sources. The non-technical issues, which have recently gained attention, include: (1) Environmental and ecological factors e.g., carbon sequestration, reforestation and revegetation. (2) Renewables as a CO_2 neutral replacement for fossil fuels. (3) Greater recognition of the importance of renewable energy, particularly modern biomass energy carriers, at the policy and planning levels. (4) Greater recognition of the difficulties of gathering good and reliable renewable energy data, and efforts to improve it. (5) Studies on the detrimental health efforts of biomass energy particularly from traditional energy users.

REFERENCES

[1] Robinson, G. Changes in construction waste management. Waste Management World p. 43-49. May-June 2007.

[2] Wu, J. and Boggess, W. The optimal allocation of conservation funds. Journal Environmental Economic Management. 1999: 38.

[3] Pernille, M. Feature: Danish lessons on district heating. Energy Resource Sustainable Management and Environmental March/April 2004: 16-17.

[4] D'Apote, S.L. IEA biomass energy analysis and projections. In: Proceedings of Biomass Energy Conference: Data, analysis and Trends, Paris: OECD; 23-24 March 1998.

[5] Hall O. and Scrase J. Will biomass be the environmentally friendly fuel of the future? Biomass and Bioenergy 1998: 15: 357-67.

[6] Sims, R.H. Not too late: IPCC identifies renewable energy as a key measure to limit climate change. Renewable Energy World 10 (4): 31-39. 2007.

[7] Omer, A.M., et al. 2003. Biogas energy technology in Sudan. Renewable Energy, 28 (3): 499-507.

[8] Bacaoui, A., Yaacoubi, A., Dahbi, C., Bennouna, J., and Mazet, A. Activated carbon production from Moroccan olive wastes-influence of some factors. Environmental Technology 19: 1203-1212. 1998.

[9] Rossi, S., Arnone, S., Lai, A., Lapenta, E., and Sonnino, A. ENEA's activities for developing new crops for energy and industry. In: Biomass for Energy and Industry (G.

Grassi, G. Gosse, G. dos Santos Eds.). Vol.1, p.107-113, Elsevier Applied Science, London and New York. 1990.

[10] Erlich, P. Forward facing up to climate change, in Global Climate Change and Life on Earth. R.C. Wyman (Ed), Chapman and Hall, London. 1991.

[11] Omer, A.M. Review: Organic waste treatment for power production and energy supply. Cells and Animal Biology 1 (2): 34-47.

In: Energy in Europe: Economics, Policy and Strategy
Editors: F. L. Magnusson and O. W. Bengtsson.
ISBN: 978-1-60456-829-5

Chapter 11

POSSIBILITY FOR ADAPTATION OF CONVENTIONAL ENERGY TO EUROPEAN ENERGY STRATEGY: CONSIDERATION OF COAL UTILIZATION BASED ON CASE STUDIES

Ryunosuke Kikuchi[1]
Department of Basic Science and Environment (CERNAS); ESAC - Polytechnic Institute of Coimbra, Bencanta, 3040-316 Coimbra, Portugal

ABSTRACT

Coal production has been in decline since 1980 in Western Europe; this decline has mainly resulted from the reduced consumption of low-quality lignite in Germany. Although some coal-fired power plants are earmarked for retirement in Europe, investment plans have been embarked on to refurbish the existing coal plants for the purpose of extending their operating lives. Coal is recognized as an indigenous energy base, and subsidies to the coal industry are authorized by the European Commission.

Since greenhouse gas, pollutant gases and dust are associated with coal utilization, the introduction of coal clean technologies is required to solve these environmental pressures. Furthermore, it is important to combine renewable energy with conventional energy (e.g. coal) in a proper manner. The EU has a binding target to increase the share of reneweable energy in total energy consumption in the EU to 20% by 2020, that is, the remaining 80% of required energy will have to come from conventional energy sources (e.g. coal).

A case study in Portugal shows that renewable energy (hydroelectricity production) is quite variable, and a model-based study in Ireland shows that the potential fuel and emissions savings are tempered by the inherent intermittence of wind and the increase in total generation costs (138 EUR per CO_2 ton).

The fluctuant supply of energy from renewable sources should be supplemented by conventional energy sources (e.g. coal-fired thermal plant) in order to meet the required demand. Indigenous coal has an important role in not only securing national energy but also measuring the variable supply of renewable energy, but the problem lies in the

[1] Tel.: +351 239 802287; fax: +351 239 802979. E-mail address: kikuchi@mail.esac.pt.

combination of renewable energy with coal-based energy production: although there is a reduction in the overall fuel bill when wind powered generation is introduced, this reduction is curtailed by the increased inefficiency of thermal power plants as a result of frequent start-ups and low loading levels. It is not advisable to separately consider renewable energy and conventional energy when determining how to realize a 20% share of renewable energy in total energy consumption in Europe. Renewable energy and coal (as a conventional energy source) are not mutually opposed but should coexist in the EU energy strategy. To this end, each new coal-fired power system is required to make up variable energy production from renewable sources, and this system is guided by the following conditions – frequent start-up with energy-saving measures, achievement of target efficiency directly after start-up, and stable efficiency at a low loading level.

INTRODUCTION

Sustainable development has been defined in many ways, including development that meets the needs of the present without compromising the ability of future generations to meet their own needs [World Commission on Environment and Development, 1987]. Sustainable development or sustainability is a long-term objective of the policies of European parties, and it is particularly stressed in Austria, the Czech Republic, the Netherlands, Portugal, Slovenia and the Russian Federation [Economic Commission for Europe, 1995].

How society meets its needs for energy services is central to the prospects for sustainable development and one of the most critical challenges facing humanity today [Science Council of Japan, 2003]. The challenge for energy policy is that of reducing the environmental costs of energy production and use, while extending access to basic energy services (e.g. refrigeration, heating and lighting), and preserving energy security. However, energy development is increasingly dominated by major concerns of air pollution, fresh water pollution, coastal pollution, deforestation, biodiversity loss, and global climate deterioration [Linor, 2007]. To prevent disastrous global consequences, it would be increasingly impossible to engage in large-scale energy-related activities without insuring their sustainability, even for European countries in which there is a perceived priority of energy development and use and power generation over their impact on the environment, society and indeed on the energy sources themselves.

The share of alternative energy in total energy consumption in Europe increased slightly during 1992-2004: (i) from 5.5% to 6.7% in Western Europe, (ii) from 10.4% to 11.9% in Central and Eastern Europe and (iii) from 2.5% to 2.8% in Eastern Europe plus the Caucasus and Central Asia (see Figure 1) [International Energy Agency, 2006]. In Western Europe, this growth was supported by a range of policy interventions, mainly aimed at stimulating the promotion of new renewable technologies; in Central and Eastern Europe, most growth came from an expansion of hydropower and waste/biomass incineration; the rate of renewable energy was low in 12 countries of Eastern Europe plus the Caucasus and Central Asia due to a decrease in energy production from combustible renewable sources and hydropower [Directorate General for Energy, 1999].

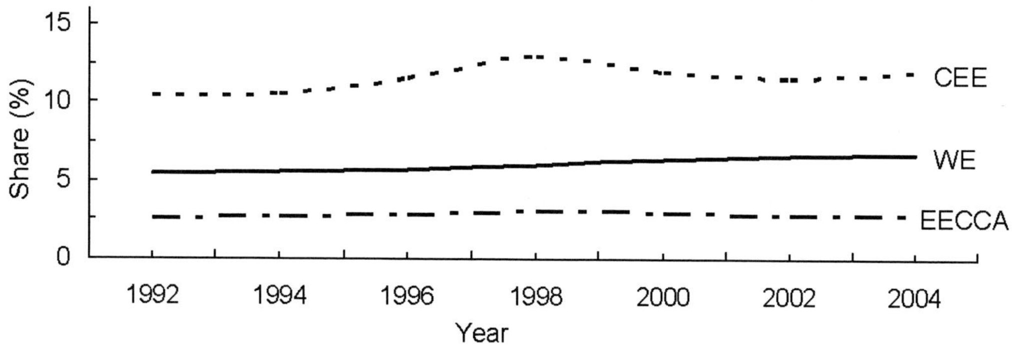

Figure 1. Contribution of renewable energy sources (including waste) to total energy consumption in Europe during 1992 to 2004 (redrawn from [International Energy Agency, 2006]). Western Europe (WE) = Austria, Belgium, Denmark, Finland, France, Germany, Greece, Ireland, Luxembourg, the Netherlands, Portugal, Spain, United Kingdom, Ireland, Liechtenstein, Norway, Switzerland, Andorra, Monaco and San Marino; Central and Eastern Europe (CEE) = Albania, Bosnia-Herzegovina, Macedonia, Hungary, Latvia, Lithuania, Poland, Romania, Serbia and Montenegro, Slovak, Slovenia, Marta and Turkey; and 12 countries of Eastern Europe, the Caucasus and Central Asia (EECCA) = Armenia, Azerbaijan, Belarus, Georgia, Moldova, Russia, Ukraine, Kazakhstan, Tajikistan, Turkmenistan and Uzbekistan.

The future contribution of renewable sources to total energy consumption has been foreseen by different types of scenarios [Directorate General for Energy, 1999]: the Conventional Wisdom (CW) scenario denotes the "business as usual" world, representing a conventional wisdom view of events; in the Battlefield (BF) scenario, the world reverts to isolationism, power blocks and protectionism; under Hypermarket (HM), the predominant themes are market forces, liberalism and free trade, so there is a minimum of intervention from government and public administration; and in Forum (FO), the world moves to consensus and a co-operative international structure with a strong role for public administration. These scenarios show that the prevailing trend of renewable energy is the growing penetration in Europe by 2020: its share of total primary energy production will increase to up to 30% in the CW scenario, 20% in the BF scenario, 25% in the HM scenario and 30% in the FO scenario [Directorate General for Energy, 1999]. In January 2007, the European Commission presented a package (COM/2006 848 final) on climate change and energy which included a new strategy within the energy sector: a target share of 12% has been set for renewable energy in total energy consumption by 2010 [Commission of the European Communities, 2007]. However, the long-term perspectives and the new strategy suggest that the remaining major portion of required energy will have to come from conventional energy sources such as coal, oil and nuclear. Focusing attention on this major portion, the present situation and the future of European energy are discussed.

Overview of European Energy

Basic information about the current situation in Europe is briefly reviewed first, followed by a description of the main discussion.

European Union

The European Union (EU) only comprised 12 countries in the early 1990s. Austria, Finland and Sweden entered the Union in 1995. From 2004 onwards, the Union expanded again as Cyprus, Malta and eight Central and East European countries entered the EU. Finally, the EU attained its current form when Romania and Bulgaria joined the community in 2007. In this case, it is practical to treat the EU as two separate entities: the 15 old member states (referred to as EU 15) which committed themselves to the Kyoto protocol as one group, and the 12 more recent arrivals (referred to as New Member States - NMS 12). The total population of the EU 27 nations was estimated at approximately 486 million in 2004 [Eurostat, 2007]. At the same time, real gross domestic product (GDP) measured in constant prices ($ = international dollar in 2000) was approximately $\$1.1 \times 10^{13}$ or around $21,800 per capita [Heston et al., 2007]. The total energy consumption was approximately 1170 MToe (million tons of oil equivalent) in 2004 [European Environment Agency, 2007].

Energy and Environment

Energy use contributes to a range of environmental pressures and is the major source of greenhouse and acid gas emissions in Europe. Options for reducing environmental pressures include using less-polluting energy sources, using energy more efficiently, and using less of the energy-consuming services such as transport, space heating and manufactured products. Figure 2 shows the energy-related emissions of greenhouse gas and air pollutants (sulfur dioxide SO_2 and nitrogen oxide NOx).

Overall emissions of energy-related greenhouse gas emissions in Europe fell considerably between 1990 and 1999: this was due mainly to the Russian Federation and Ukraine, two of the biggest energy consumers in Europe, which reduced their total emissions by 36% and 50% respectively over the period [European Environment Agency, 2003]. These reductions were mostly the result of economic difficulties and restructuring, which resulted in a substantial reduction in the energy use of these two countries over this period. However, it follows from Figure 2a that energy-related emissions in Western Europe fell by only 1.6%.

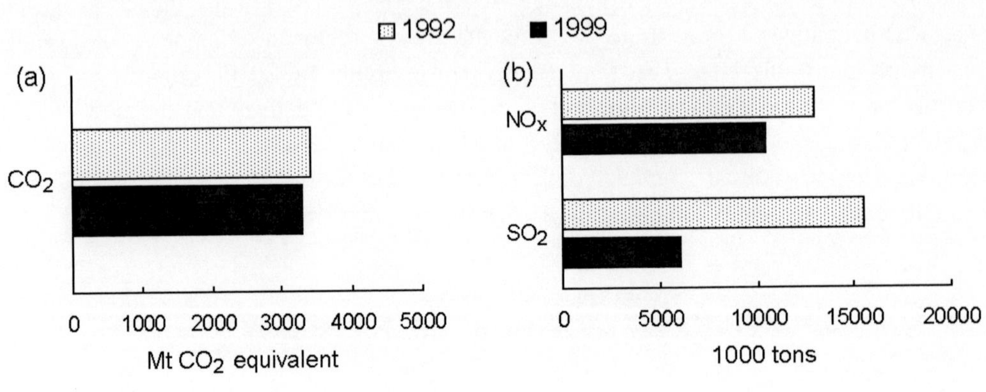

Figure 2. Energy-related emissions of (a) greenhouse gas, (b) SO_2 and NO_x in Western Europe between 1992 and 1999 [European Environment Agency, 2003].

Nevertheless, this was achieved against a background of an 18% increase in economic growth over the same period.

Energy use is the major source of sulfur dioxide (SO_2) and nitrogen oxide (NO_x) emissions, accounting for over 90% of both emissions in Europe in 1999 [European Environment Agency, 2003]. Considerable progress has been made in reducing these energy-related acid gas emissions, and this has greatly helped all European regions - Western Europe (WE), Central and Eastern Europe (CEE) and Eastern Europe, the Caucasus and Central Asia (EECCA), to be on track to achieve their aggregate targets under the Convention on Long-Range Transboundary Air Pollution (CLRTAP) which was established within the framework of the United Nations Economic Commission for Europe (UNECE). The reductions in acid gas emissions in Western Europe (see Figure 2b) were achieved mainly by direct actions including switching to lower sulfur fuels, installing flue gas clean-up systems, introducing catalytic converters in cars and modifying combustion processes [European Environment Agency, 2003].

Summary of EU Renewable Energy Policy

European leaders signed up in March 2007 to a binding EU-wide target to source 20% of their energy needs from renewables such as biomass, hydro, wind and solar power by 2020 (Energy Policy for Europe 7224/07) [Archive and Documentation Center, 2007]. The EU policy for development of renewable energy is briefly reviewed on the basis of published data [EurActiv, 2008].

1997 – the EU white paper "Energy for the future - renewable sources of energy" sets the target of increasing the share of renewable energy to 12% of total energy consumption by 2010.

2001 – EU adopts the Directive on the Promotion of Electricity produced from Renewable Energy Sources (RES-E Directive). The directive sets an EU-wide target of 21% of renewables share in electricity production by 2010.

2003 – EU adopts the Biofuels Directive setting "reference values" of 2% market share for biofuels in 2005 and 5.75% share in 2010.

2006 – European Parliament calls for 25% target for renewables in EU's energy consumption by 2020.

January 2007 – the Commission presents "Renewable Energy Roadmap" as part of its "energy-climate change package" (COM/2006 848 final).

March 2007 – the EU summit endorses the Commission roadmap (Energy Policy for Europe 7224/07): (i) a binding target to have 20% of the EU's overall energy consumption coming from renewables by 2020; and (ii) as part of the overall target, a binding minimum target for each member state to achieve at least 10% of their transport fuel consumption from biofuels.

Consumption and Sources of Energy

During the 1990s, total energy consumption in Europe decreased slightly, but it has been rising since 1999, with increases seen in three European regions (see Figure 3a). The consumption trend by region is summarized as follows [European Environment Agency, 2007a]: (i) *Western Europe* - steady growth has been seen in total energy consumption. There was a 13% increase between 1992 and 2004, lower than the growth in GDP, thereby demonstrating partial decoupling; (ii) *Central and Eastern Europe* – total energy consumption increased by 15% over the period 1992-2004. Energy consumption per unit of GDP over the period decreased by 16%, and it showed some decoupling of energy consumption from economic growth; and (iii) *Eastern Europe, the Caucasus and Central Asia* – total energy consumption decreased significantly during the 1990s, largely as a result of political transition and economic restructuring, together with limited investment in efficiency measures. In spite of the rising trend since the end of the 1990s, in 2004 it was still approximately 20% below 1992 levels.

In the coming years, total energy consumption is projected to increase steadily, with faster growth in CEE and EECCA than in WE; this is despite some further decoupling of energy consumption from economic growth in all regions, reflecting the move to services [European Environment Agency, 2007a]. Action will thus be required to minimize the environmental impacts of increasing energy usage, through greater deployment of energy-efficiency measures and greater use of less-polluting energy sources.

Coal – after a large decrease in the 1990s, reflecting a clear switch to less-polluting fuels, coal consumption in the pan-European region started to increase again (refer to Figure 3b). Despite WE coal import dependence of more than 50%, coal use is projected to increase further as high oil and gas prices, and concerns over the security of oil and gas imports, drive a return to coal-fired electricity generation in all the three regions [European Environment Agency, 2007a].

Petroleum – oil's share of total energy consumption has been increasing over the whole region since the end of the 1990s (see Figure 3b). In WE, it remained fairly stable between 1999 and 2004, large increases in the transport sector being partially offset by a switch to other fuels in other sectors. In CCE, declines in North Sea oil production have been offset by increasing imports from the Russian Federation and the Caspian basin and the use of a number of new fields and supply routes.

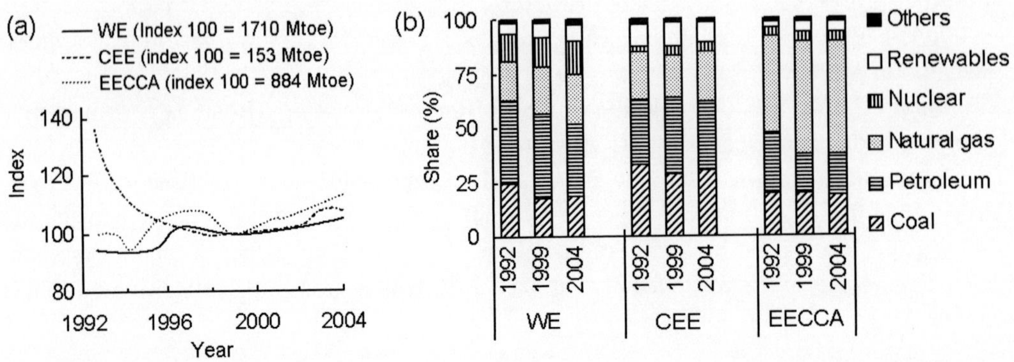

Figure 3. Energy trends and sources in Europe from 1992 to 2004. (a) total energy consumption (index 100 = 1999) and (b) energy sources (redrawn from [International Energy Agency, 2006]). See the legend of Figure 1 for region abbreviations.

Oil consumption in the EECCA region has been increasing following the period of decline during economic restructuring in the late 1990s [European Environment Agency, 2007a].

Natural gas – gas's share of total energy consumption grew in the three regions during the 1990s (see Figure 3b). Natural gas became the fuel of choice for new power generation capacity in WE at the expense of oil and coal, and this was driven by relatively low gas prices and the lower environmental impacts of gas-fired electricity generation [European Environment Agency, 2007a].

Nuclear – consumption of nuclear power increased by 22% overall between 1992 and 2004. A downward trend is expected in the EU over the coming years since several EU Member States have decided to phase out nuclear power and many existing plants will be decommissioned [European Environment Agency, 2007a].

Renewable energy – renewables showed some growth in absolute terms, increasing by 11% across the region between 1999 and 2004. The overall growth in renewable energy masked significant regional differences [European Environment Agency, 2007a]: (i) in EECCA, the 2004 consumption share of renewable energy was below 5% (see Figures 1 and 3b), due to a decrease in production from combustible renewable sources, primarily due to poor maintenance of existing plants; (ii) in CEE, renewable energy increased by 31% between 1992 and 2004, primarily from an increase in hydropower and combustible biomass; and (iii) the increase was greatest (39%) in WE, mainly driven by a strong increase in biomass use for energy production and, to a lesser extent, to increased wind capacity. Installed wind capacity in WE grew by approximately 600% between 1997 and 2003. This was largely due to the implementation of a number of fiscal and regulatory support measures, such as feed-in tariffs and obligations regarding the share of renewable energy.

CASE STUDIES

As stated in the section titled energy and environment, practices of energy consumption and production place pressures on the environment; therefore, the EU has adopted policies to reduce the environmental pressures associated with energy by means of energy saving, efficiency improvement, switching to less-polluting fuels, etc. Coal use generally exerts the greatest pressures because of the high levels of greenhouse gas, acid gas (e.g. SO_2 and NO_x) and particulate emissions [Adachi, 1997]. However, as stated above, coal consumption is showing a tendency to rise in Europe. Renewable energy resources offer clean sources of energy [Adachi, 1997], and their penetration is promoted by the new energy policy (Energy Policy for Europe 7224/07, signed in March 2007). From these viewpoints, coal and renewable energy will receive attention; these different types of energy are therefore discussed on the basis of case studies.

Renewable Energy in Portugal

Based on a national report [Red Eléctrica Nacional, 2002 and 2003], the energy situation in Portugal is outlined. Primary energy consumption registered strong growth in the period

1990-2003 (3.0% per year), although not so high in the latest years. However, primary energy consumption per capita, from 1.78 to 2.46 toe/capita between 1990 and 2003, is still the lowest per capita consumption among the EU-15 member states (average of 3.76 toe/capita). Final energy demand registered strong growth in the period 2000-2003 (2.8% per year), mainly due to consumption of oil products (1.9% per year for oil, 30.8% per year for natural gas and 4.5% per year for electricity). Of the total, 83.5% of energy supply was dependent on external resources, with 59.3% of the total primary energy consumption covered by imported oil in 2003. This means that the greatest domestic energy policy concern is associated with the security and diversification of its supply at the lowest cost, reducing dependence on oil and diversifying energy products.

Traditionally, the renewable sources which have made the greatest contribution to the total consumption of primary energy in Portugal are hydropower and biomass, particularly forest products, although there has been a large increase, more recently, in the use of wind power. Increasing the use of the potential of renewable energy sources is an essential cornerstone of a policy for sustainable development. Wind and geothermal energy accounted for a gross electricity production of 25GWh and 51GWh respectively in 1997, while these figures increased to 468 and 90GWh respectively in 2003.

Hydro has been a relevant source of electricity production, with a 34.4% share of the total gross electricity production in 2003. The annual average of hydropower capability for electricity production was 10,600GWh in 2002; it falls to 6,500GWh in a dry year and rises to 15,800GWh in a wet year [Red Eléctrica Nacional, 2002 and 2003]. Figure 4 shows the change in electricity consumption and the rate of hydroelectric production by month.

As seen in Figure 4, the problem lies in a bad balance between electricity consumption and hydroelectric production; output from a hydroelectric station is far lower than the normal level until September; in contrast, electricity consumption has 2 peaks (April and July) before September. Furthermore, hydroelectric stations generated 6,764GWh in 2002, which was down 46% from 2001 due to poor hydraulicity [Red Eléctrica Nacional, 2003]. This means that the operation of a hydroelectric system is heavily dependent on climatic conditions; to put it differently, ~35% of electricity production is dependent on climatic conditions in Portugal. Hydroelectric power generation is considered a clean technology under the 1997 Kyoto Protocol which aims to reduce greenhouse gases worldwide [Oberthnr and Ott, 1999].

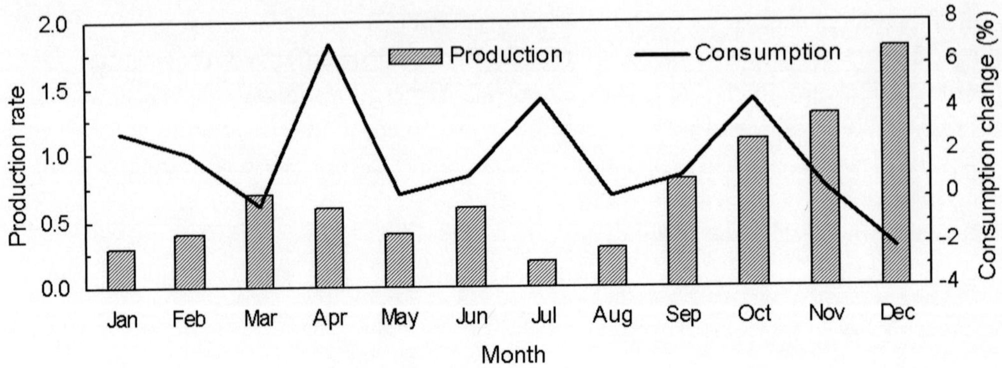

Figure 4. Variation of electricity consumption and rate of hydroelectric production in Portugal during 2002 [Red Eléctrica Nacional, 2003].

However, this Portuguese case suggests that the strong importance of the hydro component creates a quite variable contribution of renewable energy to electricity generation because there are large annual variations in the hydrologic conditions. Of course it is environmentally important to reduce greenhouse gases, but it is also socially and economically important to guarantee a stable supply of energy; there is therefore doubt as to how to maintain the balance of these two factors. As the utilization rate of hydropower increases according to the program for promoting clean energy, this balance may create a serious problem in Portugal.

Consumption, Production and Import of Coal in Western Europe

World coal consumption has been in a period of slow growth since the late 1980s, a trend that is projected to continue. Although total world consumption (5.3 billion tons) of coal in 2001 was more than 27% higher than the total in 1980, it was 1% below the 1989 peak (see Figure 5a). The reference case in Figure 5 projects growth in coal use between 2001 and 2025, at an average annual rate of 1.5%.

Coal consumption is heavily concentrated in the electricity generation sector, although significant amounts are also used for steel production. Coal accounted for 24% of total world energy consumption and for 38% of the energy consumed worldwide for electricity production in 2001 [Energy Information Administration, 2004]. Total recoverable reserves of coal around the world are estimated at 1,083 billion tons – enough to last approximately 210 years at current consumption levels [Energy Information Administration, 2001]. Although coal deposits are widely distributed, 60% of the world's recoverable reserves are located in three countries: the United States (25%), former Soviet Union (23%), and China (12%); and another four regions – Australia, India, Europe and Africa – account for an additional 29% [Energy Information Administration, 2001].

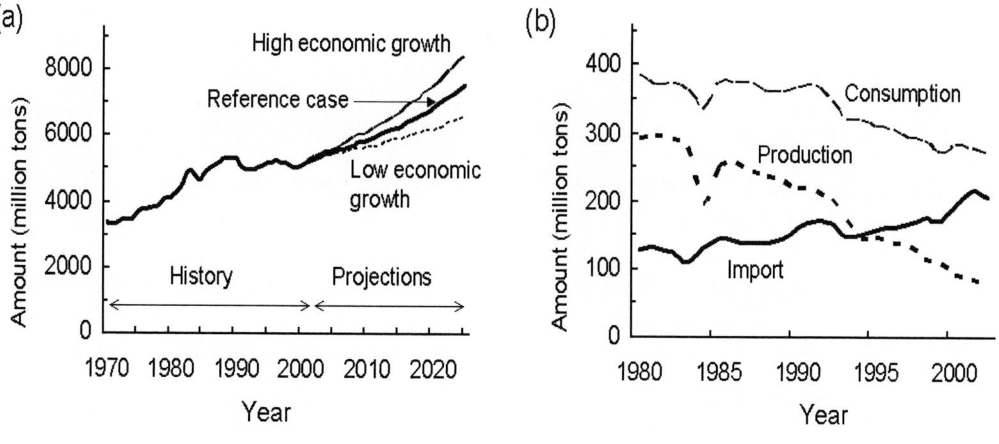

Figure 5. Outlook for coal use. (a) world consumption of coal during 1970-2025, and (b) production, consumption and import of hard coal in Western Europe during 1980-2002 (redrawn from [Energy Information Administration, 2004]).

In Western Europe, recent trends in consumption of hard coal are closely correlated with trends in its production, primarily because coal imports have increased by considerably less than production has declined (see Figure 5b). From 1980 to 2002, coal imports to Western Europe increased by 77 million tons, while hard coal production declined by 214 million tons (Figure 5b). The decline was small (30%) on a Btu basis, reflecting the fact that much of it resulted from reduced consumption of low-Btu lignite in Germany. The governments of Germany, Spain, France, and the United Kingdom currently support domestic production of hard coal through subsidies approved by the European Commission, and these authorized subsidies amounted to ~4,060 million EUR in Germany, ~1,010 million EUR in Spain, ~970 million EUR in France and ~100 million EUR in the United Kingdom in 2001 [Commission of the European Communities, 2002]. The governments in Germany and Spain plan to continue financial support for their hard coal industries, while acknowledging that future reductions in coal production are inevitable when existing mines exhaust their minable reserves [Energy Information Administration, 2004]. This means that coal will continue to play an important role in Europe's overall energy mix. Electricity producers in Germany, Spain, France, Italy, Poland, the Czech Republic, and Slovakia have revealed plans to upgrade or replace existing coal-fired generating facilities over the next two decades [International Energy Agency, 2003; Energy Information Administration, 2007].

The introduction of clean coal technology could help to limit additional pressure on the environment [World Coal Institute, 2004; European Environment Agency, 2007a]: A number of approaches can be used during refurbishment of existing plants, or their replacement, combined with the application of pollution abatement measures. For example, flue-gas treatment technologies can reduce SO_2 emissions (desulfurization systems), remove particulates (electrostatic precipitators) and reduce NO_X emissions (e.g. low NO_X burners) [World Coal Institute, 2004]. Coal gasification allows high levels of efficiency and lower emissions [World Coal Institute, 2004]. While technologies to reduce emissions of air pollutants are commercial and in widespread use in WE, their use in EECCA and CEE is limited [European Environment Agency, 2007a]. One of the most promising technologies for reduction of CO_2 emissions is carbon capture and storage (CCS), where CO_2 is captured direct from the plant and stored deep underground, either in depleted oil and gas wells, or in aquifers; however, the technology has not yet been implemented on a commercial scale [World Coal Institute, 2004].

CO_2 Reduction by Application of Wind Power in Ireland

As stated in the section presenting an overview of European energy, installed wind capacity in WE grew by approximately 600% between 1997 and 2003 [European Environment Agency, 2007a]. Since the EU has set down a binding target for the share of renewabl energy (see the section on the EU's policy summary), it is clear that a greater portion of the need for energy will be met by wind power in the future.

An Irish study shows the operational and economic impacts of wind power generation [ESB National Grid, 2004]. The quantity of CO_2 emitted as a result of meeting Ireland's electricity needs is directly related to the amount and type of fossil fuels used in conventional power stations. Producing electricity from wind reduces the consumption of fossil fuels and therefore leads to emissions savings. However, to accurately quantify the emissions saving which can be derived from wind power generation (WPG), the growing inefficiency of the conventional plant portfolio must also be taken into account. It is not sufficient to estimate the amount of energy which can be obtained from a given capacity of WPG, and to assume that the equivalent percentage of fossil fuel and therefore CO_2 can be avoided. A great amount of energy is required to start up thermal power units, and the capacity factor of a unit is a measure of the annual energy that the unit produces compared with the maximum possible production level. As WPG increases, the number of starting operations of thermal power units increases and the capacity factor decreases; in a word, this is a manifestation of the growing inefficiency of thermal generation as it is required to act as a counterbalance for WPG.

From Figure 6, it can be seen that 500MW of wind reduces the amount of CO_2 emitted by 5.4%. If there is an emission saving of 5.4% associated with the introduction of 500MW of WPG, it is to be expected that the potential fuel saving is 3 times this value – 16.2% for 1500 MW. However, as WPG increases to 1500MW, it follows from Figure 6 that the corresponding reduction in fossil fuel requirements is only 12.9%. This quantifies the growing inefficiency of WPG to curtail emissions. Based on the results obtained from this Irish study, the following findings are reported [ESB National Grid, 2004]: (i) given the intermittent nature of wind, additional or surplus capacity needs to be available to ensure security of supply when the wind is not blowing. The level of additional capacity required is conditional on the level of wind powered generation and increases as the level of wind powered generation increases. If wind powered generation increases to 1,500MW, an additional 1,160MW of conventional plant is required for generation adequacy purposes; (ii) introducing wind powered generation into the generation portfolio changes the operational regime of conventional generation and is likely to have a significant negative effect on plant reliability; (iii) the efficiency of a thermal plant declines when it is required to operate at low load levels. Although there is a reduction in the overall fuel bill when wind powered generation is introduced, this reduction is curtailed by increased inefficiencies as a result of increased start-ups and low loading levels; (iv) the total cost of meeting customer electricity demand increases as the penetration of wind generation increases.

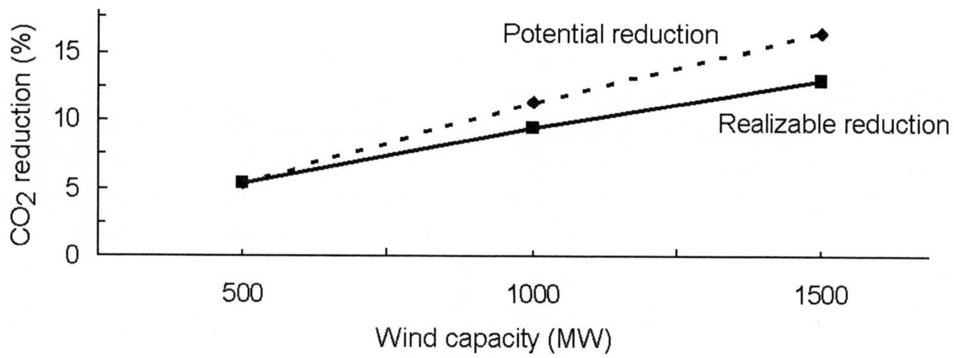

Figure 6. Reduction of CO_2 emissions in 5,000MW system according to CREEP and PROMOD models [ESB National Grid, 2004]. Note: CREEP is an acronymic name for capacity requirement evaluation by exact probability; and PROMOD is an hourly simulation program of Monte Carlo generation to determine system performance.

This is primarily due to the retention or installation of surplus conventional capacity, for system security reasons, and the operation of all installed conventional generation in a sub-optimal manner; and (v) the reduction in CO_2 emissions from increased wind power generation is tempered by increased generation costs. Although emissions levels fall by 1.42 million tons, the total generation cost increases by 196 million EUR, implying a cost per ton of 138 EUR, which appears high relative to other abatement alternatives.

SUMMARY AND FINDINGS

Since renewables energy sources (wind power, hydro power, solar energy, etc.) are clean, they help to reduce environmental preesures. The EU summit (March 2007) agreed on a binding target to increase the share of renewable energy in total energy consumption in the EU to 20% by 2020 (Energy Policy for Europe 7224/07). The output of renewable energy is quite variable, that is, its output is dependent on climatic conditions (refer to the case study of Portugal). It does not necessarily follow that adverse impacts on the environment affect energy policies. Energy is vital to social and economic well-being, so it is also important to supply stable energy which is independent of climatic conditions. Considering the intermittent nature of renewable energy (see Figure 4), supplemental power generation needs to ensure the energy supply when renewable sources are not available (i.e. less wind, less precipitation, etc.) and/or energy demand is greater that the output from the renewable energy system; this supplement cannot help counting on conventional energy.

Coal use is projected to increase in Europe (also refer to Figure 5a) because of concerns over energy import and security as well as high prices of oil and gas [European Environment Agency, 2007a]. Coal is recognized as the "indigenous energy base" by the European Commission, and coal production is financially supported by the EU state members [Commission of the European Communities, 2002]. Although the utilization of coal is responsible for emissions of GHG and air pollutants (Figure 2), the introduction of clean coal technologies will help to limit adverse environmental impacts (see the section discussing the WE situation for coal).

The problem lies in the combination of renewable energy with conventional energy production (e.g. coal-fired power plant) (see the case study of Ireland): although there is a reduction in the overall fuel bill when wind powered generation is introduced, this reduction is curtailed by the increased inefficiency of thermal power plants as a result of frequent start-ups and low loading levels; and the total cost of customer electricity demand increases as the penetration of wind generation increases. This cost increase is caused by the retention/installation of surplus conventional capacity, system security and the operation of all installed conventional generation in a sub-optimal manner.

Even if the share of renewable energy increases to up to 20%, the remaining 80% of required energy will have to come from conventional energy sources. Consideration of just the total cost and coal clean technologies is not necessarily enough to make it worthwhile to combine two different types of energy source. The following system of thermal power should be developed to adapt coal use to the future clean energy society: (i) start-up and shutdown of the system is simple; (ii) less energy is required for start-up; (iii) the desired operation efficiency can be obtained in a short time after start-up; and (iv) the desired efficiency can be obtained at low loading levels of fuel. In conclusion, it becomes important to consider how to enable renewable energy and conventional energy to coexist in a proper manner because the EU energy strategy implies that a 20% share of renewable type energy and 80% of conventional type energy should be combined to form a new energy society by 2020.

ACKNOWLEDGMENTS

Parts of this work are supported by Centro de Estudos de Recursos Naturais, Ambiente e Sociedade in Portugal. The author is grateful to Ms C. Lentfer for English review.

REFERENCES

Archive and Documentation Center (2007). *Road map for renewable energy in Europe.* Luxembourg: European Parliament, No. INI 2007 2090.

Adachi, Y. (1997). New energy technologies. Tokyo: Ohmsha, Ltd.

Commission of the European Communities (2007). *Renewable energy road map – renewable energies in the 21sth century.* Brussels: European Communities, No COM (2006) 848 final.

Commission of the European Communities (2002). Report from the commission on the application of the community rules for state aid to the coal industry in 2001. Brussels: European Communities, No. COM (2002) 176 01.

Directorate General for Energy (1999). *European energy to 2020 – a scenario approach.* Luxembourg: Office for Official Publications of the European Communities, DG XVII.

Direcção-Geral da Empresa (2005). *National reporting guidelines for CSD 14/15 thematic are – energy.* Lisbon: Ministério da Economia e da Inovaçao, national report 14CDS.

Economic Commission for Europe (1995). *Strategies and polices for air pollution abatement.* Geneva: United Nations, ECE/EB. AIR/44.

Energy Information Administration (2007). *International energy outlook 2004.* Washington DC: US Department of Energy, DOE/EIA 0484/2007.

Energy Information Administration (2004). *International energy outlook 2004.* Washington DC: US Department of Energy, DOE/EIA 0484/2004.

Energy Information Administration (2001). *International Energy Annual 2001.* Washington DC: US Department of Energy, DOE/EIA 0219/2001.

ESB National Grid (2004). *Impact of wind power generation in Ireland on the operation of conventional plant and the economic implications.* Dublin: EirGrid Plc.

EurActiv (2007). *EU renewable energy policy.* Brussels: EurActiv com PLC, No.117536.

European Environment Agency (2003). *Europe's environment – the third assessment.* Luxembourg: Office for Official Publications of the European Communities, environmental assessment report No. 10.

European Environment Agency (2007). *Annual European Community greenhouse gas inventory 1990–2005 and inventory report 2007.* Luxembourg: Office for Official Publications of the European Communities, technical report No. 7/2007.

European Environment Agency (2007a). *Europe's environment – the forth assessment.* Luxembourg: Office for Official Publications of the European Communities, environmental report No. 1/2007.

Eurostat (2007a). *Eurostat population database*. Luxembourg: Statistical Office of the European Communities.

Heston, A., Summers, R., Aten, B. (2007). *Penn world table* – version 6.2. Philadelphia: Center for International Comparisons of Production, Income and Prices.

International Energy Agency (2006). *Extended energy balances of OECD countries and non-OECD.* Paris: IEA Data Services, 2006 edition.

International Energy Agency (2003). *Coal information 2003*. Paris: IEA Data Services.

Linor, N. (2007). Energy resources and use: the present situation and possible paths to the future. *Energy*, doi:10.1016/j.energy.2007.09.009 (in press).

Oberthnr, S., Ott, H. (1999). *The Kyoto Protocol - International Climate Policy for the 21st Century*. Berlin: Springer-Verlag,

Red Eléctrica Nacional – REN, S.A. (2002). Relatório e contas 2001 (Report and accounts 2001). Lisbon: Divisão comunicação e imagem da REN (in Portuguese).

Red Eléctrica Nacional – REN, S.A. (2003). Report and accounts 2002. Lisbon: Divisão comunicação e imagem da REN.

Science Council of Japan (2003). *Energy and Sustainability Science. CD-ROM proceedings on International Conference on Science and Technology for Sustainability 2003*, 16-19 December, Tokyo.

World Coal Institute (2004). *Clean coal – building a future through technology*. London: World Coal Institute, WCI report.

World Commission on Environment and Development (1987). *Our common future*. Oxford: Oxford University Press.

In: Energy in Europe: Economics, Policy and Strategy
Editors: F. L. Magnusson and O. W. Bengtsson.
ISBN: 978-1-60456-829-5

Chapter 12

STRATEGIES AND POLICIES FOR SPEEDING UP ENERGY INNOVATIONS AND NEW TECHNOLOGIES TO MEET EUROPE'S ENERGY CHALLENGES

Peter D. Lund[1]
Helsinki University of Technology, Advanced Energy Systems
Espoo, Finland

ABSTRACT

European Union has set challenging energy targets for year 2020 including increasing the renewable energy utilization up to 20% of the final energy use. The European Commission has also announced an ambitious strategic energy technology plan. The combined effect from these market stimulating and technology development measures could lead to cost reductions of new energy technologies and enhance significantly their market penetration. With this positive starting point in mind, the paper investigates further the commercialization process of new energy technologies and particular the effects from different strategies on the support level needed. The methodological frame employed is based on a learning curve approach and the corresponding analytical equations used to quantify the effects from policies are also derived. The results indicate that finding an optimal balance between technology push and market pull policies may lead to significant savings in the support needed. If the new technology is far off from a cost breakthrough then strong early RTD actions would well pay off compared to market stimulation measures only. Closer to the breakthrough, a shift to stronger market deployment action would be motivated. In case of indicative EU targets on photovoltaic technology, an intensive 5 years RTD action could lead to savings of tens of billions of € over a long run in the total costs of reaching the breakeven point. As another example, the EU renewables target could lead to a 15-20% cost reduction of wind power by year 2020.

[1] peter.lund@tkk.fi.

INTRODUCTION

Europe confronts several energy challenges in the coming decades. If not changing the present trend, the energy imports will increase in a few decades from present 50% up to 70% of all energy, notably from Russia which has raised political concerns on the energy security. Europe is also strongly committed to reduce its CO_2 emissions in 2020 by 20%, but even more dramatic reductions of 60-80% will be necessary to fulfill the UN Climate Convention targets by 2050. Securing reasonable cost of energy is also important for the European economy, as high energy costs could put at risk the already slow economic growth in Europe compared to its global competitors. Any relevant energy policies and strategies in the European context need thus to address closely the 3 Es: energy security, economics and environment.

There are different policy options available to meet the energy challenges ahead. On a longer term in particular, creating better and cleaner energy technologies will be of vital importance as the required change from the present energy production in mitigating the global warming is so large that none of the mainstream energy options has the required properties to meet the goals. Considering the growing energy demand from rapid economic growth and industrialization in emerging economies e.g. in Asia, would mean that by mid-century clean carbon-emission free energy should be world-wide available twice the present fossil fuel production, or saving annually some 14 GtC. In their archival paper Socolow and Pacala [2004] identified several so-called clean technology wedges for reducing dramatically green house gas emissions. Such massive introduction of new energy technologies would in practice require a full energy technology revolution, which in turn implies huge efforts in technology development, commercialization, and deployment of these options. Resembling a paradigm change, would also mean large business opportunities.

Europe's position in the coming energy technology challenge has not been quite clear. On one hand many member states have strong market deployment activities for clean energy, mainly renewable energy, but Europe has been lagging to USA and Japan in research and technology development funding in many future energy technology areas. An important change in the situation may be come along the Strategic Energy Technology Plan (SET) introduced by the European Commission in 2007-2008 [Commission of the European Communities, 2007], which intends to strengthen Europe's position in energy technology and innovations through better financed concerted research and technology development (RTD) efforts.

Together with the new European energy policy targets from January 2008 for all member states, requiring in average a 2.5 times higher renewable energy utilization and 20% lower CO_2 emissions in 2020 [Commission of the European Communities, 2008a] would mean a unique opportunity for a more integrated and strategic approach to go for a true market breakthrough and massive introduction of new energy technologies. Three central issues need to be addressed in this context, namely improving technology cost-effectiveness, speeding up the commercialization process, but also policy effectiveness, i.e., costs and impacts of the measures to be formed to realize the policy targets. As a matter of fact, earlier research indicate that policies with a stronger link to the commercialization process may provide much faster market penetration and still be by far cheaper per impact unit than the traditional static policies [Lund, 2007]. Such policies would rely strongly on market mechanisms and forces.

The central theme of this chapter is on the policies for bringing new energy technologies to the European energy system in large scale. The chapter is organized as follows: description of the commercialization process and its link to policies, analysis of the EU targets in renewable energy and their implications, and finally discussion of improved strategies for accelerating the market breakthrough of new energy technologies.

COMMERCIALIZATION PROCESS

An important part of the energy innovation process is the commercialization of the new energy technology. Actually this is one of the most central questions for European energy policies today. In practice the key challenge is to improve the cost-effectiveness of the new technologies to enable market breakthrough. The whole issue is much about speeding up the commercialization process, and finding effective policies for this purpose. Basically the strategies for promoting new technologies can be divided into two main categories namely technology push measures such as RTD and market pull actions, for example market incentives or investment subsidies [Dosi, 2000; Freeman and Soete, 2000]. Figure 1 illustrates the phases of the innovation chain from research to market.

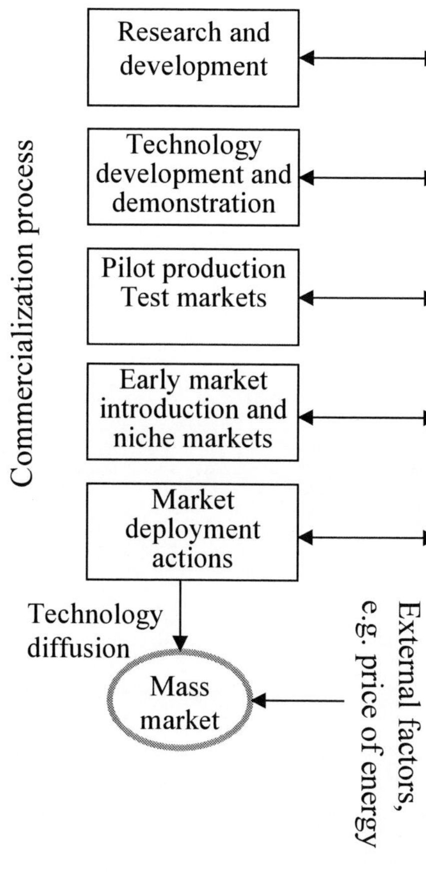

Figure 1. Commercialization process.

The basic policies used in Europe for energy innovation demonstrates a range of technology push (e.g. RTD, standardization) and market pull (e.g. feed-in-tariffs, investment subsidies) measures. Traditionally these have been separated, but in modern policy approaches they are more closely interlinked and form a policy portfolio. The contents of the portfolio should be specified for each technology separately based on their maturity, but also reflecting the progress in the commercialization. In this way advanced energy policies would support the whole commercialization process targeting for a breakthrough rather than subsidizing an energy technology product. For example, for a new technology such as photovoltaics which is still far off from a breakeven point, forming more coherent EU wide RTD policies could be highly relevant than just going for large national deployment measures. Closer to market breakthrough such as in case of wind power, the weight of the portfolio would be more on the market deployment side. When the new energy technology takes off, the deployment policies need to carefully designed not to cause supply bottlenecks and price disturbances.

The benefit of a more integrated view on the commercialization is illustrated in Figure 2. In a linear model the time from knowledge to products on the market is long due to the sequential development steps. Whereas putting these more in parallel and considering all steps part of the portfolio needed to get the new technology on the market, would lead to considerable shortening of the time-to-market and thus speeding up the market penetration.

When viewing the basic policies and strategies used to accelerate clean energy development in Europe in recent years, the emphasis have been more on creating new markets for new technologies and thus stimulate technology improvements and cost reductions through volume increases [International Energy Agecny, 2006]. Whereas compared to USA and Japan, the weight on technology development has been of less importance in Europe – for example the public RTD in energy in Europe has been lower in several strategically important future technology areas such as fuel cells and hydrogen, clean coal, photovoltaics, etc.

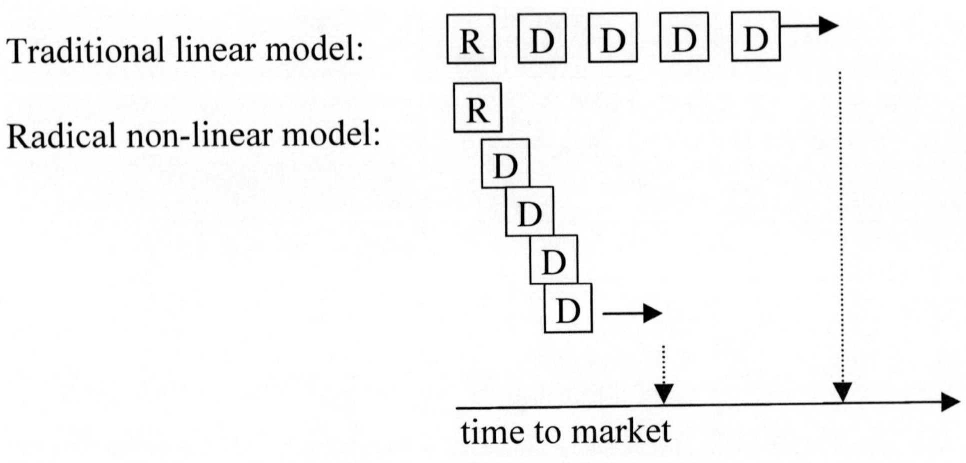

Figure 2. Examples of different profiles of the innovation chain. R=research, D1=development, D2=demonstration (pilot production), D3=dissemination, D4=deployment.

An often referred policy instrument for promoting renewable electricity is so-called feed-in-tariffs that provide a guaranteed reimbursement on produced renewable energy such as wind or solar power over a long period of time [Lewis and Wiser, 2007]. Another market stimulating instrument has been the green certificates, which allow a stronger market-based incentive system for renewable energies. Reducing the economic risks in this way, have boosted the take-up of renewables in several European countries, notably in Germany, and has lead to creation of major new industries.

The important EU strategic energy technology plan (SET-Plan) from 2007 and 2008 puts much more emphasis on integrated RTD and managing the commercialization chain as a whole. This would mean better balancing of the technology development vis-à-vis market deployment. More specifically it is anticipated that the RTD efforts could increase considerably in the coming years including actions to solve bottle-necks in the innovation chain such as larger demonstration projects or critical research resources. Important European integrated RTD and innovation instruments in coming years will be the Joint Technology Initiatives (JTI) pulling together European resources, actors, know-how and early markets for major leapfrogging in new technologies such as fuel cells and hydrogen; European Institute of Technology EIT aiming at strengthening the early part of the innovation chain but bridging effectively to industries; different networking activities such as ERA-NETs, etc.

Another interesting dimension in strategic elaboration of energy innovation policies would be the effectiveness of the measures, for example realized or perceived impacts versus extra funding needed over the reference technologies e.g. traditional energy. Previous work [Lund, 2007] indicates that policy instruments with a strong link to the innovation or commercialization process as a whole may provide much faster market penetration and still be up to a factor of 10 cheaper per impact unit than the traditional more static policies. Such policies would rely strongly on market mechanisms and powers and include catalytic measures such as different forms of procurement, use of purchasing power and competition, information measures, lead markets initiatives, guaranteed purchase prices, etc. The extra money needed would not go directly to subsidize the product but to enhancing the process bringing the product to a breakthrough.

It is a general view that there is not a single energy source that could provide the solution to the energy issues, but a mix of different technologies will be needed. However, advanced energy technologies such as nuclear Generation IV, large-scale renewable energy schemes, or fossil fuels with carbon separation and storage (CCS) could each alone provide a major contribution in the European context. None of these satisfy all policy criteria, and a full penetration would still require huge efforts being cost reductions, performance improvements, safety improvements, etc. The SET-Plan provides an important initiative in general to accelerate the development of these technologies, but in addition huge market deployment actions are under way in Europe, in particular in renewable energy sources.

Important Market Deployment Actions - European Goals on Renewable and Emissions Through

European Union made important decisions in 2007 and 2008 on its future energy directions up to year 2020 [Commission of the European Communities, 2007b]. The overall energy policy directions defined in 2007 set the following targets compared to the reference year 2005: 20% reduction in CO_2 emissions, 20% increase in renewable energy use, 20%

energy efficiency improvement, and a 10% biofuel obligation (so-called 20-20-20-10 directives). In fall 2007 the European Commission introduced the Strategic Energy Technology Plan (SET-Plan), which includes important RTD measures on new energy technologies to improve their competitiveness, and in particular to lower their costs, but also to ensure the competitiveness of European industrial positions in the field. The kind of market deployment and technology development actions together could mean a major effort for the market breakthrough of new and renewable energy technologies.

In January 2008 specific targets on renewable energy and emissions in the non-ETS (ETS = the European Emission Trading System) sector for each member of the 27 member states of EU was laid [Commission of the European Communities, 2008a]. Table 1 shows the present renewable energy use and future targets in EU-27. For countries with a low renewable energy share under 5% such as Belgium, Netherlands and UK, the new goals mean many folding the renewable energy use. Countries with considerable RES use such as Finland, Latvia and Sweden with a RES share over 25%, the relative increase us much lower, but the targets are still challenging considering the past progress and measures. In absolute terms, the largest 5 member states France, Germany, Italy, Spain and UK represents 70% of the overall goal. Thus, the type of technology solutions that these countries choose for their national implementation plans may have large impacts on the commercialization of new technologies. For example, wind power seems to be one of the renewable technologies that are strongly present in the national plans, as also some biomass forms.

Table 1. Renewable energy use in 2005 and targets for 2020 in European Union

Member States	RES 2005, %	RES 2020, %	Increase, %	RES 2020, Mtoe
Belgium	2.2	13	491	3.93
Bulgaria	9.4	16	70	0.63
Czech Republic	6.1	13	113	1.78
Denmark	17	30	76	1.99
Germany	5.8	18	210	26.58
Estonia	18	20	11	0.06
Ireland	3.1	16	416	1.60
Greece	6.9	18	161	2.30
Spain	8.7	20	130	10.97
France	10.3	23	123	20.08
Italy	5.2	17	227	15.82
Cyprus	2.9	13	348	0.17
Latvia	34.9	42	20	0.28
Lithuania	15	23	53	0.36
Luxembourg	0.9	11	1122	0.44
Hungary	4.3	13	202	1.57
Malta	0	10	∞	0.05
Netherlands	2.4	14	483	5.99
Austria	23.3	34	46	2.92
Poland	7.2	15	108	4.46

Table 1. (Continued)

Member States	RES 2005, %	RES 2020, %	Increase, %	RES 2020, Mtoe
Portugal	20.5	31	51	1.96
Romania	17.8	24	35	1.52
Slovenia	16	25	56	0.44
Slovak Republic	6.7	14	109	0.77
Finland	28.5	38	33	2.39
Sweden	39.8	49	23	3.10
United Kingdom	1.3	15	1054	20.77
EU-27	8.5	20	135	132.95

The targets represent a major market boost for renewable energy technologies in Europe. For energy policy the targets may represent just an important contribution to meet the overall policy objectives for emission reductions and energy security, but strategically such a market deployment measure could also have positive consequences on the overall competitiveness of renewable energy. A large capacity addition should definitely reduce the cost of renewables as stated in the background documents of the European Commission, but also require a significant upfront financial support as renewable energy is in most cases more expensive than the traditional energy. The EU leaves it up to the individual member states to decide on the practical measures or support schemes how to reach the goals as the local conditions, financial possibilities and potential for exploiting various renewable energy forms varies much from country to country. In practice, there are several paths to fulfill the renewable energy obligations, but in general one may envision prioritizing e.g. conversion processes with high efficiencies or those with lowest production costs.

The target will mean cumulative new investments in the order of 600 billion € by year 2020. Though renewable energy may find already today economic applications in certain technology areas and in good local conditions, in mainstream bulk energy markets renewable energy may cost even up to 50% more than traditional energy. This would mean subsidy needs amounting to even some hundred billion € to meet the targets.

Therefore from a strategic point of view considering both the large volumes and major costs associated with the European targets in RES it would be of outmost importance to formulate effective policies that both deliver required results but are cost-effective at the same time. Furthermore, in light of the other long term goals on CO_2 emission reduction which would need much more clean energy than the 2020 targets and hence also very huge investments, it would be very important to formulate such policy measures, or portfolio of measures, that could effectively contribute to the true market breakthrough of clean energy, i.e., leading to a major cost-reduction and to an economic breakeven. The role of commercializing new technologies and innovations would be vital in this context.

To address above strategic questions on optimal and effective policies for implementing European renewable targets, and to identify possible opportunities in such areas, a more detailed analysis on the final energy and renewable energy use is necessary. There has to be a matching between the form of final energy needed (heat, electricity, transport/fuel) in the different end use sectors and the delivered energy from the renewable energy source. The potential of different renewable energy sources varies in terms of final energy forms.

The primary energy use of EU-27 in 2005 corresponded to 1816 Mtoe of which 6.6% was based on renewable energy [Eurostat, 2007]. The policy target on renewable energy is, however, linked to the final energy consumption which was 1170 Mtoe. The final energy by sector and source is illustrated in Figure 3. Households, industry and road transport represent each around a quarter of the final energy use. Of the final energy forms, gas and oil correspond to two thirds emphasizing their importance as a fuel for transport and heating, correspondingly. Electricity stands for 20%.

The renewable energy share of the final energy is 8.5% and the target set for EU-27 year 2020 is 20%. The absolute amount of renewable energy to be delivered depends also on how the overall energy use develops. The Commission has set an overall target to reduce energy consumption by 20% in the same time frame. If the EU succeeds in freezing the energy use to present level then the amount of additional renewable energy to be delivered in 2020 is 133 Mtoe. But with the business-as-usual 10 years average growth of 1.15% per annum observed, the amount would increase to 156 Mtoe or 16% higher.

Heating and electricity form important final energy uses in the consumer sectors in Europe. Figures 4 and 5 show a split down of the energy sources used in these sectors [Commission of the European Communities, 2008b]. Close to 12% of the heating is based on renewable energy (RES) when also accounting for the RES share in electricity. Electricity and gas represent 70% of all heat sources. This information may have importance when introducing more RES as the heat delivery systems may put some preconditions to the use. This in turn could favor renewable that could be easily integrated into existing heating systems, for example fuel switch (oil→bio oils, gas→ biogas or bio-syngas, electricity→RES electricity) or supplementary use of RES (e.g. solar heat).

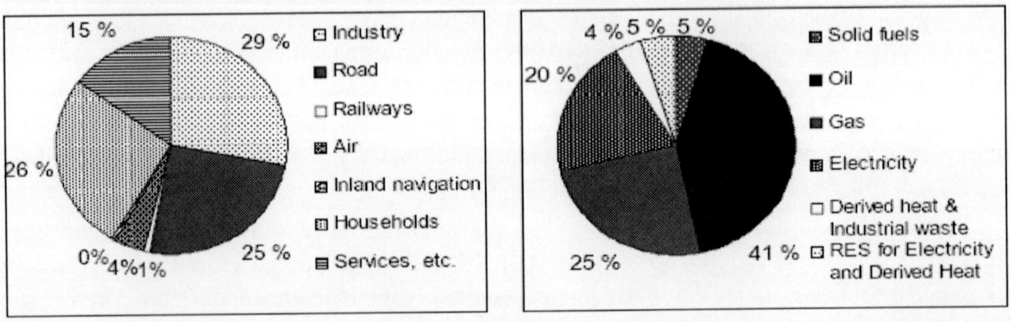

Figure 3. (a) Final energy use of EU-27 by sector (2005). (b) Final energy use in EU-27 by source (2005). Total final energy use was 1140 Mtoe [Eurostat, 2007].

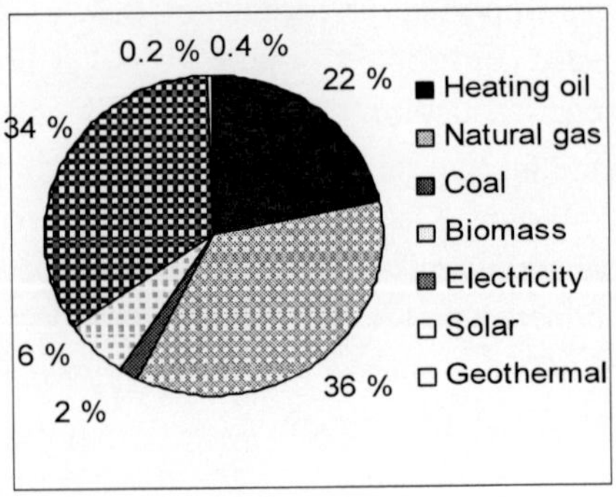

Figure 4. Sources for heating in EU-27 (2005).

Renewables represent a 14% share of the total electricity production. Two-thirds of this comes form traditional large-scale hydropower schemes, electricity from forestry biomass and wind power make up the rest with around equal shares. Future growth would, however, come mainly from new renewable energy such as wind and multiple forms of bioenergy.

Finally we show in Figure 6 the renewable targets and final energy forms for each member state [Eurostat, 2007]. Observing the dominance of oil in several member states, would mean albeit the general 10% biofuel target that renewable energy use would need to increase strongly in electricity and heating production.

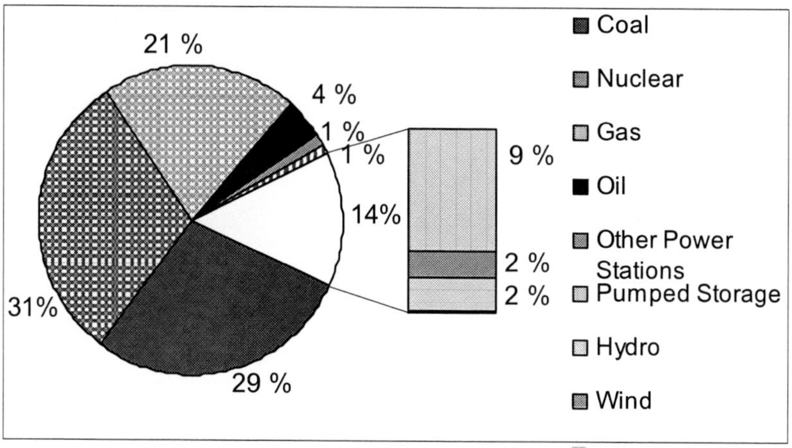

Figure 5. Sources for electricity in EU-27 (2005).

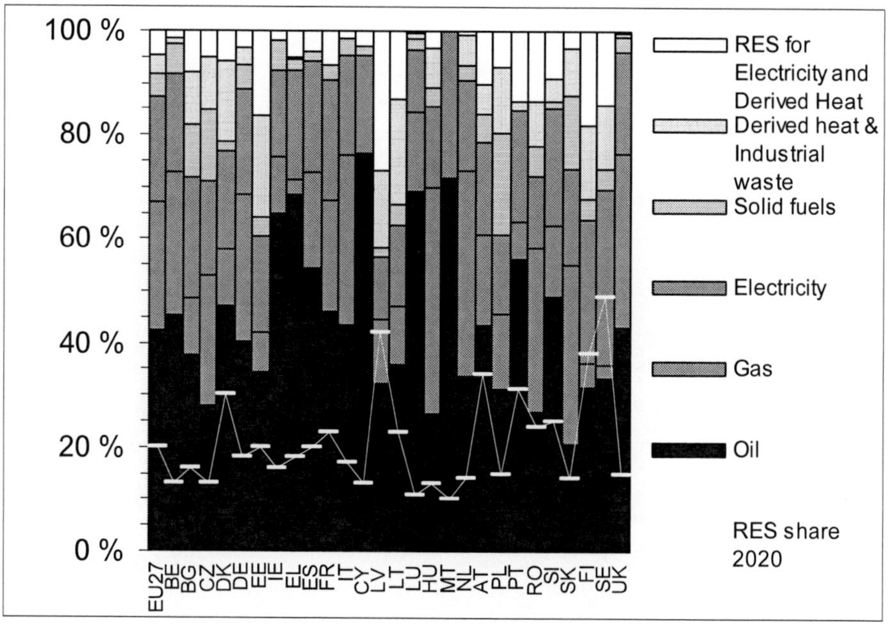

Figure 6. Final energy use in EU-27 member states and the RES targets.

Analytical Framework for Linking Policies to Commercialization and Innovation Strategies

Market deployment or technology development measures reduce the cost of the new technology. For example, the EU targets on renewable energy will not just increase the use of renewables in Europe but may have an important impact on costs as well. Thus the RES targets can be considered as kind of market deployment actions creating new demand for renewables.

One way to describe the link of targets/policies vis-à-vis costs, is to use so-called learning or experience curves [Wene, 2002]. The increase of the installed capacity leads to positive learning effects through doing and using more of the new technology. Analytically the kind of endogenous learning can be described in the following form:

$$\frac{c_V}{c_0} = \left(\frac{V}{V_0}\right)^{\alpha} \tag{1}$$

where c= unit cost of the new technology, V=cumulative volume, α= ln(1-LR)/ln2, LR=learning rate ; the subscript 0 refers to the initial state.

The unit cost drops by LR% for each doubling of the installed capacity or volume. The learning rate is typically around 15% for generic manufacturing industries and products. In new energy technologies reported values vary from 5 to 20% [Wene, 2002].

When assessing the impacts from initiatives which may cause just infinitesimal volume increases dV in relation to the cumulative volume, the corresponding decrease in unit costs can directly be obtained through differentiating Eq. (1)

$$dc_V = \alpha \frac{c_V}{V} dV \tag{2}$$

or alternatively rearranging Eq.(2) into a relative change form as follows

$$\frac{dc_V}{c_V} = \alpha \frac{dV}{V} \tag{3}$$

The coefficient α defined earlier in Eq.(1) thus determines directly the slope of the cost reduction for small volume increases. For example for a 10% volume increase using a learning ratio LR=15% or α=-0.235 that could be typical for many new energy technologies, would imply a 2.35% reduction in unit costs. For large volume changes Eq.(1) should be used to assess the cost reduction.

If the reference energy cost is denoted by c_{br} then the cumulative volume leading to a market breakthrough is given by

$$V_{br} = V_0 \left(\frac{c_{br}}{c_0} \right)^{1/\alpha} \tag{4}$$

Using the learning curve we may next assess the extra cost needed to reach the cost breakthrough by integrating Eq.(1) from V_0 to of V_{br} and subtracting the reference energy costs over the same interval. Calling the extra costs as learning investments we have

$$\text{Learning investments} = \int_{V_0}^{V_{br}} \left[c_V - c_{br} \right] \; dV$$

$$= \int \left[c_0 \left(\frac{V}{V_0} \right)^{\alpha} - c_{br} \right] dV = \frac{c_0 V_0}{1+\alpha} \left(\frac{V_{br}}{V_0} \right)^{1+\alpha} - \frac{c_0 V_0}{1+\alpha} - c_0 \left(\frac{V_{br}}{V_0} \right)^{\alpha} (V_{br} - V_0) \tag{5}$$

where V_{br} is obtained from Eq.(1).

The learning curve is basically a time-wise static equation but we may incorporate a time dependence through assuming that the production capacity increases by β each year. This gives an exponentially increasing installed capacity v_t

$$v_t = v_0 + \beta^t dv_0 \tag{6}$$

where v_0= yearly production capacity and dv_0 is the production capacity addition at $t=t_0$

The cumulative capacity V evolves then over time as follows:

$$V_t = v_0 + \sum_{i=t_0}^{t} dv_i = v_0 + \beta \frac{(1-\beta^t)}{1-\beta} dv_0 \tag{7}$$

Substituting Eq.(7) into Eq,(1) gives the unit cost as a function of time

$$c_t = \left[1 + \frac{\beta}{\beta_0} \frac{(1-\beta^t)}{1-\beta} \right]^{\alpha} \tag{8}$$

where $\beta_0 = v_0/dv_0$. From Eq. (8) we may finally estimate the time needed for reaching a cost breakthrough point:

$$t_{br} = \frac{\ln \left[1 + \beta_0 \frac{1+\beta}{\beta} \left(\left(\frac{c_{br}}{c_0} \right)^{1/\alpha} - 1 \right) \right]}{\ln \beta} \tag{9}$$

The target cost is reached with a cumulative volume of V_{br}.

Energy policy measures aim at improving the competitiveness of the new technology. In the previous analytical framework, policies may influence the learning rate (LR) and capacity increase (β). Several relevant links to policies can be identified as follows:

- balanced market deployment and technology development measures that enable a volume increase leading to the classical unit cost reduction (reference case, LR constant)
- intensified and concerted RTD efforts that lead to a steeper or even a abrupt reduction in the unit costs, for example a new innovations or technology breakthrough (LR increased, LR→1)
- if the market deployment measures becomes oversized, demand of the new technology may exceed the supply which would lead to a cost saturation or increase (LR→0 or even LR<0)

Figure 7 illustrates the cases (a) and (b) which are relevant for innovation based strategies, for example strengthened RTD efforts. The area between the solid and dotted learning curves represent the cost savings over the reference case through mainly technology intensive policy measures. Using Eqs.(1)-(9) one may analyze how the different policy options could influence the overall policy costs, acceleration of commercialization and cost breakeven, and policy effectiveness.

Economics of Different Innovation Paths

The central challenge is to bring the costs down and volumes up. The complex processes associated with bringing the costs new energy technologies down described through the commercialization process precedes the more massive market penetration. This phase is very development intensive and requires strong public support matching the technology specific needs over the whole innovation process. In this chapter, the cost issue of bringing the new energy technologies to a breakthrough is discussed.

The distance from the cost breakeven point affects the optimal balance between technology push and market pull actions. If being far away from the commercial breakthrough, focused RTD efforts to enable technology jumps could intuitively be more effective than market deployment. When reaching sensible cost levels and being closer to the market breakeven point, exercising stronger market pull measures to accelerate volume growth could be more motivated. In practice a kind of active portfolio management of the policy measures would be necessary to cope with such a dynamical frame.

Employing the analytical methodology derived in previous chapter we may crudely estimate important strategic questions related to the impacts from different policies for speeding up the market penetration of new energy technologies.

Starting with a few more general observations, the time needed for a cost breakthrough based on Eq.(9) is shown in Figure 8 as function of the yearly volume growth and using the cost gap as a parameter. The learning ratio used in this example is LR=20% and β_0=5.

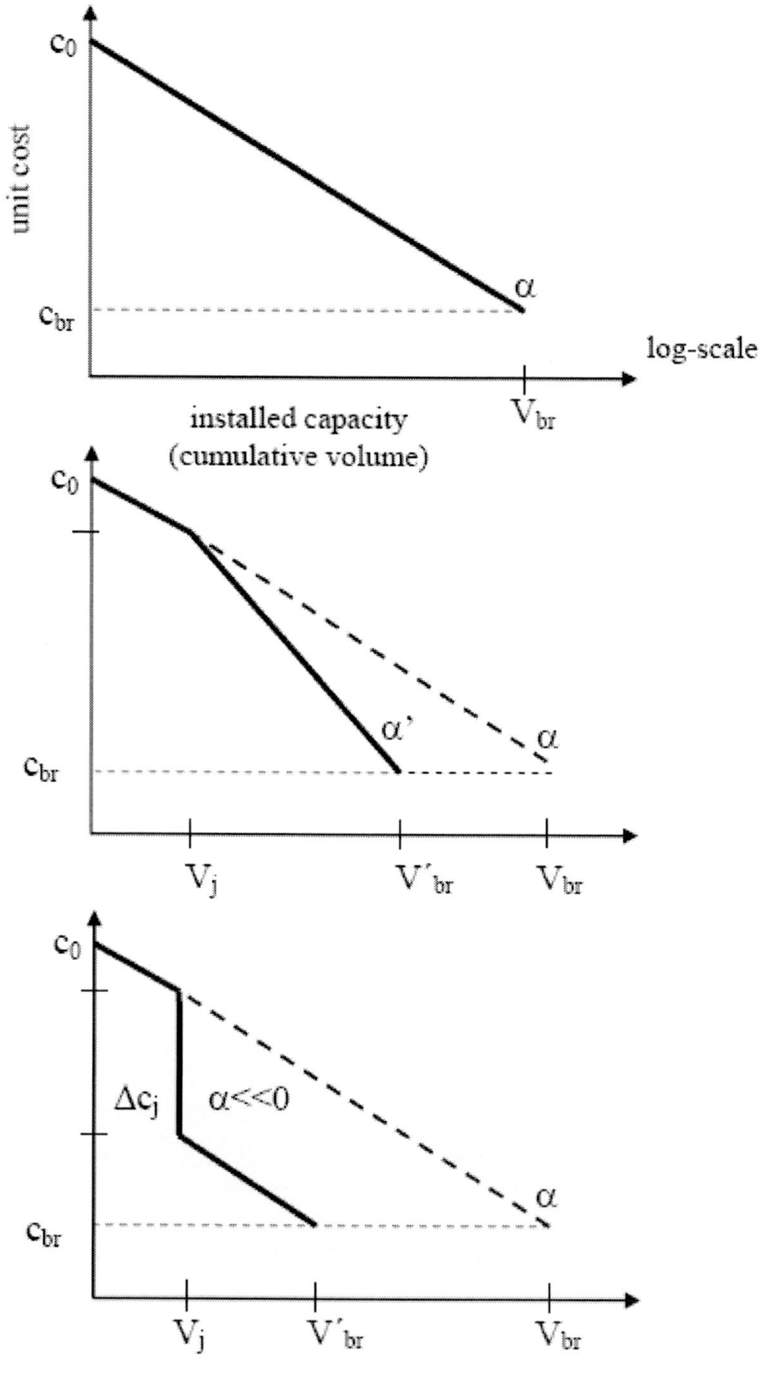

Figure 7. Illustration of the learning curve (unit cost versus cumulative installed capacity) in the reference case and for intensified RTD measures.

The results clearly indicate that if the new technology is economically immature (large cost difference) introducing generous subvention policies that increase the capacity even up to 40-50% a year, tens of years would still be required to reach the cost breakthrough point.

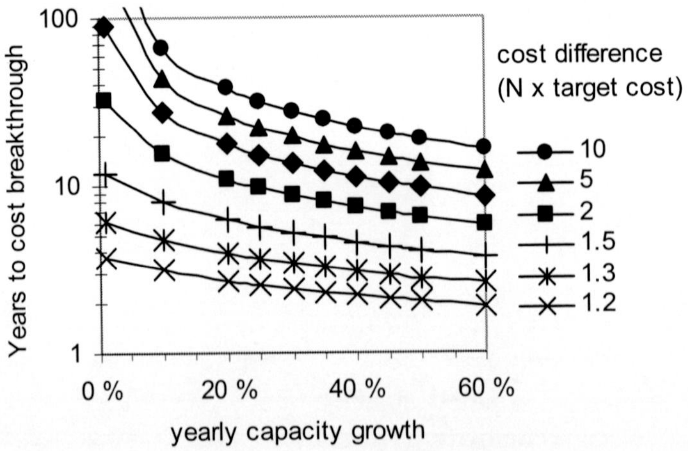

Figure 8. Time need for a cost breakthrough of the new technology.

When the cost difference in Figure 8 drops to less than a factor of 2, or the energy from the new technology is less than twice more expensive than the reference energy, accelerating the volume increase could speed up the breakthrough by several years. On the other hand, close to the breakthrough (e.g. cost gap less than 1.3) increasing the demand significantly through more financial public support would not shorten much the remaining time to the breakeven point. If using a LR value that would decrease with higher volumes instead of a constant value, then the absolute values for breakthrough in Figure 8 could increase, in particular when the cost gap becomes smaller. But the main observations explained above would remain the same emphasizing the importance of the type of support vis-à-vis the market readiness of the new energy technology.

The cost difference of the new technology contra the reference cost strongly determines the time needed to achieve a market breakthrough and also the amount of support needed. Figure 8 and the above analysis demonstrated the inadequacy of volume increasing measures to speed-up the commercialization process in all cases. Therefore we also looked on a policy case involving a strong concerted RTD action that would lead to an abrupt decrease in the cost of the new technology. Here Eq.(5) was used to estimate the savings in support spending that could be achieved with such a technology push strategy compared to a pure market pull.

Figure 9 illustrates the outcomes of the analysis for the focused RTD policy case. The parameter assumptions here are LR=15% and the new energy technology is initially twice as expensive as the reference energy meaning that the cost breakthrough would be obtained at a point where the volume is 20-fold compared to present use of the new technology. As the learning or experience curve in Eq.(1) decreases exponentially with increasing volume, it means that technology leapfrogging in the beginning of the commercialization process would lead to largest savings as clearly indicated by Figure 9. For example, a 20% cost drop at $V=V_0$ which could save more than half of the public support needed whereas close to the breakthrough point the effects from even a sudden removal of the cost gap would affect the total support need at most by 20%. This means that for new technologies that are cost-wise still far off from the cost breakthrough point, a strong RTD effort could economically be highly motivated and lead to large savings in public support but also increase the effectiveness of private investments as the new technology would become mature earlier.

Next the European Union's proposition to increase the renewable energy share to 20% of all final energy use by 2020 was analyzed in more detail with similar approach as described above. One example to fulfill this target in the electricity sector described in EU's background documents [Resch et al., 2007] is shown in Table 2. Strong increase in new renewable energy technologies such as wind and solar power is predicted. As the new technologies are more expensive than the traditional electricity (reference), the massive introduction does not come without an extra cost. For comparison, the total electricity use in EU was 3,310 TWh in 2005 of which 14% is RES [Eurostat, 2007]. In 2020, RES electricity would represent 16 (low) to 22 (high) % of all electricity if electricity consumption remains at present level, but drops to 13.5% versus 18.5% if present growth trend of electricity continues.

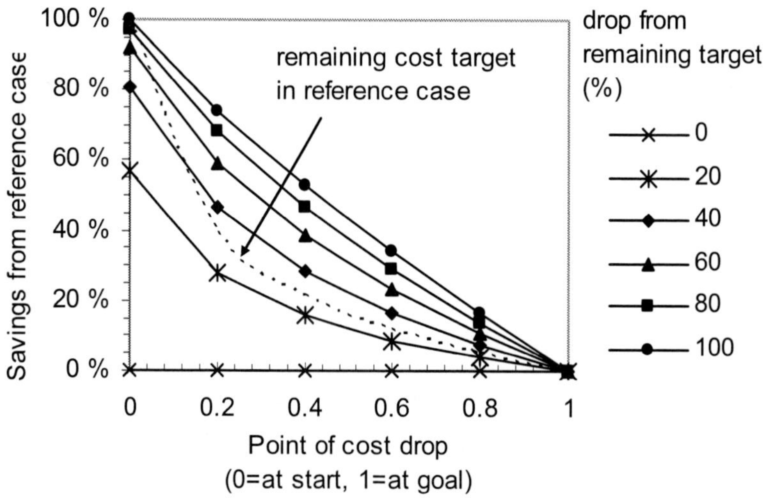

Figure 9. Cost savings in support needed for the breakthrough of new energy technology through a focused RTD effort that leads to a cost drop.

Table 2. Breakdown of new electricity generation from renewable energy sources.

Renewable electricity	Use in 2005 TWh	%	Additions by 2020 (high and low case) TWh	%	x '05	TWh	%	x '05
wind power	70.5	15.2	394	54.3	5.6	285	51.7	4.0
bioenergy	80	17.2	247	34.1	3.1	179	34.1	2.2
hydro	307	66.1	49	6.7	0.2	35	7.7	0.1
PV	1.5	0.3	16	2.2	10.6	12	2.2	7.7
other	5.4	1.2	20	2.7	3.6	14	4.3	2.6
Total	464	100	725	100	1.6	525	100	1.1

The total cumulative investments for the high scenario in Table 2 are 330 billion € [Resch et al., 2007]. Table 3 gives a more detailed split down for the largest investment objects namely wind power and solar photovoltaics (PV) that constitute over 60% of the total investments in RES electricity.

As the new technologies are often more expensive than traditional electricity, the new renewable energy technologies would need financial support to penetrate on the market. The amount of support depends on how successful the commercialization process will be. In the next, strategies for photovoltaics in the EU frame was analyzed using a coupled learning curve – diffusion model [Lund, 2006] as PV appears to be the most expensive choice in the EU RES electricity mix.

Using the European Commission's base scenario for PV as reference, the capacity and cost development would follow those shown in Figure 10. With an average volume growth of 15%/yr, the total photovoltaic electricity would correspond to around 25 TWh/yr in 2020. The cost of solar electricity would be some 30% cheaper than at the starting point, or for average European solar conditions (1.2kWh/Wp) 180 €/MWh which still exceeds the reference electricity cost (150 €/MWh).

In Figure 11, we have investigated the effects of a 5-years joint European RTD effort in PV which would lead to a 20% cost drop in addition to the fast learning rate (LR=20%, after 5 years LR=10%) in the reference case. The technology jump achieved would speed-up the cost breakthrough by 10 years, or down to year 2022 just before the PV capacity starts to take off with increasing pace. The accelerated technology development case would save the consumers close to 80% or 50 billion € from the reference case employing market deployment actions only. The extra support needed to achieve a commercial breakthrough would now be 10 billion € in total.

Table 3. Estimated investment needs for new PV and wind power by 2020 in EU.

Renewable electricity additions (high and low case)				
	TWh	billion €	TWh	billion €
wind power	393.7	161.7	285.1	104.6
PV	16.0	61.4	11.6	43.5

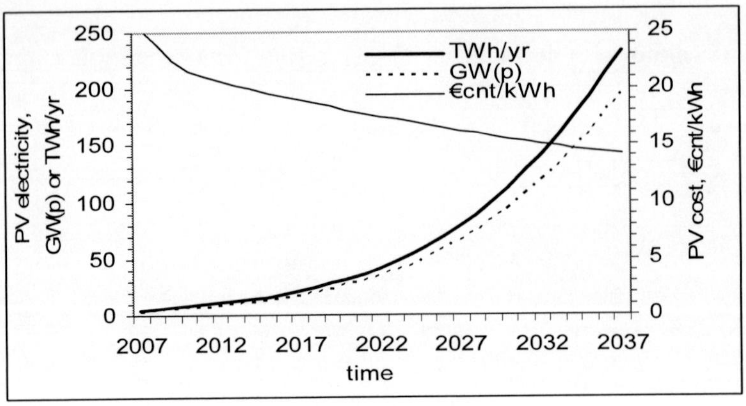

Figure 10. Estimated capacity and cost development of photovoltaics through the renewable energy targets in EU.

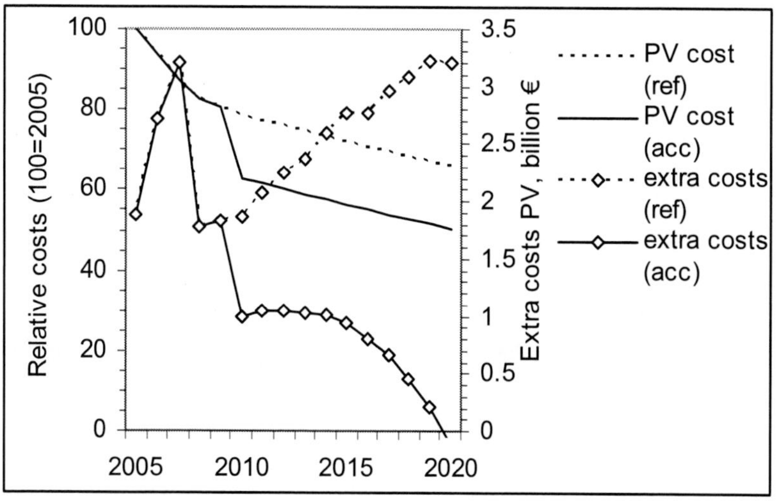

Figure 11. Support requirement in the reference and accelerated technology development cases.

Wind power offers an interesting case as it approaches the electricity production from hydropower in 2020 according to the EU scenarios. Being close to the breakthrough point, measures such as feed-in-tariffs or other guaranteed price type of mechanisms may be highly justified in this case. The wind power capacity increase from present to 2020 is 4 to 5.6 fold. Assuming a LR=10% (α=-0.152) and the present EU share of global wind power of around 70%, then the EU goals would results in a 15-20% cost reduction of wind power. For good wind regimes this could mean a cost breakeven.

CONCLUSION

The energy targets in Europe for 2020 will mean large investments in clean energy technologies and major support schemes for new energy technologies. The analysis presented in this paper demonstrates clearly that choosing optimal policy paths may lead to considerable savings in the cost of commercializing new technologies. It is important to find a correct balance between technology development and market deployment measures vis-à-vis the distance from the breakeven point. A strategy worthwhile considering would be the one that gives more emphasis on RTD actions when being further off from the cost-breakthrough point but shifting to market stimulation when getting closer to cost target.

In the European context the new policy directions on increasing renewable energy use up to 20% of final energy by 2020 has also important implications on the commercialization of new renewable energy technologies. Translating the RES targets for wind energy could mean a cost reduction of 15-20% which for optimal wind conditions could mean a cost breakthrough necessary for a massive market introduction. Large investments but smaller energy impacts are perceived through photovoltaics. In particular in PV, significant savings in the range of tens of billions € may be achieved through more intelligent policy choices over the traditional volume support schemes used in Europe. The same conclusion would apply to some other future orientated energy technologies such as fuel cells and hydrogen.

REFERENCES

Commission of the European Communities (2007a). A European Strategic Energy Technology Plan (SET-Plan). SEC(2007) 1508-1509. Brussels.

Commission of the European Communities (2007b). An energy policy for Europe, COM(2007). Brussels.

Commission of the European Communities (2008a). Proposal for a Directive of the European Parliament and of the Council on the promotion of the use of energy from renewable sources, COM(2008). Brussels.

Commission of the European Communities (2008). The support of electricity from renewable energy sources. Accompanying document to the Proposal for a Directive of the European Parliament and of the Council on the promotion of the use of energy from renewable sources, COM(2008)19. Brussels.

Dosi, G. (2000). *Innovations, organization and economic dynamics*. UK: Edward Elgar.

Eurostat (2007). Energy, Yearly statistics 2005. Luxembourg: Office for Official Publications of the European Communities.

Freeman, C., Soete, L. (2000*). Industrial innovation*. Cambridge USA: MIT Press.International Energy Agency (2006). Energy technology perspectives. Paris.

Lewis J.I., Wiser R.H. (2007). Fostering a renewable energy technology industry: An international comparison of wind industry policy support mechanisms. *Energy Policy,* 35, 1844–1857.

Lund, P. (2007). Effectiveness of policy measures in transforming the energy system. *Energy Policy*, 35, 627-639.

Lund, P. (2006). Analysis of energy technology changes and associated costs. *Energy Research*, 30, 967-984.

Pacala, R., Socolow, S. (2005). Stabilization Wedges: Solving the Climate Problem for the Next 50 Years with Current Technologies. *Science*, 305, 968-971.

Resch, G., Faber, T., Haas, R., Huber, C., Ragwitz, M., Held, A., Morthorst, P.E., Jensen, S.G., Coenraads, R., Voogt, M., Reece, G., Konstantinaviciute, I., Heyder, B. (2007). Assessment and optimisation of renewable energy support schemes in the European electricity market. Austria: Vienna University of Technology, Institute of Power Systems and Energy Economics, Energy Economics Group (EEG).

Wene, C.O. (2002). Experience curves for energy technology policy. Paris: International Energy Agency.

Reviewer: Jukka Paatero.

INDEX

A

B

C

D

E

F

G

H

I

J

K

L

M

P

Q

R

S

T

U

X

Y

Z